I0057708

Philippe Jeandet (Ed.)

Phytoalexins: Current Progress and Future Prospects

This book is a reprint of the special issue that appeared in the online open access journal *Molecules* (ISSN 1420-3049) in 2014 (available at: http://www.mdpi.com/journal/molecules/special_issues/phytoalexins-progress).

Guest Editor
Philippe Jeandet
Laboratory of Stress, Defenses and Plant Reproduction
U.R.V.V.C., UPRES EA 4707,
Faculty of Sciences, University of Reims,
PO Box. 1039,
51687 Reims cedex 02,
France

Editorial Office
MDPI AG
Klybeckstrasse 64
Basel, Switzerland

Publisher
Shu-Kun Lin

Managing Editor
Ran Dang

1. Edition 2015

MDPI • Basel • Beijing • Wuhan

ISBN 978-3-03842-058-3 (Hbk)
ISBN 978-3-03842-059-0 (PDF)

Table of Contents

About the Guest Editor

Philippe Jeandet received his doctorates in plant physiology and biochemistry in 1991 and 1996 from the University of Bourgogne (France). He started his research activities on resveratrol, a phytoalexin from the Vitaceae produced in response to fungal attacks or injury. He received an associate professor position at the University of Bourgogne (1993-1997). In 1997, Philippe Jeandet accepted a position as a professor and chairman of the laboratory of oenology and applied chemistry at the University of Reims. His research activities focussed on physico-chemistry applied to wine and microbiology. He has been respectively the director (2003-2008) and adjunct director (2008-2013) of the research unit vine and wine of Champagne and adjunct to the director of research and technology in the Champagne- Ardennes area (2004-2010). He is now co-leader of a research team on resveratrol at the laboratory of stress, defences and plant reproduction and member of the council of the Georges Chappaz research institute of vine and wine of Champagne. He has published over 250 papers in referred Journals or books (with many of them concerning phytoalexins) as well as technical papers, edited two books and two special issues and presented 240 communications to numerous symposia or congresses.

List of Contributors

Marielle Adrian: Université de Bourgogne, UMR1347 Agroécologie, ERL CNRS 6300, BP 86510, 21065 Dijon Cedex, France.

Jong Seog Ahn: Korea Research Institute of Bioscience and Biotechnology, Ochang, Chungbuk 363-883, Korea.

Yannis Andrey: School of Pharmaceutical Sciences, EPGL, University of Geneva, University of Lausanne, Quai Ernest-Ansermet 30, Geneva CH-1211, Switzerland.

Aziz Aziz: Laboratory of Stress, Defenses and Plant Reproduction, Research Unit "Vines and Wines of Champagne", UPRES EA 4707, Department of Biology and Biochemistry, Faculty of Sciences, University of Reims, P.O. Box 1039, 51687 Reims cedex 02, France.

Loïc Becker: Laboratoire de Chimie et Physique-Approche Multi échelle des Milieux Complexes (LCP-A2MC), Institut Jean Barriol (FR 2843), Université de Lorraine, ICPM 1 Boulevard Arago, F-57078 Metz, France.

Waldemar Bednarski: Institute of Molecular Physics, Polish Academy of Sciences, Smoluchowskiego 17, Poznań 60-179, Poland.

Mariana Budovska: Department of Organic Chemistry, Institute of Chemical Sciences, Faculty of Science, Pavol Jozef Safarik University, 040 80 Kosice, Slovak Republic.

Wenhao Cao: South China Sea Institute of Oceanology, Chinese Academy of Sciences, Guangzhou 510301, China.

Vincent Carré: Laboratoire de Chimie et Physique-Approche Multi échelle des Milieux Complexes (LCP-A2MC), Institut Jean Barriol (FR 2843), Université de Lorraine, ICPM 1 Boulevard Arago, F-57078 Metz, France.

Pierre-Alain Carrupt: School of Pharmaceutical Sciences, EPGL, University of Geneva, University of Lausanne, Quai Ernest-Ansermet 30, Geneva CH-1211, Switzerland.

Patrick Chaimbault: Laboratoire de Chimie et Physique-Approche Multi échelle des Milieux Complexes (LCP-A2MC), Institut Jean Barriol (FR 2843), Université de Lorraine, ICPM 1 Boulevard Arago, F-57078 Metz, France.

Malik Chalal: Université de Bourgogne, 21000 Dijon, France; Institut de Chimie Moléculaire de l'Université de Bourgogne, ICMUB UMR CNRS 6302, 9, avenue Alain Savary, 21000 Dijon, France.

Yinning Chen: Key Laboratory of Plant Resources Conservation and Sustainable Utilization, South China Botanical Garden, Chinese Academy of Sciences, Guangzhou 510650, China.

Yangrae Cho: Korea Research Institute of Bioscience and Biotechnology, Ochang, Chungbuk 363-883, Korea.

Surinder Chopra: Department of Plant Science, The Pennsylvania State University, University Park, PA 16802, USA; Intercollege Graduate Degree Program in Plant Biology, Pennsylvania State University, University Park, PA 16802, USA.

Martina Chripkova: Department of Pharmacology, Faculty of Medicine, Pavol Jozef Safarik University, 040 11 Kosice, Slovak Republic.

Sylvain Cordelier: Laboratory of Stress, Defenses and Plant Reproduction, Research Unit "Vines and Wines of Champagne", UPRES EA 4707, Department of Biology and Biochemistry, Faculty of Sciences, University of Reims, P.O. Box 1039, 51687 Reims cedex 02, France.

Jérôme Crouzet: Laboratory of Stress, Defenses and Plant Reproduction, Research Unit "Vines and Wines of Champagne", UPRES EA 4707, Department of Biology and Biochemistry, Faculty of Sciences, University of Reims, P.O. Box 1039, 51687 Reims cedex 02, France.

Dominique Delmas: Université de Bourgogne, 21000 Dijon, France; Laboratoire de Biochimie (Bio-PeroxIL) INSERM IFR 100, 6, boulevard Gabriel, Dijon, France; INSERM UMR 866, 7, boulevard Jeanne d'Arc, 21000 Dijon, France.

Rakshit Devappa: Korea Research Institute of Bioscience and Biotechnology, Ochang, Chungbuk 363-883, Korea.

Marie-Alice Deville: Champagne Deville, 13 rue Carnot, Verzy 51380, France.

Stéphan Dorey: Laboratory of Stress, Defenses and Plant Reproduction, Research Unit "Vines and Wines of Champagne", UPRES EA 4707, Department of Biology and Biochemistry, Faculty of Sciences, University of Reims, P.O. Box 1039, 51687 Reims cedex 02, France.

David Drutovic: Department of Pharmacology, Faculty of Medicine, Pavol Jozef Safarik University, 040 11 Kosice, Slovak Republic.

Abdelwahad Echairi: Welience, Maison Régionale de L'Innovation, 64 A rue de Sully, CS 77124, 21071 Dijon Cedex, France.

Magda Formela: Department of Plant Physiology, Poznań University of Life Sciences, Wołyńska 35, Poznań 60-637, Poland.

Iffa Gaffoor: Department of Plant Science, The Pennsylvania State University, University Park, PA 16802, USA.

Chenghai Gao: Guangxi Key Laboratory of Marine Environmental Science, Guangxi Academy of Sciences, Nanning 530007, China.

Katia Gindro: Station de recherche Agroscope, Institut des Sciences en Production Végétale IPV, Route de Duiller 50, P.O. Box 1012, Nyon 1260, Switzerland.

Igor V. Grigoriev: Joint Genome Institute, 2800 Mitchell Drive, Walnut Creek 94598, CA, USA.

Lee A. Hadwiger: Department of Plant Pathology, Washington State University, Pullman, WA 99164-6430, USA.

Glen L. Hartman: United States Department of Agriculture (USDA), Agricultural Research Service, University of Illinois, 1101 W. Peabody Drive, Urbana, IL 61801, USA.
Morifumi Hasegawa: College of Agriculture, Ibaraki University, 3-21-1 Chuo, Ami, Ibaraki 300-0393, Japan.
Claire Hébrard: Laboratory of Stress, Defenses and Plant Reproduction, Research Unit "Vines and Wines of Champagne", UPRES EA 4707, Department of Biology and Biochemistry, Faculty of Sciences, University of Reims, P.O. Box 1039, 51687 Reims cedex 02, France.
Curtis B. Hill: Department of Crop Sciences, University of Illinois, 1201 W. Gregory Drive, Urbana, IL 61801, USA.
Cimona Hinton: Center for Cancer Research and Therapeutic development, Department of Biological Sciences, Clark Atlanta University, Atlanta, GA 30314, USA.
Riming Huang: Key Laboratory of Plant Resources Conservation and Sustainable Utilization, South China Botanical Garden, Chinese Academy of Sciences, Guangzhou 510650, China.
Farag Ibraheem: Department of Plant Science, The Pennsylvania State University, University Park, PA 16802, USA; Intercollege Graduate Degree Program in Plant Biology, Pennsylvania State University, University Park, PA 16802, USA; Botany Department, Faculty of Science, Mansoura University, AlMansoura, 35516, Egypt.
Antonio Ippolito: Dipartimento di Scienze del Suolo, della Pianta e degli Alimenti, Università degli Studi Aldo Moro, Via G. Amendola 165/A, Bari 70126, Italy.
Takayoshi Iwai: School of Food, Agricultural and Environmental Sciences, Miyagi University, 2-2-1 Hatadate, Taihaku, Sendai, Miyagi 982-0215, Japan.
Philippe Jeandet: Laboratory of Stress, Defenses and Plant Reproduction, Research Unit "Vines and Wines of Champagne", UPRES EA 4707, Department of Biology and Biochemistry, Faculty of Sciences, University of Reims, P.O. Box 1039, 51687 Reims cedex 02, France.
M. Emília Juan: Departament de Fisiologia and Institut de Recerca en Nutrició i Seguretat Alimentària (INSA-UB), Universitat de Barcelona (UB), Av. Joan XXIII s/n, 08028 Barcelona, Spain.
Simon Kaja: Vision Research Center, Department of Ophthalmology, School of Medicine, University of Missouri—Kansas City, 2411 Holmes St., Kansas City, MO 64108, USA.
Anna Kasprowicz-Maluśki: Department of Molecular and Cellular Biology, Faculty of Biology, Adam Mickiewicz University, Umultowska 89, Poznań 60-614, Poland.
Martin Kello: Department of Pharmacology, Faculty of Medicine, Pavol Jozef Safarik University, 040 11 Kosice, Slovak Republic.
Bo Yeon Kim: Korea Research Institute of Bioscience and Biotechnology, Ochang, Chungbuk 363-883, Korea.
Agnès Klinguer: INRA, UMR1347 Agroécologie, ERL CNRS 6300, BP 86510, 21065 Dijon Cedex, France.

Peter Koulen: Vision Research Center, Department of Ophthalmology, School of Medicine, University of Missouri—Kansas City, 2411 Holmes St., Kansas City, MO 64108, USA; Department of Basic Medical Science, School of Medicine, University of Missouri—Kansas City, 2411 Holmes St., Kansas City, MO 64108, USA.

Lucia Kulikova: Department of Experimental Medicine, Faculty of Medicine, Pavol Jozef Safarik University, 040 11 Kosice, Slovak Republic.

Norbert Latruffe: Université de Bourgogne, 21000 Dijon, France; Laboratoire de Biochimie (Bio-PeroxIL) INSERM IFR 100, 6, boulevard Gabriel, Dijon, France.

Hyang Burm Lee: Division of Applied Bioscience and Biotechnology, College of Agriculture and Life Sciences, Chonnam National University, Buk-Gu, Gwangju 500-757, Korea.

Glòria Lozano-Mena: Departament de Fisiologia and Institut de Recerca en Nutrició i Seguretat Alimentària (INSA-UB), Universitat de Barcelona (UB), Av. Joan XXIII s/n, 08028 Barcelona, Spain.

Vera V. Lozovaya: Department of Crop Sciences, University of Illinois, 1201 W. Gregory Drive, Urbana, IL 61801, USA.

Anatoli V. Lygin: Department of Crop Sciences, University of Illinois, 1201 W. Gregory Drive, Urbana, IL 61801, USA.

Łukasz Marczak: Institute of Bioorganic Chemistry, Polish Academy of Sciences, Z. Noskowskiego 12/14, Poznań 61-704, Poland.

Guillaume Marti: School of Pharmaceutical Sciences, EPGL, University of Geneva, University of Lausanne, Quai Ernest-Ansermet 30, Geneva CH-1211, Switzerland.

Audrey E. McCalley: Vision Research Center, Department of Ophthalmology, School of Medicine, University of Missouri—Kansas City, 2411 Holmes St., Kansas City, MO 64108, USA.

Didier Merdinoglu: Institut National de Recherche en Agronomie (INRA) – Santé de la Vigne et Qualité du Vin (UMR 1131), 28 rue de Herrlisheim, F-68021 Colmar, France

Philippe Meunier: Université de Bourgogne, 21000 Dijon, France; Institut de Chimie Moléculaire de l'Université de Bourgogne, ICMUB UMR CNRS 6302, 9, avenue Alain Savary, 21000 Dijon, France.

Roman Mezencev: Department of Biology, Georgia Institute of Technology, Atlanta, GA 30332, USA.

Ichiro Mitsuhara: National Institute of Agrobiological Sciences, Tsukuba, Ibaraki 305-8602, Japan.

Jan Mojzis: Department of Pharmacology, Faculty of Medicine, Pavol Jozef Safarik University, 040 11 Kosice, Slovak Republic.

Iwona Morkunas: Department of Plant Physiology, Poznań University of Life Sciences, Wołyńska 35, Poznań 60-637, Poland.

Dorota Narożna: Department of Biochemistry and Biotechnology, Poznań University of Life Sciences, Dojazd 11, Poznań 60-632, Poland.

Witold Nowak: Laboratory of Molecular Biology Techniques, Faculty of Biology, Adam Mickiewicz University, Umultowska 89, Poznań 60-614, Poland.

Valerie Odero-Marah: Center for Cancer Research and Therapeutic development, Department of Biological Sciences, Clark Atlanta University, Atlanta, GA 30314, USA.

Yuko Ohashi: National Institute of Agrobiological Sciences, Tsukuba, Ibaraki 305-8602, Japan.

Robin A. Ohm: Joint Genome Institute, 2800 Mitchell Drive, Walnut Creek 94598, CA, USA.

Kazunori Okada: Biotechnology Research Center, The University of Tokyo, 1-1-1 Yayoi, Bunkyo-ku, Tokyo 113-8657, Japan.

Michelle L. Pawlowski: Department of Crop Sciences, University of Illinois, 1201 W. Gregory Drive, Urbana, IL 61801, USA.

Andrew J. Payne: Vision Research Center, Department of Ophthalmology, School of Medicine, University of Missouri—Kansas City, 2411 Holmes St., Kansas City, MO 64108, USA.

Martina Pilatova: Department of Pharmacology, Faculty of Medicine, Pavol Jozef Safarik University, 040 11 Kosice, Slovak Republic.

Joana M. Planas: Departament de Fisiologia and Institut de Recerca en Nutrició i Seguretat Alimentària (INSA-UB), Universitat de Barcelona (UB), Av. Joan XXIII s/n, 08028 Barcelona, Spain.

Alana Poloni: Department of Microbial Genetics, Institute of Applied Microbiology, Aachen Biology and Biotechnology, RWTH Aachen University, Worringerweg 1, Aachen 52074, Germany.

Anne Poutaraud: Institut National de Recherche en Agronomie (INRA) – Santé de la Vigne et Qualité du Vin (UMR 1131), 28 rue de Herrlisheim, F-68021 Colmar, France.

Diandra Randle: Center for Cancer Research and Therapeutic development, Department of Biological Sciences, Clark Atlanta University, Atlanta, GA 30314, USA.

Sławomir Samardakiewicz: Laboratory of Electron and Confocal Microscopy, Faculty of Biology, Adam Mickiewicz University, Umultowska 89, Poznań 60-614, Poland.

Marta Sánchez-González: Departament de Fisiologia and Institut de Recerca en Nutrició i Seguretat Alimentària (INSA-UB), Universitat de Barcelona (UB), Av. Joan XXIII s/n, 08028 Barcelona, Spain.

Simona M. Sanzani: Dipartimento di Scienze del Suolo, della Pianta e degli Alimenti, Università degli Studi Aldo Moro, Via G. Amendola 165/A, Bari 70126, Italy.

Leonardo Schena: Dipartimento di Agraria, Università degli Studi Mediterranea, Località Feo di Vito, Reggio Calabria 89124, Italy.

Jan Schirawski: Department of Microbial Genetics, Institute of Applied Microbiology, Aachen Biology and Biotechnology, RWTH Aachen University, Worringerweg 1, Aachen 52074, Germany.

Sylvain Schnee: Station de recherche Agroscope, Institut des Sciences en Production Végétale IPV, Route de Duiller 50, P.O. Box 1012, Nyon 1260, Switzerland.

Shigemi Seo: National Institute of Agrobiological Sciences, Tsukuba, Ibaraki 305-8602, Japan.

Chi-Ren Shyu: MU Informatics Institute, University of Missouri, Columbia, MO 65201, USA.

Claudia Simoes-Pires: School of Pharmaceutical Sciences, EPGL, University of Geneva, University of Lausanne, Quai Ernest-Ansermet 30, Geneva CH-1211, Switzerland.

Basil Smith: Center for Cancer Research and Therapeutic development, Department of Biological Sciences, Clark Atlanta University, Atlanta, GA 30314, USA.

Qixian Tan: Department of Plant Science, The Pennsylvania State University, University Park, PA 16802, USA.

Kiwamu Tanaka: Department of Plant Pathology, Washington State University, Pullman, WA 99164-6430, USA.

LeeShawn Thomas: Department of Biological Sciences, Florida A & M University, Tallahassee, FL 32307, USA.

Peter Urdzik: Department of Gynaecology and Obstetrics, Faculty of Medicine, Pavol Jozef Safarik University, 040 11 Kosice, Slovak Republic; Pasteur University Hospital, 040 11 Kosice, Slovak Republic.

Dominique Vervandier-Fasseur: Université de Bourgogne, 21000 Dijon, France; Institut de Chimie Moléculaire de l'Université de Bourgogne, ICMUB UMR CNRS 6302, 9, avenue Alain Savary, 21000 Dijon, France.

Jack M. Widholm: Department of Crop Sciences, University of Illinois, 1201 W. Gregory Drive, Urbana, IL 61801, USA.

Jean-Luc Wolfender: School of Pharmaceutical Sciences, EPGL, University of Geneva, University of Lausanne, Quai Ernest-Ansermet 30, Geneva CH-1211, Switzerland.

Hisakazu Yamane: Department of Biosciences, Teikyo University, 1-1 Toyosatodai, Utsunomiya, Tochigi 320-8551, Japan.

Tao Yan: South China Sea Institute of Oceanology, Chinese Academy of Sciences, Guangzhou 510301, China.

Olga V. Zernova: Department of Crop Sciences, University of Illinois, 1201 W. Gregory Drive, Urbana, IL 61801, USA.

Phytoalexins: Current Progress and Future Prospects

Philippe Jeandet

Reprinted from *Molecules*. Cite as: Jeandet, P. Phytoalexins: Current Progress and Future Prospects. *Molecules* **2015**, *20*, 2770-2774.

Phytoalexins are low molecular weight antimicrobial compounds that are produced by plants as a response to biotic and abiotic stresses. As such they take part in an intricate defense system which enables plants to control invading microorganisms. In the 1950s, research on phytoalexins started with progress in their biochemistry and bio-organic chemistry, resulting in the determination of their structure, their biological activity, as well as mechanisms of their synthesis and catabolism by microorganisms. Elucidation of the biosynthesis of numerous phytoalexins also permitted the use of molecular biology tools for the exploration of the genes encoding enzymes of their synthesis pathways and their regulators. This has led to potential applications for increasing plant resistance to diseases. Phytoalexins display an enormous diversity belonging to various chemical families such as for instance, phenolics, terpenoids, furanoacetylenes, steroid glycoalkaloids, sulfur-containing compounds and indoles.

Research and review papers dealing with numerous aspects of phytoalexins including modulation of their biosynthesis, molecular engineering in plants, biological activities, structure/activity relationships and phytoalexin metabolism by micro-organisms are published in this issue.

In the first paper of this special issue on phytoalexins, Jeandet *et al.* present an overview of this diverse group of molecules, namely their chemical diversity, the main biosynthetic pathways and their regulatory mechanisms, fungal metabolism, phytoalexin gene transfer in plants and their role as antifungal and bactericidal agents as well as their involvement in human health [1].

General aspects of phytoalexins from the Leguminosae and Poaceae families are also discussed in this issue. Phytoalexins from sorghum and maize are presented in details by Poloni and Schirawski [2]. Sorghum produces two distinct phytoalexins belonging to the 3-deoxyanthocyanidin chemical group, apigeninidin and luteolinidin. Their biosynthetic pathways start from the flavanone naringenin according to a scheme slightly different from that of the anthocyanin route. In maize, phytoalexins are represented by members of the terpenoid class, including zealexins and kauralexins on the one hand and benzoxazinoids on the other hand, the biosynthesis of which are fully described. Biosynthesis aspects have been linked to both the elicitation and the up-regulation mechanisms of those phytoalexins. Various applications of sorghum and maize phytoalexins in plant disease resistance and health and biomedicine are also presented.

Within the Leguminosae family, the genus *Tephrosia*, a large pantropical genus composed of more than 350 species, is a source of numerous chemical constituents possessing various biological properties, including phytoalexin-like compounds. These compounds which are reviewed in this issue by Chen *et al.*, are mainly polyphenolics (flavones, flavonols, flavononols, flavans, isoflavones and chalcones), triterpenoids and sesquiterpenes [3]. Biosynthetic pathways of a number of these compounds are described as well as some of their biological activities as estrogenic, antitumor, antimicrobial, antiprotozoal and antifeedant agents.

Elucidating the molecular mechanisms of the modulation of phytoalexin biosynthesis finds applications in plant engineering for disease resistance. In this issue, Formela *et al.* report the effects of various sugars (sucrose, glucose and fructose) acting as endogenous signals on the mechanisms regulating the biosynthesis and accumulation of the lupine phytoalexin, genistein as well as the expression of other isoflavonoid biosynthetic genes [4]. Zernova *et al.* describe the transformation of soybean hairy roots with both the peanut resveratrol synthase 3 *AhRS3* gene and the resveratrol-*O*-methyltransferase *ROMT* gene [5]. Overexpression of these two genes resulted in the production of resveratrol and its methylated derivative pterostilbene and a lower necrosis of the transformed tissues (only 0 to 7%) in response to the soybean pathogen *Rhizoctonia solani* compared to the wild-type ones which exhibited about 84% necrosis. Biosynthesis of the 3-deoxyanthocyanidin phytoalexins from sorghum is reported in transgenic maize lines expressing the MYB transcription factor *yellow seed1* (*y1*), an orthologue of the maize gene *pericarp color1* (*p1*) in the work of Ibraheem *et al.* [6]. Expression of this transcription factor leads to the production of chemically modified 3-deoxyanthocyanidins and a resistance response of *Y1*-maize plants to leaf blight (*Colletotrichum graminicola*).

It is well known that treatment of plants with various biotic or abiotic agents, the so-called elicitors, can activate complex mechanisms in the cells by altering primary and secondary metabolisms in a coordinate fashion. Elicitors are also recognized as efficient inducers of phytoalexins. In this issue, Hadwiger and Tanaka report that EDTA, used at low concentrations, is a new elicitor of pisatin, a phytoalexin indicator of non-host resistance in pea [7]. Eliciting activity of EDTA seems to be linked to induction of cell DNA damage and defense-responsive genes.

The question of the function of phytoalexins as true antifungal agents still remains unanswered. Interestingly, a study of Sanzani *et al.* underline the effectiveness both *in vitro* and *in vivo* of some polyphenolic phytoalexins, namely the coumarin, scopoletin, on the reduction of green mold symptoms caused by *Penicillium digitatum* on oranges by 40 to 85% [8]. Based on these results, the authors conclude that treatment of plants with phytoalexins may represent an interesting alternative to synthetic fungicides. In another work by Hasegawa *et al.*, the activity of two rice phytoalexins, sakuranetin and momilactone A was tested *in vitro* and *in vivo* on the blast fungus *Magnaporthe oryzae*. Sakuranetin exhibits a higher antifungal activity than does momilactone A, respectively 40%–55% and 12%–17% reduction of mycelial growth [9].

To increase the fungitoxicity of phytoalexins, design and synthesis of more active phytoalexin derivatives is needed. Chalal *et al.* report in this issue the synthesis of a series of 13 *trans*-resveratrol analogues via Wittig or Heck reactions and assess their antimicrobial activity on two different grapevine pathogens, *Plasmopara viticola* and *Botrytis cinerea* [10]. Stilbenes displayed a spectrum of activity ranging from low to high, suggesting a relationship between the chemical structures of the synthesized stilbenes (number and position of methoxy and hydroxyphenyl groups) and their antimicrobial activity.

The ability of a fungal pathogen to weaken or neutralize the toxic effects of phytoalexins is one of the essential parameters determining the outcome of the interaction between this pathogen and its host plant. The necrotrophic fungus *Alternaria brassicicola* is known to detoxify brassinin, the indolic phytoalexin from the Brassicaceae family. A transcription factor *Bdtf1* is essential for brassinin detoxification and fungal host range. In this issue, Cho *et al.* show that beside this transcriptional factor, 10 putative genes were assumed to be involved in the detoxification of brassinin using a *Bdtf1*-deletion mutant of the necrotrophic fungus *A. brassicicola* [11].

Another limitation in our knowledge of phytoalexins is the difficulty in analyzing the events occurring between the plant and the pathogen under natural conditions. Some attempts to determine the actual concentrations and the nature of phytoalexins directly in plant tissues in response to invading microorganisms have been carried out using spectroscopic methods. Becker *et al.* in this issue describe mass spectrometry (ESI-FTIR-RMS) and imaging mass spectrometry techniques to evaluate the response of grapevine leaves to *P. viticola*, the causal agent of downy mildew [12]. Most importantly, molecular mapping of grapevine leaves by laser desorption/ionization mass spectrometry reveals a specific spatial distribution of some stilbene phytoalexins produced upon the infection process. To assess modifications of the phytoalexin metabolism *in planta*, global and untargeted approaches are also needed. Here, Marti *et al.* use a Liquid Chromatography-High Resolution Mass Spectrometry-based metabolomic approach to evaluate stilbene phytoalexin modifications as a response to an abiotic stress (UV-C radiations) in leaves of three different model plant species, *Cissus Antarctica* Vent. (Vitaceae), *Vitis vinifera* L. (Vitaceae) and *Cannabis sativa* L. (Cannabaceae) [13].

Interestingly, phytoalexins have found many applications in human health and disease. For example, Lozano-Mena *et al.* review in this issue the role of maslinic acid, a pentacyclic triterpene phytoalexin-like compound present in various natural sources such as herbal remedies as well as edible vegetables and fruits, as an antitumor, antidiabetic, antioxidant, cardioprotective, neuroprotective, antiparasitic and growth-stimulating agent both in experimental and animal models [14]. This offers perspectives for this compound to be used as a nutraceutical. Moreover, other phytoalexins such as brassinin and its derivative, homobrassinin, show marked antiproliferative activities *in vitro*. In this issue, Kello *et al.* indeed report that the inhibitory effects of the phytoalexin homobrassinin in human colorectal cancer cells is associated with apoptosis, G2/M phase arrest, deregulation of tubulin expression together with the loss of mitochondrial membrane potential, caspase-3 activation and intracellular reactive oxygen species production [15]. Smith *et al.* also demonstrate that the indolic phytoalexin, camalexin, exerts antitumor activity against prostate cancer cell lines by alterations of expression and activity of a lysosomal protease, cathepsin D [16]. Immunochemical analysis reveals cathepsin D relocalization from the lysosome to the cytoplasm according to camalexin treatment which is responsible for apoptosis in those cells. One of the most promising molecules in terms of biological benefits for humans, the resveratrol, is reviewed by McCalley *et al.* regarding its effects on intracellular calcium signaling mechanisms [17]. Resveratrol's mechanisms of action are likely to be pleitropic and mediated by the interaction of this compound with key signaling proteins controlling cellular calcium homeostasis. The clinical relevance of resveratrol actions on excitable cells, transformed or cancer cells and immune cells was put in parallel with the molecular mechanisms affecting intra cellular calcium signaling proteins.

Lack of efficacy of some natural phytoalexins in reducing tumors has led to a number of investigations regarding the design and synthesis of more potent anticancer derivatives of known phytoalexins. Chalal *et al.* in this issue describe the synthesis of hydroxylated and methylated resveratrol derivatives using Wittig and Heck reactions as well as of ferrocenyl-stilbene analogs, with potent anticancer activities on human colorectal tumor SW480 cell lines [18]. However, weaker effects of the synthesized resveratrol derivatives were observed on the human hepatoblastoma HepG2 cells, showing the selectivity of those compounds for cancer treatment.

All the papers presented in this special issue thus underline the central role of phytoalexins in plant diseases as well as their involvement in human health and disease.

Acknowledgments

The Guest Editor thanks all the authors for their contributions to this special issue, all the reviewers for their work in evaluating the submitted articles and the editorial staff of *Molecules*, especially Jessica Bai, Jiahua Zhang and Wei Zhang, Assistant Editors of this journal for their kind help in making this special issue.

References

1. Jeandet, P.; Hébrard, C.; Deville, M.A.; Cordelier, S.; Dorey, S.; Aziz, A.; Crouzet, J. Deciphering the role of phytoalexins in plant-microorganism interactions and human health. *Molecules* **2014**, *19*, 18033–18056.
2. Poloni, A.; Schirawski, J. Red card for pathogens: Phytoalexins in sorghum and maize. *Molecules* **2014**, *19*, 9114–9133.
3. Chen, Y.; Tao, Y.; Gao, C.; Cao, W.; Huang, R. Natural products from the genus *Tephrosia*. *Molecules* **2014**, *19*, 1432–1458.
4. Formela, M.; Samardakiewicz, S.; Marczak, L.; Nowak, W.; Narozna, D.; Waldemar, B.; Kasprowicz-Maluski, A.; Morkunas, I. Effects of endogenous signals and *Fusarium oxysporum* on the mechanism regulating genistein synthesis and accumulation in yellow lupine and their impact on plant cell cytoskeleton. *Molecules* **2014**, *19*, 13392–13421.
5. Zernova, O.V.; Lygin, A.V.; Pawlowski, M.L.; Hill, C.B.; Hartman, G.L.; Widholm, J.M.; Lozovaya, V.V. Regulation of plant immunity through modulation of phytoalexin synthesis. *Molecules* **2014**, *19*, 7480–7496.
6. Ibraheem, F.; Gaffoor, I.; Sharma, M.; Shyu, C.R.; Chopra, S. A sorghum MYB transcription factor induces 3-deoxyanthocyanidins and enhances resistance against leaf blights in maize. *Molecules* **2015**, *20*, 2388–2404.
7. Hadwiger, L.A.; Tanaka, K. EDTA, a novel inducer of pisatin, a phytoalexin indicator of the non-host resistance in peas. *Molecules* **2015**, *20*, 24–34.
8. Sanzani, S.; Schena, L.; Ippolito, A. Effectiveness of phenolic compounds against citrus green mould. *Molecules* **2014**, *19*, 12500–12508.
9. Hasegawa, M.; Mitsuhara, I.; Seo, S.; Okada, K.; Yamane, H.; Iwai, T.; Ohashi, Y. Analysis on blast fungus-responsive characters of a flavonoid phytoalexin sakuranetin; Accumulation in infected rice leaves, antifungal activity and detoxification by fungus. *Molecules* **2014**, *19*, 11404–11418.
10. Chalal, M.; Klinguer, A.; Echairi, A.; Meunier, P.; Vervandier-Fasseur, D.; Adrian, M. Antimicrobial activity of resveratrol analogues. *Molecules* **2014**, *19*, 7679–7688.

11. Cho, Y.; Ohm, R.A.; Devappa, R.; Lee, H.B.; Grigoriev, I.V.; Kim, B.Y.; Ahn, J.S. Transcriptional responses of the *bdtf1*-deletion mutant to the phytoalexin brassinin in the necrotrophic fungus *Alternaria brassicicola*. *Molecules* **2014**, *19*, 10717–10732.
12. Becker, L.; Carré, V.; Poutaraud, A.; Merdinoglu, D.; Chaimbault, P. MALDI mass spectrometry imaging for the simultaneous location of resveratrol, pterostilbene and viniferins on grapevine leaves. *Molecules* **2014**, *19*, 10587–10600.
13. Marti, G.; Schnee, S.; Andrey, Y.; Simoes-Pires, C.; Carrupt, P.A.; Wolfender, J.L.; Gindro, K. Study of leaf metabolome modifications induced by UV-C radiations in representative *Vitis*, *Cissus* and *Cannabis* species by LC-MS based metabolomics and antioxidant assays. *Molecules* **2014**, *19*, 14004–14021.
14. Lozano-Mena, G.; Sanchez-Gonzalez, M.; Juan, M.E.; Planas, J.M. Maslinic acid, a natural phytoalexin-type triterpene from olives—A promising nutraceutical? *Molecules* **2014**, *19*, 11538–11559.
15. Kello, M.; Drutovic, D.; Chripkova, M.; Pilatova, M.; Budovska, L.; Kulikova, L.; Urdzik, P.; Mojzis, J. ROS-dependent antiproliferative effect of brassinin derivative homobrassinin in human colorectal cancer Caco2 cells. *Molecules* **2014**, *19*, 10877–10897.
16. Smith, B.; Randle, D.; Mezencev, R.; Thomas, L.; Hinton, C.; Odero-Marah, V. Camalexin-induced apoptosis in prostate cancer cells involves alterations of expression and activity of lysosomal protease cathepsin D. *Molecules* **2014**, *19*, 3988–4005.
17. McCalley, A.E.; Kaja, S.; Payne, A.J.; Koulen, P. Resveratrol and calcium signaling: molecular mechanisms and clinical relevance. *Molecules* **2014**, *19*, 7327–7340.
18. Chalal, M.; Delmas, D.; Meunier, P.; Latruffe, N.; Vervandier-Fasseur, D. Inhibition of cancer derived cell lines proliferation by synthesized hydroxylated stilbenes and new ferrocenyl-stilbene analogs. Comparison with resveratrol. *Molecules* **2014**, *19*, 7850–7868.

Deciphering the Role of Phytoalexins in Plant-Microorganism Interactions and Human Health

Philippe Jeandet, Claire Hébrard, Marie-Alice Deville, Sylvain Cordelier, Stéphan Dorey, Aziz Aziz and Jérôme Crouzet

Abstract: Phytoalexins are low molecular weight antimicrobial compounds that are produced by plants as a response to biotic and abiotic stresses. As such they take part in an intricate defense system which enables plants to control invading microorganisms. In this review we present the key features of this diverse group of molecules, namely their chemical structures, biosynthesis, regulatory mechanisms, biological activities, metabolism and molecular engineering.

Reprinted from *Molecules*. Cite as: Jeandet, P.; Hébrard, C.; Deville, M.-A.; Cordelier, S.; Dorey, S.; Aziz, A.; Crouzet, J. Deciphering the Role of Phytoalexins in Plant-Microorganism Interactions and Human Health. *Molecules* **2014**, *19*, 18033-18056.

1. Phytoalexins: A Global Survey

Phytoalexins take part in an intricate defense system used by plants against pests and pathogens [1,2]. These are low molecular weight antimicrobial compounds both synthesized by and accumulated in plants as a response to biotic and abiotic stresses. The concept of phytoalexins was first introduced over 70 years ago by Müller and Börger [3] after observing that infection of potato tubers with a strain of *Phytophthora infestans* capable of initiating hypersensitive reactions, significantly inhibited the effect of a subsequent infection with another strain of *P. infestans*. This inhibition was linked to a "principle" produced by the plant cells reacting hypersensitively that they named *phytoalexin* [4].

Most of what is known about phytoalexins derives from extensive work on a limited number of plant families: Leguminosae or Fabaceae and Solanaceae [5,6], on one hand, and investigations on one or a few species within other plant families, namely Amaryllidaceae, Euphorbiaceae, Orchidaceae, Chenopodiaceae, Compositae, Convolvulaceae, Ginkgoaceae, Poaceae, Linaceae, Moraceae, Orchidaceae, Piperaceae, Rosaceae, Rutaceae and Umbelliferae on the other hand [7]. More intensive studies recently focused on phytoalexins from plant families of significant economic importance: Poaceae (maize and rice) [8], Vitaceae [9,10] and Malvaceae (cotton) [11]. Camalexin, the main phytoalexin from Brassicaceae (Cruciferae) has also been the subject of numerous studies focusing on its biosynthetic pathway and the regulatory networks involved in its production in the model plant *Arabidopsis thaliana* [1,12]. However, the question of the ubiquity of phytoalexins throughout the plant kingdom still remains.

Phytoalexins are restricted to compounds produced from remote precursors, through *de novo* synthesis of enzymes. This peculiarity makes deciphering their biosynthesis and regulation mechanisms very complex [1,2]. Phosphorylation cascades, defense-related marker genes, calcium sensors and elicitors as well as hormone signaling are potentially important regulators for the modulation of phytoalexin production and pathogen resistance. As a corollary, knowledge of the control mechanisms of phytoalexin accumulation has served as the basis for the genetic manipulation of those compounds in engineered plants for enhanced disease resistance [1,13,14].

The question as to whether phytoalexins are active *in vivo* and play a significant role in plant defense mechanisms has long been debated addressing both the actual antimicrobial activity of phytoalexins under the conditions found within plant tissues and their localization around invading organisms [15,16]. These intractable interrogations are indeed crucial to their proposed role as microbial growth regulators in infected plant tissues. Nonetheless there is considerable evidence that these compounds exhibit *in vitro* toxicity across much of the biological spectrum, prokaryotic and eukaryotic.

The nature of the interaction between plants and pathogens largely depends on the ability of the latter to metabolize the phytoalexins to which they are exposed. Engineering of fungal genes responsible for detoxification of phytoalexins in plants has pointed out their role in the interactions between plants and pathogens [17]. In phytopathogenic fungi, ATP-Binding Cassette (ABC) transporters may also extrude plant defense products as well as fungicides. These transporters act as virulence factors providing protection against phytoalexins produced by the host. Many factors thus interplay to affect the outcome of the interaction between plants and pathogens.

It has recently been demonstrated that phytoalexins may also display health-promoting effects in humans. For instance, resveratrol produced by Vitaceae has been acclaimed for its wondrous effects and its wide range of purported healing and preventive powers as a cardioprotective, antitumor, neuroprotective and antioxidant agent as well as an antifungal and antibacterial compound [14] (see Section 8).

Work on phytoalexins has been prolific and the production of these compounds in infected tissues has become one of the most intensively studied mechanisms of disease resistance in plants. This review will focus on some of the main features of phytoalexins:

- Chemical diversity
- Main biosynthetic pathways and regulation networks
- Biological activity against microorganisms
- Molecular engineering for disease resistance in plants
- Metabolism/Transport in fungi
- Role in human health

2. Chemical Diversity of Phytoalexins

Most phytoalexins produced by the Leguminosae belong to six isoflavonoid classes: isoflavones, isoflavanones, pterocarpans, pterocarpenes, isoflavans and coumestans (Table 1) ([1] and references therein). Some pterocarpan phytoalexins are especially well known: pisatin, phaseollin, glyceollin, medicarpin and maackiain. Pisatin was the first phytoalexin to be isolated and characterized from garden pea, *Pisum sativum* [18]. Besides these compounds, a small number of legumes also produce non-isoflavonoid phytoalexins such as furanoacetylenes and stilbenes (Table 1).

Table 1. Phytoalexins from different plant families.

Plant Families (in Alphabetical Order)	Types of Phytoalexins/Examples	References
Amaryllidaceae	Flavans	[19]
Brassicaceae (Cruciferae)	Indole phytoalexins/camalexin	[20]
	Sulfur-containing phytoalexins/brassinin	[21]
Chenopodiaceae	Flavanones/betagarin Isoflavones/betavulgarin	[22]
Compositae	Polyacetylenes/safynol	[23]
Convolvulaceae	Furanosesquiterpenes/Ipomeamarone	[24]
Euphorbiaceae	Diterpenes/casbene	[25]
Poaceae	Diterpenoids:Momilactones; Oryzalexins; Zealexins; Phytocassanes; Kauralexins	[8,26]
	Deoxyanthocyanidins/luteolinidin and apigeninidin	[26,27]
	Flavanones/sakuranetin	[1]
	Phenylamides	[28]
Leguminosae	Isoflavones Isoflavanones Isoflavans Coumestans Pterocarpans/pisatin, phaseollin, glyceollin and maiackiain Furanoacetylenes/wyerone Stilbenes/resveratrol Pterocarpens	[1] and references therein
Linaceae	Phenylpropanoids/coniferyl alcohol	[29]
Malvaceae	Terpenoids naphtaldehydes/gossypol	[11]
Moraceae	Furanopterocarpans/moracins A-H	[30]
Orchidaceae	Dihydrophenanthrenes/loroglossol	[31]
Rutaceae	Methylated phenolic compounds/xanthoxylin	[32]
Umbelliferae	Polyacetylenes/falcarinol	[33]
	Phenolics: xanthotoxin	[34]
	6-methoxymellein	[35]
Vitaceae	Stilbenes/resveratrol	[9]
Rosaceae	Biphenyls/auarperin Dibenzofurans/cotonefurans	[36]
Solanaceae	Phenylpropanoid related compounds Steroid glycoalkaloids Norsequi and sesquiterpenoids Coumarins Polyacetylenic derivatives	[1] and references therein

Chen *et al.*, describe a series of compounds produced by the genus *Tephrosia*, which belongs to the Leguminosae family, and possesses phytoalexin-like activities [37]. Five main classes of phytoalexins have been reported in Solanaceae: phenylpropanoid-related compounds, steroid glycolalkaloids, norsesqui- and sesquiterpenoids, coumarins and polyacetylenic derivatives (Table 1) ([1] and references therein).

Although considerable work has been done on phytoalexins from the Leguminosae and Solanaceae families, it has been recently overshadowed by the discovery of two new phytoalexin classes from the Poaceae [8,26] and Brassicaceae families [20,21]. The main phytoalexins of Poaceae (rice, maize and sorghum) are represented by members of the labdane-related diterpenoid superfamily (zealexins, kauralexins, momilactones, oryzalexins and phytocassanes) [8,26], flavanones, an unusual group of flavonoid phytoalexins, the 3-deoxyanthocyanidins [27] and phenylamides [28] (Table 1 and references therein). The current knowledge on phytoalexins produced by sorghum (3-deoxy-anthocyanidins like luteolinidin and apigeninidin) and maize (zealexins, kauralexins) has been reviewed previously [26]. Indole compounds such as camalexin and brassinin represent the major phytoalexins from the Brassicaceae family (Table 1 and references therein).

Not unexpectedly, phytoalexins from very diverse plant families are represented by many different chemical classes. Naphtaldehyde compounds such as gossypol and its derivatives constitute the main Malvaceae phytoalexins [11] (Table 1). Antifungal polyacetylenes have been isolated as phytoalexins from the Compositae and Umbelliferae families [23,33]. Furanosesquiterpenes and diterpenes constitute the phytoalexins from the Convolvulaceae and Euphorbiaceae families [24,25]. The majority of phytoalexins found in the following plant families are phenolic compounds: flavans in Amaryllidaceae [19], flavanones and isoflavones from Chenopodiaceae [22], Linaceae phenylpropanoids [29], furanopterocarpans in Moraceae [30], dihydrophenanthrenes from Orchidaceae [31], Rutaceae methylated phenolics [32], biphenyls and dibenzofurans in Rosaceae [36], xanthotoxin and 6-methoxymellein in Umbellifereae [34,35] and finally hydroxystilbenes from Vitaceae (Table 1) [1,9,10].

3. Main Biosynthetic Pathways

Various pathways are utilized for producing different phytoalexins. As it is not our goal to describe each of these biosynthetic routes in details, we will simply outline the three most characteristic ones:

(i) The phenylpropanoic-polymalonic acid route
(ii) The methylerythritol phosphate and geranyl-geranyl diphosphate pathway
(iii) The indole phytoalexin pathway

3.1. Phytoalexins Deriving from the Phenylpropanoic-Polymalonic Acid Route

All flavonoid phytoalexins (isoflavonoids, isoflavones, pterocarpans, isoflavans, coumestans and arylbenzofurans) as well as stilbene phytoalexins and derivatives (dihydrophenanthrenes) are formed through the universal phenylpropanoic-polymalonic acid pathway. It begins with phenylalanine and the phenylalanine ammonia lyase (PAL) or to a lesser extent with tyrosine and the tyrosine ammonia lyase (TAL). The obtained *para*-coumaric acid is activated in *para*-coumaroyl-CoA by ligation to a coenzyme A by 4-coumaroyl:CoA ligase (C4L). Subsequently, chalcone synthase (CHS) on the one hand and stilbene synthase (STS) on the other hand use this same substrate and condense it with three successive units of malonyl-CoA, leading respectively to the production of naringenin chalcone, the first C15 intermediate in the flavonoid pathway and resveratrol, the precursor of all stilbenes. The possible biosynthetic routes to the main flavonoid and stilbene-like phytoalexins from the Leguminosae family are illustrated in Figure 1 [10,38,39].

Figure 1. Biosynthetic pathways to the main flavonoid and stilbenoid phytoalexins from the Leguminosae family. (adapted from [10,38,39]). The dashed arrows represent hypothetical steps and the solid arrows denote reactions for which the catalyzing enzymes have been cloned.

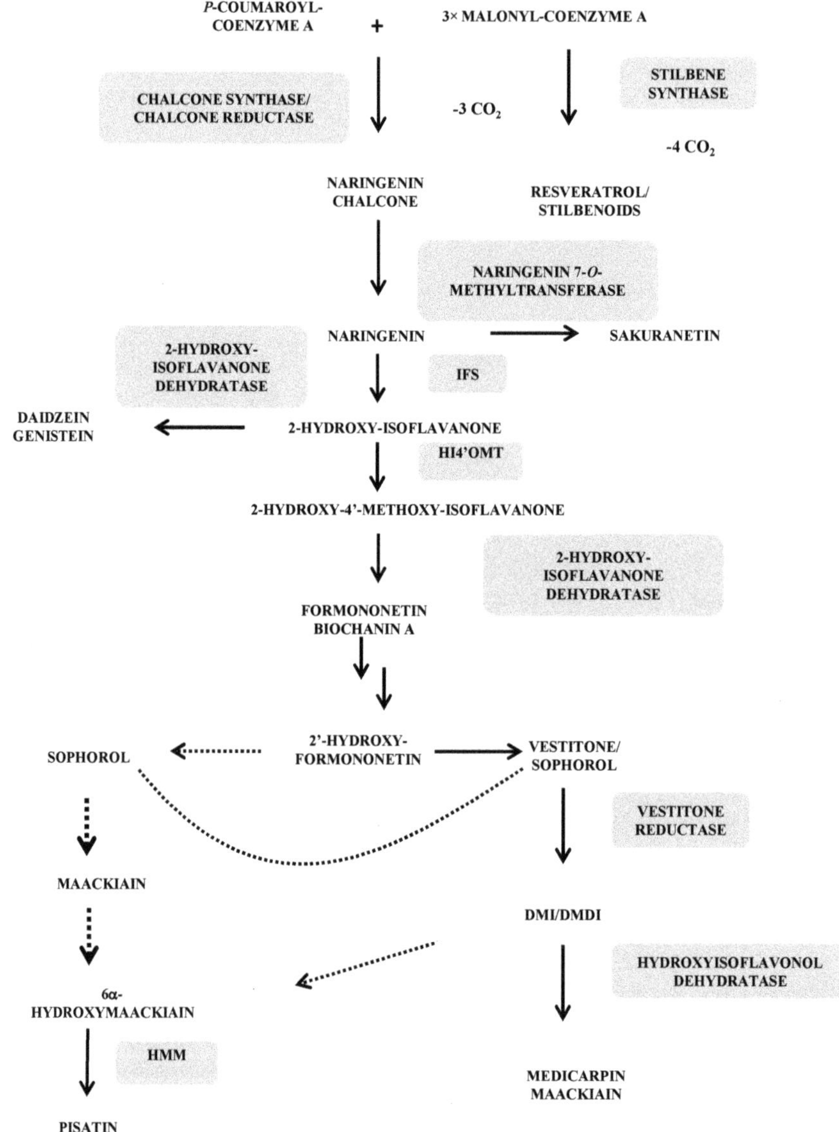

3.2. Mevalonoid-Derived Phytoalexins

These phytoalexins are represented by members of the monoterpene, sesquiterpene, carboxylic sesquiterpene and diterpene families. Specific attention will be given to the diterpene phytoalexin class [8]. This assumption has been confirmed by the observed synchronous accumulation of seven

MEP pathway gene transcripts (OsDXS3, OsDXR, OsCMS, OsCMK, OsMCS, OsHDS and OsHDR) in elicitor-induced rice (*Oryza sativa*) cells and the next steps of this biosynthesis are predicted to occur in plastids. Diterpenoids result from the subsequent action of diverse enzymes using GGDP as the starting block. Class II diterpene cyclases named copalyldiphosphate synthases (CPS) are the first to act on GGDP catalyzing the initial cyclization of the latter to copalyldiphosphate (CDP). CDP is the required substrate for class I diterpene synthases named kaurene synthase like (KSL). Sequential action of CPS and KSL produces the olefin precursors of the main diterpene phytoalexin families [8]. Stereochemically differentiated isomers are used subsequently by KSL: the *ent*-CDP in the biosynthesis of phytocassanes A-E and oryzalexins A-F and the *syn*-CDP in the construction of momilactones A and B (Figure 2). Further additions of oxygen in the formation of oryzalexins, momilactones and phytocassanes require a series of cytochrome P450 (CYPs) (Figure 2).

Figure 2. Biosynthetic pathway of diterpenoid phytoalexins.

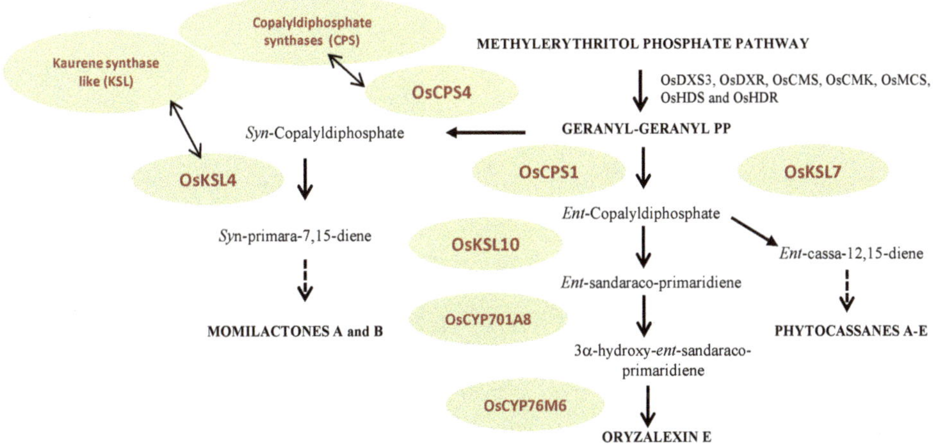

3.3. Indole Phytoalexins

Specific attention will be paid in this section to camalexin, the major phytoalexin of Arabidopsis. The indolic ring of camalexin is derived from tryptophan (Trp) which in turn arises from chorismate (Figure 3). The first step in the route from Trp to camalexin is under the control of two cytochrome P450 homologues CYP79B2 and CYP79B3, leading to indole-3-acetaldoxime. The latter is then transformed into indole-3-acetonitrile (IAN) via the cytochrome P450, CYP71A13. Subsequent conjugation of IAN with glutathione is performed by the combined action of a glutathione-S-transferase and most likely a cytochrome P450. The IAN glutathionyl derivative is then converted into IAN cysteinyl-glycine via a phytochelatin synthase or into γ-glutamyl-cysteine IAN through the action of two γ-glutamyltranspeptidases 1 and 3 [2]. Both intermediates lead to the IAN cysteine conjugate. The last steps of this biosynthesis pathway are under the control of a *CYP71B15* (*PHYTOALEXIN DEFICIENT 3, PAD 3*) gene encoding a multifunctional enzyme which forms camalexin via dihydrocamalexic acid (Figure 3).

Figure 3. Biosynthetic pathway from tryptophan to camalexin (adapted from [2]).

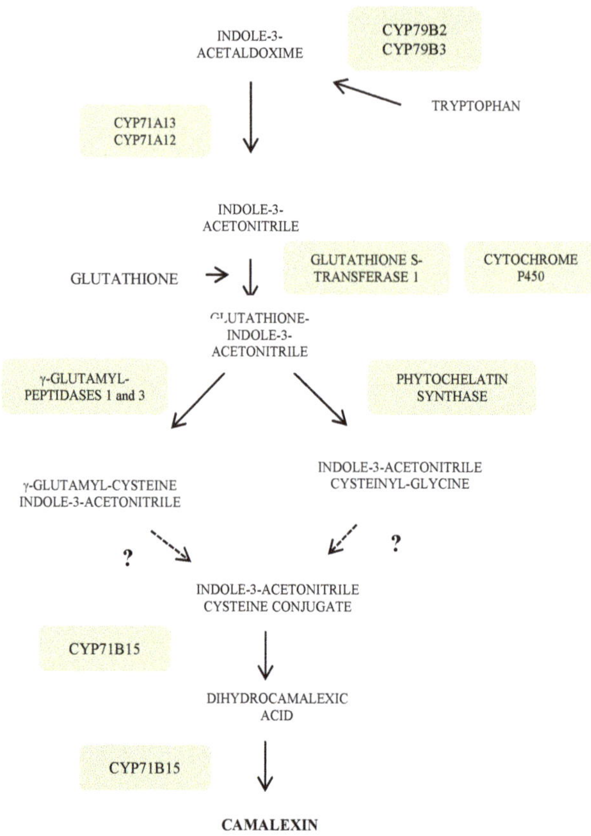

4. Regulation Networks

Phytoalexin biosynthesis is up- or downregulated by expression of many endogenous molecules such as phytohormones (jasmonic acid, salicylic acid, ethylene, auxins, abscisic acid, cytokinins and to a lesser extent gibberellins), transcriptional regulators, defense-related genes, phosphorylation relays and cascades [1,2].

Regulatory mechanisms of phytoalexin biosynthesis also depend on the nature of the infecting pathogen as well as the nature of the induced phytoalexin itself. For example, in the Arabidopsis-*Alternaria brassicicola* interaction, accumulation of camalexin was reported to be independent from jasmonic acid (JA) [40,41] though JA was involved in the regulatory signaling pathways of this phytoalexin in Arabidopsis plants challenged with the fungal pathogen *Botrytis cinerea* [42]. Similarly, existence of JA-dependent and independent pathways in the modulation of diterpenoid phytoalexins in the interaction between rice and the fungal agent *Magnaporthe oryzae* was clearly evidenced by the use of rice mutants [43]. These mutants lacking a functional allene oxide cyclase required for JA production were impaired in momilactone accumulation upon fungal infection whereas phytocassane production was not altered.

Besides, regulation of camalexin production in *Arabidopsis* is controlled by either salicylic acid (SA)-independent [44,45] or SA-dependent signaling pathways [46]. Indeed, biosynthesis of this phytoalexin was also found to be lower in SA-induction deficient mutants of *Arabidopsis* with impaired production of ethylene upon bacterial infection by *Pseudomonas syringae* [47].

Other phytohormones have been involved in the regulatory mechanisms of phytoalexin biosynthesis. Auxins and abscisic acid (ABA) generally appear to negatively regulate phytoalexin production [1]. Suppression of auxin signaling has recently been shown to increase the resistance of Arabidopsis to biotrophic pathogens and to redirect phytoalexin metabolism [48]. The biosynthesis of numerous phytoalexins is downregulated by ABA. For example, synthesis of kievitone in bean [49], synthesis of glyceollin in soybean [50,51] and production of rishitin and lubimin in potato [52] are all decreased by ABA. Tobacco mutants deficient in ABA exhibit twice as much capsidiol as wild-type plants [53]. In contrast, cytokinin overexpression was shown to enhance resistance of tobacco to *P. syringae* [54]. This increased resistance correlated well with the up-regulated synthesis of two phytoalexins, capsidiol and scopoletin.

Mitogen-Activated Protein Kinases (MAPKs) have been involved in the induction of camalexin accumulation in Arabidopsis plants upon treatment with Microbe-Associated Molecular Patterns (MAMPs) [12]. Specifically two MAP kinases, MPK3 and MPK6 take part in the up-regulation of numerous enzymes of the camalexin biosynthetic route. For example, expression of the CYP71B15 gene, which encodes the multifunctional enzyme acting at the end of the pathway showed a 400-fold increase upon overexpression of these two MAPKs [12]. In Arabidopsis *mpk3/mpk6* double mutants, camalexin production was completely abolished concomitantly with an increased susceptibility to *B. cinerea.*

Protein phosphorylation-induced phytoalexin production is also under the control of cell calcium transfers which in turn are decoded and transmitted by a toolkit of calcium binding proteins [1]. Several families of calcium sensors are indeed involved in the phytoalexin regulation networks. For instance, overexpression of two genes encoding a calcineurin B-like protein-interacting protein kinase in rice was found to induce two phytoalexin classes, phytocassanes and momilactones upon MAMP treatment [55].

Other regulators of phytoalexin biosynthesis have been identified [1]. Overexpression of Rac proteins in rice induced disease resistance to bacterial blight together with a 19- to 180-fold increase in the accumulation of the rice phytoalexin momilactone A [56]. Production of this phytoalexin is also controlled by selenium-binding protein homologues as shown in the interaction between rice and both the rice blast fungus and the rice bacterial blight [57]. Overexpression of microbial virulence factors belonging to the Nep1-like protein family in *Arabidopsis* was associated with a strong transcriptional activation of genes involved in the camalexin route [58]. Various sugars (sucrose, glucose and fructose) acting as endogenous signals, have been reported to regulate the biosynthesis and accumulation of some phytoalexins [59]. Finally, overexpression of non-expressor of pathogenesis-related genes-1 which play a critical role in the systemic acquired resistance was reported to induce the biosynthesis of the cotton phytoalexin gossypol [60]. Knowledge of the regulatory mechanisms of phytoalexin biosynthesis thus paves the way for metabolic engineering of plants for disease resistance (see Section 6).

5. Biological Activity against Microorganisms

Are phytoalexins biologically active compounds? Do phytoalexins show antibacterial activities? Over 70 years after their discovery, the actual role of phytoalexins in plant defense mechanisms is still debated. Phytoalexins are considerably less toxic than chemical fungicides. Lack of activity of isoflavonoid phytoalexins was indeed reported in comparison to classic fungicides like benomyl and mancozeb [61]. Effective doses of phytoalexins generally fall within orders of magnitude 10^{-5} to 10^{-4} M [62,63]. Phytoalexin fungitoxicity is clearly evidenced by the inhibition of germ-tube elongation, radial mycelial growth and/or mycelia dry weight increase, as best illustrated by the action of resveratrol on *B. cinerea*, the causal agent for gray mold in grapevine [63,64]. Phytoalexin antifungal activity can considerably vary from one compound to another. For example, Hasegawa *et al.*, show that the rice phytoalexin sakuranetin displays a higher activity against the blast fungus than does another rice phytoalexin, momilactone A, both *in vivo* and *in vitro* [65].

Phytoalexins may also exert some effects on the cytological, morphological and physiological characteristics of fungal cells. The activity of four phytoalexins from the Solanaceae family (rishitin, phytuberin, anhydro-β-rotunol and solavetivone) on three *Phytophthora* species resulted in loss of motility of the zoospores, rounding-up of the cells associated with some level of swelling, cytoplasmic granulation and bursting of the cell membrane [66]. The two latter are very general features of the action of phytoalexins on fungal cells ([63,64,67] and references therein). The extensive membrane damage occurring after fungal exposure to phytoalexins is reflected in substantial leakage of electrolytes and metabolites [68]. However, it has been observed that despite the presence of wyerone acid or resveratrol, surviving *B. cinerea* fungal cells could produce secondary and to a lesser extent tertiary germ tubes suggesting that some sort of escape from phytoalexin damage could take place [63,69]. Asymetric growth of the germ tube resulting in the production of "curved-germ tubes" has also been observed in *B. cinerea* conidia treated with sub-lethal doses of resveratrol [63]. This cytological abnormality suggests that stilbenic compounds may interact with tubulin polymerization, the mode of action of many synthetic fungicides and anticancer agents [70]. Moreover, phytoalexins may affect glucose uptake by fungal cells as reported in the interactions between phaseollin or kievitone/and *Rhizoctonia solani* [68]. Observations of *B. cinerea* conidia showed a complete disorganization of mitochondria and disruption of the plasma membrane upon treatment with the stilbene phytoalexins, resveratrol and pterostilbene [63,64,67]. Pterostilbene especially led to a rapid and complete cessation of respiration in *B. cinerea* conidia which can be explained by its activity as an uncoupling agent of electron transport and phosphorylation [67]. Camalexin has recently been involved in the induction of fungal apoptotic programmed cell death in *B. cinerea* [71]. The efficaciousness *in vivo* of some phytoalexins, namely the coumarin phytoalexin, scopoletin on the reduction of green mold symptoms caused by *Penicillium digitatum* on oranges was shown [72]. In the same way, phenolic phytoalexins (resveratrol, scopoletin, scoparone and umbelliferone) were shown to significantly inhibit the growth of *Penicillium expansum* and patulin accumulation in apples [73]. To increase the fungitoxicity of phytoalexins, design and synthesis of more active phytoalexin derivatives is needed [74,75].

Beside their antifungal activity, phytoalexins possess some antibacterial activity. Rishitin for instance decreased the viability of cells of *Erwinia atroseptica* by around 100% at a dose of 360 μg/L [76]. Resveratrol also exerts some activity against numerous bacteria affecting humans: *Chlamydia*, *Helicobacter*, *Staphylococcus*, *Enterococcus*, *Pseudomonas* and *Neisseria* ([14] and references therein). It is thus clear that phytoalexins exhibit toxicity across much of the biological spectrum, prokaryotic and eukaryotic.

6. Engineering of Phytoalexins and Role in Plant Defense Mechanisms

Gain- or loss-of-function genetic approaches addressing phytoalexin production for disease resistance have provided direct and indirect proofs of their implication in plant/microorganism interactions. Relatively simple genetic constructs involving the introduction of a single gene in plants are required in the case of the grapevine phytoalexin resveratrol, synthesis of which is controlled by the stilbene synthase (STS). The first report of increased disease resistance resulting from foreign phytoalexin expression in a novel plant was brought by the group of Kindl with the transfer of two grapevine STS genes (*Vst 1* and *Vst 2*) into tobacco [77]. Introduction of these two genes was shown to confer higher resistance to *B. cinerea*. From that point, a number of transformations were then completed in alfalfa, rice, barley, wheat, tomato, papaya and Arabidopsis using the same STS genes or STS genes from other plant origins, conferring resistance to various pathogens [1]. All these results clearly showed that phytoalexins could act as determinant factors in the expression of the defense mechanisms of plants against phytopathogenic microorganisms though there are rare examples of STS overexpression not being associated with disease resistance [1].

Following the works on stilbene phytoalexins, other genetic transformations were achieved with other phytoalexin genes. Surprisingly, engineering phytoalexins seems to have been limited to exploiting only a few phytoalexin biosynthetic genes. This has mainly concerned the genetic manipulation of phytoalexin glycosylation by the use of a tobacco glucosyltransferase acting on scopoletin [78]. Overexpression of the isoflavonoid-7-*O-methyltransferase* in alfalfa, an enzyme with a crucial role in the biosynthesis of the phytoalexin maiackiain, was also linked to an increased resistance of that plant to *Phoma medicaginis* [79]. Transformation of soybean hairy roots with both the peanut resveratrol synthase 3 *AhRS3* gene and resveratrol-*O-methyltransferase ROMT* gene catalyzing the transformation of resveratrol to pterostilbene [80] resulted in the resistance of that plant to *Rhizoctonia solani* [81]. In many cases, engineering the entire phytoalexin biosynthetic pathway is not feasible and the problem researchers are facing is to choose the right enzyme catalyzing the limitant step of this pathway.

Loss-of-function genetic approaches clearly evidenced the role played by phytoalexins in plant-microorganism interactions. In almost all experiments, mutants impaired in phytoalexin production showed increased susceptibility to pathogens. Reduced amounts of pisatin in hairy roots of pea transformed with antisense 6-α-hydroxymaiackiain-3-*O*-methyltransferase were associated with a decreased resistance to the fungal pathogen *Nectria haematococca* [17]. RNAi silencing of isoflavone synthase or chalcone reductase in soybean suppressed by 90% the accumulation of daidzein and glyceollin as well as disease resistance to *P. sojae* [82]. Loss-of-function alleles of the *yellow seed1* gene encoding CHS, chalcone isomerase, dihydroflavonol reductase and flavonoid-3'-

hydroxylase induced deficiency in the accumulation of 3-deoxyanthocyanidin associated with severe symptoms of the anthracnose disease in sorghum [83]. Effect of the *PHYTOALEXIN DEFICIENT* mutation on camalexin in Arabidopsis was found to be dependent on the infecting pathogen. This mutation was not associated with increased susceptibility to *P. syringae*, *Perenospora parasitica*, *Erysiphae oronti* and *B. cinerea* though it markedly affected its susceptibilty to *A. brassicicola* ([1] and references therein). Finally loss-of-function genetic approaches have underlined the role of phytoalexin glycosylation in plant-pathogen interactions. Transgenic tobacco leaves downregulated for a tobacco specific phenylpropanoid glucosyltransferase saw their scopolin content decreased by 70% to 75% associated with a 63% increase in TMV lesion surfaces [84].

Indirect modulation of phytoalexin levels through manipulation of hormone signaling, phosphorylation cascades or defense-related marker genes also demonstrated the role of phytoalexins in plant defense mechanisms. For instance cytokinin overexpression in tobacco led to increased resistance to *P. syringae* which strongly correlated with up-regulated synthesis of two phytoalexins, capsidiol and scopoletin [54]. Mutations in two MAP kinases MPK3 and MPK6 impaired camalexin production and disease resistance to *B. cinerea* in Arabidopsis [12].

Though phytoalexin engineering seems to have been limited to exploiting only a few genes mainly stilbene and isoflavonoid ones, indirect modulation of phytoalexin accumulation employing transcriptional regulators or components of upstream regulatory pathways becomes a useful approach to improve plant disease resistance [1].

7. Fungal Metabolism and Transporters

The interaction between a plant and its pathogen can be envisaged as a balance between phytoalexin production by the host and phytoalexin metabolism or inactivation via transporters (mainly ATP Binding Cassette, ABC transporters) by the second actor of this interaction. Modification of any of the factors contributing to this balance could modify the outcome of the interaction. We will further see that in phytopathogenic fungi, ABC transporters act as virulence factors, conferring protection against defense compounds produced by the host. In several plant-fungus interactions, it has become evident that the ability to weaken or neutralize the effects of phytoalexins is one of the essential determinants of fungal/host coupling. Detoxification processes of phytoalexins by fungi are far from being clearly understood [85].

It is rather difficult to derive any comprehensive generalizations from the existing data on phytoalexin metabolism *per se*. The known catabolic pathways of phytoalexins by fungi may involve monoxygenation, reduction, hydration, oxidation, oxidative dimerization, glycosylation and demethylation reactions. Since most phytoalexins are lipophilic compounds that efficiently penetrate cell membrane structures, phytoalexin metabolism usually involves their conversion to more polar products. Creation of new hydroxyl groups by oxygenation, demethylation, reduction of aldehydes and ketones or hydration of double bonds as well as glycosylation increase the degree of polarity of phytoalexins. Detoxification of the cruciferous phytoalexins, brassinin, 1-methoxybrassinin and cyclobrassinin, by the stem rot fungus *Sclerotinia sclerotiorum* indeed requires a brassinin glucosyltransferase [86]. Methylated phytoalexins are well known to be more fungitoxic than the non-methylated ones owing to the fact that phytoalexin methylation enhances their lipophilic character. Moreover the presence of methylated groups or any other electron-attracting groups on the aromatic ring of some phytoalexins plays an important role in the formation of charge transfer complexes, favoring contact and affinity with (membrane) proteins and acting as uncoupling agents of electron transport and photophosphorylation. A cytochrome P450 pisatin demethylase transforming pisatin into 6-α-hydroxymaiackiain and 3-hydroxymaackiain-isoflavan was characterized from the fungal pathogen *Nectria haematococca* [87]. Interestingly, fungal isolates with the highest pisatin demethylating activity were shown to be the most virulent on pea [87]. In addition, overexpression of this pisatin demethylating activity in hairy roots of pea resulted in reduced amounts of this phytoalexin in the plant tissues upon infection by *N. haematococca* with a correlated decreased resistance to this pathogen [17]. Two hydratases, a kievitone hydratase from *Fusarium solani* [88] and one inducible hydrolase of *Leptosphaeria maculans* acting on brassinin [89] were implicated in the detoxification process of these two phytoalexins.

Oxidation and oxidative dimerization processes can also take place in the metabolism of phytoalexins. A brassinin detoxifying oxidase with a molecular mass of 57 kDa has been characterized and purified from the blackleg fungus *L. maculans*. This oxidase transforms the cruciferous phytoalexin brassinin into the less fungitoxic compound indol-3-carboxaldehyde [90]. Interestingly, this pathogen was unable to metabolize camalexin, another major phytoalexin from crucifers, conferring this plant family protection against *L. maculans*. Simple stilbenes produced by members of the Vitaceae family may undergo oxidative dimerization by a laccase-like stilbene oxidase from *B. cinerea* with a molecular mass of 32 kDa [91]. This process involves the 4'-hydroxyphenyl group of one resveratrol unit leading to a dehydrodimer with a dihydrobenzofuran structure named δ-viniferin [92,93]. Importantly, *B. cinerea* isolates possessing the highest oxidative activity were found to be the most virulent on grapevine [94]. Very recently, the stilbene-type phytoalexin astringin produced by Norway spruce in the interaction with the bark beetle (*Ips* spp.) and its fungal associate, *Ceratocystis polonica* was shown to undergo metabolism by the latter [95]. *C. polonica* converted astringin to ring-opened lactones, aglycones and dehydrodimers *in vitro*. In this study, the virulence of the fungal pathogen on Norway spruce correlated well with differential usage of the various pathways for stilbene biotransformation.

Phytopathogenic fungi evolved mechanisms of insensitivity or resistance to protect themselves against phytoalexins. One of them involves extruding toxic compounds out of the cell through transporters, conferring them protection against plant defense products. In fungi, the role of ATP-binding cassette (ABC) transporters in the efflux of natural and synthetic toxicants is well known [96–98]. In addition, several genes encoding ABC transporters have been shown to be involved in fungal virulence on plant hosts [99–104]. A number of phytoalexins and other toxic compounds induced expression of these fungal transporters [99,100,105–107]. However, only a few studies have demonstrated the ability of these transporters to confer tolerance to a known phytoalexin. In *B. cinerea*, the BcatrB ABC transporter has been shown to be a virulence factor that increases tolerance of the pathogen towards phytoalexins. Indeed, *BcatrB* replacement mutants revealed increased sensitivity to resveratrol and reduced virulence on grapevine leaves [100]. Moreover, a *B. cinerea* strain lacking functional BcatrB was more sensitive to camalexin *in vitro* and less virulent on *A. thaliana* wild-type plants, but was fully virulent on camalexin-deficient *A. thaliana* mutants [103]. In the same way, ABC transporters related to *BcatrB*, such as GpABC1 from *Gibberella pulicaris*, act as virulence factors on potato. GpABC1 provides tolerance to rishitin, while a *GpABC1* mutant is essentially non-pathogenic [101]. In *N. haematococca*, the *NhABC1* gene is induced after treatment with pisatin *in vitro* and during infection of pea plants. Mutation in *NhABC1* gene rendered the fungus even more sensitive to pisatin and led to lower pathogenicity on pea, indicating that *NhABC1* contributes to the tolerance to pisatin and acts as a virulence factor [102]. The substrate range of ABC transporters can vary from a single compound to a wide spectrum of molecules with no identified common feature. BcatrB from *B. cinerea* has a wide substrate range, comprising mainly aromatic compounds [100,103,107–109] such as the phytoalexins eugenol, resveratrol and camalexin, the fungicides fenpiclonil and fludioxonil, as well as the antibiotics phenazine-1-carboxylic acid and phenazine-1-carboxamide. The closest identified homologue of BcatrB, AtrB from *Aspergillus nidulans*, shows a similar function in multidrug resistance [110]. Two other homologues with similar substrate ranges are MgAtr5 from *Mycosphaerella graminicola* [111] and PMR5 from *Penicillium digitatum* [112], indicating that other homologues of BcatrB may function also in multidrug resistance and pathogenesis. Taken together, it becomes evident that ABC transporters can be essential for the development of phytopathogenic fungi, providing protection against phytoalexins produced by the host plant and acting as virulence factors.

8. Role of Phytoalexins in Human Health

Phytoalexins may display health-promoting effects in humans. A few of them have been reported to exert antioxidant, anticarcinogenic and cardiovascular protective activities. Maslinic acid, a natural phytoalexin-type triterpene from olives exerts a wide range of biological activities as an antitumor, antidiabetic, neuroprotective, cardioprotective, antiparasitic and growth-stimulating agent, providing evidence of the potential of this molecule as a nutraceutical [113]. Health benefit properties of Brassicaceae were attributed in part to their phytoalexins, camalexin and related indolic compounds [114]. Camalexin namely is able to induce apoptosis in prostate cancer cells [115]. 3-deoxyanthocyanidins, flavonoid phytoalexins produced by members of the Poaceae family, are helpful in reducing the incidence of gastro-intestinal cancer [116]. Moreover, other indolic phytoalexins such as brassinin and its derivative, homobrassinin, show marked antiproliferative activities in human colorectal cancer cells *in vitro* [117]. The most promising results in this area were obtained with resveratrol, the phytoalexin from grapevine. This compound is indeed considered to be an antiproliferative agent exerting antitumor activity either as a cytostatic or a cytotoxic agent in various cancers [118]. The first report of the cancer chemoprotective activity of a phytoalexin is the study of the group of Pezzuto [119]. This pioneering work was then confirmed on many other human cancer models. The most frequently described mode of antitumor action for phytoalexins concerns apoptosis which may be via the inhibition of antiapoptotic molecules such as survivin [120] or, for instance as reported in this issue, alterations of expression and activity of lysosomal protease cathepsin D [115].

Resveratrol exerts antitumor activities *in vivo*, namely in skin cancers by topical applications. It does not seem to be very effective in inhibiting leukemia despite displaying antileukemic activity *in vitro* and shows anticancer effects in experimentally-induced breast cancers only at high doses.

Resveratrol presents some anticancer activities in hepatoma, lung carcinoma and intestinal tumors ([14] and references therein). However lack of efficacy of natural phytoalexins in reducing tumors has led to a number of investigations regarding the design and synthesis of more potent anticancer derivatives of known phytoalexins such as brassinin [121], methoxybrassinol [122] and resveratrol [123].

There are also several studies providing evidence of the cardioprotective activity of phytoalexins such as indoles and stilbenes [14,114]. Resveratrol namely was proven to inhibit LDL peroxidation in *ex vivo* rat heart studies, to have a potent role in preventing atherosclerosis and to block platelet aggregation from high-cholesterol-fed rabbits ([14] and references therein). Besides, this compound has an effect in neurological diseases such as cerebral ischemia, Parkinson's disease, pain and cognitive impairment in rats, spinal cord lesion in rabbits and finally brain edema and tumors in human cells ([14] and references therein). Interestingly, additional studies have also demonstrated that resveratrol increases lifespan in lower organisms (yeast, metazoans) and higher organisms through the activation of the sirtuin proteins [124,125]. Resveratrol's mechanisms of action are likely to be pleitropic and mediated by the interaction of this compound with key signaling proteins controlling cellular calcium homeostasis [126]. Interestingly, quercetin and umbelliferone were also reported to reduce mycotoxin accumulation in apple fruits by *P. expansum* by down-regulating relative expression of genes encoding patulin biosynthesis [127].

Some other phytoalexins like the steroid glycoalkaloids from potato or the dimeric sesquiterpene gossypol from cotton display a certain level of toxicity for humans explaining the crucial interest in engineering those plants for abolishing production of these undesirable compounds [11,128].

9. Concluding Remarks

Works on phytoalexins from diverse chemical families have generated a lot of data regarding basic aspects of plant defenses and their regulatory mechanisms. As a result, engineering of phytoalexins has arisen as a new area in the development of useful approaches to disease control. Nonetheless, while a variety of genetic transfers were carried out in order to investigate the potential of stilbene and flavonoid phytoalexin biosynthetic genes in conferring disease resistance, strategies focusing on the other phytoalexin chemical families did not [1].

Some studies have attempted to determine the actual concentration and the nature of phytoalexins directly in plant tissues in response to invading microorganisms using spectroscopic methods [129,130]. However, our general knowledge remains limited by the difficulty to analyse the events occurring under natural conditions between the plant and the pathogen.

On the other hand, the potential value of several phytoalexins on a therapeutic point of view has made their large-scale production a necessity. Engineering yeast and bacteria, may represent valuable means for the production of phytoalexins at an industrial scale [14,131]. However, their tailoring is needed as they do not possess the genes encoding phytoalexin biosynthesis. Another approach is large-scale production of phytoalexins using plant cell suspensions in bioreactors. Some experiments are underway to optimize stilbene phytoalexin production in bioreactors [132,133].

Although considerable work has already been done on phytoalexins, the ways in which they act against microorganisms and the mechanisms the latter have developed to counteract their action are still poorly understood keeping this subject an active field of research even after over 70 years.

Author Contributions

Philippe Jeandet wrote the paper. Marie-Alice Deville edited the whole manuscript. Aziz Aziz, Stephan Dorey and Sylvain Cordelier checked the regulatory and molecular engineering sections. Claire Hébrard and Jérôme Crouzet wrote the ABC transporter section.

Abbreviation

IFS: 2-hydroxy isoflavanone synthase; DMI: 7,2'-dihydroxy-4'-methoxy-isoflavanol; DMDI: 7,2'-dihydroxy-4',5'-methylenedioxy-isoflavanol; HMM: 6α-hydroxymaackiain 3-*O*-methyltransferase; HI4'OMT: SAM: 2,7,4'-trihydroxy-isoflavanone 4'-*O*-methyltransferase; STS: stilbene synthase; CHS: chalcone synthase.

Conflicts of Interest

The authors declare no conflict of interest.

References

1. Jeandet, P.; Clément, C.; Courot, E.; Cordelier, S. Modulation of phytoalexin biosynthesis in engineered plants for disease resistance. *Int. J. Mol. Sci.* **2013**, *14*, 14136–14170.
2. Ahuja, I.; Kissen, R.; Bones, A.M. Phytoalexins in defense against pathogens. *Trends Plant Sci.* **2012**, *17*, 73–90.
3. Müller, K.O.; Börger, H. Experimentelle Untersuchungen über die *Phytophthora* Resistenz der Kartoffel. *Arb. Biol. Reichsanst. Land Forstwirtsch.* **1940**, *23*, 189–231.
4. Deverall, B.J. Introduction. In *Phytoalexins*; Bailey, J.A., Mansfield, J.W., Eds.; Blackie: Glasgow/London, UK, 1982; pp. 1–20.
5. Ingham, J.L. Phytoalexins from the Leguminosae. In *Phytoalexins*; Bailey, J.A., Mansfield, J.W., Eds.; Blackie: Glasgow/London, UK, 1982; pp. 21–80.
6. Kuc, J. Phytoalexins from the Solanaceae. In *Phytoalexins*; Bailey, J.A., Mansfield, J.W., Eds.; Blackie: Glasgow/London, UK, 1982; pp. 81–105.
7. Coxon, D.T. Phytoalexins from other plant families. In *Phytoalexins*; Bailey, J.A., Mansfield, J.W., Eds.; Blackie: Glasgow/London, UK, 1982; pp. 106–132.
8. Schmelz, E.A.; Huffaker, A.; Sims, J.W.; Christensen, S.A.; Lu, X.; Okada, K.; Peters, R.J. Biosynthesis, elicitation and roles of monocot terpenoid phytoalexins. *Plant J.* **2014**, *79*, 659–678.
9. Langcake, P.; Pryce, R.J. The production of resveratrol by *Vitis vinifera* and other members of the *Vitaceae* as a response to infection or injury. *Physiol. Plant Pathol.* **1976**, *9*, 77–86.

10. Jeandet, P.; Delaunois, B.; Conreux, A.; Donnez, D.; Nuzzo, V.; Cordelier, S.; Clément, C.; Courot, E. Biosynthesis, metabolism, molecular engineering and biological functions of stilbene phytoalexins in plants. *BioFactors* **2010**, *36*, 331–341.
11. Sunilkumar, G.; Campbell, L.M.; Pukhaber, L.; Stipanovic, R.D.; Rathore, K.S. Engineering cottonseed for use in human nutrition by tissue-specific reduction of toxic gossypol. *Proc. Natl. Acad. Sci. USA* **2006**, *103*, 18054–18059.
12. Ren, D.; Liu, Y.; Yang, K.Y.; Han, L.; Mao, G.; Glazebrook, J.; Zhang, S. A fungal-responsive MAPK cascade regulates phytoalexin biosynthesis in *Arabidopsis*. *Proc. Natl. Acad. Sci. USA* **2008**, *105*, 5638–5643.
13. Delaunois, B.; Cordelier, S.; Conreux, A.; Clément, C.; Jeandet, P. Molecular engineering of resveratrol in plants. *Plant Biotechnol. J.* **2009**, *7*, 2–12.
14. Jeandet, P.; Delaunois, B.; Aziz, A.; Donnez, D.; Vasserot, Y.; Cordelier, S.; Courot, E. Metabolic engineering of yeast and plants for the production of the biologically active hydroxystilbene, resveratrol. *J. Biomed. Biotechnol.* **2012**, doi:10.1155/2012/579089.
15. Mansfield, J.W.; Bailey, J.A. Phytoalexins: Current problems and future prospects. In *Phytoalexins*; Bailey, J.A., Mansfield, J.W., Eds.; Blackie: Glasgow/London, UK, 1982; pp. 319–323.
16. Hammerschmidt, R. Phytoalexins: What have we learned after 60 years? *Annu. Rev. Phytopathol.* **1999**, *37*, 285–306.
17. Wu, Q.; VanEtten, H.D. Introduction of plant and fungal genes into pea (*Pisum sativum* L.) hairy roots reduces their ability to produce pisatin and affects their response to a fungal pathogen. *Mol. Plant-Microbe Interact.* **2004**, *17*, 798–804.
18. Cruickshank, I.A.M.; Perrin, D.R. Isolation of a phytoalexin from *Pisum sativum* L. *Nature* **1960**, *187*, 799–800.
19. Coxon, D.T.; O'Neill, T.M.; Mansfield, J.W.; Porter, A.E.A. Identification of three hydroxyflavan phytoalexins from daffodil bulbs. *Phytochemistry* **1980**, *19*, 889–891.
20. Browne, L.M.; Conn, K.L.; Ayert, W.A.; Tewari, J.P. The camalexins: New phytoalexins produced in the leaves of *Camelia sativa* (Cruciferae). *Tetrahedron* **1991**, *47*, 3909–3914.
21. Pedras, M.S.C.; Okanga, F.I.; Zaharia, I.L.; Khan, A.G. Phytoalexins from crucifers: Synthesis, biosynthesis and biotransformation. *Phytochemistry* **2000**, *53*, 161–176.
22. Geigert, J.; Stermitz, F.R.; Johnson, G.; Maag, D.D.; Johnson, D.K. Two phytoalexins from sugarbeet (*Beta vulgaris*) leaves. *Tetrahedron* **1973**, *29*, 2703–2706.
23. Allen, E.H.; Thomas, C.A. Trans-trans-3,11-tridecadiene-5,7,9-triyne-1,2-diol, an antifungal polyacetylene from diseased safflower (*Carthamus tinctorius*). *Phytochemistry* **1971**, *10*, 1579–1582.
24. Uritani, I.; Uritani, M.; Yamada, H. Similar metabolic alterations induced in sweet potato by poisonous chemicals and by *Ceratostomella fimbriata*. *Phytopathology* **1960**, *50*, 30–34.
25. Sitton, D.; West, C.A. Casbene: An antifungal diterpene produced in cell-free extracts of *Ricinus communis* seedlings. *Phytochemistry* **1975**, *14*, 1921–1925.
26. Poloni, A.; Schirawski, J. Red card for pathogens: Phytoalexins in sorghum and maize. *Molecules* **2014**, *19*, 9114–9133.

27. Lo, S.C.; de Verdier, K.; Nicholson, R. Accumulation of 3-deoxyanthocyanidin phytoalexins and resistance to *Colletotrichum sublineolum* in sorghum. *Physiol. Mol. Plant Pathol.* **1999**, *55*, 263–273.
28. Lin Park, H.; Lee, S.W.; Jung, K.H.; Hahn, T.R.; Cho, M.H. Transcriptomic analysis of UV-treated rice leaves reveals UV-induced phytoalexin biosynthetic pathways and their regulatory networks in rice. *Phytochemistry* **2013**, *96*, 57–71.
29. Keen, N.T.; Littlefield, L.J. The possible association of phytoalexins with resistant gene expression in flax to *Melamspora lini*. *Physiol. Plant Pathol.* **1975**, *14*, 275–280.
30. Takasugi, M.; Nagao, S.; Masamune, T.; Shirata, A.; Takahashi, K. Structures of moracins E, F, G and H, new phytoalexins from diseased mulberry. *Tetrahedron Lett.* **1979**, *28*, 4675–4678.
31. Ward, E.W.B.; Unwin, C.H.; Stoessel, A. Loroglossol: An orchid phytoalexin. *Phytopathology* **1975**, *65*, 632–633.
32. Hartmann, G.; Nienhaus, F. The isolation of xanthoxylin from the bark of *Phytophthora*- and *Hendersonula*-infected *Citrus lemon* and its fungitoxic effect. *Phytopathol. Z.* **1974**, *81*, 97–113.
33. Harding, V.K.; Heale, J.B. The accumulation of inhibitory compounds in the induced resistance response of carrot root slices to *Botrytis cinerea*. *Physiol. Plant Pathol.* **1981**, *18*, 7–15.
34. Johnson, C.; Brannon, D.R.; Kuc, J. Xanthotoxin: A phytoalexin of *Pastinaca sativa* root. *Phytochemistry* **1973**, *12*, 2961–2962.
35. Condon, P.; Kuc, J.; Draudt, H.N. Production of 3-methyl-6-methoxy-8-hydroxy-3,4-dihydroisocoumarin by carrot root tissue. *Phytopathology* **1963**, *53*, 1244–1250.
36. Kokubun, T.; Harborne, J.B. Phytoalexin induction in the sapwood of plants of the Maloideae (Rosaceae): Biphenyls or dibenzofurans. *Phytochemistry* **1995**, *40*, 1649–1654.
37. Chen, Y.; Tao, Y.; Gao, C.; Cao, W.; Huang, R. Natural products from the genus *Tephrosia*. *Molecules* **2014**, *19*, 1432–1458.
38. Deavours, B.E.; Dixon, R.A. Metabolic engineering of isoflavonoid biosynthesis in alfalfa. *Plant Physiol.* **2005**, *138*, 2245–2259.
39. Kaimoyo, E.; VanEtten, H.D. Inactivation of pea genes by RNAi supports the involvement of two similar *O*-methyltransferases in the biosynthesis of (+)-pisatin and of chiral intermediates with a configuration opposite that found in (+)-pisatin. *Phytochemistry* **2008**, *69*, 76–87.
40. Thomma, B.P.H.J.; Nelissen, I.; Eggermont, K.; Broekaert, W.F. Deficiency in phytoalexin production causes enhanced susceptibilty of *Arabidopsis thaliana* to the fungus *Alternaria brassicicola*. *Plant J.* **1999**, *19*, 163–117.
41. Van Wees, S.C.; Chang, H.S.; Zhu, T.; Glazebrook, J. Characterization of the early response of Arabidopsis to *Alternaria brassicicola* infection using expression profiling. *Plant Physiol.* **2003**, *132*, 606–617.
42. Rowe, H.C.; Walley, J.W.; Corwin, J.; Chan, E.K.F.; Dehesh, K.; Kliebenstein, D.J. Deficiencies in jasmonate-mediated plant defense reveal quantitative variation in *Botrytis cinerea* pathogenesis. *PLoS Pathog.* **2010**, *6*, e1000861.
43. Riemann, M.; Haga, K.; Shimizu, T.; Okada, K.; Ando, S.; Mochizuki, S.; Nishizawa, Y.; Yamanouchi, U.; Nick, P.; Yano, M.; *et al.* Identification of rice *Allene Oxide Cyclase* mutants

and the function of jasmonate for defence against *Magnaporthe oryzae*. *Plant J.* **2013**, *74*, 226–238.

44. Nawrath, C.; Métraux, J.P. Salicylic acid induction-deficient mutants of *Arabidopsis* express *PR-2* and *PR-5* and accumulate high levels of camalexin after pathogen inoculation. *Plant Cell* **1999**, *11*, 1393–1404.
45. Roetschi, A.; Si-Ammour, A.; Belbahri, L.; Mauch, F.; Mauch-Mani, B. Characterization of an *Arabidopsis-Phytophthora* pathosystem: Resistance requires a functional PAD2 gene and is independent of salicylic acid, ethylene and jasmonic acid signalling. *Plant J.* **2001**, *28*, 293–305.
46. Denby, K.J.; Jason, L.J.M.; Murray, S.L.; Last, R.L. *ups1*, an *Arabidopsis thaliana* camalexin accumulation mutant defective in multiple defence signalling pathways. *Plant J.* **2005**, *41*, 673–684.
47. Heck, S.; Grau, T.; Buchala, A.; Métraux, J.P.; Nawrath, C. Genetic evidence that expression of NahG modifies defence pathways independent of salicylic acid biosynthesis in the *Arabidopsis-Pseudomonas syringae* pv. tomato interaction. *Plant J.* **2003**, *36*, 342–352.
48. Robert-Seilaniantz, A.; MacLean, D.; Jikumaru, Y.; Hill, L.; Yamaguchi, S.; Kamiya, Y.; Jones, J.D.G. The microRNA miR393 re-directs secondary metabolite biosynthesis away from camalexin and towards glucosinates. *Plant J.* **2011**, *67*, 218–231.
49. Goosens, J.F.V.; Vendrig, J.C. Effects of abscissic acid, cytokinins, and light on isoflavonoid phytoalexin accumulation in *Phaseolus vulgaris*. *Planta* **1982**, *154*, 441–446.
50. Ward, E.W.; Cahill, D.M.; Bhattacharyya, M.K. Abscisic acid suppression of phenylalanine ammonia-lyase activity and mRNA, and resistance of soybeans to *Phytophthora megasperma* f.s.p. *glycinea*. *Plant Physiol.* **1989**, *91*, 23–27.
51. Mohr, P.; Cahill, D.M. Relative roles of glyceollin, lignin and the hypersensitive response and the influence of ABA in compatible and incompatible interactions of soybeans with *Phytophthora sojae*. *Physiol. Mol. Plant Pathol.* **2001**, *58*, 31–41.
52. Henfling, J.W.D.M.; Bostock, R.M.; Kuc, J. Effect of abscisic acid on rishitin and lubimin accumulation and resistance to *Phytophthora infestans* and *Cladosporium cucumerinum* in potato tuber tissue slices. *Phytopathology* **1980**, *70*, 1074–1078.
53. Mialoundama, A.S.; Heintz, D.; Debayle, D.; Rahier, A.; Camara, B.; Bouvier, F. Abscisic acid negatively regulates elicitor-induced synthesis of capsidiol in wild tobacco. *Plant Physiol.* **2009**, *150*, 1556–1566.
54. Grosskinsky, D.K.; Naseem, M.; Abdelmoshem, U.A.; Plickert, N.; Engelke, T.; Griebel, T.; Zeier, J.; Novak, O.; Strand, M.; Pfeifhofer, H.; *et al.* Cytokinins mediate resistance against *Pseudomonas syringae* in tobacco through increased antimicrobial phytoalexin synthesis independent of salicylic acid signaling. *Plant Physiol.* **2011**, *157*, 815–830.
55. Kurusu, T.; Hamada, J.; Nokajima, H.; Kitagawa, Y.; Kiyoduka, M.; Takahashi, A.; Hanamata, S.; Ohno, R.; Hayashi, T.; Okada, K.; *et al.* Regulation of microbe-associated molecular pattern-induced hypersensitive cell death, phytoalexin production, and defense gene expression by calcineurin B-like protein-interacting protein kinases, OsCIPK14/15, in rice cultured cells. *Plant Physiol.* **2010**, *153*, 678–692.

56. Ono, E.; Wong, H.L.; Kawasaki, T.; Hasegawa, M.; Kodama, O.; Shimamoto, K. Essential role of the small GTPase Rac in disease resistance of rice. *Proc. Natl. Acad. Sci. USA* **2001**, *98*, 759–764.
57. Sadawa, K.; Hasegawa, M.; Tokuda, L.; Kameyama, J.; Kodama, O.; Kohchi, T.; Yoshida, K.; Shinmyo, A. Enhanced resistance to blast fungus and bacterial blight in transgenic rice constitutively expressing *OsSBP*, a rice homologue of mamalian selenium-binding proteins. *Biosci. Biotechnol. Biochem.* **2004**, *68*, 873–880.
58. Rauhut, T.; Luberacki, B.; Seitz, H.U.; Glawischnig, E. Inducible expression of a Nep1-like protein serves as a model trigger system of camalexin biosynthesis. *Phytochemistry* **2009**, *70*, 185–189.
59. Formela, M.; Samardakiewicz, S.; Marczak, L.; Nowak, W.; Narozna, D.; Waldemar, B.; Kasprowicz-Maluski, A.; Morkunas, I. Effects of endogenous signals and *Fusarium oxysporum* on the mechanism regulating genistein synthesis and accumulation in yellow lupine and their impact on plant cell cytoskeleton. *Molecules* **2014**, *19*, 13392–13421.
60. Parkhi, V.; Kumar, V.; Campbell, L.M.; Bell, A.A.; Shah, J.; Rathore, K.S. Resistance against various fungal pathogens and reniform nematode in transgenic cotton plants expressing Arabidopsis *NRP1*. *Transgenic Res.* **2010**, *19*, 959–975.
61. Rathmell, W.G.; Smith, D.A. Lack of activity of selected isoflavonoid phytoalexins as protectant fungicides. *Pestic. Sci.* **1980**, *11*, 568–572.
62. Ingham, J.L.; Deverall, B.J.; Kuc, J.; Coxon, D.T.; Stoessl, A.; VanEtten, H.D.; Matthews, D.E.; Smith, D.A. Toxicity of phytoalexins. In *Phytoalexins*; Bailey, J.A., Mansfield, J.W., Eds.; Blackie: Glasgow/London, UK, 1982; pp. 218–252.
63. Adrian, M.; Jeandet, P.; Veneau, J.; Weston, L.A.; Bessis, R. Biological activity of resveratrol, a stilbenic compound from grapevines, against *Botrytis cinerea*, the causal agent for gray mold. *J. Chem. Ecol.* **1997**, *23*, 1689–1702.
64. Adrian, M.; Jeandet, P. Effects of resveratrol on the ultrastructure of *Botrytis cinerea* conidia and biological significance in plant/pathogen interactions. *Fitoterapia* **2012**, *83*, 1345–1350.
65. Hasegawa, M.; Mitsuhara, I.; Seo, S.; Okada, K.; Yamane, H.; Iwai, T.; Ohashi, Y. Analysis on blast fungus-responsive characters of a flavonoid phytoalexin sakuranetin; Accumulation in infected rice leaves, antifungal activity and detoxification by fungus. *Molecules* **2014**, *19*, 11404–11418.
66. Harris, J.E.; Dennis, C. The effect of post-infectional potato tuber metabolites and surfactants on zoospores of Oomycetes. *Physiol. Plant Pathol.* **1977**, *11*, 163–169.
67. Pezet, R.; Pont, V. Ultrastructural observations of pterostilbene fungitoxicity in dormant conidia of *Botrytis cinerea* Pers. *J. Phytopathol.* **1990**, *129*, 29–30.
68. VanEtten, H.D.; Bateman, D.F. Studies on the mode of action of the phytoalexin phaseollin. *Phytopathology* **1971**, *61*, 1363–1372.
69. Rossall, S.; Mansfield, J.W.; Huston, R.A. Death of *Botrytis cinerea* and *B. fabae* following exposure to wyerone derivatives *in vitro* and during infection development in broad bean leaves. *Physiol. Plant Pathol.* **1980**, *16*, 135–146.

70. Woods, J.A.; Hafield, J.A.; Pettit, G.R.; Fox, B.W.; McGown, A.T. The interaction with tubulin of a series of stilbenes based on combretastatin A-4. *Br. J. Cancer* **1995**, *71*, 705–711.
71. Shlezinger, N.; Minz, A.; Gur, Y.; Hatam, I.; Dagdas, Y.F.; Talbot, N.J.; Sharon, A. Anti-apoptotic machinery protects the necrotrophic fungus *Botrytis cinerea* from host-induced apoptotic-like cell death during plant infection. *PLoS Pathog.* **2011**, *7*, e1002185.
72. Sanzani, S.; Schena, L.; Ippolito, A. Effectiveness of phenolic compounds against citrus green mould. *Molecules* **2014**, *19*, 12500–12508.
73. Sanzani, S.M.; de Girolamo, A.; Schena, L.; Solfrizzo, M.; Ippolito, A. Visconti, A. Control of *Penicillium expansum* and patulin accumulation on apples by quercetin and umbelliferone. *Eur. Food Res. Technol.* **2009**, *228*, 381–389.
74. Pont, V.; Pezet, R. Relation between the chemical structure and biological activity of hydroxystilbenes against *Botrytis cinerea. J. Phytopathol.* **1990**, *130*, 1–8.
75. Chalal, M.; Klinguer, A.; Echairi, A.; Meunier, P.; Vervandier-Fasseur, D.; Adrian, M. Antimicrobial activity of resveratrol analogues. *Molecules* **2014**, *19*, 7679–7688.
76. Lyon, G.D.; Bayliss, C.E. The effect of rishitin on *Erwinia carotovora* var. *atroseptica* and other bacteria. *Physiol. Plant Pathol.* **1975**, *6*, 177–186.
77. Hain, R.; Reif, H.J.; Krause, E.; Langebartels, R.; Kindl, H.; Vornam, B.; Wiese, W.; Schmelzer, E.; Schreier, P.; Stöcker, R.; *et al.* Disease resistance results from foreign phytoalexin expression in a novel plant. *Nature* **1993**, *361*, 153–156.
78. Gachon, C.; Baltz, R.; Saindrenan, P. Over-expression of a scopoletin glucosyltransferase in *Nicotiana tabacum* leads to precocious lesion formation during the hypersensitive response to tobacco mosaic virus but does not affect virus resistance. *Plant Mol. Biol.* **2004**, *54*, 137–146.
79. He, X.Z.; Dixon, R.A. Genetic manipulation of isoflavone 7-*O-methyltransferase* enhances biosynthesis of 4'-*O*-methylated isoflavonoid phytoalexins and disease resistance in alfalfa. *Plant Cell* **2000**, *12*, 1689–1702.
80. Schmidlin, L.; Poutaraud, A.; Claudel, P.; Mestre, P.; Prado, E.; Santos-Rosa, M.; Wiedemann-Merdinoglu, S.; Karst, F.; Merdinoglu, D.; Hugueney, P. A stress-inducible resveratrol *O-methyltransferase* involved in the biosynthesis of pterostilbene in grapevine. *Plant Physiol.* **2008**, *148*, 1630–1639.
81. Zernova, O.V.; Lygin, A.V.; Pawlowski, M.L.; Hill, C.B.; Hartman, G.L.; Widholm, J.M.; Lozovaya, V.V. Regulation of plant immunity through modulation of phytoalexin synthesis. *Molecules* **2014**, *19*, 7480–7496.
82. Graham, T.L.; Graham, M.Y.; Subramanian, S.; Yu, O. RNAi silencing of genes for elicitation or biosynthesis of 5-deoxyisoflavonoids suppresses race-specific resistance and hypersensitive cell death in *Phytophthora sojae* infected tissues. *Plant Physiol.* **2007**, *144*, 728–740.
83. Ibraheem, F.; Gaffoor, I.; Chopra, S. Flavonoid phytoalexin-dependent resistance to anthracnose leaf blight requires a functional *yellow seed1* in *Sorghum bicolor*. *Genetics* **2010**, *184*, 915–926.
84. Chong, J.; Baltz, R.; Schmitt, C.; Beffa, R.; Fritig, B.; Saindrenan, P. Downregulation of a pathogen-responsive tobacco UDP-Glc:phenylpropanoid glucosyltransferase reduces

scopoletin glucoside accumulation, enhances oxidative stress, and weakens virus resistance. *Plant Cell* **2002**, *14*, 1093–1107.

85. Cho, Y.; Ohm, R.A.; Devappa, R.; Lee, H.B.; Grigoriev, I.V.; Kim, B.Y.; Ahn, J.S. Transcriptional responses of the *bdtf1*-deletion mutant to the phytoalexin brassinin in the necrotrophic fungus *Alternaria brassicicola*. *Molecules* **2014**, *19*, 10717–10732.
86. Pedras, M.S.C.; Ahiahonu, P.W.K.; Hossain, M. Detoxification of the cruciferous phytoalexin brassinin in *Sclerotinia sclerotiorum* requires an inducible glucosyltransferase. *Phytochemistry* **2004**, *65*, 2685–2694.
87. Delserone, L.M.; McCluskey, K.; Matthews, D.E.; VanEtten, H.D. Pisatin demethylation by fungal pathogens and non pathogens on pea: Association with pisatin tolerance and virulence. *Physiol. Mol. Plant Pathol.* **1999**, *55*, 317–326.
88. Li, D.; Chung, K.R.; Smith, D.A.; Schardl, C.L. The *Fusarium solani* gene encoding kievitone hydratase, a secreted enzyme that catalyzes detoxification of a bean phytoalexin. *Mol. Plant-Microbe Interact.* **1995**, *8*, 388–397.
89. Pedras, M.S.C.; Gadagi, R.S.; Jha, M.; Sarma-Mamillapalle, V.K. Detoxification of the phytoalexin brassinin by isolates of *Leptosphaeria maculans* pathogenic on brown mustard involves an inducible hydrolase. *Phytochemistry* **2007**, *68*, 1572–1578.
90. Pedras, M.S.C.; Minic, Z.; Jha, M. Brassinin oxidase, a fungal detoxifying enzyme to overcome a plant defense—purification, characterization and inhibition. *FEBS J.* **2008**, *275*, 3691–3705.
91. Pezet, R. Purification and characterization of a 32-kDa laccase-like stilbene oxidase produced by *Botrytis cinerea* Pers.: Fr. *FEMS Microbiol. Lett.* **1998**, *167*, 203–208.
92. Breuil, A.C.; Adrian, M.; Pirio, N.; Meunier, P.; Bessis, R.; Jeandet, P. Metabolism of stilbene phytoalexins by *Botrytis cinerea*: Characterization of a resveratrol dehydrodimer. *Tetrahedron Lett.* **1998**, *39*, 537–540.
93. Breuil, A.C.; Jeandet, P.; Chopin, F.; Adrian, M.; Pirio, N.; Meunier, P.; Bessis, R. Characterization of a pterostilbene dehydrodimer produced by laccase of *Botrytis cinerea*. *Phytopathology* **1999**, *89*, 298–302.
94. Sbaghi, M.; Jeandet, P.; Bessis, R.; Leroux, P. Metabolism of stilbene type-phytoalexins in relation to the pathogenicity of *Botrytis cinerea* to grapevines. *Plant Pathol.* **1996**, *45*, 139–144.
95. Hammerbacher, A.; Schmidt, A.; Wadke, N.; Wright, L.P.; Schneider, B.; Bohlmann, J.; Brand, W.A.; Fenning, T.M.; Gershenzon, J.; Paetz, C. A common fungal associate of the spruce bark beetle metabolizes the stilbene defenses of Norway spruce. *Plant Physiol.* **2013**, *162*, 1324–1336.
96. Del Sorbo, G.; Schoonbeek, H.J.; de Waard, M.A. Fungal transporters involved in efflux of natural toxic compounds and fungicides. *Fungal Genet. Biol.* **2000**, *30*, 1–15.
97. De Ward, M.A.; Andrade, A.C.; Hayashi, K.; Schoonbeek, H.J.; Stergiopoulos, I.; Zwiers, L.H. Impact of fungal drug transporters on fungicide sensitivity, multidrug resistance and virulence. *Pest Manag. Sci.* **2006**, *62*, 195–207.
98. Coleman, J.J.; Mylonakis, E. Efflux in fungi: La pièce de résistance. *PLoS Pathog.* **2009**, *5*, e1000486.

99. Urban, M.; Bhargava, T.; Hamer, J.E. An ATP-driven efflux pump is a novel pathogenicity factor in rice blast disease. *EMBO J.* **1999**, *18*, 512–521.
100. Schoonbeek, H.J.; del Sorbo, G.; de Waard, M.A. The ABC transporter BcatrB affects the sensitivity of *Botrytis cinerea* to the phytoalexin resveratrol and the fungicide fenpiclonil. *Mol. Plant-Microbe Interact.* **2001**, *14*, 562–571.
101. Fleissner, A.; Sopalla, C.; Weltring, K.M. An ATP-binding cassette multidrug-resistance transporter is necessary for tolerance of *Gibberella pulicaris* to phytoalexins and virulence on potato tubers. *Mol. Plant-Microbe Interact.* **2002**, *15*, 102–108.
102. Del Sorbo, G.; Ruocco, M.; Schoonbeek, H.J.; Scala, F.; Pane, C.; Vinale, F.; de Waard, M.A. Cloning and functional characterization of BcatrA, a gene encoding an ABC transporter of the plant pathogenic fungus *Botryotinia fuckeliana* (*Botrytis cinerea*). *Mycol. Res.* **2008**, *112*, 737–746.
103. Stefanato, F.L.; Abou-Mansour, E.; Buchala, A.; Kretschmer, M.; Mosbach, A.; Hahn, M.; Bochet, C.G.; Métraux, J.P.; Schoonbeek, H.J. The ABC transporter BcatrB from *Botrytis cinerea* exports camalexin and is a virulence factor on *Arabidopsis thaliana. Plant J.* **2009**, *58*, 499–510.
104. Coleman, J.J.; White, G.J.; Rodriguez-Carres, M.; VanEtten, H.D. An ABC transporter and a cytochrome P450 of *Nectria haematococca* MPVI are virulence factors on pea and are the major tolerance mechanisms to the phytoalexin pisatin. *Mol. Plant-Microbe Interact.* **2010**, *24*, 368–376.
105. Zwiers, L.H.; de Waard, M.A. Characterization of the ABC transporter genes *MgAtr1* and *MgAtr2* from the wheat pathogen *Mycosphaerella graminicola. Fungal Genet. Biol.* **2000**, *30*, 115–125.
106. Stergiopoulos, I.; van Nistelrooy, J.G.M.; Kema, G.H.J.; de Waard, M.A. Multiple mechanisms account for variation in base-line sensitivity to azole fungicides in field isolates of *Mycosphaerella graminicola. Pest Manag. Sci.* **2003**, *59*, 1333–1343.
107. Schoonbeek, H.J.; van Nistelrooy, J.G.M.; de Waard, M.A. Functional analysis of ABC transporter genes from *Botrytis cinerea* identifies BcatrB as a transporter of eugenol. *Eur. J. Plant Pathol.* **2003**, *109*, 1003–1011.
108. Schoonbeek, H.J.; Raaijmakers, J.M.; de Waard, M.A. Fungal ABC transporters and microbial interactions in natural environments. *Mol. Plant-Microbe Interact.* **2002**, *15*, 1165–1172.
109. Vermeulen, T.; Schoonbeek, H.J.; de Waard, M.A. The ABC transporter BcatrB from *Botrytis cinerea*is a determinant of the activity of the phenylpyrrole fungicide fludioxonil. *Pest Manag. Sci.* **2001**, *57*, 393–402.
110. Andrade, A.C.; del Sorbo, G.; van Nistelrooy, J.G.M.; de Waard, M.A. The ABC transporter AtrB from *Aspergillus nidulans* mediates resistance to all major classes of fungicides and some natural toxic compounds. *Microbiology* **2000**, *146*, 1987–1997.
111. Zwiers, L.H.; Stergiopoulos, I.; Gielkens, M.M.C.; Goodall, S.D.; de Waard, M.A. ABC transporters of the wheat pathogen *Mycosphaerella graminicola* function as protectants against biotic and xenobiotic toxic compounds. *Mol. Genet. Genomics* **2003**, *269*, 499–507.

112. Nakaune, R.; Hamamoto, H.; Imada, J.; Akutsu, K.; Hibi, T. A novel ABC transporter gene, *PMR5*, is involved in multidrug resistance in the phytopathogenic fungus *Penicillium digitatum*. *Mol. Genet. Genomics* **2002**, *267*, 179–185.
113. Lozano-Mena, G.; Sanchez-Gonzalez, M.; Juan, M.E.; Planas, J.M. Maslinic acid, a natural phytoalexin-type triterpene from olives—A promising nutraceutical? *Molecules* **2014**, *19*, 11538–11559.
114. Jahangir, M.; Kim, H.K.; Choi, Y.H.; Verpoorte, R. Health-affecting compounds in Brassicaceae. *Compr. Rev. Food Sci. Food Saf.* **2009**, *8*, 31–43.
115. Smith, B.; Randle, D.; Mezencev, R.; Thomas, L.; Hinton, C.; Odero-Marah, V. Camalexin-induced apoptosis in prostate cancer cells involves alterations of expression and activity of lysosomal protease cathepsin D. *Molecules* **2014**, *19*, 3988–4005.
116. Yang, L.; Browning, J.D.; Awika, J.M. Sorghum 3-deoxyanthocyanidins possess strong phase II enzyme inducer activity and cancer cell growth inhibition properties. *J. Agric. Food Chem.* **2009**, *57*, 1797–1804.
117. Kello, M.; Drutovic, D.; Chripkova, M.; Pilatova, M.; Budovska, L.; Kulikova, L.; Urdzik, P.; Mojzis, J. ROS-dependent antiproliferative effect of brassinin derivative homobrassinin in human colorectal cancer Caco2 cells. *Molecules* **2014**, *19*, 10877–10897.
118. Athar, M.; Back, J.H.; Tang, X.; Kim, K.H.; Kopelovich, L.; Bickers, D.R.; Kim, A.L. Resveratrol: A review of preclinical studies for human cancer protection. *Toxicol. Appl. Pharmacol.* **2007**, *224*, 274–283.
119. Jang, M.; Cai, L.; Udeani, G.O.; Slowing, K.V.; Thomas, C.F.; Beecher, C.W.; Fong, H.H.; Farnsworth, N.R.; Kinghorn, A.D.; Metha, R.G.; *et al.* Cancer chemopreventive activity of resveratrol, a natural product derived from grapes. *Science* **1997**, *275*, 218–220.
120. Fuda, S.; Debatin, K.M. Sensitization for tumor necrosis factor-related apotosis-inducing ligand-induced apoptosis by the chemopreventive agent resveratrol. *Cancer Res.* **2004**, *64*, 337–346.
121. Budovska, M.; Pilatova, M.; Varinska, L.; Mojzis, J.; Mezncev, R. The synthesis and anticancer activity of analogs of the indole phytoalexins brassinin, 1-methoxyspirobrassinol methyl eher and cyclobrassinin. *Bioorg. Med. Chem.* **2013**, *21*, 6623–6633.
122. Mezencev, R.; Kutschy, P.; Salayova, A.; Updegrove, T.; McDonald, J.F. The design, synthesis and anticancer activity of new nitrogen mustard derivatives of natural indole phytoalexin 1-methoxybrassinol. *Neoplasma* **2009**, *56*, 321–330.
123. Chalal, M.; Delmas, D.; Meunier, P.; Latruffe, N.; Vervandier-Fasseur, D. Inhibition of cancer derived cell lines proliferation by synthesized hydroxylated stilbenes and new ferrocenyl-stilbene analogs. Comparison with resveratrol. *Molecules* **2014**, *19*, 7850–7868.
124. Wood, J.G.; Rogina, B.; Lavu, S.; Howitz, K.; Helfand, S.L.; Tatar, M; Sinclair, D. Sirtuin activators mimic caloric restriction and delay ageing in metazoans. *Nature* **2004**, *430*, 686–689.

125. Barger, J.L.; Kayo, T.; Vann, J.M.; Arias, E.B.; Wang, J.; Hacker, T.A.; Wang, Y.; Raederstorff, D.; Morrow, J.D.; Leeuwenburgh, C.; *et al.* A Low dose of dietary resveratrol partially mimics caloric restriction and retards aging parameters in mice. *PLoS One* **2008**, *3*, e2264.
126. McCalley, A.E.; Kaja, S.; Payne, A.J.; Koulen, P. resveratrol and calcium signaling: Molecular mechanisms and clinical relevance. *Molecules* **2014**, *19*, 7327–7340.
127. Sanzani, S.M.; Schena, L.; Nigro, F.; de Girolamo, A.; Ippolito, A. Effect of quercetin and umbelliferone on the transcript level of *Penicillium expansum* genes involved in patulin biosynthesis. *Eur. J. Plant Pathol.* **2009**, *125*, 223–233.
128. Matthews, D.; Jones, H.; Gans, P.; Coates, S.; Smith, L.M.J. Toxic secondary metabolite production in genetically modified potatoes in response to stress. *J. Agric. Food Chem.* **2005**, *53*, 7766–7776.
129. Becker, L.; Carré, V.; Poutaraud, A.; Merdinoglu, D.; Chaimbault, P. MALDI mass spectrometry imaging for the simultaneous location of resveratrol, pterostilbene and viniferins on grapevine leaves. *Molecules* **2014**, *19*, 10587–10600.
130. Marti, G.; Schnee, S.; Andrey, Y.; Simoes-Pires, C.; Carrupt, P.A.; Wolfender, J.L.; Gindro, K. Study of leaf metabolome modifications induced by UV-C radiations in representative *Vitis*, *Cissus* and *Cannabis* species by LC-MS based metabolomics and antioxidant assays. *Molecules* **2014**, *19*, 14004–14021.
131. Donnez, D.; Jeandet, P.; Clément, C.; Courot, E. Bioproduction of resveratrol and stilbene derivatives by plant cells and microorganisms. *Trends Biotechnol.* **2009**, *27*, 706–713.
132. Jeandet, P.; Vasserot, Y.; Chastang, T.; Courot, E. Engineering microbial cells for the biosynthesis of natural compounds of pharmaceutical significance. *BioMed Res. Int.* **2013**, doi:10.1155/2013/780145.
133. Jeandet, P.; Clément, C.; Courot, E. Resveratrol production at large scale using plant cell suspensions. *Eng. Life Sci.* **2014**, doi:10.1002/elsc.201400022.

Red Card for Pathogens: Phytoalexins in Sorghum and Maize

Alana Poloni and Jan Schirawski

Abstract: Cereal crop plants such as maize and sorghum are constantly being attacked by a great variety of pathogens that cause large economic losses. Plants protect themselves against pathogens by synthesizing antimicrobial compounds, which include phytoalexins. In this review we summarize the current knowledge on phytoalexins produced by sorghum (luteolinidin, apigeninidin) and maize (zealexin, kauralexin, DIMBOA and HDMBOA). For these molecules, we highlight biosynthetic pathways, known intermediates, proposed enzymes, and mechanisms of elicitation. Finally, we discuss the involvement of phytoalexins in plant resistance and their possible application in technology, medicine and agriculture. For those whose world is round we tried to set the scene in the context of a hypothetical football game in which pathogens fight with phytoalexins on the different playing fields provided by maize and sorghum.

Reprinted from *Molecules*. Cite as: Poloni, A.; Schirawski, J. Red Card for Pathogens: Phytoalexins in Sorghum and Maize. *Molecules* **2014**, *19*, 9114-9133.

1. What's at Stake: Maize and Sorghum

The world population increases by around 1% every year [1]. The rising number of people necessitates an ongoing expansion in food production. Currently, the largest part of food supply stems from the production of cereal crops such as maize (*Zea mays*) and sorghum (*Sorghum biolor*) [2]. To be sufficient to feed the population, studies indicate that global crop production needs to double by 2050. However, the production of maize, the most cultivated cereal in the world, is only increasing at a rate of 1.6% per year, while the rate increase necessary to match world population growth would be 2.4% [3].

Maize is the most cultivated cereal in the world, with 875 million tons produced in 2012 [2] and a worldwide consumption of more than 116 million tons. Maize ranks highest in net energy content and lowest in protein and fiber content relative to other cereals [2]. The plant is utilized mainly for human and animal livestock feed, but is also used for non-food products and for generation of bioenergy, for example in agricultural biogas production [4].

Sorghum is the fifth most highly produced crop and total production reached 58 million tons in 2012 [2]. The predominantly cultivated sorghum plant (*Sorghum bicolor*) exists in several subspecies or races with different morphological and physiological characteristics [5]. Among the main advantages of this plant are its high draught and heat tolerance, its high sugar content and the high yields of forage biomass that can be obtained per unit of land. Due to these advantages, sorghum is cultivated especially in hot and arid regions, like Nigeria and India, as well as USA and China [2]. Although sorghum is also used for industrial purposes, such as the generation of fiber, paper and ethanol, its main use is still for feed and food. Recently, the plant became even more important for the food industry, because it can be used to produce gluten-free products [6].

2. Setting the Game: Pathogens Attack

Despite the importance of sorghum and maize for agriculture and industry, a part of the harvest is lost either during cultivation or storage. Losses during cultivation are mainly due to the action of different pathogens, like bacteria, fungi, oomycetes, nematodes, parasitic plants and viruses. Pathogen attack affect the amount and quality of the grains, making them unsuitable for consumption. Microbial pathogens are difficult to control since they have genetically diverse populations that quickly adapt to changing environments and can readily break resistance of potential host plants [7]. However, although many pathogens exist, generally only a few species or a few strains of a given species are able to successfully infect a certain plant host. For example, the smut fungus *Ustilago maydis* causes tumors in leaves and inflorescences of maize, while the close relatives *Ustilago hordei* and *Sporisorium scitamineum* attack barley and sugarcane, respectively [8,9]. A more extreme example of host specificity occurs in the smut fungus *Sporisorium reilianum*, where two *formae speciales* exhibit different host preferences: *S. reilianum f. sp. reilianum* is an efficient sorghum pathogen, while *S. reilianum f. sp. zeae* is a maize pathogen that is unable to cause disease on sorghum [10].

In order to deal with or avoid the damaging effects of the pathogens, plants possess intricate defense mechanisms. Defense responses include the generation of reactive oxygen species and callose deposition at the point of entry, lignification of colonized plant tissues, activation of defense genes and production of antimicrobial substances [11–13]. Among the plant-produced antimicrobials that are induced upon pathogen attack are the phytoalexins, which are compounds of low molecular weight that are induced by stress [14–16].

In spite of our increased understanding of plant health and disease, millions of dollars' worth of harvest are still lost every year due to plant pathogens. In the search for effective plant protection measures, there is a renewed interest in the study of phytoalexins, since they are natural compounds with the potential to effectively protect our crops against pathogen attack. In this review, we focus on phytoalexins known to be produced by sorghum and maize. We summarize what is known about their biosynthetic pathways and their mechanisms of elicitation. Finally, we discuss the use of phytoalexins in agriculture, human health and industry.

3. The Fullback: Phytoalexins in Sorghum

Sorghum produces two distinct 3-deoxyanthocyanidin phytoalexins, known as apigeninidin (2-(4-hydroxyphenyl)benzopyrilium chloride) and luteolinidin (2-(3,4-dihydroxyphenyl)chromenylium-5,7-diol) (Figure 1A), in addition to a variety of derivatives, like 5-methoxy-luteolinidin, caffeic acid ester of arabinosyl-5-O-apigeninidin, and 7-methoxyapigeninidin [17,18]. A mixture of these characteristic reddish- and orange-colored compounds are known to be synthesized in the cytoplasm of epidermal sorghum cells infected with *Colletotrichum sublineolum* where they accumulate in initially colorless inclusion bodies. These inclusion bodies migrate to the infection zone, where they first accumulate and become pigmented, then lose their spherical shape and release their red contents at the infection site [19,20]. Accumulation of the 3-deoxyanthocyanidinsoccurs much faster in pathogen-challenged cells of resistant cultivars than of susceptible ones [21,22], suggesting that early phytoalexin accumulation is important to prevent proliferation and spread of fungal hyphae.

Figure 1. Structural formula of phytoalexins produced by sorghum and maize. (**A**) 3-Deoxyanthocyanidins produced by sorghum; (**B–D**) Phytoalexins produced by maize; (**B**) Zealexins; (**C**) Kauralexins; (**D**) Benzoxazinoids. Structures adapted from [23,24].

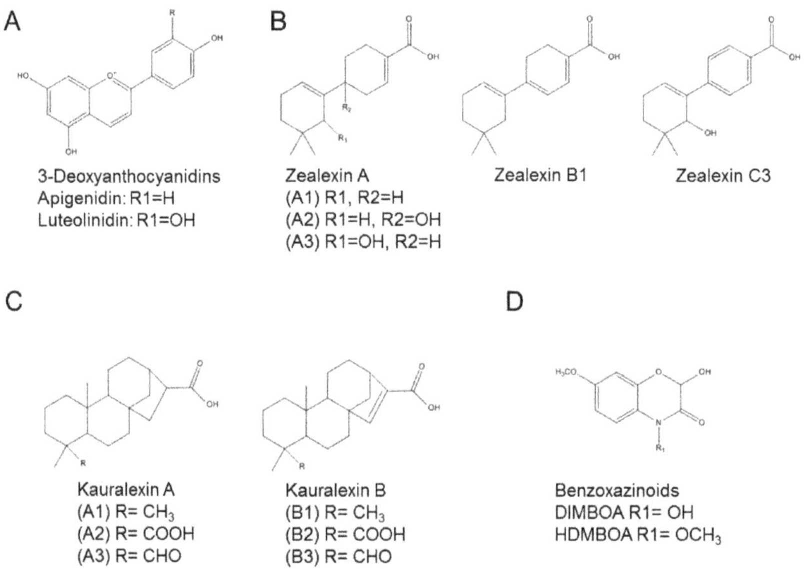

The measurement of the phytoalexin precursors naringenin and eriodictyol shows that they do not accumulate to measurable constitutive levels in unchallenged *S. bicolor* [25]. The low concentration or the rapid turnover of precursor compounds suggests *de novo* synthesis upon challenge of the plant by a fungal intruder [25]. In line with this hypothesis, genes of the 3-deoxyanthocyanidin biosynthesis phenylalanine ammonia lyase, chalcone synthase, and dihydroflavonol 4-reductase were induced when sorghum was challenged with the maize pathogens *Bipolaris maydis* or *S. reilianum f. sp. zeae* but not when challenged with the sorghum pathogen *S. reilianum f. sp. reilianum* [10,26].

4. Additional Defense Players: Versatile Plant Defense Responses

There is evidence that 3-deoxyanthocyanidin induction is not the only level of defense used by sorghum upon pathogen attack. In sorghum seedlings inoculated with *Fusarium proliferatum* and *Fusarium thapsinum*, the increase in apigeninidin and luteolinidin levels was accompanied by increased concentrations of peroxidases, beta-1,3-glucanases and chitinases [27]. In sorghum seedlings inoculated with *Cochliobolus heterostrophus*, in addition to phytoalexin accumulation, a fast and coordinated accumulation of *PR-10* and chalcone synthase transcripts was observed. This accumulation was delayed in sorghum seedlings inoculated with the sorghum pathogen *C. sublineolum* [28,29]. Recently, the whole transcriptome of sorghum inoculated with the necrotroph *Bipolaris sorghicola* was analyzed. In addition to the up-regulation of genes encoding key enzymes for phytoalexin biosynthesis, many other plant genes with a suspected role in defense were up-regulated, which included genes encoding plant receptors, genes involved in MAPK cascades and Calcium signaling, transcription factors and genes involved in downstream responses (peroxidases, PR proteins and genes implicated in biosynthesis of lignin) [30,31].

5. Preparation Phase: Phytoalexin Biosynthesis

Biosynthesis of the 3-deoxyanthocyanidin phytoalexins is independent of light and occurs in the dark, in contrast to the biosynthesis of anthocyanins that is light dependent [32]. Biosynthesis of the 3-deoxyanthocyanidins luteolinidin and apigeninidin, of the flavones luteolin and apigenin, and the leucoanthocyanidins and anthocyanins occurs via common and specific pathway steps [17,33,34]. Commons steps include the formation of *p*-coumaryl CoA, that is generated from phenylalanine via the action of the enzymes phenylalanine ammonia lyase (PAL) to synthesize cinnamic acid, cinnamate-4-hydroxylase (C4H) to synthesize *p*-coumaric acid, and coumaryl CoA ligase (CCL) for generation of *p*-coumaryl CoA. *p*-Coumaryl CoA is the substrate of the enzyme chalcone synthase (CHS) [33], which catalyzes the condensation of *p*-coumaryl CoA and three molecules of malonyl CoA form naringenin chalcone that is converted to naringenin by a chalcone isomerase (CHI) (Figure 2).

It is from the flavanone naringenin that the biosynthesis pathways of anthocyanin, flavone and 3-deoxyanthocyanidin split [33]. The flavones apigenin and luteolin are generated from naringenin and the related flavanon eriodictyol that is likely generated from naringenin via a flavonoid-3'-hydroxylase (F3'H). It is the enzyme flavon synthase (FNS) that catalyzes both hydroxylation at C-2 and abstraction of water [33]. In the anthocyanin pathway, naringenin and related flavanones are hydroxylated at C-3 by flavanone-3-hydroxylase (F3H), followed by an NADPH-dependent reduction of the C-4 carbonyl group by dihydroflavonol 4-reductase (SbDFR1), and the action of the anthocyanidin synthase (ANS), that abstracts water leaving a double bond between C-3 and C-4. The unstable anthocyanidins are then converted to the stable anthocyanins by a flavonol 3-*O*-glucosyltransferase (3GT) that attaches a glucose molecule to the C-3 hydroxyl group [33]. In contrast, for biosynthesis of the 3-deoxyanthocyanidins, naringenin and eriodictyol are direct targets of NADPH-dependent reduction of the C-4 carbonyl group by a dihydroflavonol 4-reductase (SbDFR3), that has been shown to be a different enzyme than the dihydroflavonol 4-reductase SbDFR1 involved in anthocyanin biosynthesis [33]. The generated luteoferol and apiferol (that might also be generated from luteoferol via an F3'H) are likely reduced by an unidentified anthocyanidin synthase that removes the C-4 hydroxyl group leaving a double bond between C-3 and C-4 and creating the 3-deoxyanthocyanidins apigeninidin and luteolinidin [33].

Figure 2. Biosynthetic pathway of 3-deoxyanthocyanidins in sorghum. Structures of intermediates and products are shown. Where known, enzyme classes are indicated. In many cases the specific enzyme has not been identified yet. ANS, anthocyanidin synthase; C4H, cinnamate-4-hydroxylase; CCL, coumaryl-CoA ligase; CHI, chalcone isomerase; F3'H, flavanone-3'-hydroxylase; F3H, flavanone-3-hydroxylase; NCS, Naringenin chalcone synthase; PAL, phenylalanine ammonia lyase; SbDFR1, dihydroflavonol 4-reductase 1; SbDFR3, dihydroflavonol 4-reductase 3; SbFNS2, flavone synthase 2. Pathway adapted from [33,35].

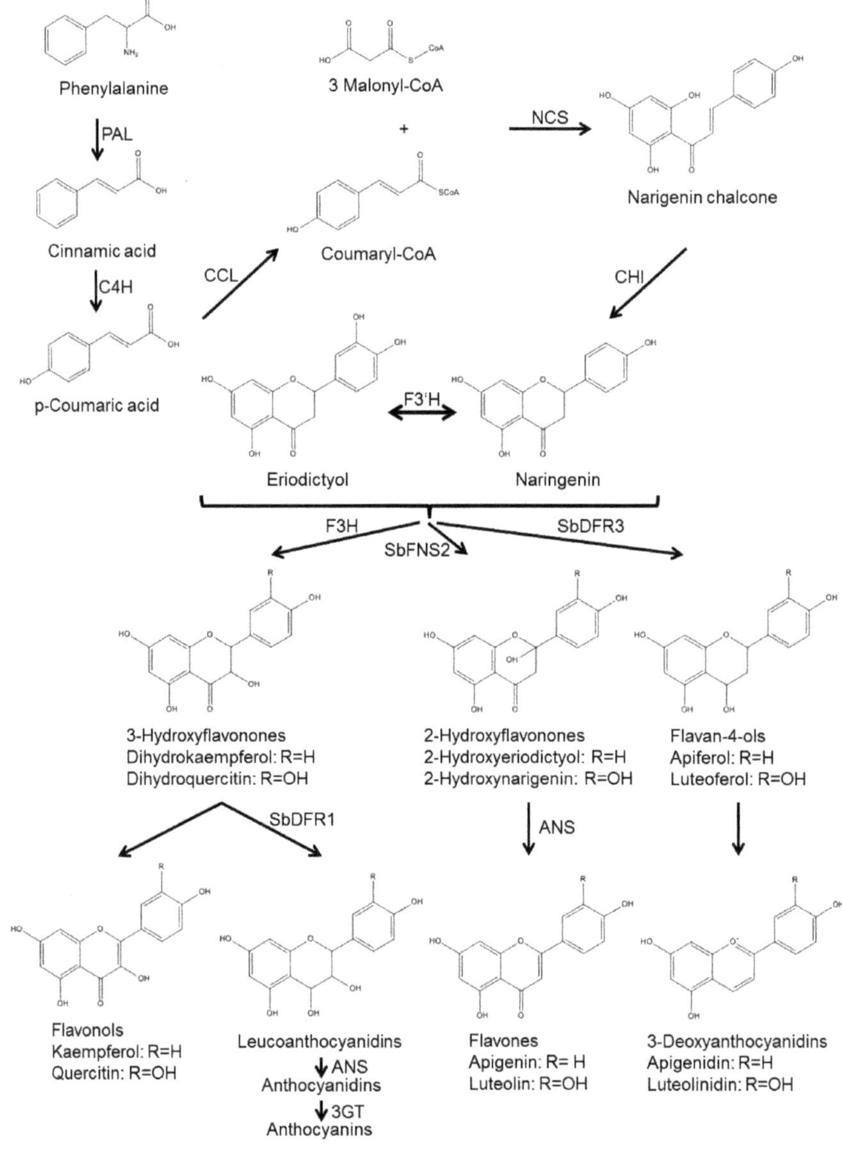

A candidate gene, *Sb06g029550*, has been identified as induced upon *B. sorghicola* infection of sorghum [30], which might correspond to the unidentified anthocyanidin synthase responsible for generation of apigeninidin and luteolinidin from apiferol and luteoferol. Gene expression of this enzyme coincided with the accumulation of apigeninidin detected during *B. sorghicola* infection [30]. Fungal inoculation with the non-sorghum pathogen *Cochliobolus heterostrophus* shifts metabolic flux away from anthocyanin synthesis towards 3-deoxyanthocyanidin synthesis [36]. This is achieved by induction of PAL and CHS presumably leading to an increased synthesis of naringenin, as well as SbDFR3 and the ANS involved in this pathway [33,36], and simultaneous repression of the anthocyanin biosynthesis genes *F3H*, *SbDFR1* and *ANS*. Both SbDFR1 and SbDFR3 are able to convert flavanones to flavan-4-ols *in vitro* [33]. However, only *SbDFR1* is up-regulated under light-induced anthocyanin biosynthesis, while only *SbDFR3* is up-regulated during 3-deoxyanthocyanidin biosynthesis [33]. Transcriptome data of sorghum infected with *B. sorghicola* identified four putative paralogs of *DFR* genes, but only one was up-regulated and had a similar sequence as *SbDFR3*, suggesting that this was the one involved in the reduction of the C-4 group of narigenin [30].

6. Training with New Methods: Unknown Biosynthesis Genes

Although the enzyme class necessary for the specific biosynthetic steps can be easily predicted, the redundancy of the enzyme complement makes it difficult to predict which of the *S. bicolor* genes encode the relevant one for a specific catalytic step. For example, first analysis of the genome sequence of *S. bicolor* identified eight genes potentially encoding chalcone synthases [37]. Seven of these (SbCHS1 to SbCHS7) are highly conserved, while one (SbCHS8) shows an amino acid identity of only about 82% to the other seven enzymes [38]. Of the eight *CHS* genes, *SbCHS8* was overrepresented in a cDNA library prepared from *C. sublineolum*-inoculated sorghum, suggesting that SbCHS8 was involved in 3-deoxyanthocyanidin biosynthesis [38]. However, *SbCHS8* was found to encode a functional stilbene synthase (STS) that is activated during both host and non-host responses, being therefore renamed to *SbSTS1* [39]. In contrast, *SbCHS2* could be shown to encode a typical chalcone synthase that is able to synthesize naringenin chalcone *in vitro* [39]. It is possible that more than one of the remaining seven chalcone synthases of sorghum is responsible for production of naringenin chalcone during pathogen attack and that there are even more chalcone synthases present. Transcriptome sequencing of sorghum infected with *B. sorghicola* revealed nine *SbCHS* genes, of which six were up-regulated [30].

The genome of *S. bicolor* shows presence of three unique sorghum flavonoid 3'-hydroxylases [37]. Of these, *SbF3'H1* expression was induced during light-induced anthocyanin accumulation, while *SbF3'H2* expression was induced during pathogen-specific 3-deoxyanthocyanidin synthesis. Expression of *SbF3'H3* was not detected under these conditions, leaving its potential *in vivo* function unexplained [23]. Identification of the specific flavone synthase involved in flavone biosynthesis was more straightforward. Investigation of pathogen-inducible gene expression identified a cytochrome P450 protein that was shown to generate flavones from flavanones via formation of 2-hydroxyflavanones. The gene is located in single-copy on chromosome 2 and was named *SbFNS2* [35].

7. Kick-Off: Phytoalexin Induction

The 3-deoxyanthocyanidins are synthesized in sorghum in response to different stimuli [40]. The study of phytoalexin induction by fungal intruders was pioneered by Nicholson and coworkers [19]. The authors inoculated sorghum leaves with the non-sorghum pathogen *Helminthosporium maydis* and the sorghum pathogen *Colletotrichum graminicola.* They tested two sorghum cultivars, BR54 (resistant) and P721N (susceptible) and identified the phytoalexins apigenidin and luteolinidin, of which apigenidin accumulated in both cultivars, while luteolinidin was present only in the resistant one. Both compounds were active in inhibiting germling development in *H. maydis* and elongation of *C. graminicola* germ-tubes *in vitro* [19]. Apigeninidin was also shown to inhibit growth of the gram-positive bacteria *Bacillus cereus*, *Staphylococcus aureus*, *Staphylococcus epidermidis*, and *Streptococcus faecalis*, as well as the Gram-negative bacteria *Escherichia coli*, *Serratia marcescens*, and *Shigella flexneri* [41].

The order and timing of appearance of the different phytoalexins in sorghum seems to be carefully choreographed. Sorghum leaves inoculated with the maize pathogen *B. maydis* show detectable levels of apigeninidin first at 10 h post inoculation (hpi), of luteolinidin and apigeninidin-5-*O*-arabinoside at 14 hpi, of luteolinidin-5-methylether at 18 hpi and apigeninidin-7-methylether at 20 hpi. At 24 hpi, the levels of apigeninidin and luteolinidin were similar, but by 48 hpi the amounts of luteolinidin reached the double of apigenidin [17].

Differential induction of phytoalexins may support host specificity of the two *formae speciales* of *S. reilianum*. Inoculation of sorghum with the maize-pathogenic S. *reilianum f. sp. zeae* resulted in a strong deposition of luteolinidin and apigeninidin (Figure 3A). In contrast, no phytoalexins were generated when sorghum was inoculated with the sorghum-pathogenic *S. reilianum f. sp. reilianum* [10]. Quantitative RT-PCR confirmed increased expression of the phytoalexin biosynthesis gene *SbDFR3* in samples infected with *S. reilianum f. sp. zeae*, while in samples infected with *S. reilianum f. sp. reilianum* the levels were similar to control samples (Figure 3B). Phytoalexins were visible on leaves at 3 dpi as dark red-colored stains which increased in number and size at later time points (Figure 3C). *In vitro*-assays demonstrated that luteolinidin but not apigeninidin was able to slow growth of *S. reilianum* [10]. The concentration of phytoalexins in infected host cells was estimated to be between 0.48 and 1.20 ng of luteolindin and 0.24 to 0.91 ng of apigeninidin per cell [42], which is more than what is necessary for *in vitro* toxicity. Interestingly, *in vitro* growth of both *S. reilianum f. sp. zeae* and *S. reilianum f. sp. reilianum* was equally affected by luteolinidin [10], suggesting that the active phytoalexin is not induced when sorghum is colonized by *S. reilianum f. sp. reilianum.*

8. Changing the Pitch: Phytoalexins in Maize

In maize, phytoalexins are represented by terpenoids, which include zealexins and kauralexins and benzoxazinoids, represented by DIMBOA and HDMBOA. Zealexins were recently identified as a group of acidic sesquiterpenoids that are related to β-macrocarpene (4',5,5-trimethyl-1,1'-bis(cyclo-hexane)-1,3'-diene) [43]. The group contains at least five different compounds, zealexin A1, A2, A3, B1 and C3 (Figure 1B), that accumulate to very high levels of about 800 μg/g in *Fusarium graminearum*-infected maize [43]. Nine additional related compounds have been detected by expanded GC/(+)CI-MS but their exact identity is not yet known [43]. High concentrations of zealexins were also found in maize challenged with other fungal pathogens, such as *Aspergillus flavus* and *Rhizopus microsporus* [43]. In vitro toxicity tests showed that zealexin A1 and A3 (but not A2) inhibited growth of *A. flavus* and *F. graminearum*, and that A1 was also effective against *R. microsporus* [43].

Figure 3. Deposition of 3-deoxyanthocyanidins in sorghum during interaction with the smut fungus *Sporisorium reilianum*. (**A**) Sorghum leaves infiltrated with water (control), infected with *S. reilianum f. sp. zeae* or infiltrated with chitin. The latter treatments lead to appearance of spots with a characteristic red color indicating phytoalexin production; (**B**) Quantitative RT-PCR of sorghum samples inoculated with water (Mock), *S. reilianum f. sp. reilainum* (SRS) or *S. reilianum f. sp. zeae* (SRZ). Sorghum leaves were collected at 0.5, 1, 2 and 3 days post infection (dpi). Up-regulation of the gene *SbDFR3* was observed only for samples infected with SRZ; (**C**) Sorghum leaves infected with SRZ showing the emergence of red color at 3 dpi, which gets more intense with time.

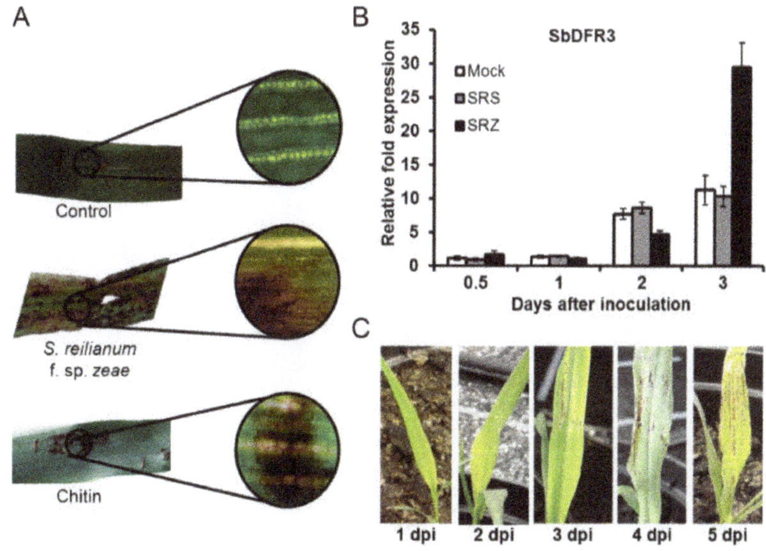

Zealexin biosynthesis likely involves terpene synthases 6 and 11 (TPS6 and TPS11). These very similar enzymes were shown to convert farnesyl pyrophosphate to (*S*)-β-macrocarpene via an (*S*)-β-bisabolene intermediate [44,45]. Accordingly, *TPS6* and *TPS11* are transcriptionally induced prior to the rise in zealexin concentration and figure among the most highly induced maize genes upon *F. graminearum* infection [43]. *TPS6* and *TPS11* were also found to be highly induced in maize leaves colonized by the maize smut fungus *U. maydis* [46,47] and in maize inflorescences colonized by the head smut fungus *S. reilianum f. sp. zeae* [48]. Co-silencing of *TPS6* and *TPS11* led to increased susceptibility to *U. maydis* [49]. This indicates that the zealexins are terpenoid phytoalexins produced by maize as part of its defense against fungal intruders.

In addition to zealexin, maize can produce kauralexins that are *ent*-kaurane related diterpenoid phytoalexins. So far, six kauralexins are known: kauralexin A1 (*ent*-kauran-17-oic acid), A2 (*ent*-kauran-17,19-dioic acid), A3 (*ent*-kaur-19-al-17-oic acid), B1 (*ent*-kaur-15-en-17-oic acid), B2 (*ent*-kaur-15-en-17,19-dioic acid), and B3 (*ent*-kaur-15-en-19-al-17-oic acid) (Figure 1C) [50]. Before the accumulation of kauralexins, a strong up-regulation of the *AN2* gene that encodes the enzyme *ent*-copalyl diphosphate synthase anther ear 2 was observed in maize infected with *F. graminearum* [51]. This gene was identified as an ortholog of rice genes encoding ent-copalyl diphosphate synthases that supply precursors to diterpenoid phytoalexins [50,51]. Kauralexins also accumulated in maize stems colonized by the pathogens *R. microsporus* and *C. graminicola* [50]. From the six detected *ent*-kaurane–related diterpenoids, kauralexin A3 and B3 presented antimicrobial activity against *R. microsporus* and *C. graminicola* when used in relevant concentrations [50].

Kauralexin biosynthesis seems to be regulated by phytohormones. Application of a combination of jasmonic acid (JA) and ethylene to maize plants was sufficient to induce kauralexin generation [50]. The gene *AN2* was highly expressed at 24 h after attack by the insect *Ostrinia nubilalis*, which was accompanied by an increase in the concentrations of JA and ethylene together with the expression of allene oxide synthase and 1-aminocyclopropane-1-carboxylic acid oxidase genes, key enzymes involved in biosynthesis of these hormones [52]. At 48 h post attack, plants accumulated kauralexins at higher concentration than at 24 h [53].

Microarray analysis in *Fusarium graminearum*-infected maize showed that not only *AN2* was up-regulated, but also *TPS6* and *TPS11* that are involved in zealexin biosynthesis [43]. In maize roots infected with the oomycete *Phytophthora cinnamori, TPS11* and *AN2* were two of the most highly up-regulated genes [54]. These results demonstrate that zealexins and kauralexins are co-inducted and co-produced in maize [43].

In addition to zealexins and kauralexins, benzoxazinoid hydroxamic acids can be produced in maize leaves [55], namely HDMBOA-Glc (2-hydroxy-4,7-dimethoxy-1,4-benzoxazin-3-one-glucoside) and DIMBOA-Glc (4-dihydroxy-7-methoxy-1,4-benzoxazin-3-one-glucoside, Figure 1D), in addition to precursors and variants that include HMBOA-Glc (2-hydroxy-7-methoxy-1,4-benzoxazin-3-one glucoside), DIM2BOA-Glc (2,4-dihydroxy-7,8-dimethoxy-1,4-benzoxazin-3-one glucoside), HBOA (2-hydroxy-2*H*-1,4-benzoxazin-3(4*H*)-one), DIBOA (2,4-dihydroxy-2*H*-1,4-benzoxazin-3(4*H*)-one) and TRIBOA (2,4,7-trihydroxy-2*H*-1,4-benzoxazin-3(4H)-one). DIMBOA-Glc is predominant in maize seedlings, and its levels decrease as the plants get older [56]. Therefore, these compounds have been described as phytoanticipins. However, additional DIMBOA can be synthesized upon pathogen infection [57], and this pool of DIMBOA has to be considered as phytoalexins. The benzoxazinoid biosynthetic pathway has been elucidated and involves the generation of indole by an indole-3-glycerol phosphate lyase (BENZOXAZINE-DEFICIENT1 [BX1]), and subsequent production of indolin-2-one by BX2, 3-hydroxyindolin-2-one by BX3, HBOA by BX4, DIBOA by BX5, DIBOA-Glc by BX8 and BX9, TRIBOA-Glc by BX6 and finally DIMBOA-Glc by BX7 [58,59].

Insect feeding triggers the conversion of DIMBOA-Glc to HDMBOA-Glc and increases plant resistance against some pathogens, but can also be associated with susceptibility in selected examples [52,53,55]. Maize inoculation with the mycorrhizal fungus *Glomus mosseae* increased the production of DIMBOA and reduced disease caused by *Rhizoctonia solani*. The defense genes

PR2a, *PAL*, and *AOS* were up regulated, together with *BX9*, one of the key genes in DIMBOA biosynthesis [60]. During herbivory with *O. nubilalis*, a decrease of DIMBOA-Glc was observed, while both HDMBOA-Glc and plant resistance increased [52]. The insect *Spodoptera littoralis* triggered accumulation of DIMBOA that was attributed to *de novo* synthesis [61]. HDMBOA-Glc also accumulated in response to treatment with JA, pathogen infection, and herbivory [62]. The smut fungus *U. maydis* induced DIMBOA in maize, but the fungus was resistant to this compound [46]. Another experiment using *U. maydis* detected differences in gene expression for *BX1*, *BX2*, *BX5* and *BX8*, but none for *BX3*, *BX4*, *BX6* and *BX7* [58]. HDMBOA-Glc accumulated in tissues infected with *F. graminearum* in wild-type and benzoxazine-deficient1 (bx1) mutant lines, suggesting that the Bx1 gene and the presence of DIMBOA-Glc were not necessary for HDMBOA-Glc biosynthesis [43]. A gene similar to *BX1*, called *IGL*, also encodes an indole-3-glycerol phosphate lyase, and could be a potential candidate for the synthesis of HDMBOA-Glc in *bx1* plants [59].

9. Know the Rules of the Game: Elicitation and Regulation of Phytoalexin Biosynthesis

Phytoalexins in maize and sorghum are induced during pathogen infection, which suggests that molecules originating from the pathogen or generated during host-pathogen interaction act as elicitors [15]. Although non-pathogens and even several pathogens induce phytoalexins in maize and sorghum, only very little is known about the eliciting molecules. Specific elicitors may include avirulence proteins or effectors that are produced by the pathogen during infection, while pathogen associated molecular patterns (PAMPs) such as conserved proteins, glycoproteins, oligosaccharides and fatty acids could serve as more general elicitors. PAMPs, including flagellin and lipopolysaccharide of bacteria, and chitin, chitosan and β-glucan of fungi, are documented as elicitors of phytoalexins in several plant species, like tobacco, rice, soybean, lemon and *Arabidopsis* [63,64]. In maize, treatment of wounded stems with the PAMP polygalacturonase from *Rhizopus sp.* resulted in kauralexin accumulation within 24 h [50]. Fungal β-1,3-glucan also serves as PAMP, as shown by over-expression of the glucan synthase CgGLS1 of *C. graminicola* in maize, which led to up-regulation of the terpene synthase genes potentially involved in phytoalexin biosynthesis as well as to a reduction in pathogen spread [65]. Interestingly, *C. graminicola* decreases its β-1,3-glucan production during the first hours after plant penetration, which presumably leads to avoidance of PAMP-elicited plant defense responses during its biotrophic growth phase [65].

In sorghum, infiltration of leaves with a chitin solution led to a strong induction of phytoalexins, revealing the potential of this PAMP as an elicitor (Figure 3A; Poloni and J. Schirawski, unpublished). Different preparations of *Saccharomyces cerevisiae* generated by heat inactivation and extracted with ethanol were also able to induce phytoalexins when incubated with sorghum seedlings, and the response increased with higher concentrations of proteins in the elicitor sample [66]. Carbohydrates and peptides extracted from conidia of *C. graminicola* also induced phytoalexin deposition in sorghum mesocotyls [67].

While in *Arabidopsis* and rice several signaling components involved in phytoalexin biosynthesis have been identified (for example, camalexin production in *Arabidopsis* is regulated by mitogen-activated protein kinases (MAPK) AtMPK3, AtMPK6 and AtMPK4, and diterpenoid phytoalexins in rice are regulated via the MAPKs OsMPK3 and OsMPK6 [68], very little is known about regulation of phytoalexin biosynthesis in maize and sorghum. In sorghum, the *Y1* (*YELLOW SEED1*) gene that encodes a MYB transcription factor involved in regulation of phlobaphene biosynthesis [69] was also shown to regulate biosynthesis of 3-deoxyanthocyanidins [70]. *Y1* null alleles do not accumulate 3-deoxyanthocyanidins when challenged with the non-pathogenic fungus *C. heterostrophus*, and also show greater susceptibility to the pathogenic fungus *C. sublineolum*, demonstrating that the accumulation of 3-deoxyanthocyanidins and resistance to *C. sublineolum* in sorghum require a functional *Y1* gene [70].

In maize, a number of genes have been identified to be involved in regulation of anthocyanin biosynthesis. Plants with a mutated *PAC1* gene (*PALE ALEURONE COLOR1*) exhibited reduced levels of anthocyanins and reduced transcript levels of the anthocyanin biosynthetic genes, while transcript levels of the regulatory genes *B* and *C1* did not decrease [71]. The product of the *R* gene induces phytoalexins through regulation of a CHS encoded by *C2*, a DFR encoded by *A1*, and 3GT encoded by *Bz1*. For activation of anthocyanin biosynthesis, one protein of the bHLH-transcription factors B and R, and one protein of the Myb-transcription factors C1 and P1 needs to be expressed [72]. Maize cells overexpressing C1 and R accumulated anthocyanins, while cells overexpressing P accumulated 3-deoxyflavonoids [73]. In sorghum, anthocyanin biosynthesis has not been elucidated. Because of partial overlap in precursors, generation of anthocyanins and 3-deoxyanthocyanidin phytoalexins in sorghum may be regulated by the same proteins. As homolog of the maize *P1* gene, the sorghum *Y1* gene (see above) was identified. The encoded MYB-type regulatory protein controls expression of a F3'H [74]. However, in sorghum leaves infected with *B. sorghicola*, *Y1* was completely suppressed in both infected and control samples, while the genes encoding *CHS*, *CHI*, and *F3'H* were differentially expressed [30]. This suggests that additional regulators are active in sorghum.

Phytoalexin biosynthesis in sorghum is regulated via hormones. In sorghum roots, JA stimulates phytoalexin deposition while salyclic acid (SA) had an inhibitory function [33]. In contrast, microarray analysis of sorghum gene expression in response to SA and methyl jasmonate (MeJA) identified many phytoalexin biosynthesis genes as induced by both SA and MeJA. These included genes encoding PAL, C4H, cinnamyl alcohol dehydrogenase, cinnamoyl-CoA reductase, CHS, chalcone-flavanone isomerase, flavanone 3-hydroxylase, dihydroflavonal-4-reductase, isoflavone reductase, and leucoanthocyanidin dioxygenase [75].

10. Scoring Goals on other Fields: Applications of Phytoalexins

In addition to up-regulation of phytoalexin biosynthesis genes together with other plant defense genes, a positive relationship between the presence of phytoalexins and non-virulence of certain pathogens was established and was supported by *in vitro* inhibition tests [10,20,35,76,77]. However, in some plant pathogen interactions, fungal strategies for overcoming the deleterious effects of phytoalexins were discovered and the responsible genes identified. For example, the fungal pathogens *Leptosphaeria maculans* and *Alternaria brassicicola* detoxify the phytoalexin brassinin present in crucifers using brassinin oxidases that hydrolyze the dithiocarbamate group of brassinin to generate the non-toxic (1H-indol-3-yl)methanamine [78,79]. Similarly, the pea pathogen *Nectria haematococca* detoxifies the phytoalexin pisatin using pisatin demethylase [80,81]. In potato,

Gibberella pulicaris is successful during infection due to the ability to detoxify rishitin [82]. This knowledge could be used to develop inhibitors of phytoalexin-detoxifying enzymes as part of a phytosanitary treatment based on supporting self-defense of the plant. Alternatively, resistant plants that produce several different phytoalexins could be generated by either classical breeding or biotechnological methods. However, this approach requires additional knowledge on the genes and regulators involved in phytoalexin biosynthesis.

Identification of genes involved in phytoalexin generation is facilitated by the elucidation of the maize and sorghum genome sequences [37,83]. The sequence information now allows prediction of target genes that then need to be functionally characterized. One of the challenges lies in the identification of genes specific for phytoalexin production that do not lead to the generation of unplanned derivatives with a potentially adverse effect on human health.

To characterize gene function and study the impact of phytoalexins in plant resistance, the generation of plants that either do not express or overexpress phytoalexins will be a great help. Generation of recombinant plants is still a major challenge in research on maize and sorghum gene function. However, several techniques have been successfully employed and may accelerate the gain of knowledge. In addition to the laborious search for mutants in plant libraries generated by chemical mutagenesis using TILLING [84,85], Virus induced gene silencing (VIGS) was successfully applied to transiently silence genes in maize [49] and sorghum [86]. Sorghum transformation via microparticle bombardment of undifferentiated cells has been done [87], as well as new tools developed for the generation of targeted gene knockouts. Among these promising tools are the use of zinc finger nucleases (ZFNs) [88], TAL effector nucleases (TALENs) [88,89] and, most recently, the use of the clustered regulatory interspersed short palindromic repeat (CRISPR)/CRISPR-associated protein (Cas) system [90,91]. The newly generated plants would need to be subsequently tested for pathogen resistance under real-life conditions to learn how environmental stress affects plant protection by phytoalexins. In addition to new tools for generation of targeted gene disruption lines, mRNA sequencing of infected plants has been successfully used in maize and sorghum to discover target genes contributing to plant defense [30,31,92,93].

The 3-deoxyanthocyanidins have been proposed as medical agents against proliferation of several human cancer cell lines [94] and have been shown to induce apoptosis, inhibit cell proliferation, metastasis, and angiogenesis and sensitize tumor cells to therapeutic-induced cytotoxicity [95]. The flavone luteolin is well characterized for its antioxidant and anti-inflammatory activities both *in vitro* and *in vivo* [96]. Luteolinidin in a concentration of 200 μM reduced the viability of HL-60 cells by 90% and HepG2 by 50% [94]. Experiments using 3-deoxyanthocyanins against colon cancer stem cells showed a reduction of proliferation and apoptosis in these cells, in which luteolinidin was more effective than apigeninidin [97]. The compounds also presented effect against breast cancer MCF 7 cells [98] and were effective preventing the oxidation of LDL [99]. Moreover, sorghum extracts rich in 3-deoxyanthocyanidins were demonstrated to induce phase II enzymes [100], which are considered indicators of protection against carcinogens in animal cells [101]. An increased knowledge of the mode of action of phytoalexins and related compounds will help to explore their potential for use in human health.

In addition to their potential use in cancer treatment, other beneficial uses for the sorghum-specific phytoalexins have been proposed. These include the use as natural and persistant hair and food colorants [102,103], since they are natural products that have high color stability at various pH values, temperatures and light intensities [104]. In order to facilitate industrial production, a new sorghum variety *REDforGREEN* (*RG*) was recently developed using mutagenesis-assisted breeding. This variety overexpresses 3-deoxyanthocyanidins in leaf tissues, which highly increases the yield of pigments [103].

11. Striving for the Trophy: Challenges Ahead

Great efforts have resulted in a wealth of information on the identity, inducing conditions, and the biosynthesis genes of the major phytoalexins in both sorghum and maize. However, for some of the compounds, biosynthesis pathways are not completely elucidated. Although genes can be predicted, their involvement in phytoalexin biosynthesis is unclear. Once their involvement is assured, the enzymes can be tested for their catalytic abilities to know how they contribute to the pool of different phytoalexins. In addition to the identification of the remaining biosynthesis genes, their regulation will need to be studied. Different regulators and inducing molecules are expected to function in orchestrating sequential phytoalexin accumulation. Finally, we need to learn more about the mechanism of toxicity towards fungi, bacteria or insects, in order to be able to use phytoalexins for the creation of resistant crop plants. Future research on these promising compounds will help to preserve world nutrition and improve world economy in many different aspects.

Acknowledgments

We wish to thank Theresa Wollenberg for critical comments on the manuscript, the German Academic Exchange Service (DAAD), the RWTH Aachen University and the German Science Foundation (DFG) for funding.

Author Contributions

Alana Poloni and Jan Schirawski reviewed the literature, designed figures, and wrote the paper.

Conflicts of Interest

The authors declare no conflict of interest.

References

1. Worldometers. Available online: http://www.worldometers.info/ (accessed on 12 May 2014).
2. Food and Agriculture Organization of the United Nations (FAO). Available online: http://faostat.fao.org/ (accessed on 12 May 2014).
3. Ray, D.K.; Mueller, N.D.; West, P.C.; Foley, J.A. Yield trends are insufficient to double global crop production by 2050. *PLoS One* **2013**, *8*, e66428.

4. Schittenhelm, S. Chemical composition and methane yield of maize hybrids with contrasting maturity. *Eur. J. Agron.* **2008**, *29*, 72–79.
5. Zeller, F.J. Sorghum (*Sorghum bicolor* L. Moench): Utilization, genetics, breeding. *Bodenkultur* **2000**, *51*, 71–85.
6. Pontieri, P.; Mamone, G.; de Caro, S.; Tuinstra, M.R.; Roemer, E.; Okot, J.; de Vita, P.; Ficco, D.B.M.; Alifano, P.; Pignone, D.; *et al.* Sorghum, a healthy and gluten-free food for celiac patients as demonstrated by genome, biochemical, and immunochemical analyses. *J. Agric. Food Chem.* **2013**, *61*, 2565–2571.
7. Strange, R.N.; Scott, P.R. Plant disease: A threat to global food security. *Annu. Rev. Phytopathol.* **2005**, *43*, 83–116.
8. Laurie, J.D.; Ali, S.; Linning, R.; Mannhaupt, G.; Wong, P.; Guldener, U.; Munsterkotter, M.; Moore, R.; Kahmann, R.; Bakkeren, G.; *et al.* Genome comparison of barley and maize smut fungi reveals targeted loss of RNA silencing components and species-specific presence of transposable elements. *Plant Cell* **2012**, *24*, 1733–1745.
9. Munkacsi, A.B.; Stoxen, S.; May, G. Domestication of maize, sorghum, and sugarcane did not drive the divergence of their smut pathogens. *Evolution* **2007**, *61*, 388–403.
10. Zuther, K.; Kahnt, J.; Utermark, J.; Imkampe, J.; Uhse, S.; Schirawski, J. Host specificity of *Sporisorium reilianum* is tightly linked to generation of the phytoalexin luteolinidin by *Sorghum bicolor*. *Mol. Plant Microbe Interact.* **2012**, *25*, 1230–1237.
11. Flors, V.; Ton, J.; Jakab, G.; Mauch-Mani, B. Abscisic Acid and Callose: Team players in defence against pathogens? *J. Phytopathol.* **2005**, *153*, 377–383.
12. Van Loon, L.C.; Rep, M.; Pieterse, C.M.J. Significance of inducible defense-related proteins in infected plants. *Annu. Rev. Phytopathol.* **2006**, *44*, 135–162.
13. O'Brien, J.A.; Daudi, A.; Butt, V.S.; Bolwell, G.P. Reactive oxygen species and their role in plant defence and cell wall metabolism. *Planta* **2012**, *236*, 765–779.
14. Ahuja, I.; Kissen, R.; Bones, A.M. Phytoalexins in defense against pathogens. *Trends Plant Sci.* **2012**, *17*, 73–90.
15. Hammerschmidt, R. Phytoalexins: What have we learned after 60 years? *Annu. Rev. Phytopathol.* **1999**, *37*, 285–306.
16. Jeandet, P.; Clement, C.; Courot, E.; Cordelier, S. Modulation of Phytoalexin Biosynthesis in Engineered Plants for Disease Resistance. *Int. J. Mol. Sci.* **2013**, *14*, 14136–14170.
17. Wharton, P.S.; Nicholson, R.L. Temporal synthesis and radiolabelling of the sorghum 3-deoxyanthocyanidin phytoalexins and the anthocyanin, cyanidin 3-dimalonyl glucoside. *New Phytol.* **2000**, *145*, 457–469.
18. Nicholson, R.; Wood, K. Phytoalexins and secondary products, where are they and how can we measure them? *Physiol. Mol. Plant Pathol.* **2001**, *59*, 63–69.
19. Nicholson, R.L.; Kollipara, S.S.; Vincent, J.R.; Lyons, P.C.; Cadena-Gomez, G. Phytoalexin synthesis by the sorghum mesocotyl in response to infection by pathogenic and nonpathogenic fungi. *Proc. Natl. Acad. Sci. USA* **1987**, *84*, 5520–5524.
20. Snyder, B.A.; Nicholson, R.L. Synthesis of phytoalexins in sorghum as a site-specific response to fungal ingress. *Science* **1990**, *248*, 1637–1639.

21. Wharton, P.; Julian, A. A cytological study of compatible and incompatible interactions between *Sorghum bicolor* and *Colletotrichum sublineolum*. *New Phytol.* **1996**, *134*, 25–34.
22. Basavaraju, P.; Shetty, N.P.; Shetty, H.S.; de Neergaard, E.; Jørgensen, H.J.L. Infection biology and defence responses in sorghum against *Colletotrichum sublineolum*. *J. Appl. Microbiol.* **2009**, *107*, 404–415.
23. Shih, C.H.; Chu, I.K.; Yip, W.K.; Lo, C. Differential expression of two flavonoid 3'-hydroxylase cDNAs involved in biosynthesis of anthocyanin pigments and 3-deoxyanthocyanidin phytoalexins in sorghum. *Plant Cell Physiol.* **2006**, *47*, 1412–1419.
24. Schmelz, E.A.; Huffaker, A.; Sims, J.W.; Christensen, S.A.; Lu, X.; Okada, K.; Peters, R.J. Biosynthesis, elicitation and roles of monocot terpenoid phytoalexins. *Plant J.* **2014**, doi:10.1111/tpj.12436.
25. Stafford, H.A. Teosinte to maize—Some aspects of missing biochemical and physiological data concerning regulation of flavonoid pathways. *Phytochemistry* **1998**, *49*, 285–293.
26. Cui, Y.; Magill, J.; Frederiksen, R.; Magill, C. Chalcone synthase and phenylalanine ammonia-lyase mRNA levels following exposure of sorghum seedlings to three fungal pathogens. *Physiol. Mol. Plant Pathol.* **1996**, *49*, 187–199.
27. Huang, L.D.; Backhouse, D. Effects of *Fusarium* species on defence mechanisms in sorghum seedlings. *N. Z. Plant Prot.* **2004**, *57*, 121–124.
28. Lo, S.C.; Hipskind, J.D.; Nicholson, R.L. cDNA cloning of a sorghum pathogenesis-related protein (PR-10) and differential expression of defense-related genes following inoculation with *Cochliobolus heterostrophus* or *Colletotrichum sublineolum*. *Mol. Plant Microbe Interact.* **1999**, *12*, 479–489.
29. Lo, S.C.; de Verdier, K.; Nicholson, R.L. Accumulation of 3-deoxyanthocyanidin phytoalexins and resistance to *Colletotrichum sublineolum* in sorghum. *Physiol. Mol. Plant Pathol.* **1999**, *55*, 263–273.
30. Mizuno, H.; Kawahigashi, H.; Kawahara, Y.; Kanamori, H.; Ogata, J.; Minami, H.; Itoh, T.; Matsumoto, T. Global transcriptome analysis reveals distinct expression among duplicated genes during sorghum-interaction. *BMC Plant Biol.* **2012**, *12*, 121.
31. Yazawa, T.; Kawahigashi, H.; Matsumoto, T.; Mizuno, H. Simultaneous transcriptome analysis of sorghum and *Bipolaris sorghicola* by using RNA-seq in combination with de novo transcriptome assembly. *PLoS One* **2013**, *8*, e62460.
32. Weiergang, I.; Hipskind, J.D.; Nicholson, R.L. Synthesis of 3-deoxyanthocyanidin phytoalexins in sorghum occurs independent of light. *Physiol. Mol. Plant Pathol.* **1996**, *49*, 377–388.
33. Liu, H.; Du, Y.; Chu, H.; Shih, C.H.; Wong, Y.W.; Wang, M.; Chu, I.K.; Tao, Y.; Lo, C. Molecular dissection of the pathogen-inducible 3-deoxyanthocyanidin biosynthesis pathway in sorghum. *Plant Cell Physiol.* **2010**, *51*, 1173–1185.
34. Winkel-Shirley, B. Flavonoid biosynthesis. A colorful model for genetics, biochemistry, cell biology, and biotechnology. *Plant Phisiol.* **2001**, *126*, 485–493.

35. Du, Y.; Chu, H.; Wang, M.; Chu, I.K.; Lo, C. Identification of flavone phytoalexins and a pathogen-inducible flavone synthase II gene (SbFNSII) in sorghum. *J. Exp. Bot.* **2010**, *61*, 983–994.
36. Lo, S.C.; Nicholson, R.L. Reduction of light-induced anthocyanin accumulation in inoculated sorghum mesocotyls. Implications for a compensatory role in the defense response. *Plant Physiol.* **1998**, *116*, 979–989.
37. Paterson, A.H.; Bowers, J.E.; Bruggmann, R.; Dubchak, I.; Grimwood, J.; Gundlach, H.; Haberer, G.; Hellsten, U.; Mitros, T.; Poliakov, A.; *et al.* The *Sorghum bicolor* genome and the diversification of grasses. *Nature* **2009**, *457*, 551–556.
38. Lo, C.; Coolbaugh, R.C.; Nicholson, R.L. Molecular characterization and in silico expression analysis of a chalcone synthase gene family in *Sorghum bicolor*. *Physiol. Mol. Plant Pathol.* **2002**, *61*, 179–188.
39. Yu, C.K.Y.; Springob, K.; Schmidt, J.; Nicholson, R.L.; Chu, I.K.; Yip, W.K.; Lo, C. A stilbene synthase gene (SbSTS1) is involved in host and nonhost defense responses in sorghum. *Plant Physiol.* **2005**, *138*, 393–401.
40. Nicholson, R.L.; Hammerschmidt, R. Phenolic compounds and their role in disease resistance. *Annu. Rev. Phytopathol.* **1992**, *30*, 369–389.
41. Stonecipher, L.L.; Hurley, P.S.; Netzly, D.H. Effect of apigeninidin on the growth of selected bacteria. *J. Chem. Ecol.* **1993**, *19*, 1021–1027.
42. Snyder, B.A.; Leite, B.; Hipskind, J.; Butler, L.G.; Nicholson, R.L. Accumulation of sorghum phytoalexins induced by *Colletotrichum graminicola* at the infection site. *Physiol. Mol. Plant Pathol.* **1991**, *39*, 463–470.
43. Huffaker, A.; Kaplan, F.; Vaughan, M.M.; Dafoe, N.J.; Ni, X.; Rocca, J.R.; Alborn, H.T.; Teal, P.E.A.; Schmelz, E.A. Novel acidic sesquiterpenoids constitute a dominant class of pathogen-induced phytoalexins in maize. *Plant Physiol.* **2011**, *156*, 2082–2097.
44. Köllner, T.G.; Schnee, C.; Li, S.; Svatos, A.; Schneider, B.; Gershenzon, J.; Degenhardt, J. Protonation of a neutral (S)-beta-bisabolene intermediate is involved in (S)-beta-macrocarpene formation by the maize sesquiterpene synthases TPS6 and TPS11. *J. Biol. Chem.* **2008**, *283*, 20779–20788.
45. Köllner, T.G.; Schnee, C.; Gershenzon, J.; Degenhardt, J. The variability of sesquiterpenes emitted from two *Zea mays* cultivars is controlled by allelic variation of two terpene synthase genes encoding stereoselective multiple product enzymes. *Plant Cell* **2004**, *16*, 1115–1131.
46. Basse, C.W. Dissecting defense-related and developmental transcriptional responses of maize during *Ustilago maydis* infection and subsequent tumor formation. *Plant Physiol.* **2005**, *138*, 1774–1784.
47. Dohlemann, G.; Wahl, R.; Horst, R.J.; Voll, L.M.; Usadel, B.; Poree, F.; Stitt, M.; Pons-Kuhnemann, J.; Sonnewald, U.; Kahmann, R.; *et al.* Reprogramming a maize plant: Transcriptional and metabolic changes induced by the fungal biotroph *Ustilago maydis*. *Plant J.* **2008**, *56*, 181–195.

48. Ghareeb, H.; Becker, A.; Iven, T.; Feussner, I.; Schirawski, J. *Sporisorium reilianum* infection changes inflorescence and branching architectures of maize. *Plant Physiol.* **2011**, *156*, 2037–2052.

49. Van der Linde, K.; Doehlemann, G. Utilizing virus-induced gene silencing for the functional characterization of maize genes during infection with the fungal pathogen *Ustilago maydis*. *Methods Mol. Biol.* **2013**, *975*, 47–60.

50. Schmelz, E.A.; Kaplan, F.; Huffaker, A.; Dafoe, N.J.; Vaughan, M.M.; Ni, X.; Rocca, J.R.; Alborn, H.T.; Teal, P.E. Identity, regulation, and activity of inducible diterpenoid phytoalexins in maize. *Proc. Natl. Acad. Sci. USA* **2011**, *108*, 5455–5460.

51. Harris, L.J.; Saparno, A.; Johnston, A.; Prisic, S.; Xu, M.; Allard, S.; Kathiresan, A.; Ouellet, T.; Peters, R.J. The maize AN2 gene is induced by *Fusarium* attack and encodes an ent-copalyl diphosphate synthase. *Plant Mol. Biol.* **2005**, *59*, 881–894.

52. Dafoe, N.J.; Huffaker, A.; Vaughan, M.M.; Duehl, A.J.; Teal, P.E.; Schmelz, E.A. Rapidly induced chemical defenses in maize stems and their effects on short-term growth of *Ostrinia nubilalis*. *J. Chem. Ecol.* **2011**, *37*, 984–991.

53. Dafoe, N.J.; Thomas, J.D.; Shirk, P.D.; Legaspi, M.E.; Vaughan, M.M.; Huffaker, A.; Teal, P.E.; Schmelz, E. A European corn borer (*Ostrinia nubilalis*) induced responses enhance susceptibility in maize. *PLoS One* **2013**, *8*, e73394.

54. Allardyce, J.A.; Rookes, J.E.; Hussain, H.I.; Cahill, D.M. Transcriptional profiling of *Zea mays* roots reveals roles for jasmonic acid and terpenoids in resistance against *Phytophthora cinnamomi*. *Funct. Integr. Genomics* **2013**, *13*, 217–228.

55. Glauser, G.; Marti, G.; Villard, N.; Doyen, G.A.; Wolfender, J.L.; Turlings, T.C.J.; Erb, M. Induction and detoxification of maize 1,4-benzoxazin-3-ones by insect herbivores. *Plant J.* **2011**, *68*, 901–911.

56. Meihls, L.N.; Handrick, V.; Glauser, G.; Barbier, H.; Kaur, H.; Haribal, M.M.; Lipka, A.E.; Gershenzon, J.; Buckler, E.S.; Erb, M.; *et al.* Natural variation in maize aphid resistance is associated with a DIMBOA-Glc methyltransferase. *Plant Cell* **2013**, *25*, 2341–2355.

57. VanEtten, H.D.; Mansfield, J.W.; Bailey, J.A.; Farmer, E.E. Two classes of plant antibiotics: Phytoalexins *versus* "Phytoanticipins". *Plant Cell* **1994**, *6*, 1191–1192.

58. Tanaka, S.; Brefort, T.; Neidig, N.; Djamei, A.; Kahnt, J.; Vermerris, W.; Koenig, S.; Feussner, K.; Feussner, I.; Kahmann, R. A secreted *Ustilago maydis* effector promotes virulence by targeting anthocyanin biosynthesis in maize. *Elife* **2014**, *3*, e01355.

59. Frey, M.; Schullehner, K.; Dick, R.; Fiesselmann, A.; Gierl, A. Benzoxazinoid biosynthesis, a model for evolution of secondary metabolic pathways in plants. *Phytochemistry* **2009**, *70*, 1645–1651.

60. Song, Y.Y.; Cao, M.; Xie, L.J.; Liang, X.T.; Zeng, R.S.; Su, Y.J.; Huang, J.H.; Wang, R.L.; Luo, S.M. Induction of DIMBOA accumulation and systemic defense responses as a mechanism of enhanced resistance of mycorrhizal corn (*Zea mays L.*) to sheath blight. *Mycorrhiza* **2011**, *21*, 721–731.

61. Erb, M.; Balmer, D.; de Lange, E.S.; von Merey, G.; Planchamp, C.; Robert, C.A.; Röder, G.; Sobhy, I.; Zwahlen, C.; Mauch-Mani, B.; *et al.* Synergies and trade-offs between insect and pathogen resistance in maize leaves and roots. *Plant Cell Environ.* **2011**, *34*, 1088–1103.
62. Oikawa, A.; Ishihara, A.; Tanaka, C.; Mori, N.; Tsuda, M.; Iwamura, H. Accumulation of HDMBOA-Glc is induced by biotic stresses prior to the release of MBOA in maize leaves. *Phytochemistry* **2004**, *65*, 2995–3001.
63. Angelova, Z.; Georgiev, S.; Roos, W.; Bulgaria, S. Elicitation of plants. *Biotechnol. Equip.* **2006**, *20*, 72–83.
64. Millet, Y.A.; Danna, C.H.; Clay, N.K.; Songnuan, W.; Simon, M.D.; Werck-Reichhart, D.; Ausubel, F.M. Innate immune responses activated in Arabidopsis roots by microbe-associated molecular patterns. *Plant Cell* **2010**, *22*, 973–990.
65. Oliveira-Garcia, E.; Deising, H.B. Infection structure-specific expression of β-1,3-glucan synthase is essential for pathogenicity of *Colletotrichum graminicola* and evasion of β-glucan-triggered immunity in maize. *Plant Cell* **2013**, *25*, 2356–2378.
66. Wulff, N.A.; Pascholati, S.F. Preparacoes de *Saccharomyces cerevisiae* elicitoras de fitoalexinas em mesocotilos de sorgo. *Sci. Agric.* **1988**, *55*, 1.
67. Yamaoka, N.; Lyons, P.C.; Hipskind, J.; Nicholson, R.L. Elicitor of sorghum phytoalexin synthesis from *Colletotrichum graminicola*. *Physiol. Mol. Plant Pathol.* **1990**, *37*, 255–270.
68. Kishi-Kaboshi, M.; Takahashi, A.; Hirochika, H. MAMP-responsive MAPK cascades regulate phytoalexin biosynthesis. *Plant Signal. Behav.* **2010**, *5*, 1653–1656.
69. Chopra, S.; Brendel, V.; Zhang, J.; Axtell, J.D.; Peterson, T. Molecular characterization of a mutable pigmentation phenotype and isolation of the first active transposable element from *Sorghum bicolor*. *Proc. Natl. Acad. Sci. USA* **1999**, *96*, 15330–15335.
70. Chopra, S.; Gevens, A.; Svabek, C.; Wood, K.V.; Peterson, T.; Nicholson, R.L. Excision of the Candystripe1 transposon from a hyper-mutable Y1-cs allele shows that the sorghum Y1 gene controls the biosynthesis of both 3-deoxyanthocyanidin phytoalexins and phlobaphene pigments. *Physiol. Mol. Plant Pathol.* **2002**, *60*, 321–330.
71. Selinger, D.A.; Chandler, V.L. A mutation in the PALE ALEURONE COLOR1 gene identifies a novel regulator of the maize anthocyanin pathway. *Plant Cell* **1999**, *11*, 5–14.
72. Lesnick, M.L.; Chandler, V.L. Activation of the maize anthocyanin gene A2 is mediated by an element conserved in many anthocyanin promoters. *Plant Physiol.* **1998**, *117*, 437–445.
73. Grotewold, E.; Chamberlin, M.; Snook, M.; Siame, B.; Butler, L.; Swenson, J.; Maddock, S.; St Clair, G.; Bowen, B. Engineering secondary metabolism in maize cells by ectopic expression of transcription factors. *Plant Cell* **1998**, *10*, 721–740.
74. Ibraheem, F.; Gaffoor, I.; Chopra, S. Flavonoid phytoalexin-dependent resistance to anthracnose leaf blight requires a functional YELLOW SEED1 in *Sorghum bicolor*. *Genetics* **2010**, *184*, 915–926.
75. Salzman, R.A.; Brady, J.A.; Finlayson, S.A.; Buchanan, C.D.; Summer, E.J.; Sun, F.; Klein, P.E.; Klein, R.R.; Pratt, L.; Cordonnier-Pratt, M.M.; *et al.* Transcriptional profiling of sorghum induced by methyl jasmonate, salicylic acid, and aminocyclopropane carboxylic acid reveals cooperative regulation and novel gene responses. *Plant Physiol.* **2005**, *138*, 352–368.

76. Lo, S.C.; Weiergang, I.; Bonham, C.; Hipskind, J.; Wood, K.; Nicholson, R.L. Phytoalexin accumulation in sorghum: Identification of a methylether of luteolinidin. *Physiol. Mol. Plant Pathol.* **1996**, *49*, 21–31.
77. Schutt, C.; Netzly, D. Effect of apiferol and apigeninidin on growth of selected fungi. *J. Chem. Ecol.* **1991**, *17*, 2261–2266.
78. Pedras, M.S.C.; Minic, Z.; Jha, M. Brassinin oxidase, a fungal detoxifying enzyme to overcome a plant defense—Purification, characterization and inhibition. *FEBS J.* **2008**, *275*, 3691–3705.
79. Pedras, M.S.C.; Minic, Z.; Sarma-Mamillapalle, V.K. Substrate specificity and inhibition of brassinin hydrolases, detoxifying enzymes from the plant pathogens *Leptosphaeria maculans* and *Alternaria brassicicola. FEBS J.* **2009**, *276*, 7412–7428.
80. Schäfer, W.; Straney, D.; Ciuffetti, L.; van Etten, H.D.; Yoder, O.C. One enzyme makes a fungal pathogen, but not a saprophyte, virulent on a new host plant. *Science* **1989**, *246*, 247–249.
81. George, H.L.; van Etten, H.D. Characterization of pisatin-inducible cytochrome p450s in fungal pathogens of pea that detoxify the pea phytoalexin pisatin. *Fungal Genet. Biol.* **2001**, *33*, 37–48.
82. Desjardins, A.E.; Gardner, H.W. Virulence of *Gibberella pulicaris* on potato tubers and its relationship to a gene for rishitin metabolism. *Phytopathology* **1991**, *81*, 429–435.
83. Schnable, P.S.; Ware, D.; Fulton, R.S.; Stein, J.C.; Wei, F.; Pasternak, S.; Liang, C.; Zhang, J.; Fulton, L.; Graves, T.A.; *et al.* The B73 maize genome: Complexity, diversity, and dynamics. *Science* **2009**, *326*, 1112–1115.
84. Stemple, D.L. TILLING—a high-throughput harvest for functional genomics. *Nat. Rev. Genet.* **2004**, *5*, 145–150.
85. Sikora, P.; Chawade, A.; Larsson, M.; Olsson, J.; Olsson, O. Mutagenesis as a tool in plant genetics, functional genomics, and breeding. *Int. J. Plant Genomics* **2011**, *2011*, 314829.
86. Martin, T.; Biruma, M.; Fridborg, I.; Okori, P.; Dixelius, C. A highly conserved NB-LRR encoding gene cluster effective against *Setosphaeria turcica* in sorghum. *BMC Plant Biol.* **2011**, *11*, 151.
87. Casas, A.M.; Kononowicz, A.K.; Haan, T.G.; Zhang, L.; Tomes, D.T.; Bressan, R.A.; Hasegawa, P.M. Transgenic sorghum plants obtained after microprojectile bombardment of immature inflorescences. *In Vitro Cell. Dev. Biol. Plant* **1997**, *33*, 92–100.
88. Gaj, T.; Gersbach, C.A.; Barbas, C.F. ZFN, TALEN, and CRISPR/Cas-based methods for genome engineering. *Trends Biotechnol.* **2013**, *31*, 397–405.
89. Chen, K.; Gao, C. TALENs: Customizable molecular DNA scissors for genome engineering of plants. *J. Genet. Genomics* **2013**, *40*, 271–279.
90. Jiang, W.; Zhou, H.; Bi, H.; Fromm, M.; Yang, B.; Weeks, D.P. Demonstration of CRISPR/Cas9/sgRNA-mediated targeted gene modification in Arabidopsis, tobacco, sorghum and rice. *Nucleic Acids Res.* **2013**, doi:10.1093/nar/gkt780.
91. Belhaj, K.; Chaparro-Garcia, A.; Kamoun, S.; Nekrasov, V. Plant genome editing made easy: Targeted mutagenesis in model and crop plants using the CRISPR/Cas system. *Plant Methods* **2013**, *9*, 39.

92. Campos-Bermudez, V.A.; Fauguel, C.M.; Tronconi, M.A.; Casati, P.; Presello, D.A.; Andreo, C.S. Transcriptional and metabolic changes associated to the infection by *Fusarium verticillioides* in maize inbreds with contrasting ear rot resistance. *PLoS One* **2013**, *8*, e61580.
93. O'Connell, R.J.; Thon, M.R.; Hacquard, S.; Amyotte, S.G.; Kleemann, J.; Torres, M.F.; Damm, U.; Buiate, E.A.; Epstein, L.; Alkan, N.; *et al.* Lifestyle transitions in plant pathogenic *Colletotrichum* fungi deciphered by genome and transcriptome analyses. *Nat. Genet.* **2012**, *44*, 1060–1065.
94. Shih, C.H.; Siu, S.O.; Ng, R.; Wong, E.; Chiu, L.C.M.; Chu, I.K.; Lo, C. Quantitative analysis of anticancer 3-deoxyanthocyanidins in infected sorghum seedlings. *J. Agric. Food Chem.* **2007**, *55*, 254–259.
95. Lin, Y.; Shi, R.; Wang, X.; Shen, H.M. Luteolin, a flavonoid with potential for cancer prevention and therapy. *Curr. Cancer Drug Targets* **2008**, *8*, 634–646.
96. Seelinger, G.; Merfort, I.; Wölfle, U.; Schempp, C.M. Anti-carcinogenic effects of the flavonoid luteolin. *Molecules* **2008**, *13*, 2628–2651.
97. Massey, A.R.; Reddivari, L.; Vanamala, J. The dermal layer of sweet sorghum (*Sorghum bicolor*) stalk, a byproduct of biofuel production and source of unique 3 deoxyanthocyanidins, has more antiproliferative and proapoptotic activity than the pith in p53 variants of HCT116 and colon cancer stem cells. *J. Agric. Food Chem.* **2014**, *62*, 3150–3159.
98. Suganyadevi, P.; Saravanakumar, K.M.; Mohandas, S. The antiproliferative activity of 3-deoxyanthocyanins extracted from red sorghum (*Sorghum bicolor*) bran through P53-dependent and Bcl-2 gene expression in breast cancer cell line. *Life Sci.* **2013**, *92*, 379–382.
99. Carbonneau, M.A.; Cisse, M.; Mora-Soumille, N.; Dairi, S.; Rosa, M.; Michel, F.; Lauret, C.; Cristol, J.P.; Dangles, O. Antioxidant properties of 3-deoxyanthocyanidins and polyphenolic extracts from Côte d'Ivoire's red and white sorghums assessed by ORAC and *in vitro* LDL oxidisability tests. *Food Chem.* **2014**, *145*, 701–7090.
100. Yang, L.; Browning, J.D.; Awika, J.M. Sorghum 3-deoxyanthocyanins possess strong phase II enzyme inducer activity and cancer cell growth inhibition properties. *J. Agric. Food Chem.* **2009**, *57*, 1797–1804.
101. Gao, J.; Kashfi, K.; Liu, X.; Rigas, B. NO-donating aspirin induces phase II enzymes *in vitro* and *in vivo*. *Carcinogenesis* **2006**, *27*, 803–810.
102. Patent application WO2012175720 A1. Available online: http://www.google.com/patents/WO2012175720A1 (accessed on 12 May 2014).
103. Petti, C.; Kushwaha, R.; Tateno, M.; Harman-Ware, A.E.; Crocker, M.; Awika, J.; Debolt, S. Mutagenesis breeding for increased 3-deoxyanthocyanidin accumulation in leaves of *Sorghum bicolor* (L.) Moench: A source of natural food pigment. *J. Agric. Food Chem.* **2014**, *62*, 1227–1232.
104. Awika J.M.; Rooney, L.W.; Waniska, R.D. Properties of 3-deoxyanthocyanins from sorghum. *J. Agric. Food Chem.* **2004**, *52*, 4388–4394.

Natural Products from the Genus *Tephrosia*

Yinning Chen, Tao Yan, Chenghai Gao, Wenhao Cao and Riming Huang

Abstract: The genus *Tephrosia*, belonging to the Leguminosae family, is a large pantropical genus of more than 350 species, many of which have important traditional uses in agriculture because they possess the bioactivity of phytoalexins. This review not only outlines the sources, chemistry and biological evaluations of natural products from the genus *Tephrosia* worldwide that have appeared in literature from 1910 to December 2013, but also covers work related to proposed biosynthetic pathways and synthesis of some natural products from the genus *Tephrosia*, with 105 citations and 168 new compounds.

Reprinted from *Molecules*. Cite as: Chen, Y.; Yan, T.; Gao, C.; Cao, W.; Huang, R. Natural Products from the Genus *Tephrosia*. *Molecules* **2014**, *19*, 1432-1458.

1. Introduction

The genus *Tephrosia*, belonging to the Leguminosae family, is a large pantropical genus of more than 350 species, many of which have important traditional uses [1,2]. Phytochemical investigations have revealed the presence of glucosides, rotenoids, isoflavones, chalcones, flavanones, flavanols, and prenylated flavonoids [1–9] of chemotaxonomic importance in the genus [10]. Moreover, bioactivity has been studied extensively, indicating that chemical constituents and extracts of the genus *Tephrosia* exhibited diverse bioactivities, such as insecticidal [11], antiviral [12], antiprotozoal [13], antiplasmodial [14] and cytotoxic [15] activities.

So far, the reviews on natural products isolated from the genus *Tephrosia* are limited [16]. To gain a comprehensive and systematic understanding of this genus, this review outlines the chemistry, proposed biosynthetic pathways, synthesis, and biological evaluations of natural products from the genus *Tephrosia* worldwide that have appeared in literature from 1971 to December 2013, with 105 citations and 168 new compounds from them.

2. Chemical Constituents

The chemical constituents of the genus *Tephrosia* reported since 1910 (compounds **1–168**) are shown in Table 1 and Figures 1–10 below with their names, and their biological sources. As listed in the table and Figures 1–7, flavonoids are the predominant constituents of this genus.

Figure 1. Flavones from genus *Tephrosia*.

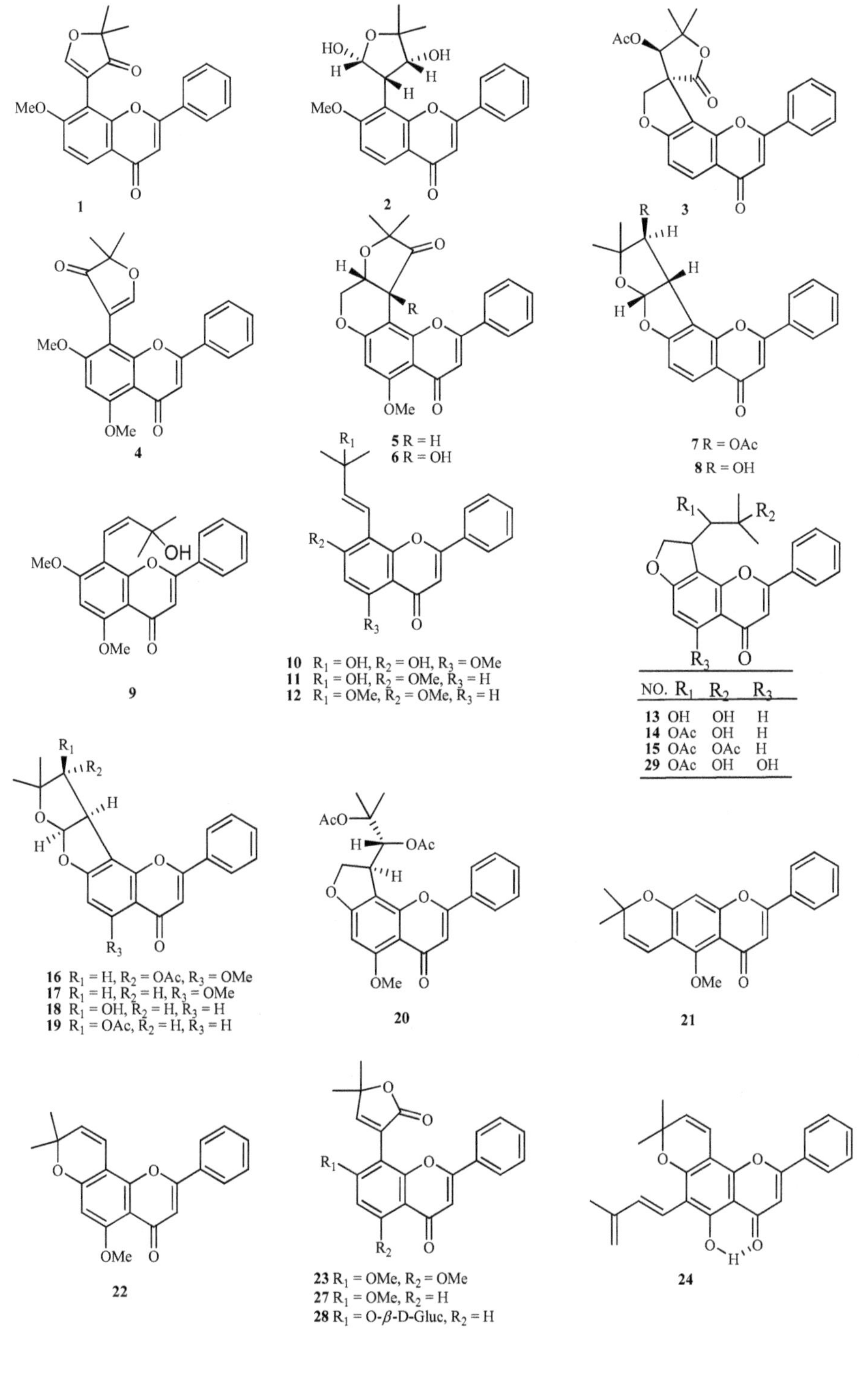

1 **2** **3**

4

5 R = H
6 R = OH

7 R = OAc
8 R = OH

9

10 R_1 = OH, R_2 = OH, R_3 = OMe
11 R_1 = OH, R_2 = OMe, R_3 = H
12 R_1 = OMe, R_2 = OMe, R_3 = H

NO.	R_1	R_2	R_3
13	OH	OH	H
14	OAc	OH	H
15	OAc	OAc	H
29	OAc	OH	OH

16 R_1 = H, R_2 = OAc, R_3 = OMe
17 R_1 = H, R_2 = H, R_3 = OMe
18 R_1 = OH, R_2 = H, R_3 = H
19 R_1 = OAc, R_2 = H, R_3 = H

20

21

22

23 R_1 = OMe, R_2 = OMe
27 R_1 = OMe, R_2 = H
28 R_1 = O-β-D-Gluc, R_2 = H

24

Figure 1. *Cont.*

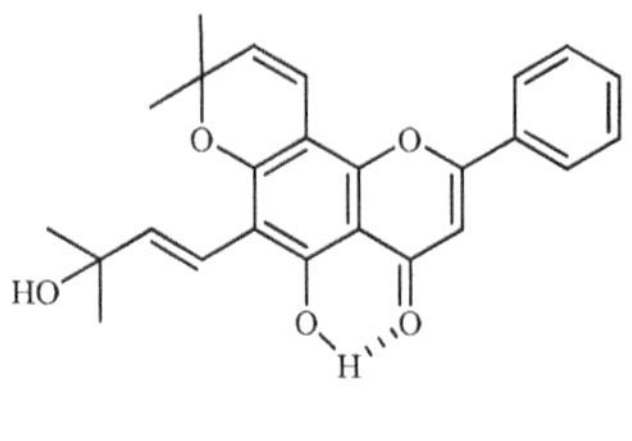

Figure 2. Flavonols from genus *Tephrosia*.

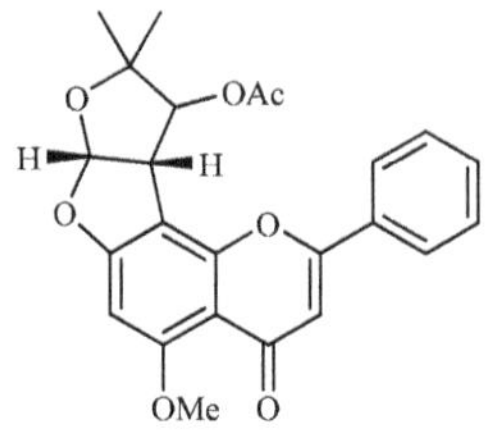

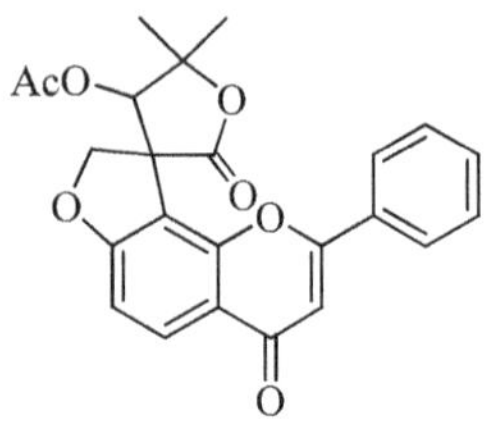

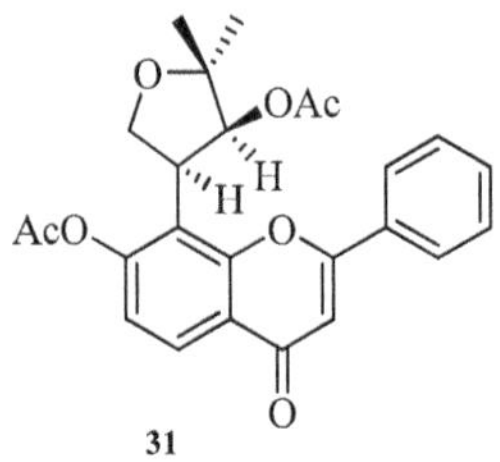

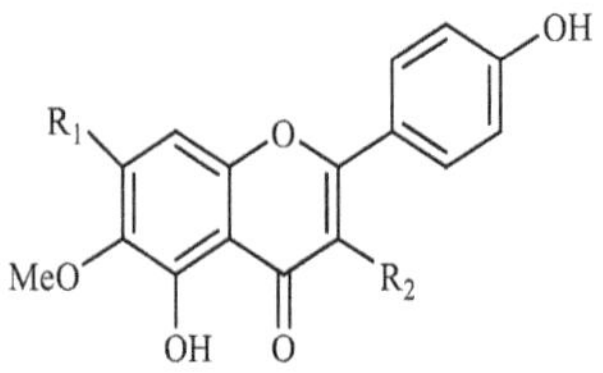

32 R_1 = O-(α-Rha), R_2 = O-(α-Rha-(1→2)-[α-Rha-(1→6)]-β-Gal-

33 R_1 = O-(α-Rha), R_2 = O-(α-Rha-(1→6)-β-Gal-

34 R_1 = OH, R_2 = O-(α-Rha-(1→2)-[α-Rha-(1→6)]-β-Gal-

35 R_1 = OH, R_2 = O-(α-Rha-(1→2)-[(3-*O*-*E*-feruloyl)-α-Rha-(1→6)]-β-Gal-

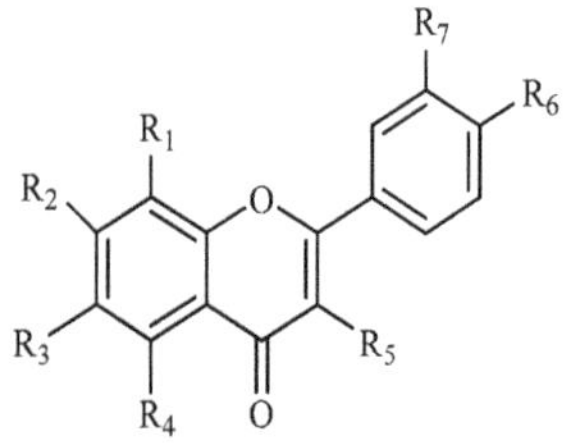

NO.	R_1	R_2	R_3	R_4	R_5	R_6	R_7
36	H	ORha	OH	OH	ORha	OMe	H
37	H	OMe	OMe	OMe	OH	OH	H
38	OMe	H	OMe	OH	OMe	OH	H
39	H	OEt	H	H	OH	OH	OH

Figure 3. Flavanonols from genus *Tephrosia*.

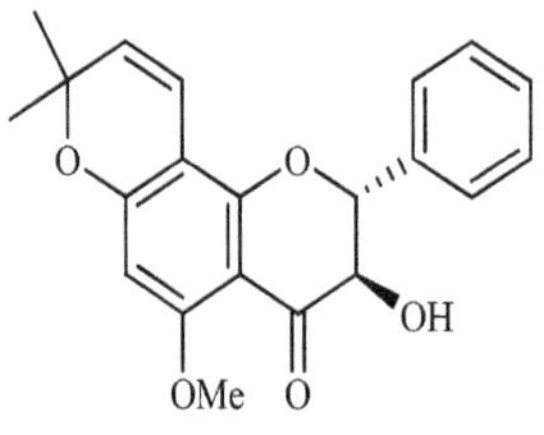

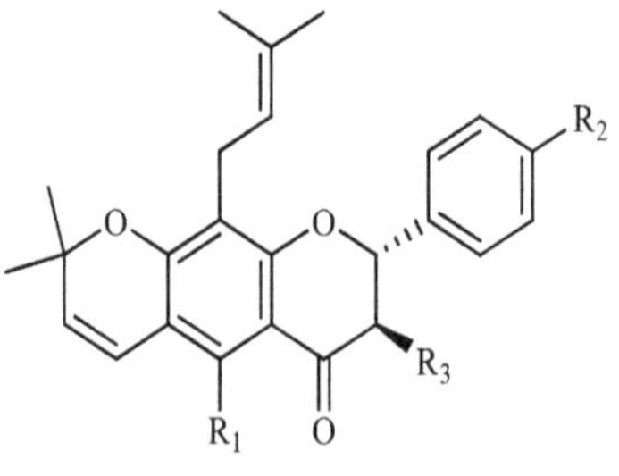

41 R_1 = OH, R_2 = OH, R_3 = OH

42 R_1 = OAc, R_2 = OAc, R_3 = OAc

Figure 4. Flavans from genus *Tephrosia*.

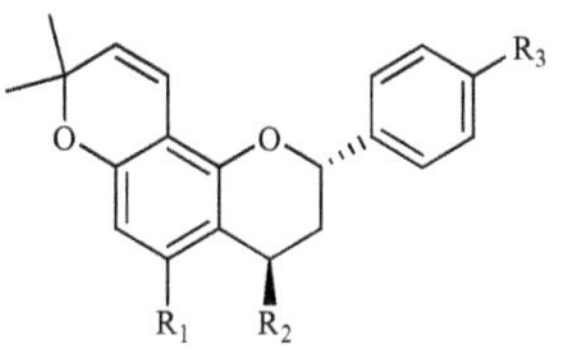

43 R_1 = OMe, R_2 = O, R_3 = OH
61 R_1 = OH, R_2 = O, R_3 = H
62 R_1 = OMe, R_2 = O, R_3 = H
63 R_1 = OMe, R_2 = OMe, R_3 = H
64 R_1 = OMe, R_2 = OH, R_3 = H

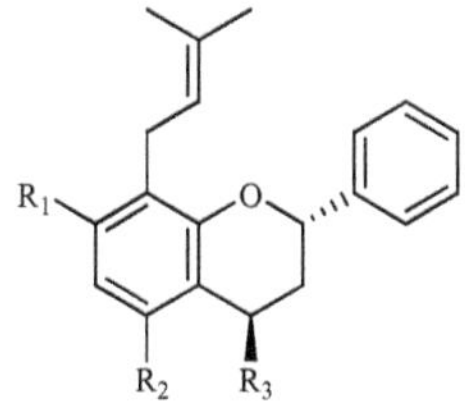

44 R_1 = OH, R_2 = OMe, R_3 = O
69 R_1 = OMe, R_2 = OH, R_3 = O
70 R_1 = OMe, R_2 = OMe, R_3 = H
71 R_1 = OMe, R_2 = OMe, R_3 = OH

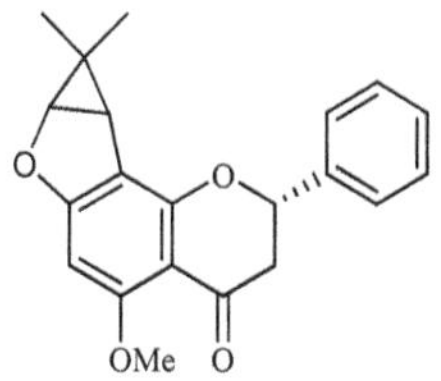

45

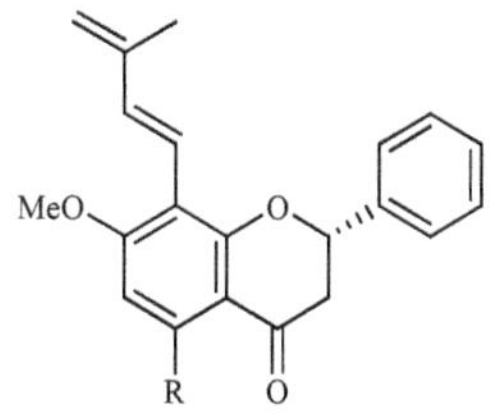

46 R = OMe
84 R = OH

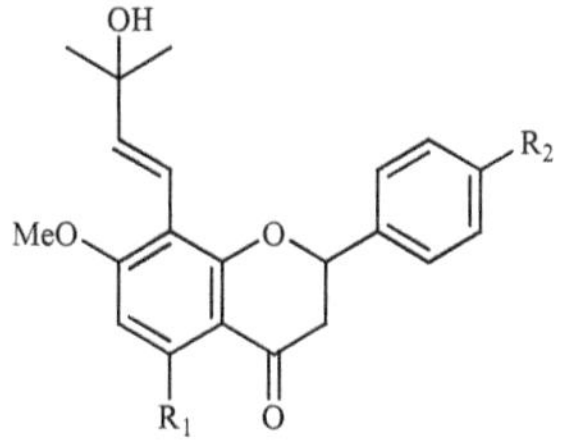

47 R_1 = OMe, R_2 = H
48 R_1 = OMe, R_2 = OH
77 R_1 = H, R_2 = H

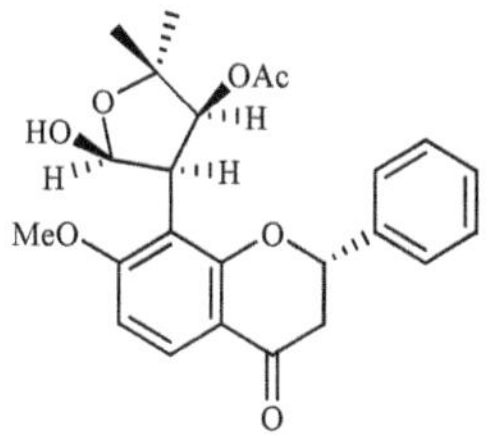

49

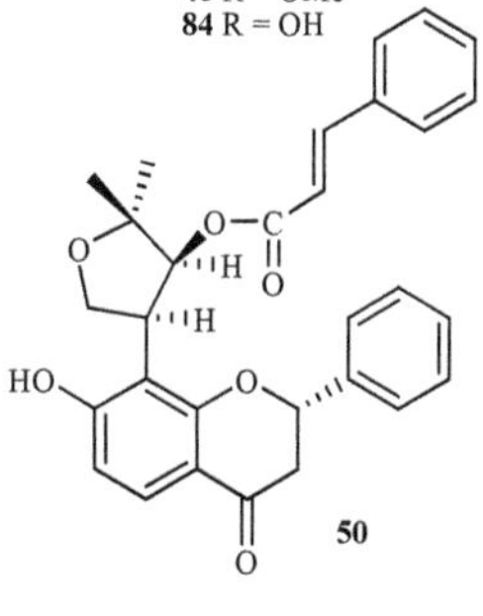

50

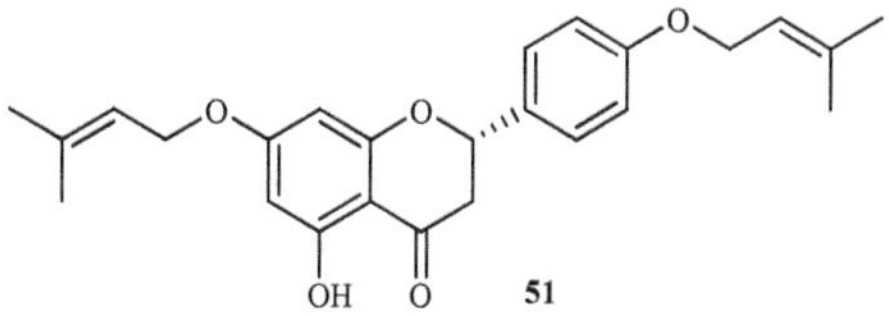

51

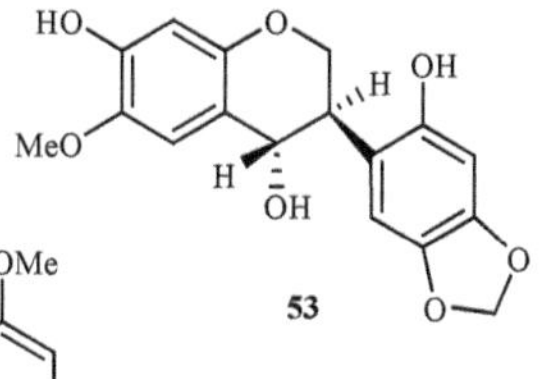

53

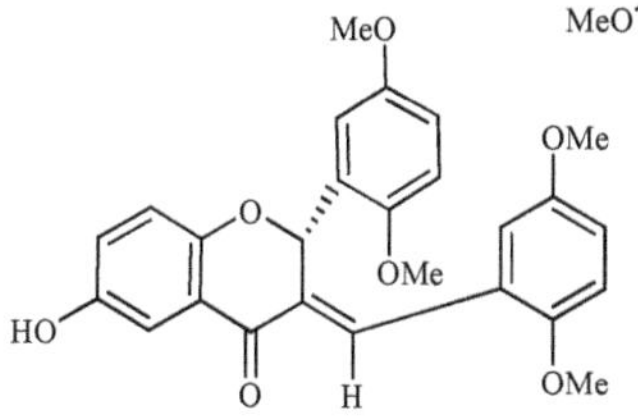

52

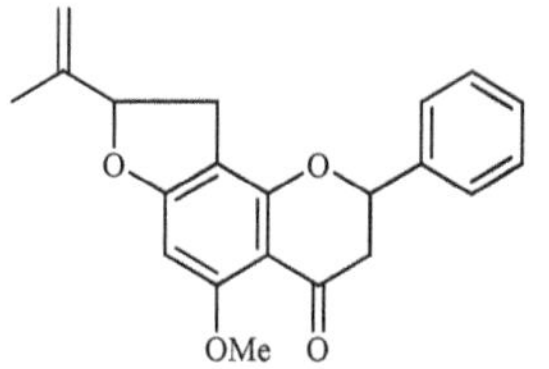

54

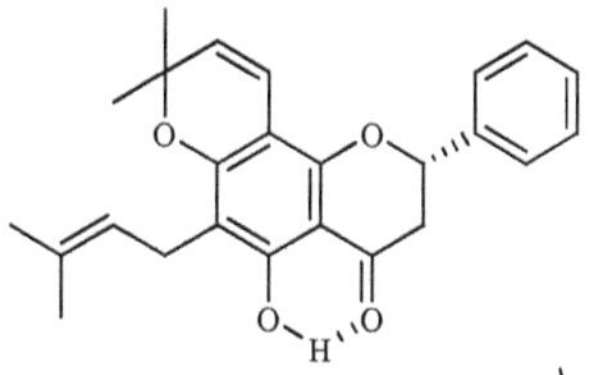

57

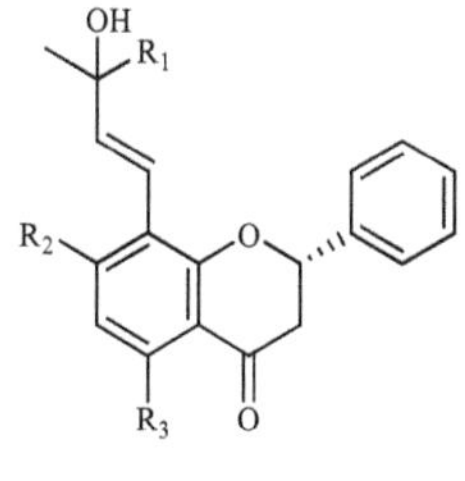

55 R_1 = Me, R_2 = OH, R_3 = H
56 R_1 = CH_2OAc, R_2 = OH, R_3 = H
81 R_1 = OH, R_2 = OMe, R_3 = OMe
83 R_1 = Me, R_2 = OMe, R_3 = OH

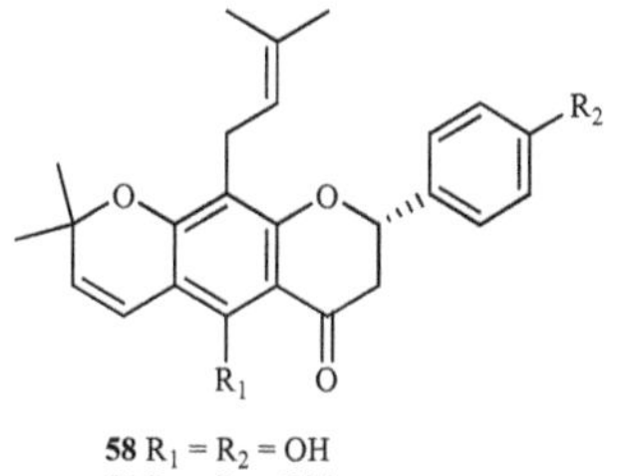

58 R_1 = R_2 = OH
59 R_1 = R_2 = OMe
60 R_1 = R_2 = OAc

Figure 4. *Cont.*

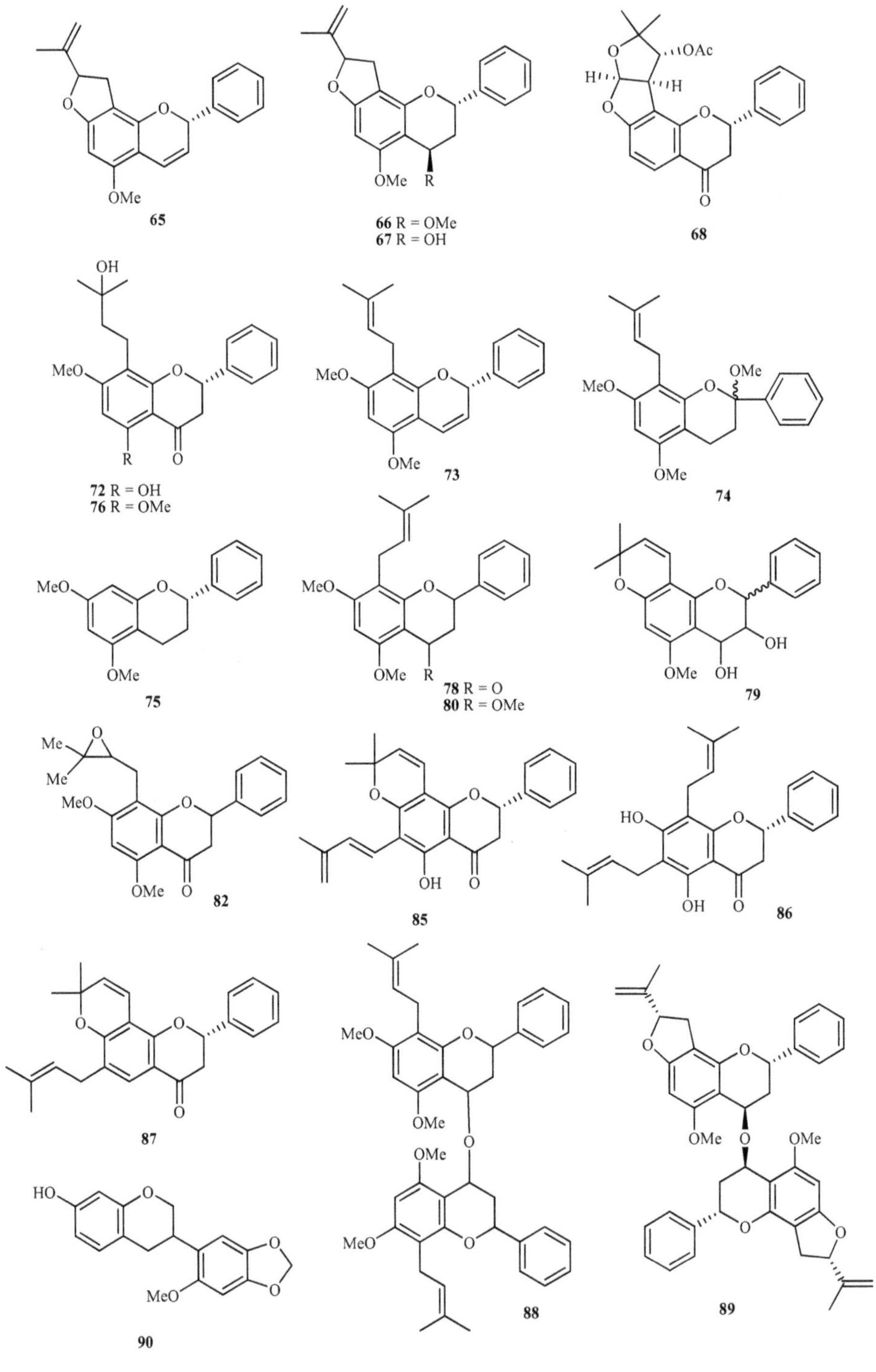

Figure 4. *Cont.*

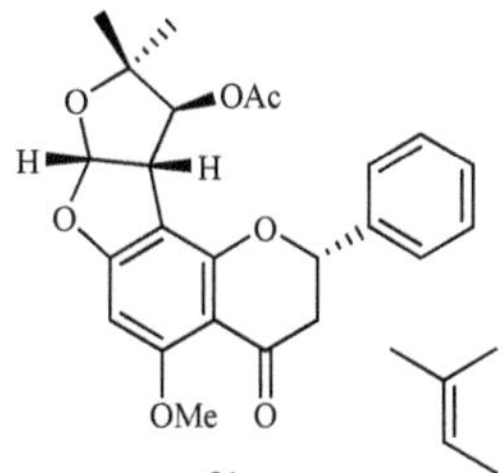

91

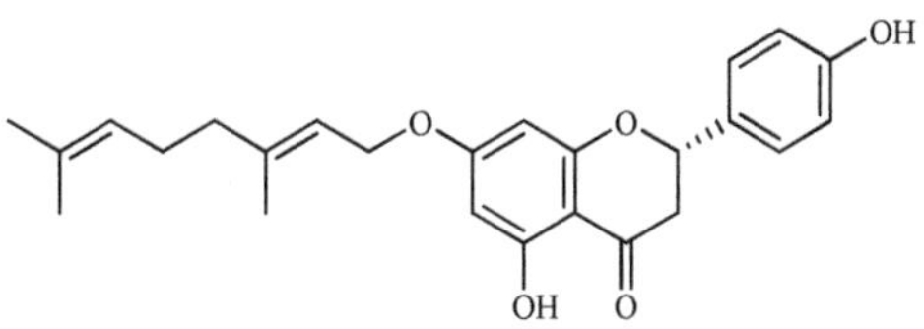

92

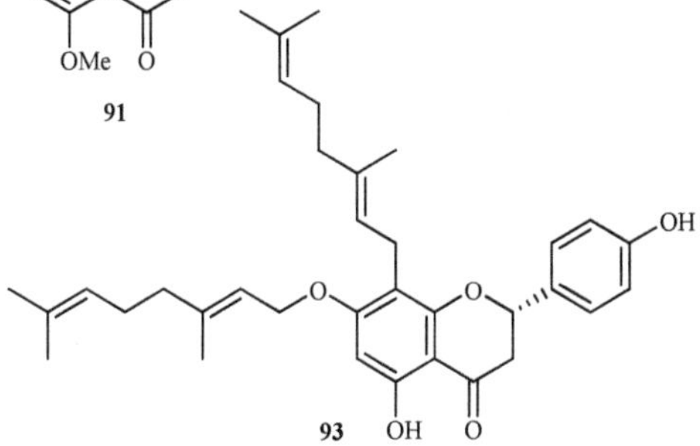

93

Figure 5. Isoflavones from genus *Tephrosia*.

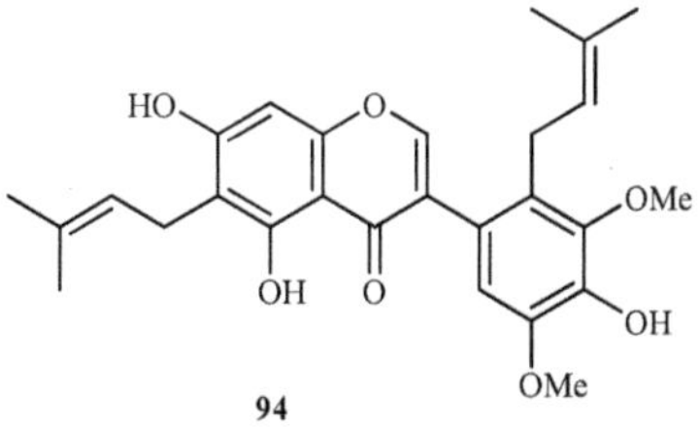

94

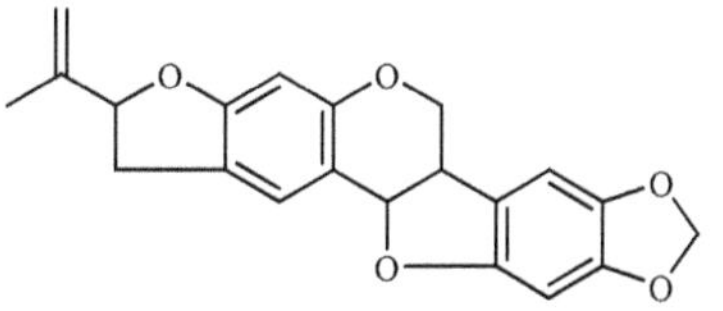

95

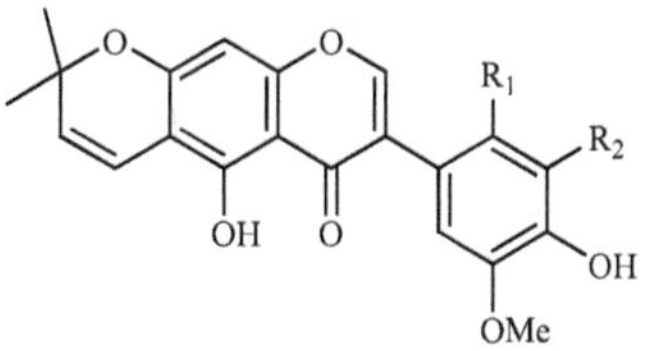

96 R_1 = OMe, R_2 = H
97 R_1 = H, R_2 = OMe

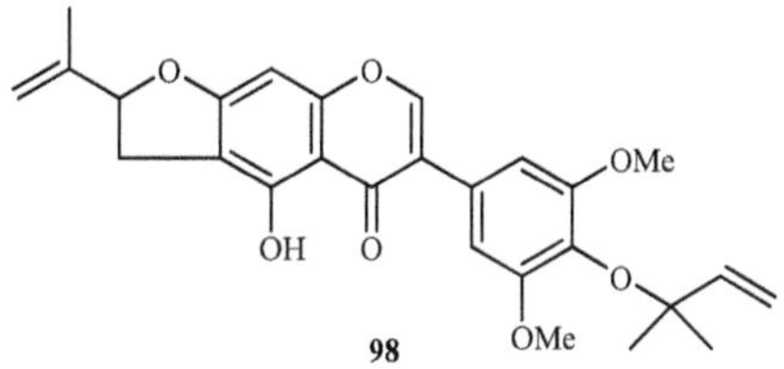

98

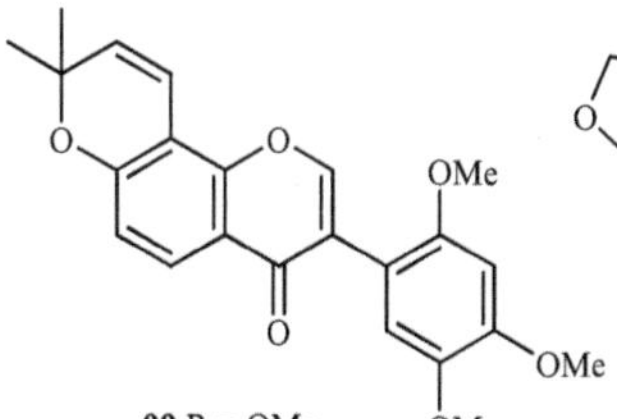

99 R = OMe
100 R = OH

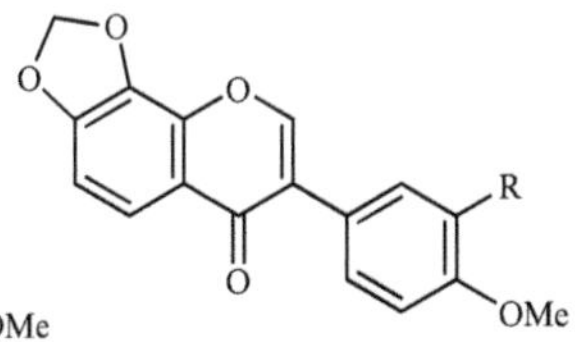

101 R = OMe
102 R = H

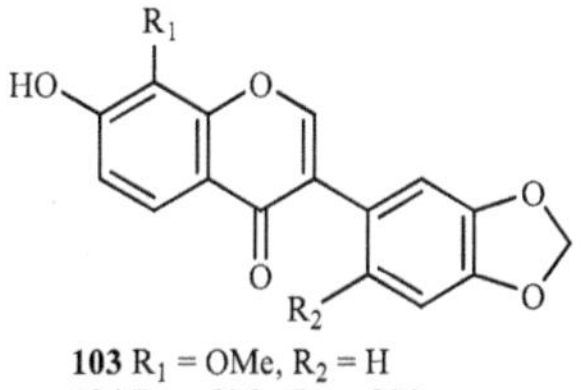

103 R_1 = OMe, R_2 = H
104 R_1 = OMe, R_2 = OMe
105 R_1 = H, R_2 = OMe

Figure 5. *Cont.*

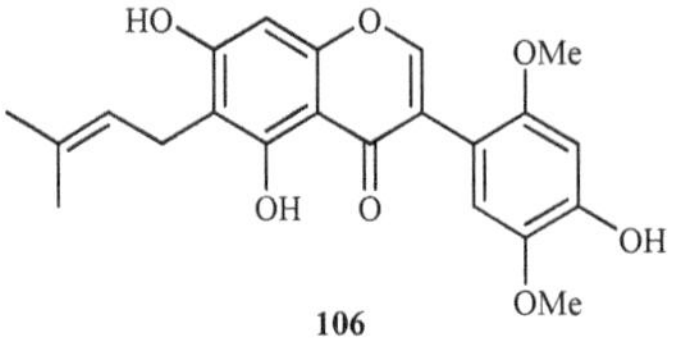

106

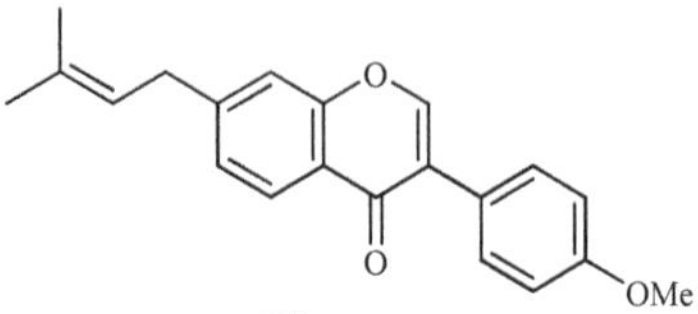

107

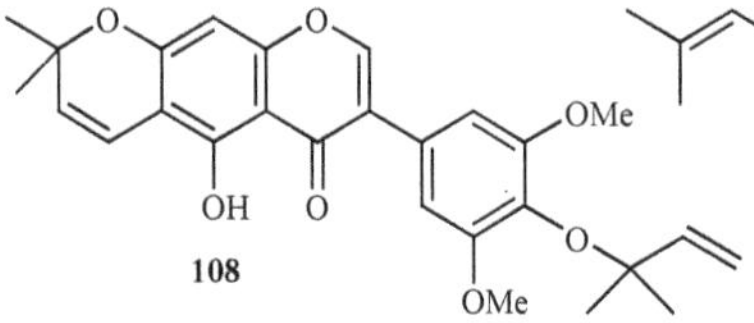

108

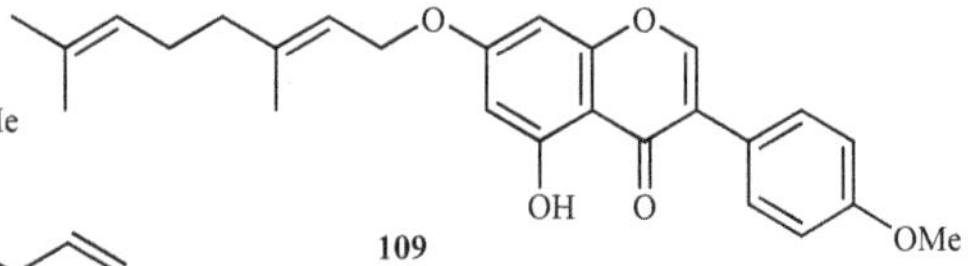

109

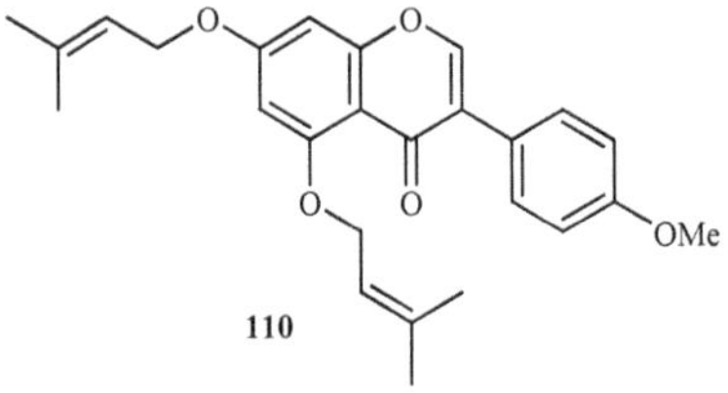

110

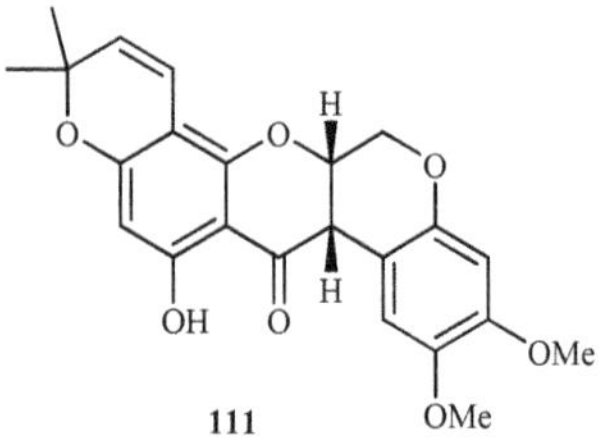

111

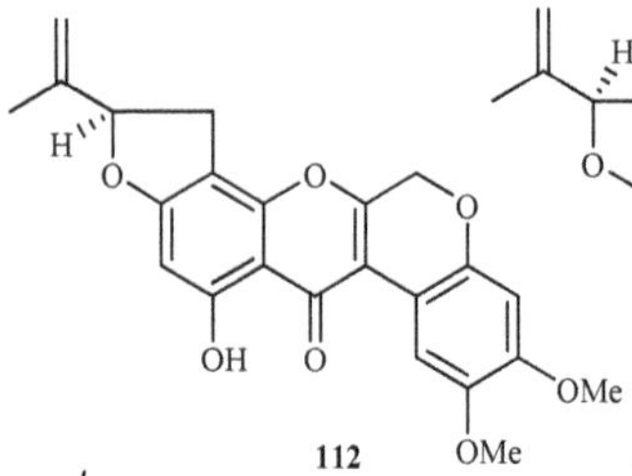

112

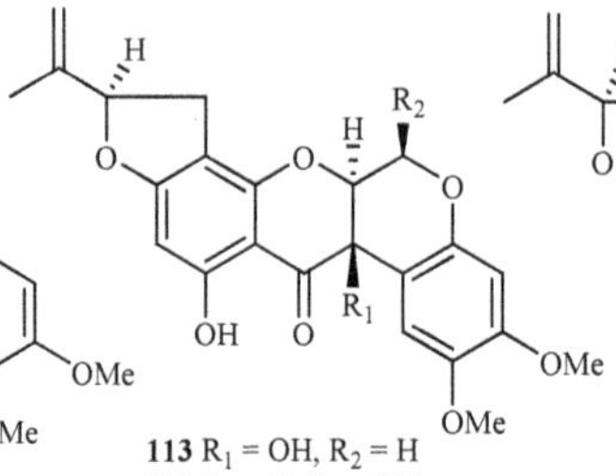

113 R_1 = OH, R_2 = H
114 R_1 = H, R_2 = OH
115 R_1 = OH, R_2 = OH

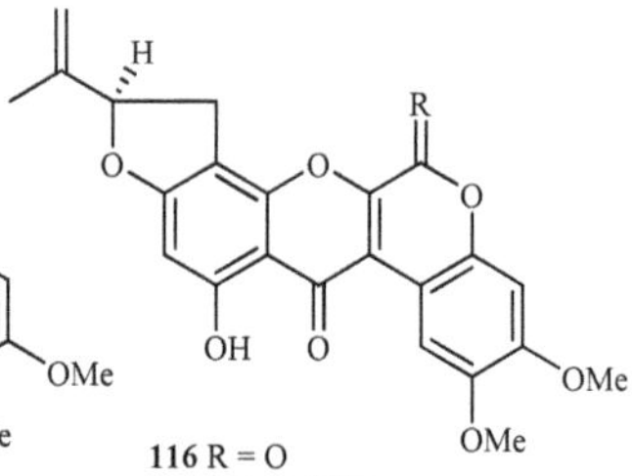

116 R = O
117 R = H, OMe

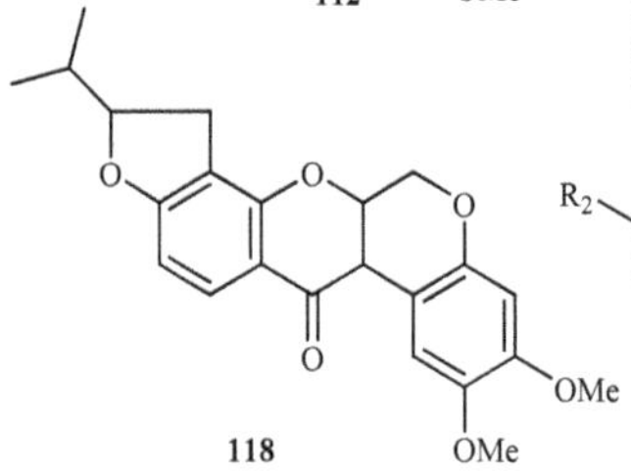

118

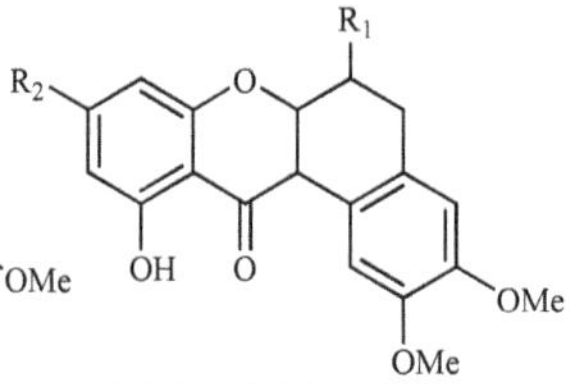

119 R_1 = OH, R_2 = OMe
120 R_1 = OH, R_2 = OH
121 R_1 = OAc, R_2 = OMe

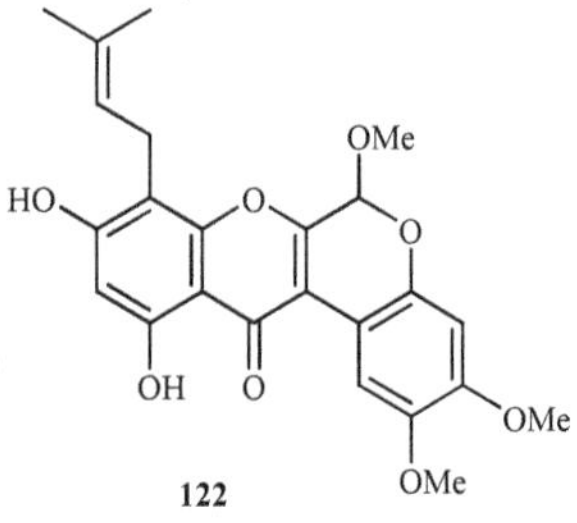

122

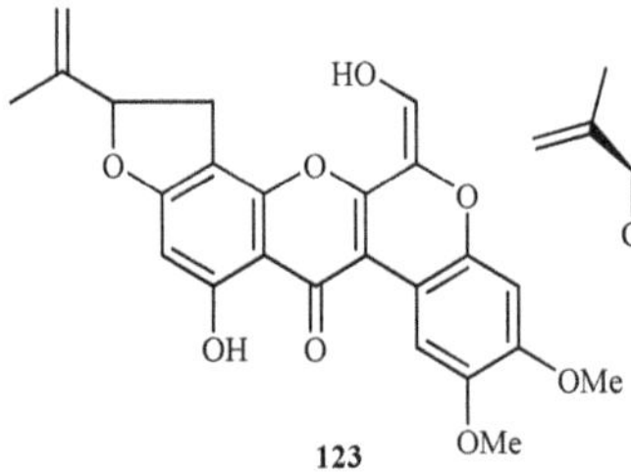

123

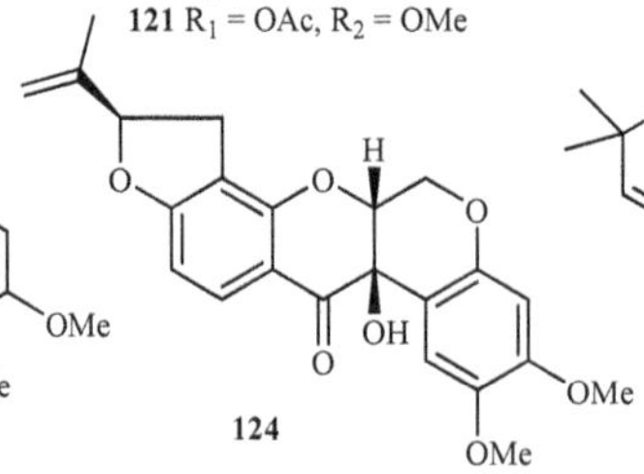

124

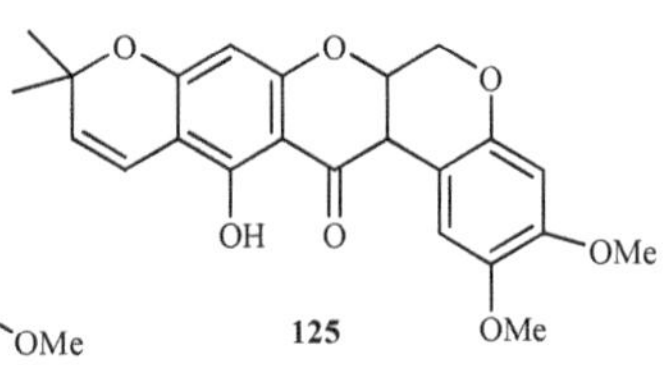

125

Figure5. *Cont.*

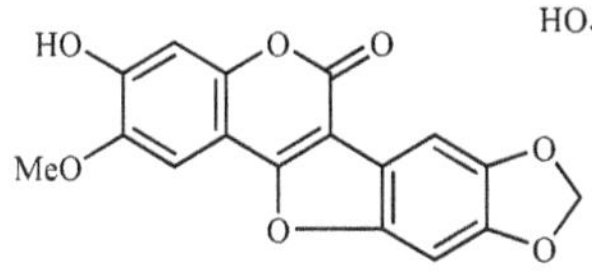

HO MeO O O O O O O 126

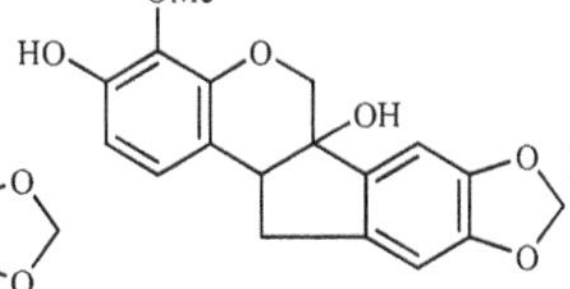

OMe HO O OH O O 127

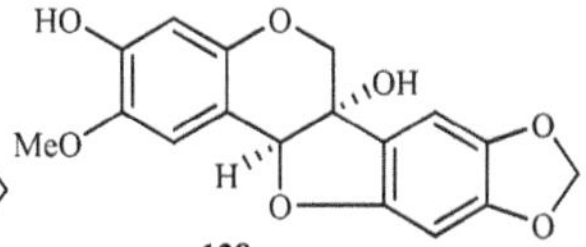

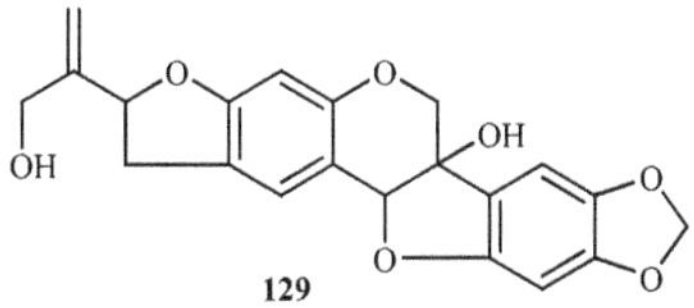

HO MeO O OH H O O O 128

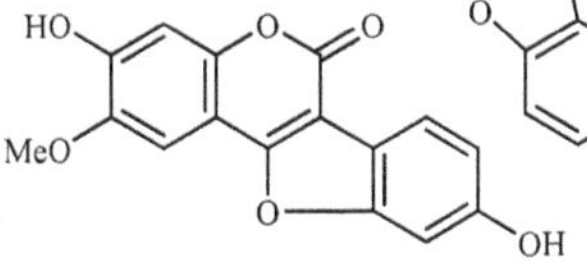

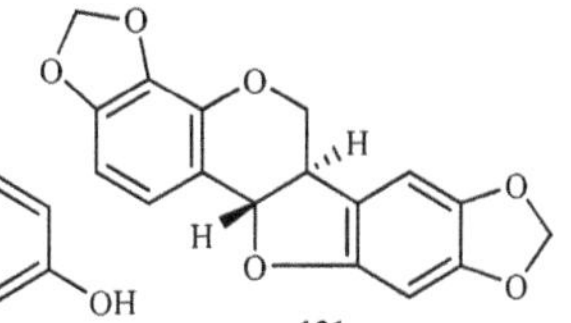

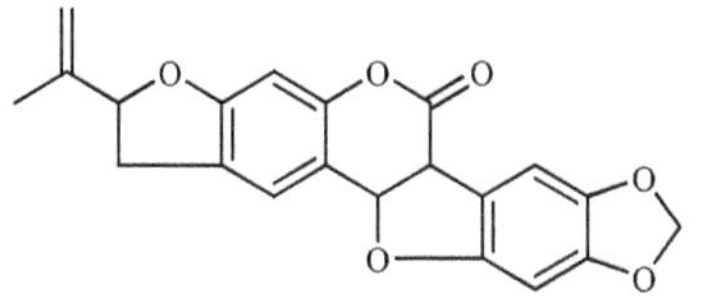

O OH O O OH O O O 129

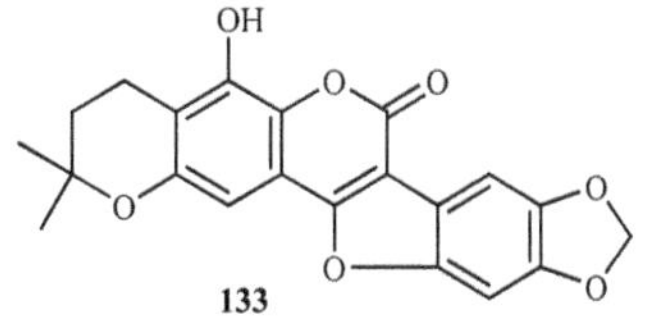

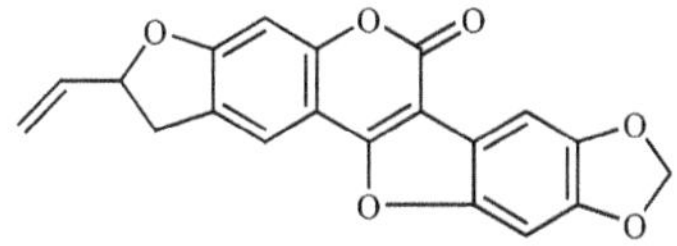

HO MeO O O O OH 130

O O O H H O O O 131

O O O O O O O 132

OH O O O O O O O 133

O O O O O O O 134

OH O O O O O O O 135

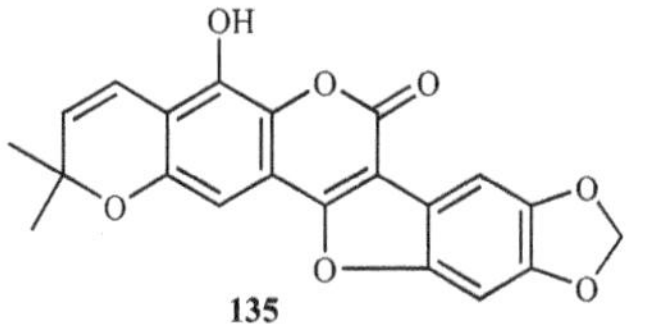

Figure 6. Chalcones from genus *Tephrosia*.

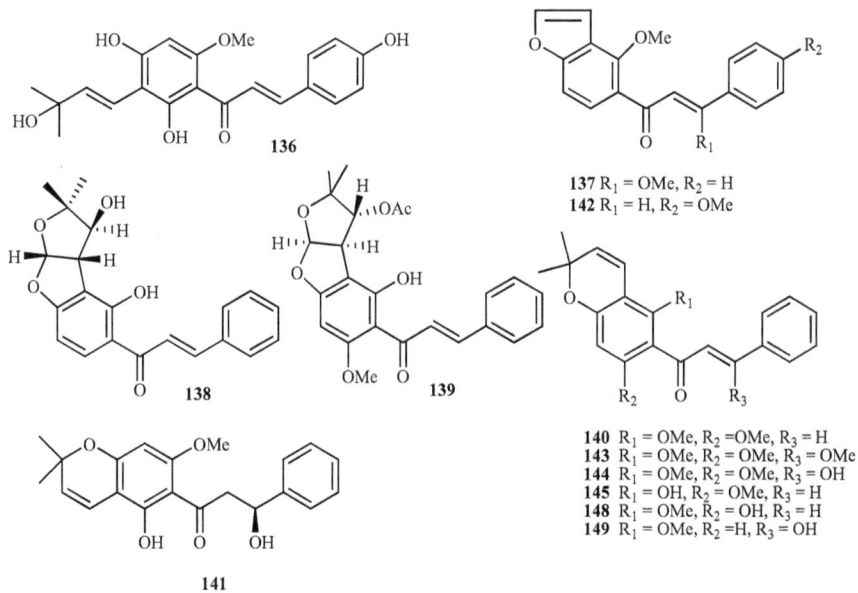

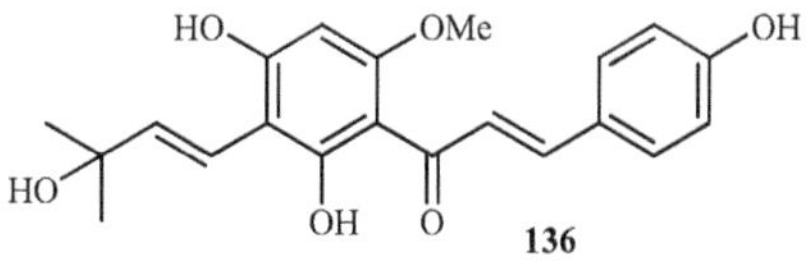

HO OMe OH HO OH O 136

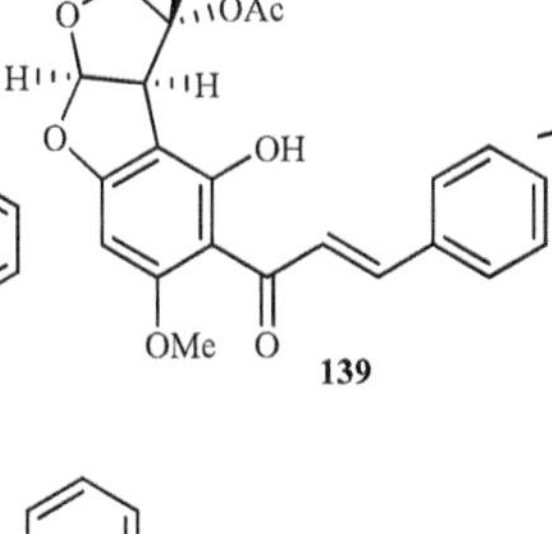

O OMe R_2 O R_1

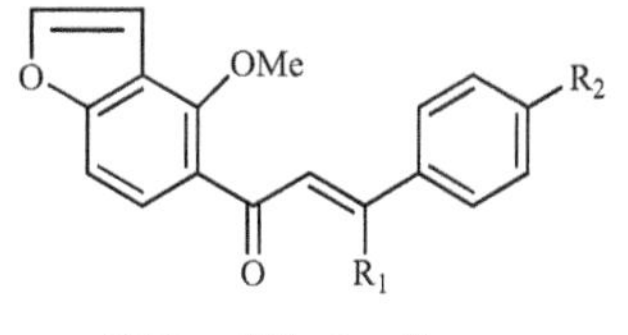

137 R_1 = OMe, R_2 = H
142 R_1 = H, R_2 = OMe

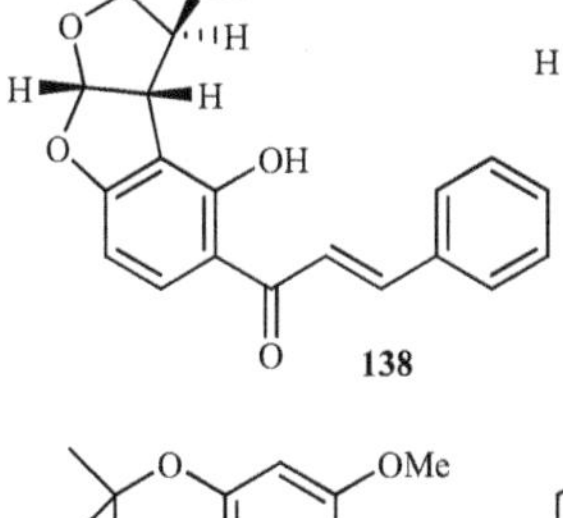

O OH H H H O OH O 138

H O OAc H H O OH OMe O 139

O R_1 R_2 O R_3

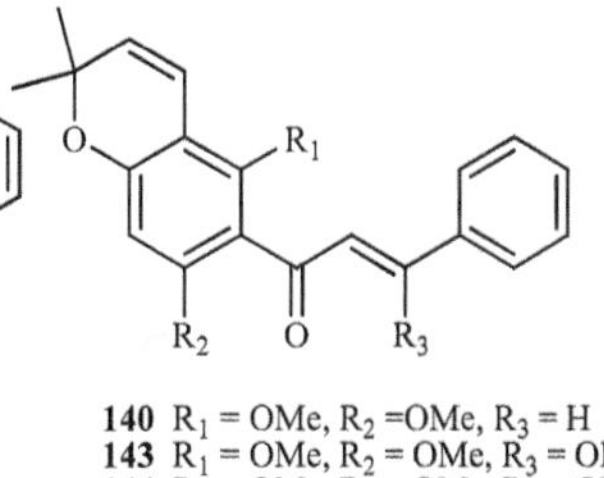

140 R_1 = OMe, R_2 =OMe, R_3 = H
143 R_1 = OMe, R_2 = OMe, R_3 = OMe
144 R_1 = OMe, R_2 = OMe, R_3 = OH
145 R_1 = OH, R_2 = OMe, R_3 = H
148 R_1 = OMe, R_2 = OH, R_3 = H
149 R_1 = OMe, R_2 =H, R_3 = OH

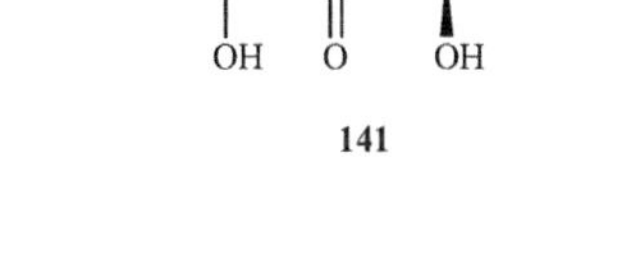

O OMe OH O OH 141

Figure 6. *Cont.*

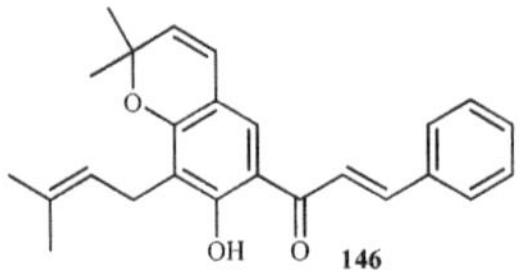

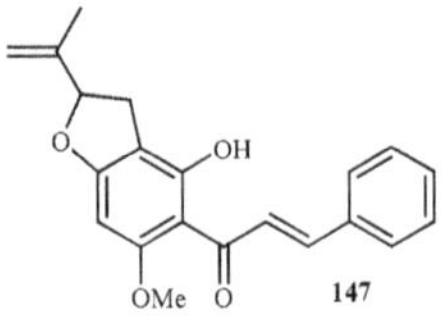

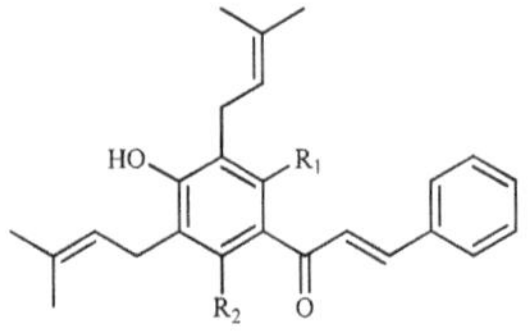

151 R_1 = H, R_2 = OH
153 R_1 = OH, R_2 = H

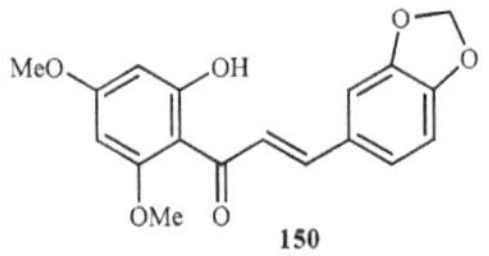

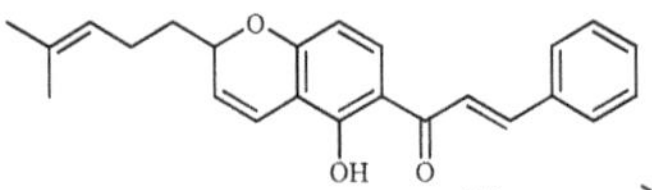

Figure 7. Other flavonoids from genus *Tephrosia*.

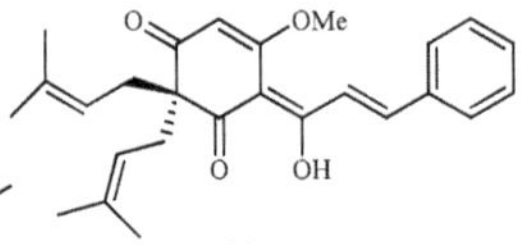

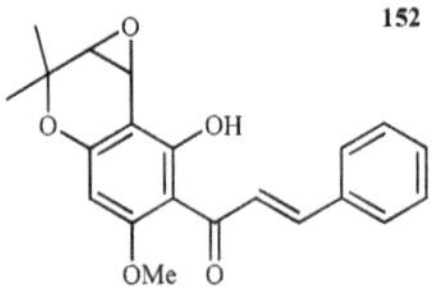

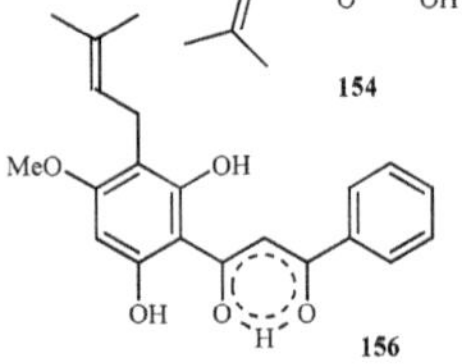

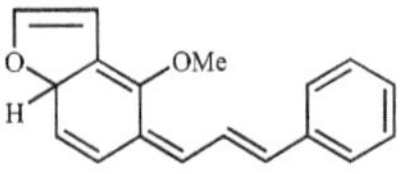

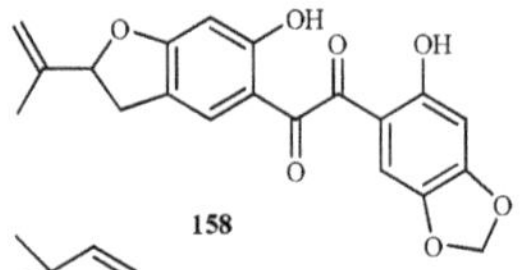

Figure 8. Triterpenoid from genus *Tephrosia*.

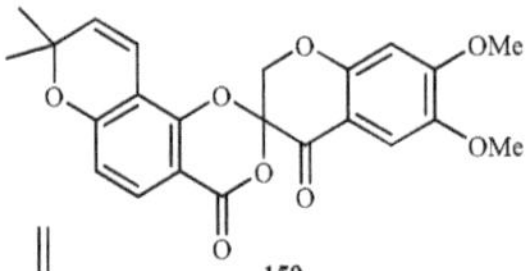

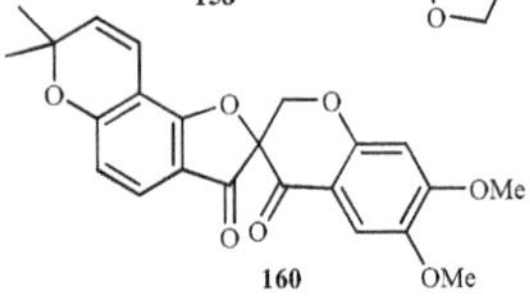

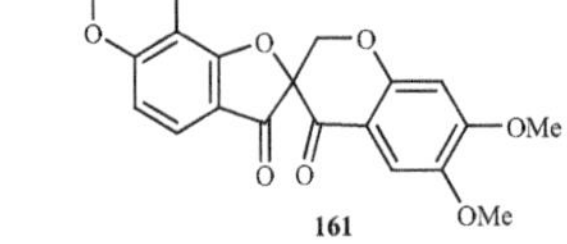

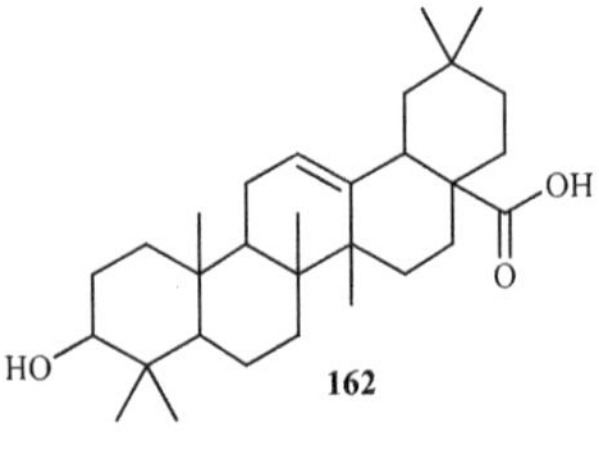

Figure 9. Sesquiterpenes from genus *Tephrosia*.

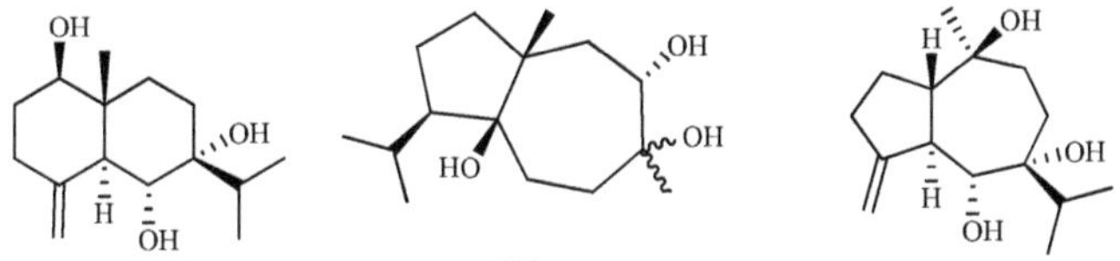

163 164 165

Figure 10. Other compounds from genus *Tephrosia*.

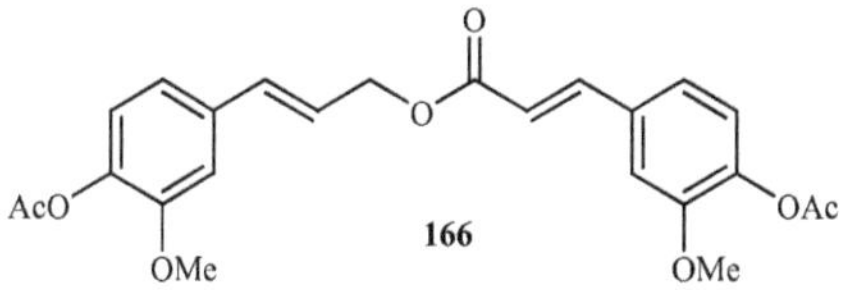

166 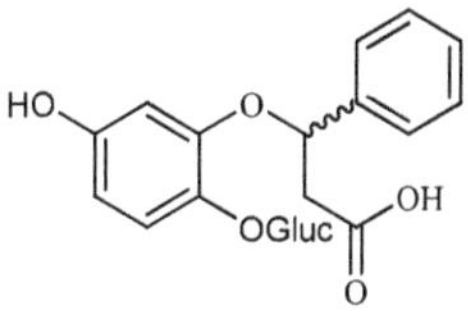 167 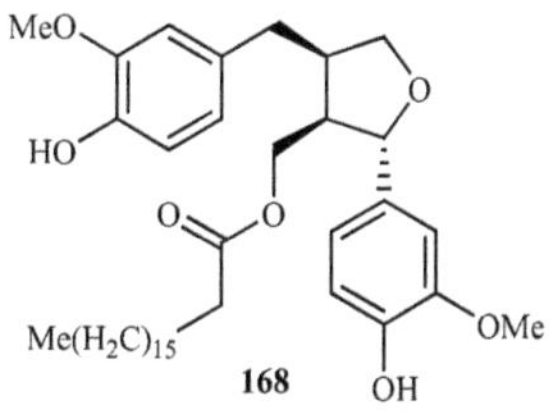 168

Table 1. Chemical constituents from the genus *Tephrosia*.

No.	Compound class and name	Source	Ref.
	Flavones		
1	tephroglabrin	*T. purpurea*	[3]
2	tepurindiol	*T. purpurea*	[3]
3	glabratephrin	*T. apollinea*	[10]
4	tachrosin	*Tephrosia polystachyoides*	[17]
5	staohyoidin	*T. polystachyoides*	[18]
6	tephrodin	*T. polystachyoides*	[18]
7	semiglabrin	*T. semiglabra, T. apollinea*	[19,20]
8	semiglabrinol	*T. semiglabra, T. apollinea*	[10,19]
9	tephrostachin	*T. polystachyoides*	[21]
10	emoroidone	*T. emoroides*	[22]
11	tephroapollin C	*T. apollinea*	[23]
12	tephroapollin D	*T. apollinea*	[23]
13	tephroapollin E	*T. apollinea*	[23]
14	tephroapollin F	*T. apollinea*	[23]
15	tephroapollin G	*T. apollinea*	[23]
16	multijugin	*T. multijuga*	[24]
17	multijuninol	*T. multijuga*	[24]
18	pseudosemiglabrinol	*T. apollinea*	[25]
19	(−)-pseudosemiglabrin	*T. semiglabra*	[26]

Table 1. *Cont.*

No.	Compound class and name	Source	Ref.
20	polystachin	*T. polystachya*	[27]
21	5-methoxy-6,6-dimethylpyrano[2,3:7,6]flavone	*T. praecans*	[28]
22	candidin	*T. candida*	[29]
23	hookerianin	*T. hookeriana*	[30]
24	fulvinervin B	*T. fulvinervis*	[31]
25	fulvinervin C	*T. fulvinervis*	[32]
26	enantiomultijugin	*T. viciodes*	[33]
27	apollinine	*T. purpurea*	[34]
28	demethylapollinin 7-*O*-*β*-D-glucopyranoside	*T. cinerea*	[35]
29	tephropurpulin A	*T. apollinea*, *T. purpurea*	[36,37]
30	isoglabratephrin	*T. purpurea*	[37]
31	terpurinflavone	*T. purpurea*	[38]
	Flavonols		
32	6-hydroxykaempferol 6-methyl ether 3-*O*-*α*-rhamnopyranosyl(7→6)-*β*-galactopyranoside-7-*O*-*α*-rhamnopyranoside	*T. vogelii*	[1]
33	6-hydroxykaempferol 6-methyl ether 3-*O*-*α*-rhamnopyranosyl(1→2)[*α*-rhamnopyranosyl(1→6)-*β*-galactopyranoside	*T. vogelii*	[1]
34	6-hydroxykaempferol 6-methyl ether 3-*O*-*α*-rhamnopyranosyl(1→2)[*α*-rhamnopyranosyl(1→ 6)]-*β*-galactopyranoside-7-*O*-*α*-rhamnopyranoside	*T. vogelii*	[1]
35	6-hydroxykaempferol 6-methyl ether 3-*O*-*α*-rhamnopyranosyl (1→2)[(3-*O*-*E*-feruloyl)-*α*-rhamnopyranosyl(1→6)]-*β*-galacto-pyranosides	*T. vogelii*	[1]
36	6-hydroxykaempferol 4'-methyl ether	*T. candida*	[39]
37	candidol		[40]
38	candirone	*T. candida*	[41,42]
39	7-ethoxy-3,3',4'-trihydroxyflavone	*T. procumbens*	[43]
	Flavanonols		
40	(2*R*,3*R*)-3-hydroxy-5-methoxy-6",6"-dimethylpyrano-[2",3":7,8]flavanone	*T. vogelii*	[1]
41	lupinifolinol	*T. lupinifolia*	[44]
42	lupinifolinol triacetate	*T. lupinifolia*	[44]
	Flavans		
43	(2*S*)-4'-hydroxy-5-methoxy-6",6"-dimethylpyrano[2",3":7,8]-flavanone	*T. vogelii*	[1]
44	(2*S*)-7-hydroxy-5-methoxy-8-prenylflavanone	*T. vogelii*	[1]
45	(2*S*)-5-methoxy-6",6"-dimethy1-4",5"-dihydrocyclopropa-[4",5"]furano[2",3":7,8]flavanone	*T. vogelii*	[1]
46	(2*S*)-5,7-dimethoxy-8-(3-methylbut-1,3-dienyl)flavanone	*T. vogelii*	[1]
47	tephrocandidin A	*T. candida*	[2]
48	tephrocandidin B	*T. candida*	[2]
49	(+)-tephrorin A	*T. purpurea*	[4]

Table 1. *Cont.*

No.	Compound class and name	Source	Ref.
50	(+)-tephrorin B	*T. purpurea*	[4]
51	(2*S*)-5-hydroxy-7,4'-di-*O*-(*γ*,*γ*-dimethylallyl)flavanone	*T. calophylla*	[6]
52	6-hydroxy-*E*-3-(2,5-dimethoxybenzylidine)-2',5'-dimethoxyflavanone	*T. calophylla*	[6]
53	pumilanol	*T. pumila*	[13]
54	emoroidenone	*T. emoroides*	[22]
55	tephroapollin A	*T. apollinea*	[23]
56	tephroapollin B	*T. apollinea*	[23]
57	fulvinervin A	*T. fulvinervis*	[30]
58	lupinifolin	*T. lupinifolia*	[44]
59	5,4'-*O*,*O*-dimethyl-lupinifolin	*T. lupinifolia*	[44]
60	lupinifolin diacelate	*T. lupinifolia*	[44]
61	obovatin	*T. obovata*	[45]
62	obovatin methyl-ether	*T. obovata*	[45]
63	methylhildardtol B	*T. hildebrandtii*	[46]
64	hildgardtol B	*T. hildebrandtii*	[46]
65	hildgardtene	*T. hildebrandtii*	[46]
66	methylhildgardtol A	*T. hildebrandtii*	[46]
67	hildgardtol A	*T. hildebrandtii*	[46]
68	purpurin	*T. purpurea*	[47]
69	tephrinone	*T. villosa*	[48]
70	5,7-dimethoxy-8-prenylflavan	*T. madrensis*	[49]
71	tephrowatsin A	*T. watsoniana*	[50]
72	tephrowatsin C	*T. watsoniana*	[50]
73	tephrowatsin B	*T. watsoniana*	[50]
74	tephrowatsin D	*T. watsoniana*	[50]
75	tephrowatsin E	*T. watsoniana*	[50]
76	nitenin	*T. nitens*	[51]
77	falciformin	*T. falciformis*	[52]
78	candidone	*T. candida*	[53]
79	quercetol A	*T. quercetorum*	[54]
80	quercetol B	*T. quercetorum*	[54]
81	quercetol C	*T. quercetorum*	[54]
82	5,7-dimethoxy-8-(2,3-epoxy-3-methylbutyl)-flavanone	*T. hamiltonii*	[55]
83	tephroleocarpin A	*T. leiocarpa*	[56]
84	tephroleocarpin B	*T. leiocarpa*	[56]
85	spinoflavanone A	*T. spinosa*	[57]
86	spinoflavanone B	*T. spinosa*	[57]
87	maxima flavanone A	*T. maxima*	[58]
88	tepicanol A	*T. tepicana*	[59]
89	crassifolin	*T. crassifolia*	[60]
90	astraciceran	*T. strigosa*	[61]
91	(+)-apollineanin	*T. apollinea*	[62]
92	(2*S*)-5,4'-dihydroxy-7-*O*-[*E*-3,7-dimethyl-2,6-octadienyl]flavanone	*T. villosa*	[63]

Table 1. *Cont.*

No.	Compound class and name	Source	Ref.
	Isoflavones		
93	(2*S*)-5,4'-dihydroxy-7-*O*-[*E*-3,7-dimethyl-2,6-octa-dienyl]-8-*C*-[*E*-3,7-dimethyl-2,6-octadienyl]flavanone	*T. villosa*	[63]
94	7,4'-dihydroxy-3',5'-dimethoxyisoflavone	*T. purpurea*	[5]
95	emoroidocarpan	*T. emoroides*	[22]
96	elongatin	*T. elongate*	[64]
97	pumilaisoflavone D	*T. pumila*	[65]
98	pumilaisoflavone C	*T. pumila*	[65]
99	barbigerone	*T. barbigera*	[66]
100	4'-demethyltoxicarol isoflavone	*T. polyphylla*	[67]
101	maxima isoflavone D	*T. maxima*	[68]
102	maxima isoflavone E	*T. maxima*	[68]
103	maxima isoflavone F	*T. maxima*	[68]
104	maxima isoflavone G	*T. maxima*	[68]
105	viridiflorin	*T. viridiflora*	[69]
106	maxima isoflavone J	*T. maxima*	[70]
107	pumilaisoflavone A	*T. pumila*	[71]
108	pumilaisoflavone B	*T. pumila*	[71]
109	7-*O*-geranylbiochanin A	*T. tinctoria*	[72]
110	5,7-di-*O*-prenylbiochanin A	*T. tinctoria*	[73]
111	toxicarol	*T. toxicaria*	[74]
112	villosinol	*T. villosa*	[75]
113	villosol	*T. villosa*	[75]
114	villosin	*T. villoss*	[76]
115	villol	*T. villoss*	[76]
116	villosone	*T. villoss*	[76]
117	villinol	*T. villoss*	[76]
118	dehydrodihydrorotenone	*T. candida*	[77]
119	dihydrostemonal	*T. pentaphylla*	[78]
120	9-demethyldihydrostemonal	*T. pentaphylla*	[78]
121	6-acetoxydihydrostemonal	*T. pentaphylla*	[78]
122	6a,12a-dehydro-2,3,6-trimethoxy-8-(3',3'-dimethylallyl)-9,11-dihydroxyrotenone	*T. villosa*	[79]
123	12a-dehydro-6-hydroxysumatrol	*T. villosa*	[80]
124	12a-hydroxyrotenone	*T. uniflora*	[81]
125	12a-hydroxy-β-toxicarol	*T. candida*	[82]
126	tephrosol	*T. villosa*	[83]
127	tephrocarpin	*T. bidwilli*	[84]
128	hildecarpin	*T. hildebrandtii*	[85,86]
129	hildecarpidin	*T. hildebrandtii*	[87]
130	2-methoxy-3,9-dihydroxy coumestone	*T. hamiltonii*	[88]
131	3,4:8,9-dimethylenedioxypterocarpan	*T. aequilata*	[89]
132	tephcalostan	*T. calophylla*	[90]
133	tephcalostan B	*T. calophylla*	[91]

Table 1. *Cont.*

No.	Compound class and name	Source	Ref.
	Chalcones		
134	tephcalostan C	*T. calophylla*	[91]
135	tephcalostan D	*T. calophylla*	[91]
136	candidachalcone	*T. candida*	[2]
137	*O*-methylpongamol	*T. purpurea*	[3]
138	(+)-tephrosone	*T. purpurea*	[4]
139	(+)-tephropurpurin	*T. purpurea*	[5]
140	2',6'-dimethoxy-4',5'-(2"2"dimethyl)-pyranochalcone	*T. pulcherrima*	[7]
141	(*S*)-elatadihydrochalcone	*T. elata*	[14]
142	purpuritenin	*T. purpurea*	[15]
143	praecansone A	*T. praecans*	[28]
144	praecansone B	*T. praecans*	[28]
145	obovatachalcone	*T. obovata*	[45]
146	spinochalcone C	*T. spinosa*	[57]
147	crassichalone	*T. crassifolia*	[60]
148	oaxacacin	*T. woodii*	[92]
149	6'-demethoxypraecansone B	*T. purpurea*	[93]
150	tephrone	*T. candida*	[94]
151	spinochalcone A	*T. spinosa*	[95]
152	spinochalcone B	*T. spinosa*	[95]
153	3',5'-diisopentenyl-2',4'-dihydroxychalcone	*T. spinosa*	[96]
154	tunicatachalcone	*T. tunicate*	[97]
155	epoxyobovatachalcone	*T. carrollii*	[98]
156	2',6'-dihydroxy-3'-prenyl-4'-methoxy-β-hydroxychalcone	*T. major*	[99]
	Other Flavonoids		
157	purpureamethied	*T. purpurea*	[15]
158	calophione A	*T. calophylla*	[91]
159	tephrospirolactone	*T. candida*	[100]
160	tephrospiroketone I	*T. candida*	[100]
161	tephrospiroketone II	*T. candida*	[100]
	Triterpenoid		
162	oleanolic acid	*T. strigosa*	[61]
	Sesquiterpenes		
163	1β-hydroxy-6,7α-dihydroxyeudesm-4(15)-ene	*T. candida*	[2]
164	linkitriol	*T. purpurea*	[34]
165	1β,6α,10α-guai-4(15)-ene-6,7,10-triol	*T. vogelii*	[101]
	Others		
166	2-propenoic acid, 3-(4-(acetyloxy) -3-methoxypheny)-3(4-actyloxy)-3-methoxyphenyl)-2-propenyl ester	*T. purpurea*	[34]
167	cineroside A	*T. cinerea*	[35]
168	(+)-lariciresinol-9'-stearate	*T. vogelii*	[101]

2.1. Flavonoids

Flavonoids were the most main constituents of the genus *Tephrosia*, even of the Leguminosae family. From the year of 1971, 161 flavonoids isolated from the genus *Tephrosia* are divided into several categories depending on their skeletons (Figures 1–7).

2.1.1. Flavones

Thirty-one flavones (**1**–**31**), were isolated from *T. polystachyoides*, *T. semiglabra*, *T. multijuga*, *T. polystachya*, *T. praecans*, *T. apollinea*, *T. candida*, *T. purpurea*, *T. fulvinervis*, *T. viciodes*, *T. emoroids* and *T. hookeriana* [3,10,17–38].

2.1.2. Flavonols

Eight flavonols (**32**–**39**), were isolated, four, *i.e.*, **32**–**34** were obtained from *T. vogelii* [1], one, *i.e.*, **35**–**38**, from *T. candida* [39–42] and **39** from *T. procumbens* [43].

2.1.3. Flavanonols

Only three flavanonols, **40**, **41** and **42** were isolated from *T. vogelii* and *T. lupinifolia*, respectively [1,44].

2.1.4. Flavans

Fifty-one flavans, **43**–**93**, were isolated from twenty-three species of the genus Tephrosia, *i.e.*, *T. obovata*, *T. villosa*, *T. madrensis*, *T. nitens*, *T. watsoniana*, *T. hildebrandtii*, *T. falciformis*, *T. hamiltonii*, *T. quercetorum*, *T. leiocarpa*, *T. spinosa*, *T. maxima*, *T. emoroides*, *T. tepicana*, *T. crassifolia*, *T. strigosa*, *T. pumila*, *T. calophylla*, *T. vogelii*, *T. apollinea*, *T. candida*, *T. purpurea* and *T. fulvinervis* [1,2,4,6,13,22,23,44–63].

2.1.5. Isoflavones

Forty-two isoflavones, **94**–**135**, have been isolated and identified from this genus [5,22,64–91]. Among them, **111**–**125** were identified as rotenoids [74–82], **94** and **126**–**135** were identified as coumestan derivatives [22,83–91].

2.1.6. Chalcones

Twenty-one chalcones, **136**–**156**, isolated from twelve species of genus *Tephrosia*, *i.e.*, *T. obovata*, *T. praecans*, *T. purpurea*, *T. candida*, *T. woodii*, *T. spinosa*, *T. crassifolia*, *T. tunicate*, *T. carrollii*, *T. major*, *T. pulcherrima* and *T. elata* [2–5,7,14,15,28,45,57,60,92–99].

2.1.7. Other Flavonoids

157 was isolated from *T. purpurea* seeds [15]. **158** was isolated from *T. calophylla* [91]. **159–161** were isolated from *T. candida* [100].

2.2. Triterpenoid

Only one triterpenoid has been isolated from this genus, that is **162** from *T. strigosa* [61].

2.3. Sesquiterpenes

Three sesquiterpenes, **163**, **164** and **165** were isolated from *T. candida* [2], *T. purpurea* [33] and *T. vogelii* [101], respectively.

2.4. Others

166–168 have been isolated from *T. purpurea* [34], *T. cinerea* [35] and *T. vogelii* [101], respectively.

3. Proposed Biosynthetic Pathways and Synthesis

8-Substituted isoflavonoids such as toxicarol isoflavone and rotenoids are well known [3]. Compounds **4–6** from *T. polystachyoides* could be explained to be evolved biogenetically from naturally occurring chrysins (A) as illustrated in the Scheme 1 [102]. It would appear that the complex substituents at C-8 arise from the ability of *Tephrosia* species to oxidise a 7-OMe group to a $-O^{+}=CH_2$ group (Scheme 2), in the same way that closely related species of Leguminosae oxidise the 2'-OMe group of isoflavonoids to yield rotenoids [103]. A pattern that explains the various C-8 substituents in *T. purpurea* and *T. apollinea* is shown in Scheme 3. In *T. polystachoides* this process is taken even further and the carbon of yet another 7-OMe group is incorporated into the additional rings attached to C-7 and C-8 (Scheme 4) [3]. We could confirm the structures of compounds **7** and **8** by their conversion into semiglabrinone, isoemiglabrinone and tephroglabrin (**3**) as shown in Scheme 5 [3]. Purpuritenin (**142**) was isolated from *T. purpurea* has been synthetized as showed in Scheme 6 [104].

Scheme 1. Possible biogenetic pathway of compounds **4–6** of *T. polystachyoides*.

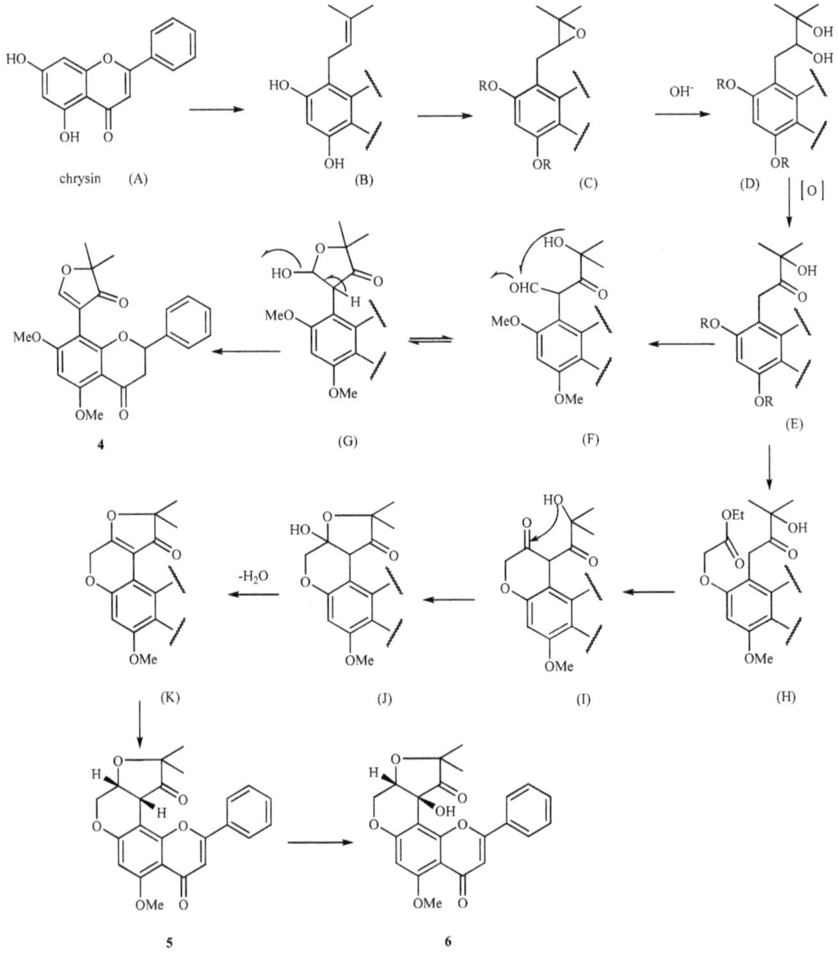

Scheme 2. Possible biogenetic pathway of compounds **8** and **11**.

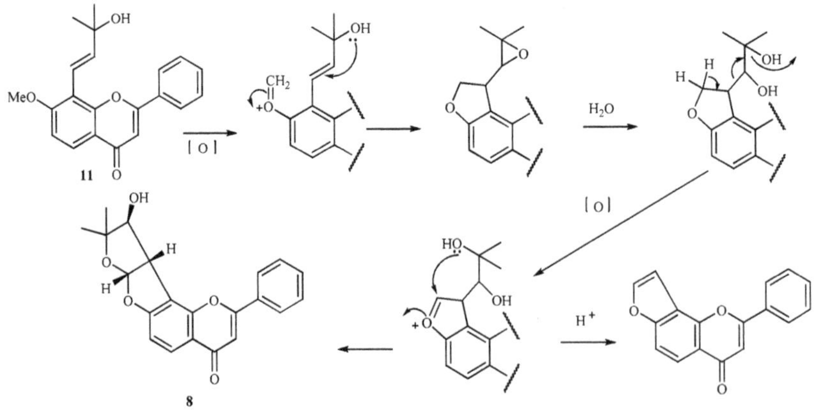

Scheme 3. Possible biogenetic pathway of compounds **3**, **8**, **11**, **27** and **137**.

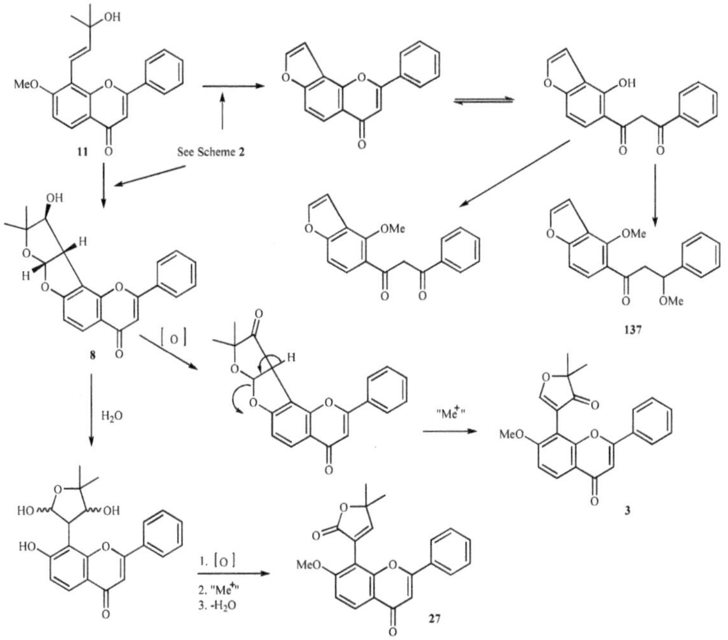

Scheme 4. Possible biogenetic pathway of compounds **4**, **5** and **137**.

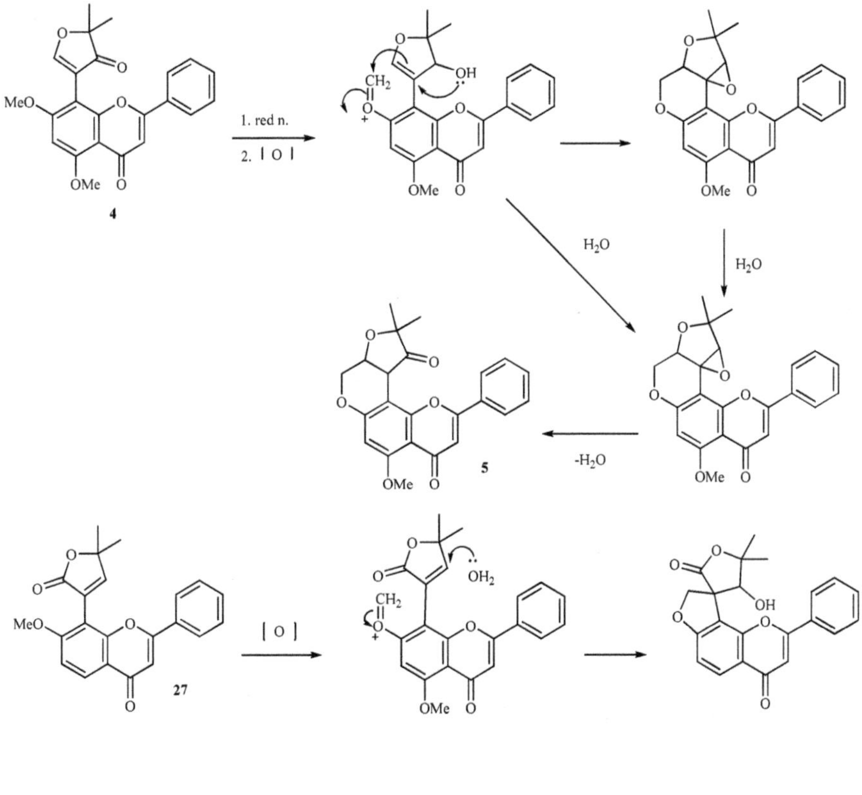

Scheme 5. Transform of compounds **3** and **8**.

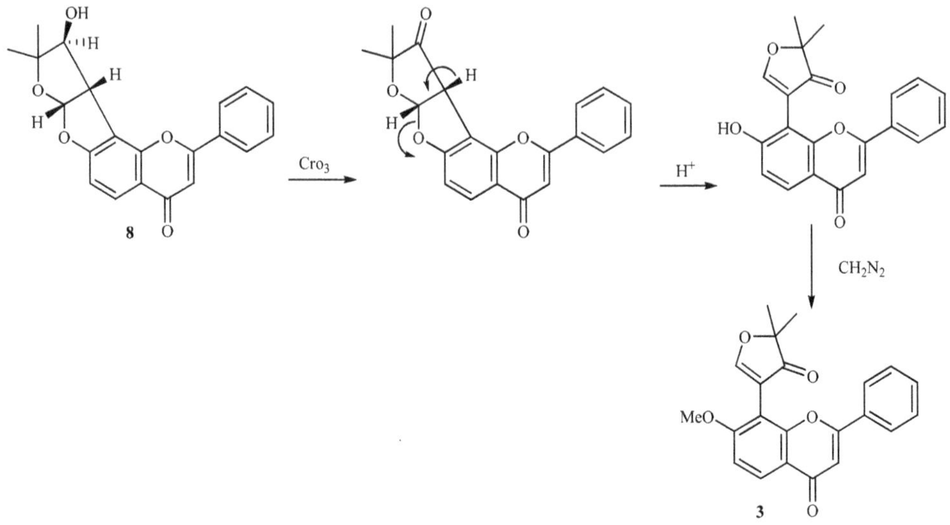

Scheme 6. The synthesis of **144**.

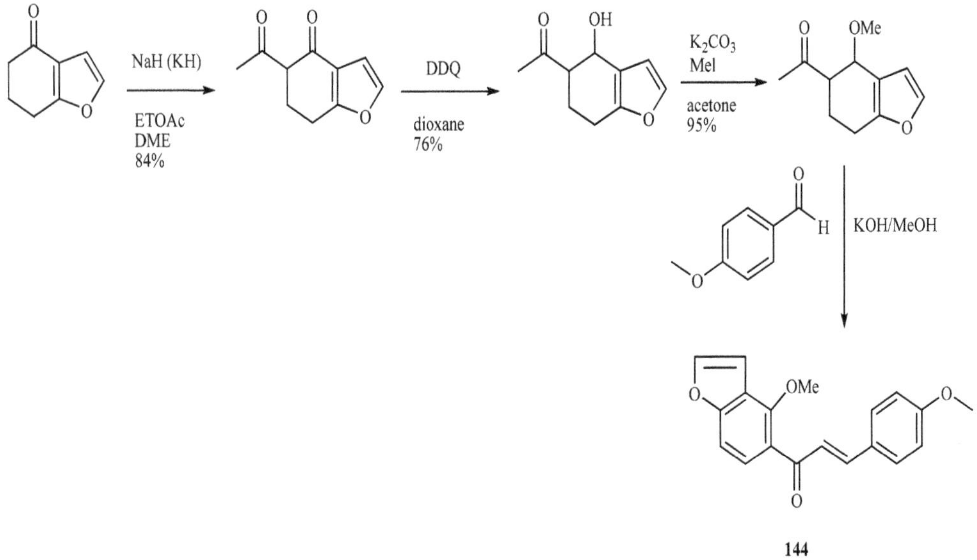

4. Biological Activities

The chemical constituents from the genus *Tephrosia* have been shown to exhibit various bioactivities, such as estrogenic, antitumor, antimicrobial, antiprotozoal, antifeedant activities [2,105].

4.1. Estrogenic Activity

Candidachalcone (**136**) isolated from *T. candida* exhibited estrogenic activity with IC_{50} value of 80 μM, compared with 18 μM for the natural steroid 17*β*-estradiol [2].

4.2. Antitumor Activities

Calophione A (**158**) and tephcalostans B–D (**133–135**) from *T. calphylla* were evaluated for cytotoxicity against RAW (mouse macrophage cells) and HT-29 (colon cancer cells) cancer cell lines. **158** exhibited significant cytotoxicity with IC_{50} of 5.00 (RAW) and 2.90 μM (HT-29), respectively, while **133–135** showed moderated cytotoxicity against both RAW and HT-29 cell lines [91]. (+)-Tephrorins A (**49**) and B (**50**), and (+)-tephrosone (**138**) isolated from *T. purpurea* were evaluated for their potential cancer chemopreventive properties using a cell-based quinone reductase induction assay [4]. 7,4'-dihydroxy-3',5'-dimethoxyisoflavone (**94**), and (+)-tephropurpurin (**139**), were obtained as active compounds from *T. purpurea*, using a bioassay based on the induction of quinone reductase (QR) activity with cultured Hepa 1c1c7 mouse hepatoma cells [5].

4.3. Antimicrobial Activities

2',6'-Dimethoxy-4',5'-(2'',2''-dimethyl)-pyranochalcone (**140**) from *T. pulcherrima* showed significant antimicrobial activity when tested against a series of micro-organisms [7]. 3,4:8,9-Dimethylenedioxypterocarpan (**131**) from *T. aequilata* exhibited low activity against gram-positive bacteria, *Bacillus subtilis* and *Micrococcus lutea* [89]. Hildecarpin (**128**) from *T. hildebrandtii* had exhibited antifungal activity against *Cladosporium cucumerinum* [85,86].

4.4. Antiprotozoal Activities

Terpurinflavone (**31**) isolated from *T. purpurea* showed the highest antiplasmodial activity against the chloroquine-sensitive (D6) and chloroquine-resistant (W2) strains of *Plasmodium falciparum* with IC_{50} values of 3.12 ± 0.28 μM (D6) and 6.26 ± 2.66 μM (W2) [38]. The crude extract of the seedpods of *T. elata* showed antiplasmodial activities against D6 and W2 strains of *P. falciparum* with IC_{50} values of 8.4 ± 0.3 and 8.6 ± 1.0 μg/mL, respectively [14]. Obovatin (**61**) and obovatin methyl ether (**62**) from *T. obovata* [45] showed antiplasmodial activities against D6 and W2 strains of *P. falciparum* with IC_{50} values of 4.9 ± 1.7 and 6.4 ± 1.1 μg/mL, and 3.8 ± 0.3 and 4.4 ± 0.6 μg/mL, respectively [14]. (*S*)-Elatadihydrochalcone (**141**) from *T. elata* exhibited good antiplasmodial activity against the D6 and W2 strains of *P. falciparum* with IC_{50} values of 2.8 ± 0.3 (D6) and 5.5 ± 0.3 μg/mL (W2), respectively [14]. Tephcalostans C (**134**) and D (**135**) from *T. calphylla* were found to be weakly antiprotozoal activity *in vitro* [91]. Pumilanol (**53**) from *T. pumila* exhibited significant antiprotozoal activity against *T. rhodensiense*, *T. cruzi* and *L. donovani* with IC_{50} of 3.7, 3.35 and 17.2 μg/mL, respectively, but displayed high toxicity towards L-6 (IC_{50} of 17.12 μg/mL) rat skeletal myoblasts [13]. Tephrinone (**69**) from *T. villosa* [48] also exhibited high degree of activity and selectivity against both *T. b. rhodensiense*, *T. cruzi* and *L. donovani* with IC_{50} of 3.3 and 16.6 μg/mL [13].

4.5. Antifeedant Activities

Emoroidenone (**54**) from *T. emoroides* showed strong feeding deterrent activity against *Chilo partellus* larvae with a mean percentage deterrence of 66.1% at a dose of 100 μg/disc [22]. Hildecarpin (**128**) from *T. hildebrandtii* had exhibited insect antifeedant activity against the legume pod-borer *Maruca testulalis*, and important pest of cowpea (*Vigna*) [85,86].

4.6. Other Activities

(−)-Pseudosemiglabrin (**19**) from *T. semiglabra* displayed *in vitro* inhibitory effects on human platelet aggregation [26]. Obovatin (**61**), obovatin methyl-ether (**62**) and obovatachalcone (**145**) from *T. obovata* displayed moderate piscicidal activity against loach fish *Misgurnus angullicaudatus*. The TLm (median tolerance limit) values of **61**, **62** and **145** were 1.25, 1.55 and 1.35 ppm, respectively [45]. Toxicarol (**111**) was a constituent of the South American fish poison *T. toxicaria* [74].

5. Conclusions

The genus *Tephrosia*, including *ca.* 400 species, with *ca.* 52 species being investigated worldwide, was reported to possess various chemical constituents and to display diverse bioactivities, especially antiplasmodial, estrogenic, antitumor, antimicrobial, antiprotozoal, antifeedant activities. Plants of the genus *Tephrosia* have important traditional uses in agriculture, because they possess the bioactivity of phytoalexins. Some compounds isolated from these plants also have the bioactivity of phytoalexins according to the reported literature, which we have list in the part of the manuscript "Biological Activities". We think that there will be many phytoalexin-type compounds isolated from plants of the genus *Tephrosia.* Although the number of natural compounds was isolated from this genus, there are still many *Tephrosia* species that received no little attention further, phytochemical and biological studies on this genus are needed in the future. In addition, the biosynthetic pathways and synthesis of these bioactive molecules in the genus remained largely unexplored. Thus, much more chemical, biosynthetic, synthetic and biological studies should be carried out on natural compounds in *Tephrosia* species in order to disclose their potency, selectivity, toxicity, and availability.

Acknowledgments

We thank the authors of all the references cited herein for their valuable contributions. Financial supported for this work by grants from National Natural Science Foundation of China (No. 31100260, 31200246), Knowledge Innovation Program of Chinese Academy of Sciences (KSCX2-EW-J-28), Program of Guangzhou City (No. 12C14061559), Foundation of Key Laboratory of Plant Resources Conservation and Sustainable Utilization, South China Botanical Garden, Chinese Academy of Sciences (No. 201210ZS).

Author Contributions

In this paper, Yinning Chen was in charge of writing the manuscript; Tao Yan was responsible for drawing the structures of the compounds; Chenghai Gao was in charge of correcting the revised manuscript; Wenhao Cao was responsible for searching for the literature; Riming Huang is the corresponding author who was responsible for arranging, checking and revising the manuscript.

Conflicts of Interest

The authors declare no conflict of interest.

References

1. Stevenson, P.C.; Kite, G.C.; Lewis, G.P.; Forest, F.; Nyirenda, S.P.; Belmain, S.R.; Sileshi, G.W.; Veitch, N.C. Distinct chemotypes of *Tephrosia vogelii* and implications for their use in pest control and soil enrichment. *Phytochemistry* **2012**, *78*, 135–146.
2. Hegazy, M.E.F.; Mohamed, A.E.H.; El-Halawany, A.M.; Djemgou, P.C.; Shahat, A.A.; Pare, P.W. Estrogenic activity of chemical constituents from *Tephrosia candida*. *J. Nat. Prod.* **2011**, *74*, 937–942.
3. Pelter, A.; Ward, R.S.; Rao, E.V.; Raju, N.R. 8-Substituted flavonoids and 3'-substituted 7-oxygenated chalcones from *Tephrosia purpurea*. *J. Chem. Soc. Perkin Trans. 1* **1981**, *9*, 2491–2498.
4. Chang, L.C.; Chavez, D.; Song, L.L.; Farnsworth, N.R.; Pezzuto, J.M.; Kinghorn, A.D. Absolute configuration of novel bioactive flavonoids from *Tephrosia purpurea*. *Org. Lett.* **2000**, *2*, 515–518.
5. Chang, L.C.; Gerhauser, C.; Song, L.; Farnsworth, N.R.; Pezzuto, J.M.; Kinghorn, A.D. Activity-guided isolation of constituents of *Tephrosia purpurea* with the potential to induce the phase II enzyme, quinone reductase. *J. Nat. Prod.* **1997**, *60*, 869–873.
6. Reddy, R.V.N.; Khalivulla, S.I.; Reddy, B.A.K.; Reddy, M.V.B.; Gunasekar, D.; Deville, A.; Bodo, B. Flavonoids from *Tephrosia calophylla*. *Nat. Prod. Commun.* **2009**, *4*, 59–62.
7. Ganapaty, S.; Srilakshmi, G.V.K.; Pannakal, S.T.; Laatsch, H. A pyranochalcone and prenylflavanones from *Tephrosia pulcherrima* (Baker) drumm. *Nat. Prod. Commun.* **2008**, *3*, 49–52.
8. Kassem, M.E.S.; Sharaf, M.; Shabana, M.H.; Saleh, N.A.M. Bioactive flavonoids from *Tephrosia purpurea*. *Nat. Prod. Commun.* **2006**, *1*, 953–955.
9. Clarke, G.; Banerjee, S.C. A glucoside from *Tephrosia purpurea*. *J. Chem. Soc.* **1910**, *97*, 1833–1837.
10. Waterman, P.G.; Khalid, S.A. The major flavonoids of the seed of *Tephrosia apollinea*. *Phytochemistry* **1980**, *19*, 909–915.
11. Kole, R.K.; Satpathi, C.; Chowdhury, A.; Ghosh, M.R.; Adityachaudhury, N. Isolation of amorpholone, a potent rotenoid insecticide from *Tephrosia candida*. *J. Agric. Food Chem.* **1992**, *40*, 1208–1210.
12. Sanchez, I.; Gomez-Garibay, F.; Taboada, J.; Ruiz, B.H. Antiviral effect of flavonoids on the dengue virus. *Phytother. Res.* **2000**, *14*, 89–92.
13. Ganapaty, S.; Pannakal, S.T.; Srilakshmi, G.V.K.; Lakshmi, P.; Waterman, P.G.; Brun, R. Pumilanol, an antiprotozoal isoflavanol from *Tephrosia pumila*. *Phytochem. Lett.* **2008**, *1*, 175–178.

14. Muiva, L.M.; Yenesew, A.; Derese, S.; Heydenreich, M.; Peter, M.G.; Akala, H.M.; Eyase, F.; Waters, N.C.; Mutai, C.; Keriko, J.M.; *et al.* Antiplasmodial beta-hydroxydihydrochalcone from seedpods of *Tephrosia elata*. *Phytochem. Lett.* **2009**, *2*, 99–102.
15. Sinha, B.; Natu, A.A.; Nanavati, D.D. Prenylated flavonoids from *Tephrosia purpurea* seeds. *Phytochemistry* **1982**, *21*, 1468–1470.
16. Touqeer, S.; Saeed, M.A.; Ajaib, M. A review on the phytochemistry and pharmacology of genus *Tephrosia*. *Phytopharmacology* **2013**, *4*, 598–637.
17. Smalberg, T.M.; Vleggaar, R.; de Waal, H.L. Tachrosin: A new flavone from *Tephrosia polystachyoides* Bak F. *S. Afr. J. Chem.* **1971**, *24*, 1–8.
18. Vleggaar, R.; Smalberg, T.M.; de Waal, H.L. Two new flavones from *Tephrosia polystachyoides* Bakf 2. *Tetrahedron Lett.* **1972**, *8*, 703–704.
19. Smalberg, T.M.; van den Berg, A.J.; Vleggaar, R. Flavonoids from *Tephrosia*—VI: The structure of semiglabrin and semiglabrinol. *Tetrahedron* **1973**, *29*, 3099–3104.
20. Vleggaar, R.; Kruger, G.J.; Smalberger, T.M.; van den Berg, A.J. Flavonoids from *Tephrosia*. XI1. Structure of glabratephrin. *Tetrahedron* **1978**, *34*, 1405–1408.
21. Vleggaar, R.; Smalberg, T.M.; de Waal, H.L. Flavonoids from *Tephrosia*. V. Structure of tephrostachin. *S. Afr. J. Chem.* **1973**, *26*, 71–73.
22. Machocho, A.K.; Lwande, W.; Jondiko, J.I.; Moreka, L.V.C.; Hassanali, A. Three new flavonoids from the root of *Tephrosia emoroides* and their antifeedant activity against the larvae of the spotted stalk Borer Chilo-Partellus Swinhoe. *Pharmaceut. Biol.* **1995**, *33*, 222–227.
23. El-Razek, M.H.A.; Mohamed, A.E.H.H.; Ahmed, A. Prenylated flavonoids, from *Tephrosia apollinea*. *Heterocycles* **2007**, *71*, 2477–2490.
24. Vleggaar, R.; Smalberger, T.M.; van den Berg, A.J. Flavonoids from *Tephrosia*. IX. Structure of multijugin and multijuginol. *Tetrahedron* **1975**, *31*, 2571–2573.
25. Ahmad, S. Natural occurrence of *Tephrosia* flavones. *Phytochemistry* **1986**, *25*, 955–958.
26. Jonathan, L.T.; Gbeassor, M.; Che, C.T.; Fong, H.H.S.; Farnsworth, N.R.; Lebreton, G.C.; Venton, D.L. Pseudosemiglabrin, a platelet-aggregation inhibitor from *Tephrosia semiglabra*. *J. Nat. Prod.* **1990**, *53*, 1572–1574.
27. Vleggaar, R.; Smalberger, T.M.; van Aswegen, J.L. Flavonoids from *Tephrosia*. X. Structure of polystachin. *S. Afr. J. Chem.* **1978**, *31*, 47–50.
28. Camele, G.; Dellemonache, F.; Dellemonache, G.; Marinibettolo, G.B. Three new flavonoids from *Tephrosia praecans*. *Phytochemistry* **1980**, *19*, 707–709.
29. Chibber, S.S.; Dutt, S.K. Candidin, a pyranoflavone from *Tephrosia candida* seeds. *Phytochemistry* **1981**, *20*, 1460–1460.
30. Prabhakar, P.; Vanangamudi, A.; Gandhidasan, R.; Raman, P.V. Hookerianin: A flavone from *Tephrosia hookeriana*. *Phytochemistry* **1996**, *43*, 315–316.
31. Rao, E.V.; Venkataratnam, G.; Vilain, C. Flavonoids from *Tephrosia fulvinervis*. *Phytochemistry* **1985**, *24*, 2427–2430.
32. Venkataratnam, G.; Rao, E.V.; Vilain, C. Fulvinervin C, a flavone from *Tephrosia fulvinervis*. *Phytochemistry* **1986**, *25*, 1507–1508.

33. Gomezgaribay, F.; Quijano, L.; Hernandez, C.; Rios, T. Flavonoids from *Tephrosia* species. IX. Enantiomultijugin, a flavone from *Tephrosia viciodes*. *Phytochemistry* **1992**, *31*, 2925–2926.
34. Khalafalah, A.K.; Yousef, A.H.; Esmail, A.M.; Abdelrazik, M.H.; Hegazy, M.E.; Mohamed, A.E. Chemical constituents of *Tephrosia purpurea*. *Pharmacogn. Res.* **2010**, *2*, 72–75.
35. Maldini, M.; Montoro, P.; Macchia, M.; Pizza, C.; Piacente, S. Profiling of phenolics from *Tephrosia cinerea*. *Planta Med.* **2011**, *77*, 1861–1864.
36. Khalafallah, A.K.; Suleiman, S.A.; Yousef, A.H.; El-kanzi, N.A.A.; Mohamed, A.E.H.H. Prenylated flavonoids from *Tephrosia apollinea*. *Chin. Chem. Lett.* **2009**, *20*, 1465–1468.
37. Hegazy, M.E.F.; Abd El-Razek, M.H.; Nagashima, F.; Asakawa, Y.; Pare, P.W. Rare prenylated flavonoids from *Tephrosia purpurea*. *Phytochemistry* **2009**, *70*, 1474–1477.
38. Juma, W.P.; Akala, H.M.; Eyase, F.L.; Muiva, L.M.; Heydenreich, M.; Okalebo, F.A.; Gitu, P.M.; Peter, M.G.; Walsh, D.S.; Imbuga, M.; *et al.* Terpurinflavone: An antiplasmodial flavone from the stem of *Tephrosia purpurea*. *Phytochem. Lett.* **2011**, *4*, 176–178.
39. Sarin, J.P.S.; Singh, S.; Garg, H.S.; Khanna, N.M.; Dhar, M.M. Flavonol glycoside with anticancer activity from *Tephrosia candida*. *Phytochemistry* **1976**, *15*, 232–234.
40. Dutt, S.K.; Chibber, S.S. Candidol, a flavonol from *Tephrosia candida*. *Phytochemistry* **1983**, *22*, 325–326.
41. Parmar, V.S.; Jain, R.; Simonsen, O.; Boll, P.M. Isolation of candirone—A novel pentaoxygenation pattern in a naturally-occurring 2-phenyl-4H-1-benzopyran-4-one from *Tephrosia candida*. *Tetrahedron* **1987**, *43*, 4241–4247.
42. Horie, T.; Kawamura, Y.; Kobayashi, T.; Yamashita, K. Revised structure of a natural flavone from *Tephrosia candida*. *Phytochemistry* **1994**, *37*, 1189–1191.
43. Venkataratnam, G.; Rao, E.V.; Vilain, C. Flavonoids of *Tephrosia procumbens*—Revised structure for praecansone A and conformation of praecansone B. *J. Chem. Soc. Perkin Trans. 1* **1987**, *12*, 2723–2727.
44. Smalberg, T.M.; Vleggaar, R.; Weber, J.C. Flavonoids from *Tephrosia*. VII: Constitution and absolute-configuration of lupinifolin and lupinifolinol, two flavanones from *Tephrosia lupinifolia* Burch (Dc). *Tetrahedron* **1974**, *30*, 3927–3931.
45. Chen, Y.L.; Wang, Y.S.; Lin, Y.L.; Munakata, K.; Ohta, K. Obovatin, obovatin methyl-ether and obovatachalcone, new piscicidal flavonoids from *Tephrosia obovata*. *Agric. Biol. Chem. Tokyo* **1978**, *42*, 2431–2432.
46. Dellemonache, F.; Labbiento, L.; Marta, M.; Lwande, W. 4-*β*-substituted flavans from *Tephrosia hildebrandtii*. *Phytochemistry* **1986**, *25*, 1711–1713.
47. Gupta, R.K.; Krishnamurti, M.; Parthasarathi, J. Purpurin, a new flavanone from *Tephrosia purpurea* seeds. *Phytochemistry* **1980**, *19*, 1264–1264.
48. Rao, P.P.; Srimannarayana, G. Tephrinone, a new flavanone from *Tephrosia villosa*. *Curr. Sci. India* **1981**, *50*, 319–320.
49. Gomez, F.; Quijano, L.; Garcia, G.; Calderon, J.S.; Rios, T. A prenylated flavan from *Tephrosia madrensis*. *Phytochemistry* **1983**, *22*, 1305–1306.
50. Gomez, F.; Quijano, L.; Calderon, J.S.; Rodriquez, C.; Rios, T. Prenylflavans from *Tephrosia watsoniana*. *Phytochemistry* **1985**, *24*, 1057–1059.

51. Gomez, F.; Calderon, J.; Quijano, L.; Cruz, O.; Rios, T. Nitenin—A new flavan from *Tephrosia nitens* Beth. *Chem. Ind.* **1984**, *17*, 632–632.
52. Khan, H.A.; Chandrasekharan, I.; Ghanim, A. Falciformin, a flavanone from pods of *Tephrosia falciformis*. *Phytochemistry* **1986**, *25*, 767–768.
53. Ganguly, A.; Bhattacharyya, P.; Bhattacharyya, A.; Adityachaudhury, N. Synthesis of Candidone—A new flavanone isolated from *Tephrosia candida*. *Indian J. Chem. B* **1988**, *27*, 462–463.
54. Gomezgaribay, F.; Quijano, L.; Calderon, J.S.; Morales, S.; Rios, T. Flavonoids from *Tephrosia* species. VI. Prenylflavanols from *Tephrosia quercetorum*. *Phytochemistry* **1988**, *27*, 2971–2973.
55. Hussaini, F.A.; Shoeb, A. A new epoxyflavanone from *Tephrosia hamiltonii*. *Planta Med.* **1987**, *2*, 220–221.
56. Gomezgaribay, F.; Quijano, L.; Rios, T. Flavonoids from *Tephrosia* species. VII. Flavanones from *Tephrosia leiocarpa*. *Phytochemistry* **1991**, *30*, 3832–3834.
57. Rao, E.V.; Prasad, Y.R. Prenylated flavonoids from *Tephrosia spinosa*. *Phytochemistry* **1993**, *32*, 183–185.
58. Rao, E.V.; Prasad, Y.R.; Murthy, M.S.R. A prenylated flavanone from *Tephrosia maxima*. *Phytochemistry* **1994**, *37*, 111–112.
59. Gomez-Garibay, F.; Calderon, J.S.; Quijano, L.; Tellez, O.; Olivares, M.D.; Rios, T. Flavonoids from *Tephrosia* species part 8—An unusual prenyl biflavanol from *Tephrosia tepicana*. *Phytochemistry* **1997**, *46*, 1285–1287.
60. Gomez-Garibay, F.; Calderon, J.S.; Arciniega, M.D.; Cespedes, C.L.; Tellez-Valdes, O.; Taboada, J. Flavonoids from *Tephrosia* species part 9—An unusual isopropenyldihydrofuran biflavanol from *Tephrosia crassifolia*. *Phytochemistry* **1999**, *52*, 1159–1163.
61. Rao, E.V.; Sridhar, P. Chemical examination of *Tephrosia strigosa*. *Indian J. Chem. B* **1999**, *38*, 872–873.
62. Hisham, A.; John, S.; Al-Shuaily, W.; Asai, T.; Fujimoto, Y. (+)-Apollineanin: A new flavanone from *Tephrosia apollinea*. *Nat. Prod. Res.* **2006**, *20*, 1046–1052.
63. Madhusudhana, J.; Reddy, R.V.N.; Reddy, B.A.K.; Reddy, M.V.B.; Gunasekar, D.; Deville, A.; Bodo, B. Two new geranyl flavanones from *Tephrosia villosa*. *Nat. Prod. Res.* **2010**, *24*, 743–749.
64. Smalberger, T.M.; Vleggaar, R.; Weber, J.C. Flavonoids from *Tephrosia*. VIII: Structure of elongatin, an isoflavone from *Tephrosia elongata* E Mey. *Tetrahedron* **1975**, *31*, 2297–2301.
65. Yenesew, A.; Dagne, E.; Waterman, P.G. Flavonoids from the seed pods of *Tephrosia pumila*. *Phytochemistry* **1989**, *28*, 1291–1292.
66. Vilain, C. Barbigerone, a new pyranoisoflavone from seeds of *Tephrosia barbigera*. *Phytochemistry* **1980**, *19*, 988–989.
67. Dagne, E.; Mammo, W.; Sterner, O. Flavonoids of *Tephrosia polyphylla*. *Phytochemistry* **1992**, *31*, 3662–3663.
68. Rao, E.V.; Murthy, M.S.R.; Ward, R.S. Nine isoflavones from *Tephrosia maxima*. *Phytochemistry* **1984**, *23*, 1493–1501.

69. Gomez, F.; Calderon, J.S.; Quijano, L.; Dominguez, M.; Rios, T. Viridiflorin, an isoflavone from *Tephrosia viridiflora*. *Phytochemistry* **1985**, *24*, 1126–1128.
70. Murthy, M.S.R.; Rao, E.V. Maxima isoflavone J: A new O-prenylated isoflavone from *Tephrosia maxima*. *J. Nat. Prod.* **1985**, *48*, 967–968.
71. Dagne, E.; Dinku, B.; Gray, A.I.; Waterman, P.G. Pumilaisoflavone A and Pumilaisoflavone B from the seed pods of *Tephrosia pumila*. *Phytochemistry* **1988**, *27*, 1503–1505.
72. Reddy, B.A.K.; Khalivulla, S.I.; Gunasekar, D. A new prenylated isoflavone from *Tephrosia tinctoria*. *Indian J. Chem. B* **2007**, *46*, 366–369.
73. Khalivulla, S.I.; Reddy, B.A.K.; Gunasekar, D.; Blond, A.; Bodo, B.; Murthy, M.M.; Rao, T.P. A new di-O-prenylated isoflavone from *Tephrosia tinctoria*. *J. Asian Nat. Prod. Res.* **2008**, *10*, 953–955.
74. Clark, E.P. Toxicarol. A constituent of the South American fish poison Cracca (*Tephrosia*) toxicaria. *J. Am. Chem. Soc.* **1930**, *52*, 2461–2464.
75. Sarma, P.N.; Srimannarayana, G.; Rao, N.V.S. Constitution of villosol and villosinol, two new rotenoids from *Tephrosia villosa* (Linn) pods. *Indian J. Chem. B* **1976**, *14*, 152–156.
76. Krupadanam, G.L.D.; Sarma, P.N.; Srimannarayana, G.; Rao, N.V.S. New C-6 oxygenated rotenoids from *Tephrosia villosa*—Villosin, villosone, villol and villinol. *Tetrahedron Lett.* **1977**, *24*, 2125–2128.
77. Roy, M.; Bhattacharya, P.K.; Pal, S.; Chowdhuri, A.; Adityachaudhury, N. Dehydrodihydrorotenone and flemichapparin B in *Tephrosia candida*. *Phytochemistry* **1987**, *26*, 2423–2424.
78. Dagne, E.; Yenesew, A.; Waterman, P.G. Flavonoids and isoflavonoids from *Tephrosia fulvinervis* and *Tephrosia pentaphylla*. *Phytochemistry* **1989**, *28*, 3207–3210.
79. Prashant, A.; Krupadanam, G.L.D. A new prenylated dehydrorotenoid from *Tephrosia villosa* seeds. *J. Nat. Prod.* **1993**, *56*, 765–766.
80. Prashant, A.; Krupadanam, G.L.D. Dehydro-6-hydroxyrotenoid and lupenone from *Tephrosia villosa*. *Phytochemistry* **1993**, *32*, 484–486.
81. Abreu, P.M.; Luis, M.H. Constituents of *Tephrosia uniflora*. *Nat. Prod. Lett.* **1996**, *9*, 81–86.
82. Andrei, C.C.; Viera, P.C.; Fernandes, J.B.; daSilva, M.F.D.F.; Fo, E.R. Dimethylchromene rotenoids from *Tephrosia candida*. *Phytochemistry* **1997**, *46*, 1081–1085.
83. Rao, P.P.; Srimannarayana, G. Tephrosol, a new coumestone from the roots of *Tephrosia villosa*. *Phytochemistry* **1980**, *19*, 1272–1273.
84. Ingham, J.L.; Markham, K.R. Tephrocarpin, a pterocarpan phytoalexin from *Tephrosia bidwilli* and a structure proposal for acanthocarpan. *Phytochemistry* **1982**, *21*, 2969–2972.
85. Lwande, W.; Bentley, M.D.; Hassanali, A. The structure of hildecarpin, an insect antifeedant 6a-hydroxypterocarpan from the roots of *Tephrosia hildebrandtii* Vatke. *Int. J. Trop. Insect Sci.* **1986**, *7*, 501–503.
86. Lwande, W.; Hassanali, A.; Njoroge, P.W.; Bentley, M.D.; Delle Monache, F.; Jondiko, J.I. A new 6a-hydroxypterocarpan with insect antifeedant and antifungal properties from the roots of *Tephrosia hildebrandtii* Vatke. *Int. J. Trop. Insect Sci.* **1985**, *6*, 537–541.

87. Lwande, W.; Bentley, M.D.; Macfoy, C.; Lugemwa, F.N.; Hassanali, A.; Nyandat, E. A new pterocarpan from the roots of *Tephrosia hildebrandtii*. *Phytochemistry* **1987**, *26*, 2425–2426.
88. Rajani, P.; Sarma, P.N. A coumestone from the roots of *Tephrosia hamiltonii*. *Phytochemistry* **1988**, *27*, 648–649.
89. Tarus, P.K.; Machocho, A.K.; Lang'at-Thoruwa, C.C.; Chhabra, S.C. Flavonoids from *Tephrosia aequilata*. *Phytochemistry* **2002**, *60*, 375–379.
90. Kishore, P.H.; Reddy, M.V.B.; Gunasekar, D.; Murthy, M.M.; Caux, C.; Bodo, B. A new coumestan from *Tephrosia calophylla*. *Chem. Pharm. Bull. (Tokyo)* **2003**, *51*, 194–196.
91. Ganapaty, S.; Srilakshmi, G.V.K.; Pannakal, S.T.; Rahman, H.; Laatsch, H.; Brun, R. Cytotoxic benzil and coumestan derivatives from *Tephrosia calophylla*. *Phytochemistry* **2009**, *70*, 95–99.
92. Dominguez, X.A.; Tellez, O.; Ramirez, G. Mixtecacin, a prenylated flavanone and oaxacacin its chalcone from the roots of *Tephrosia woodii*. *Phytochemistry* **1983**, *22*, 2047–2049.
93. Rao, E.V.; Raju, N.R. Two flavonoids from *Tephrosia purpurea*. *Phytochemistry* **1984**, *23*, 2339–2342.
94. Chibber, S.S.; Dutt, S.K. Tephrone, a new chalcone from *Tephrosia candida* seeds. *Curr. Sci. India* **1982**, *51*, 933–934.
95. Rao, E.V.; Prasad, Y.R. Two chalcones from *Tephrosia spinosa*. *Phytochemistry* **1992**, *31*, 2121–2122.
96. Sharma, V.M.; Rao, P.S. A prenylated chalcone from the roots of *Tephrosia spinosa*. *Phytochemistry* **1992**, *31*, 2915–2916.
97. Andrei, C.C.; Ferreira, D.T.; Faccione, M.; de Moraes, L.A.B.; de Carvalho, M.G.; Braz, R. C-prenylflavonoids from roots of *Tephrosia tunicata*. *Phytochemistry* **2000**, *55*, 799–804.
98. Gomez-Garibay, F.; Arciniega, M.D.O.; Cespedes, C.L.; Taboada, J.; Calderon, J.S. Chromene chalcones from *Tephrosia carrollii* and the revised structure of oaxacacin. *Z. Naturforsch. C* **2001**, *56*, 969–972.
99. Gomez-Garibay, F.; Tellez-Valdez, O.; Moreno-Torres, G.; Calderon, J.S. Flavonoids from *Tephrosia major*. A new prenyl-β-hydroxychalcone. *Z. Naturforsch. C* **2002**, *57*, 579–583.
100. Andrei, C.C.; Vieira, P.C.; Fernandes, J.B.; da Silva, M.F.; Rodrigues Fo, E. New spirorotenoids from *Tephrosia candida*. *Z. Naturforsch. C* **2002**, *57*, 418–422.
101. Wei, H.H.; Xu, H.H.; Xie, H.H.; Xu, L.X.; Wei, X.Y. Sesquiterpenes and lignans from *Tephrosia vogelii*. *Helv. Chim. Acta* **2009**, *92*, 370–374.
102. Jain, A.C.; Gupta, R.C., Possible biogenesis of novel type of flavones from *Tephrosia polystachyoides*. *Curr. Sci. India* **1978**, *47*, 770–770.
103. Crombie, L.; Dewick, P.M.; Whiting, D.A. Biosynthesis of rotenoids—Chalcone, isoflavone, and rotenoid stages in formation of amorphigenin by *Amorpha fruticosa* seedlings. *J. Chem. Soc. Perkin Trans. 1* **1973**, *12*, 1285–1294.
104. Lee, Y.R.; Morehead, A.T. A new route for the synthesis of furanoflavone and furanochalcone natural products. *Tetrahedron* **1995**, *51*, 4909–4922.
105. Belmain, S.R.; Amoah, B.A.; Nyirend, S.P.; Kamanula, J.F. Stevenson, P.C. Highly variable insect control efficacy of *Tephrosia vogelii* Chemotypes. *J. Agric. Food Chem.* **2012**, *60*, 10055–10063.

Effects of Endogenous Signals and *Fusarium oxysporum* on the Mechanism Regulating Genistein Synthesis and Accumulation in Yellow Lupine and Their Impact on Plant Cell Cytoskeleton

Magda Formela, Sławomir Samardakiewicz, Łukasz Marczak, Witold Nowak, Dorota Narożna, Waldemar Bednarski, Anna Kasprowicz-Maluśki and Iwona Morkunas

Abstract: The aim of the study was to examine cross-talk interactions of soluble sugars (sucrose, glucose and fructose) and infection caused by *Fusarium oxysporum* f.sp. *lupini* on the synthesis of genistein in embryo axes of *Lupinus luteus* L.cv. Juno. Genistein is a free aglycone, highly reactive and with the potential to inhibit fungal infection and development of plant diseases. As signal molecules, sugars strongly stimulated accumulation of isoflavones, including genistein, and the expression of the isoflavonoid biosynthetic genes. Infection significantly enhanced the synthesis of genistein and other isoflavone aglycones in cells of embryo axes of yellow lupine with high endogenous sugar levels. The activity of β-glucosidase, the enzyme that releases free aglycones from their glucoside bindings, was higher in the infected tissues than in the control ones. At the same time, a very strong generation of the superoxide anion radical was observed in tissues with high sugar contents already in the initial stage of infection. During later stages after inoculation, a strong generation of semiquinone radicals was observed, which level was relatively higher in tissues deficient in sugars than in those with high sugar levels. Observations of actin and tubulin cytoskeletons in cells of infected embryo axes cultured on the medium with sucrose, as well as the medium without sugar, showed significant differences in their organization.

Reprinted from *Molecules*. Cite as: Formela, M.; Samardakiewicz, S.; Marczak, Ł.; Nowak, W.; Narożna, D.; Bednarski, W.; Kasprowicz-Maluśki, A.; Morkunas, I. Effects of Endogenous Signals and *Fusarium oxysporum* on the Mechanism Regulating Genistein Synthesis and Accumulation in Yellow Lupine and Their Impact on Plant Cell Cytoskeleton. *Molecules* **2014**, *19*, 13392-13421.

1. Introduction

Genistein ($C_{15}H_{10}O_5$, 5,7,4'-trihydroxyisoflavone) is a fascinating molecule, attracting the attention of researchers due to its broad spectrum of its biological activity not only in plants, but also in the human organism. Plant-origin genistein in animals and human body cells acts as an anticancer, antibacterial and spasmolytic substance, but it also has antioxidant and hypotensive effects [1]. Moreover, genistein as a natural isoflavonoid phytoestrogen, is a strong inhibitor of protein tyrosine kinases [2]. Increased or aberrant expression of tyrosine kinases impact on tumor development and progression. This bioactive aglycone can also affect metabolism, *i.e.*, the kinetics of insulin binding to cell membranes [3], leptin secretion–an important factor regulating the energetic status of the whole organism [4].

In plant cells the physiological function of genistein has been established, it is a free aglycone of isoflavonoid origin capable of inhibiting the development of infections and diseases caused by pathogenic fungi [5]. Genistein may function as a phytoalexin due to its antimicrobial and fungistatic activity [6–8]. Results of *in vitro* tests showed also that genistein is a strong inhibitor of growth of a pathogenic fungus [9]. Additionally, it serves as a signalling compound that initiates the nodulation process in the legume-rhizobia symbiosis [10,11]. Plants from the family Leguminosae are rich sources of genistein [12,13]. It was shown that genistein is found in lupine (*Lupinus luteus, Lupinus albus*) [9,14,15], soy (*Glycine* sp.) [16,17], clover (*Trifolium* sp.) [18,19], in lucerne (*Medicago sativa*) [20,21] and in kudzu roots (*Pueraria lobata*) [22]. Genistein can be also secreted from roots of legumes [13]. Elucidation of the biosynthesis of numerous phytoalexins, including genistein, has facilitated the use of molecular biology tools in the exploration of the genes encoding enzymes of their synthesis pathways and their regulators [23].

Genetic modifications on the biosynthesis of phytoalexins was investigated in the context of plant resistance to pathogens. The effect of sucrose was demonstrated to improve the production of secondary metabolites, including flavonoids in plants [24,25]. Recent literature reports have documented both the essential role of sucrose as a donor of carbon skeletons for the metabolism of phenylpropanoid metabolism and that of the signalling molecules up-regulating expression of biosynthesis of phenylpropanoid [26–29]. Apart from sucrose, also glucose and fructose are recognized as signalling molecules in plants [30,31]. Sugar signals may contribute to immune responses against pathogens and probably function as priming molecules against pathogens [32]. The novel concept of “sweet priming” predicts specific key roles for saccharides in perceiving, mediating and counteracting both biotic and abiotic stresses.

The aim of the present study was to examine effects of sucrose and monosaccharides (glucose and fructose) as endogenous signals, and a hemibiotrophic fungus *Fusarium oxysporum* f.sp. *lupini* on the mechanism regulating genistein synthesis and its accumulation in embryo axes of *Lupinus luteus* L. cv. Juno. Therefore, apart from the estimation of levels of genistein and other isoflavone free aglycones, the expression of genes encoding phenylalanine ammonia-lyase (PAL), chalcone synthase (CHS), chalcone isomerase (CHI) and isoflavone synthase (IFS) was analyzed in the non-infected and *F. oxysporum*-infected embryo axes. Within this study we also analyzed changes in the activity of β-glucosidase, an enzyme which releases free aglycones from their glucoside bindings. At the exogenous addition of sucrose, glucose or fructose the endogenous levels of these sugars were determined in tissues. Moreover, generation of free radicals was estimated in view of the varied level of sucrose and monosaccharides, which may be incorporated in the defence responses of embryo axes, for example stimulation of phytoalexin synthesis or sealing of the cell walls. It was particularly interesting and important to examine post-infection changes in the actin and tubulin cytoskeleton cells of embryo axes of yellow lupine at varied levels of sucrose.

Available literature does not provide information regarding the importance of monosaccharides, *i.e.*, glucose and fructose, for the mechanism regulating genistein synthesis, generation of the superoxide anion radical and semiquinone radicals in plant response to infection. It should be stressed that the aforementioned aspects of the research, as well as monitoring of post-infection changes in actin and tubulin cytoskeletons in cells of yellow lupine embryo axes with different levels of sucrose after inoculation by *F. oxysporum*, are novel problems.

2. Results and Discussion

2.1. The Effect of Sucrose, Glucose and Fructose on Accumulation of Genistein and Other Isoflavones in Embryo Axes Infected with F. oxysporum

Exogenous addition of sucrose, glucose or fructose to the medium as a rule caused higher concentrations of genistein and other isoflavones in non-inoculated embryo axes of yellow lupine cv. Juno (+Sn, +Gn and +Fn) than in non-inoculated axes cultured at carbohydrate deficit (−Sn) (Figure 1A). However, the level of genistein in these tissues was higher in axes cultured *in vitro* on the medium with sucrose (+Sn) than in axes cultured with monosaccharides (+Gn or +Fn). In turn, infection of embryo axes with a hemibiotrophic fungus *F. oxysporum* considerably enhanced the accumulation of genistein. Already at 24 h after inoculation the level of genistein was 4 times greater in embryo axes cultured on the medium with glucose (+Gi) than in non-inoculated axes (+Gn). Moreover, post-infection accumulation of genistein was also found at successive time points after inoculation in axes cultured on the medium with sucrose and monosaccharides, *i.e.*, glucose and fructose. However, we need to focus particularly on the very strong accumulation of genistein in these tissues at 96 h after inoculation. Moreover, the post-infection accumulation of genistein was also observed in axes inoculated with *F. oxysporum* cultured under carbohydrate deficit. The highest level of this free aglycone in these tissues was recorded at 48 h after inoculation, while at successive time points this level was much lower than in inoculated tissues with high levels of sugars (+Si, +Gi and +Fi). It is of interest that starting from 48 h of culture the level of 2'-hydroxygenistein was many times higher in non-inoculated axes cultured on the medium with sucrose, glucose or fructose (+Sn, +Gn or +Fn) than in the other experimental variants (−Sn, −Si, +Si, +Gi, +Fi) (Figure 1B).

Figure 1. The effect of sucrose, glucose and fructose on accumulation of genistein (**A**) and other isoflavones, *i.e.*, 2'-hydroxygenistein (**B**), wighteone (**C**) and luteone (**D**), in *in vitro* cultured embryo axes of *Lupinus luteus* L. cv. Juno infected with *Fusarium oxysporum* f.sp. *lupine*. Statistical significance of differences between the average values of each pairs for accumulation of genistein and other isoflavones. Statistically significant differences (*p*-value) was assumed at $p < 0.05$.

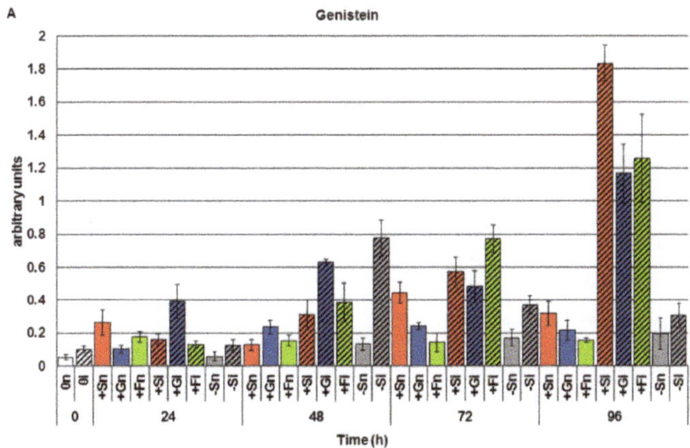

Figure 1. *Cont.*

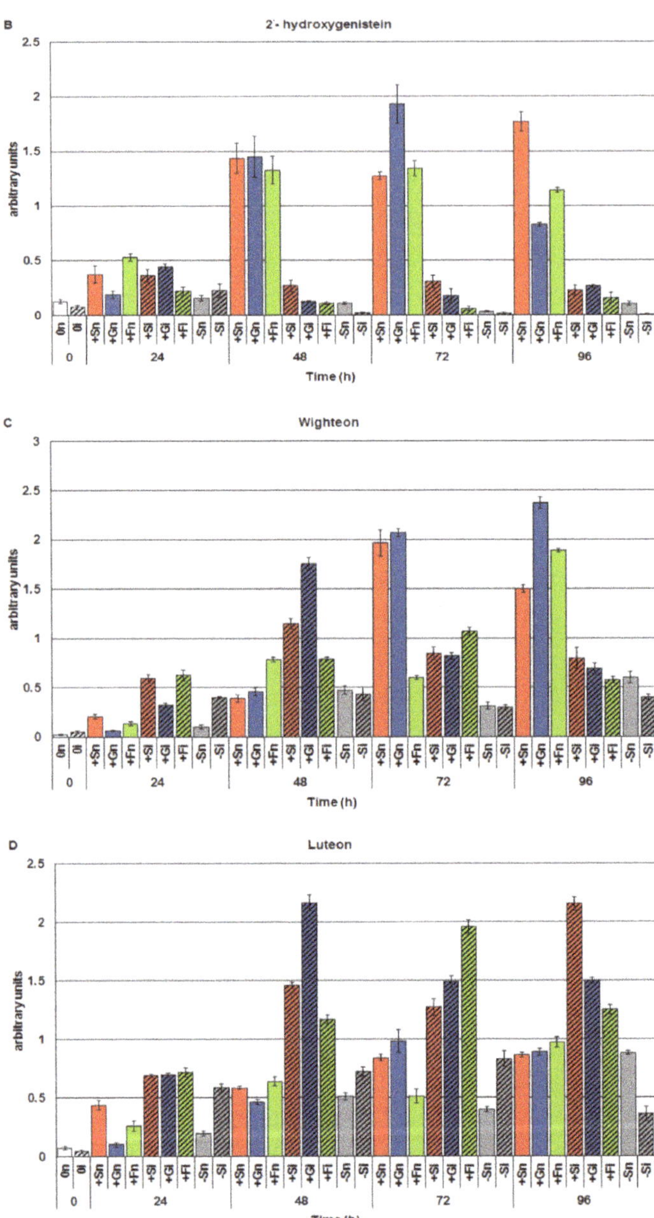

Infection caused a very strong reduction of the 2'-hydroxygenistein level in all inoculated embryo axes, occasionally so substantial that it reached the detection limit for this metabolite. As a result of inoculation with *F. oxysporum*, concentrations of another free aglycone, which is a prenylated derivative of 2'-hydroxygenistein, *i.e.*, luteone, increased many times in inoculated embryo axes with high levels of sugars (+Si, +Gi and +Fi) (Figure 1D). Moreover, in the period from 24 to 72 h after inoculation the level of luteone was higher in inoculated axes with carbohydrate deficit (−Si) than in non-inoculated ones (−Sn). In turn, the level of wighteone, a prenylated derivative of genistein, decreased considerably at 72 and 96 h after inoculation in axes with high levels of sucrose, glucose and fructose, while the concentration of genistein at these time points after inoculation was increased (Figure 1C).

2.2. The Effect of Sucrose, Glucose and Fructose on Expression Levels of Genistein Biosynthesis Pathway Genes in Response to Infection with F. oxysporum

Real-time PCR analyses of the level of mRNA encoding enzymes involved in the synthesis of isoflavones, including genistein, revealed post-infection accumulation of mRNA in embryo axes of yellow lupine cv. Juno (Figure 2). In embryo axes infected with *F. oxysporum* cultured *in vitro* on the medium with sucrose, glucose and fructose (+Si, +Gi, +Fi) the level of mRNA for PAL, CHS, CHI and IFS was higher than in non-infected axes (+Sn, +Gn, +Fn) in the period from 0 to 96 h. Moreover, up to 72 h after inoculation in inoculated axes with a sugar deficit (−Si) a higher level of mRNA encoding PAL, CHS, CHI and IFS was observed post infection in relation to non-inoculated axes (−Sn). At the next time point after inoculation, *i.e.*, at 96 h in infected axes cultured at carbohydrate deficit (−Si) a very strong reduction was recorded in the mRNA level, while at 96 h in infected axes with a high level of carbohydrates (+Si, +Gi and +Fi) the level of mRNA encoding the above mentioned enzymes was the highest. It needs to be stressed that a very high post-infection level of mRNA was recorded for enzymes of the specific isoflavone synthesis pathway, *i.e.*, CHS and IFS (Figure 2B,D) in embryo axes inoculated with *F. oxysporum*, being much higher than for PAL and CHI (Figure 2A,C).

Figure 2. The effect of sucrose, glucose and fructose on expression levels of isoflavone biosynthetic pathway genes, *i.e.*, phenylalanine ammonia-lyase (PAL) (**A**), chalcone synthase (CHS) (**B**), chalcone isomerase (CHI) (**C**) and isoflavone synthase (IFS) (**D**) in *in vitro* cultured embryo axes of *L. luteus* infected with *F. oxysporum.* Statistical significance of differences between the average values of each pairs for expression levels of isoflavone biosynthetic pathway genes. Statistically significant differences (p-value) was assumed at $p < 0.05$.

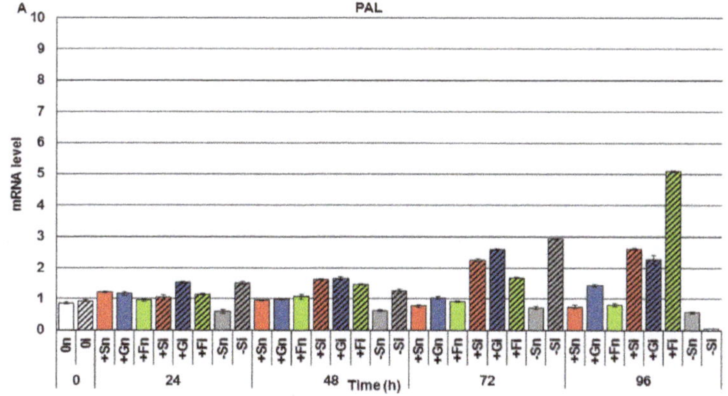

Figure 2. *Cont.*

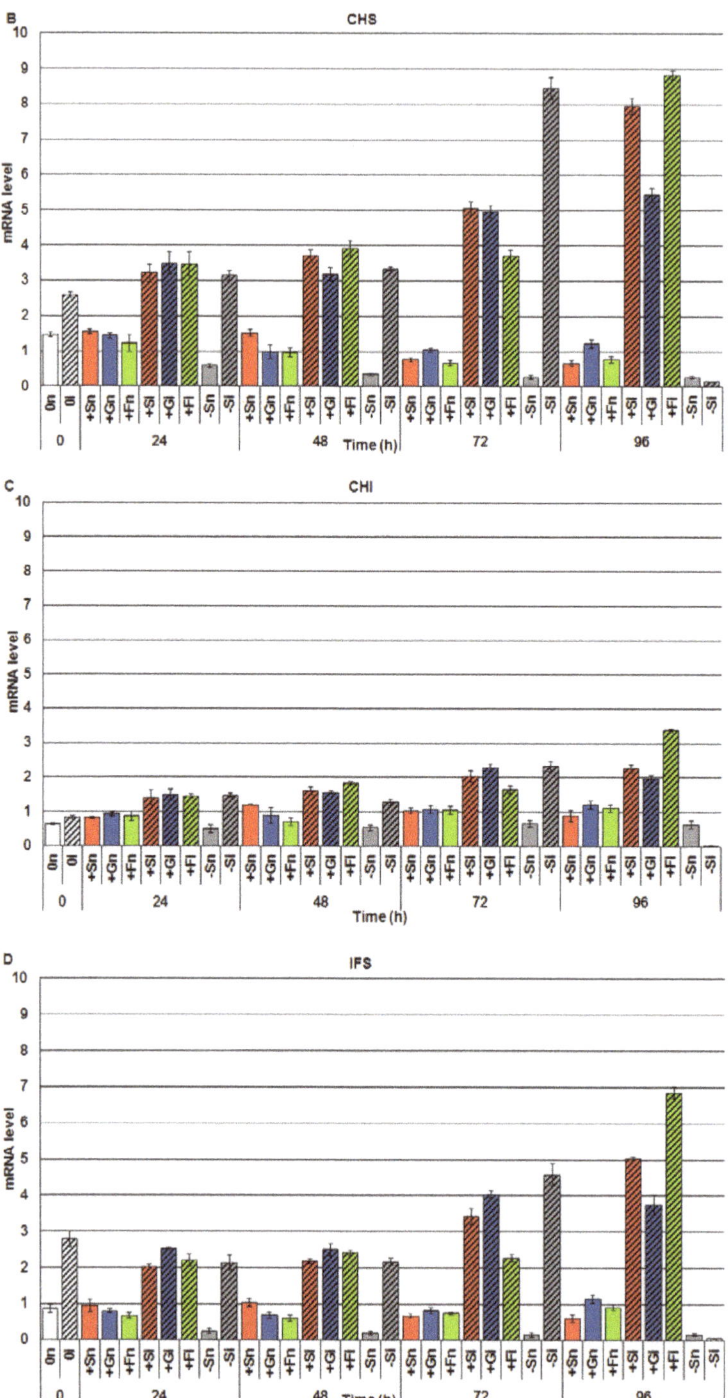

2.3. *The Effect of Exogenous Sucrose, Glucose and Fructose on Endogenous Levels of Soluble Sugars and Their Changes in Response to Infection with F. oxysporum*

GC-MS analyses of the carbohydrate levels showed that after exogenous administration of sucrose, glucose and fructose the endogenous level of these sugars increased strongly in embryo axes of yellow lupine cv. Juno (Figure 3). Levels of sucrose and monosaccharides (glucose and fructose) in embryo axes of yellow lupine cultured *in vitro* with an addition of sucrose (+Sn), glucose (+Gn) and fructose (+Fn) was much higher than in embryo axes cultured at carbohydrate deficit (−Sn). A very strong endogenous decrease in the levels of sucrose (Figure 3A) and monosaccharides (glucose and fructose) (Figure 3B,C) was observed at 72 h after infection in embryo axes cultured in the presence of these carbohydrates. At 96 h after infection the level of sucrose decreased even further (Figure 3A), while the content of fructose in +Fi tissues increased considerably and was higher than in the other experimental variants (Figure 3C).

What is more, already starting from 24 h after infection in embryo axes with carbohydrate deficit (−Si) the level of sugars decreased and it was lower than in non-inoculated embryo axes (−Sn) at all-time points. The level of carbohydrates in −Si and −Sn tissues was very low in relation to the other experimental variants.

Figure 3. The effect of sucrose, glucose and fructose on the endogenous level of soluble sugars *i.e.*, sucrose (**A**), glucose (**B**), fructose (**C**) in *in vitro* cultured embryo axes of *L. luteus* and their changes in response to infection with *F. oxysporum*. Statistical significance of differences between the average values of each pairs for endogenous level of sugars. Statistically significant differences (p-value) was assumed at $p < 0.05$.

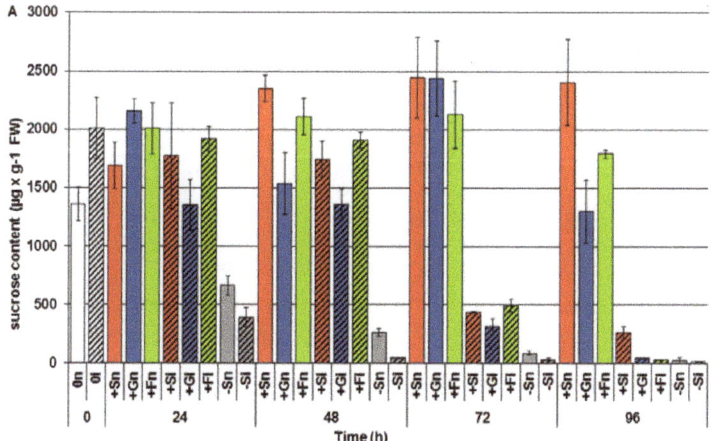

Figure 3. *Cont.*

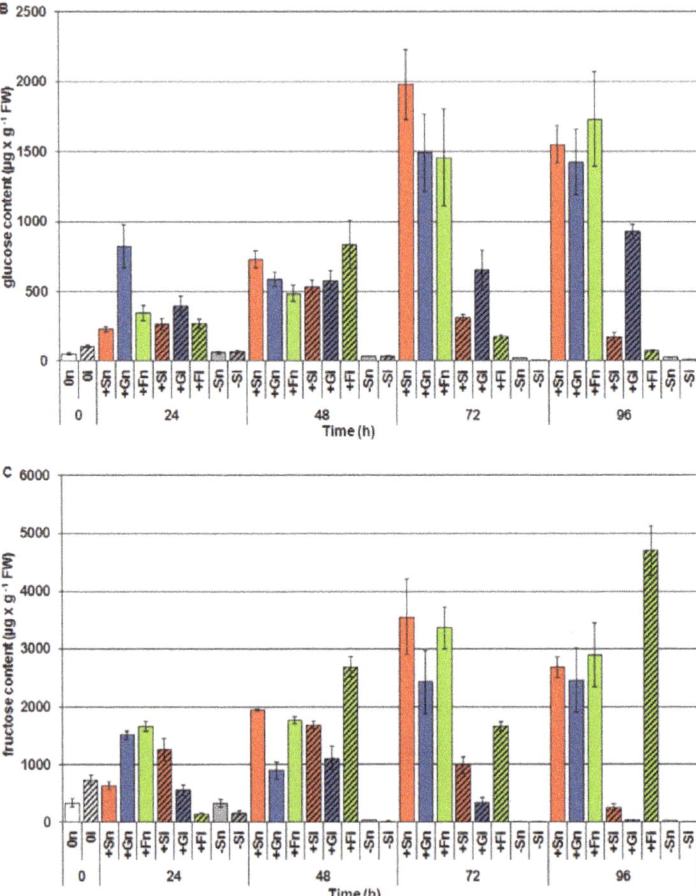

2.4. The Effect of Sucrose, Glucose and Fructose on β-Glucosidase Activity in Response to Infection with F. oxysporum

Analyses of β-glucosidase activity in the period from 0 to 96 h of culture showed a post-infection increase in the activity of this enzyme in inoculated embryo axes both cultured in the presence of sucrose, glucose and fructose, and under carbohydrate deficit (Figure 4). However, the activity of this enzyme was higher in embryo axes infected with *F. oxysporum* cultured at carbohydrate deficit (−Si) than in the infected embryo axes cultured in the presence of sugars, *i.e.*, (+Si, +Gi, +Fi). The highest activity of β-glycosidase was recorded at 96 h after inoculation.

2.5. The Effect of Sucrose, Glucose and Fructose on Superoxide Anion Radical ($O_2^{\bullet -}$) Generation in Response to Infection with F. oxysporum

In the period from 24 to 96 h after inoculation a strong generation of the superoxide anion radical was observed in embryo axes infected with *F. oxysporum* (Figure 5).

Figure 4. The effect of sucrose, glucose and fructose on β-glucosidase activity in *in vitro* cultured embryo axes of *L. luteus* infected with *F. oxysporum*. Statistical significance of differences between the average values of each pairs for β-glucosidase activity. Statistically significant differences (*p*-value) was assumed at $p < 0.05$.

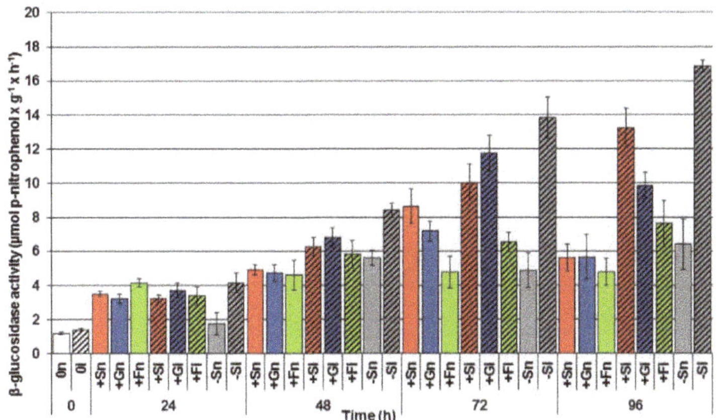

Figure 5. The effect of sucrose, glucose and fructose on superoxide anion ($O_2^{\bullet-}$) generation in *in vitro* cultured embryo axes of *L. luteus* infected with *F. oxysporum*. Statistical significance of differences between the average values of each pairs for superoxide anion radical generation. Statistically significant differences (*p*-value) was assumed at $p < 0.05$.

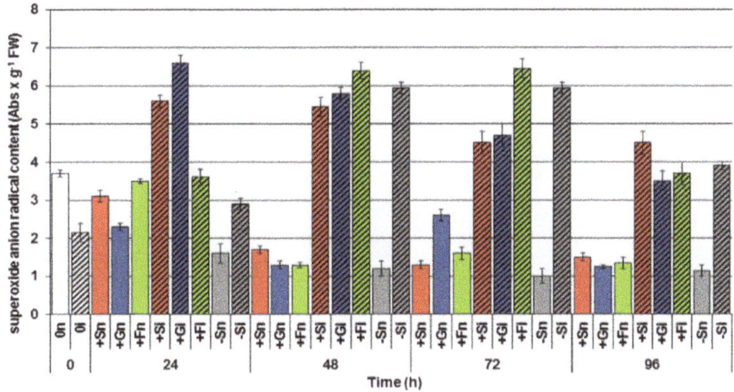

At 24 h after inoculation a particularly interesting finding was connected with the high level of $O_2^{\bullet-}$ generation in infected embryo axes with a high level of sugars (+Si, +Gi and +Fi), with this level being considerably higher than in embryo axes with the deficit (−Si). At successive time points after inoculation intensity of $O_2^{\bullet-}$ generation in −Si embryo axes was similarly as high as in the infected embryo axes with a high endogenous level of carbohydrates (+Si, +Gi and +Fi).

2.6. The Effect of Sucrose, Glucose and Fructose on Semiquinone Radical Generation in Response to Infection with F. oxysporum

Starting from 48 h culture the level of generation of semiquinone radicals with the Landé g factor $g_{\|} = 2.0036(0.005)$ $g_{\perp} = 2.0053\ (0.003)$ was higher in embryo axes with high endogenous levels of sucrose, glucose and fructose than in those with sugar deficits (Figure 6). At that time point it was observed that infection with *F. oxysporum* resulted in enhanced generation of these radicals in infected embryo axes with a carbohydrate deficit (−Si) in relation to those non-infected (−Sn). At further time points after inoculation with *F. oxysporum*, *i.e.*, at 72 and 96 h, concentrations of these radicals increased considerably, both in −Si and +Si, +Gi and +Fi embryo axes and it was much greater than in non-infected embryo axes (−Sn, +Sn, +Gn, +Fn) (Figure 6A). However, at 96 h after inoculation the concentration of these radicals was the highest in embryo axes with carbohydrate deficit (−Si), being higher than in infected embryo axes with a high endogenous carbohydrate level.

2.7. The Effect of Sucrose on Actin and Tubulin Cytoskeleton Organization in Response to Infection with F. oxysporum

The actin cytoskeleton in cells of non-inoculated axes with a high endogenous level of sucrose (+Sn) was formed of long and thick actin cables surrounding the cells and their branches, creating a dense actin meshwork extending in various directions (Figure 7A).

Figure 6. The effect of sucrose, glucose and fructose on semiquinone radical concentration in *in vitro* cultured embryo axes of *L. luteus* infected with *F. oxysporum* (**A**) and typical wide-scan EPR spectra at 72 h (**B**) and 96 h (**C**). Statistical significance of differences between the average values of each pairs for semiquinone radical concentration. Statistically significant differences (*p*-value) was assumed at $p < 0.05$.

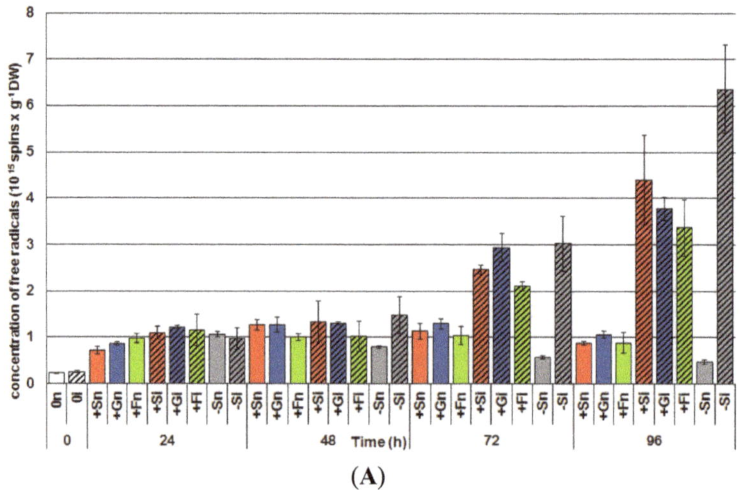

(**A**)

Figure 6. *Cont.*

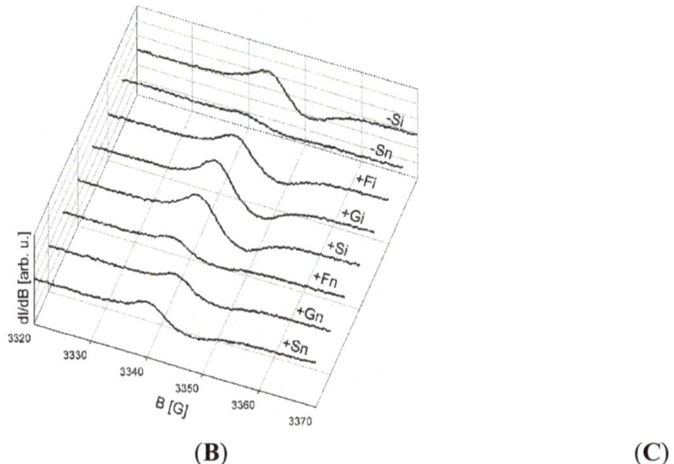

(B) (C)

Figure 7. The effect of sucrose and *F. oxysporum* on actin (**A**) (scale = 20 µm) and tubulin (**B**) (green fluorescence; scale = 10 µm) cytoskeleton organization *in vitro* cultured embryo axes of *L. luteus*. The red arrow indicates actin cables, red arrowhead-meshwork of microfilaments and white star-actin cytoskeleton around the nucleus.

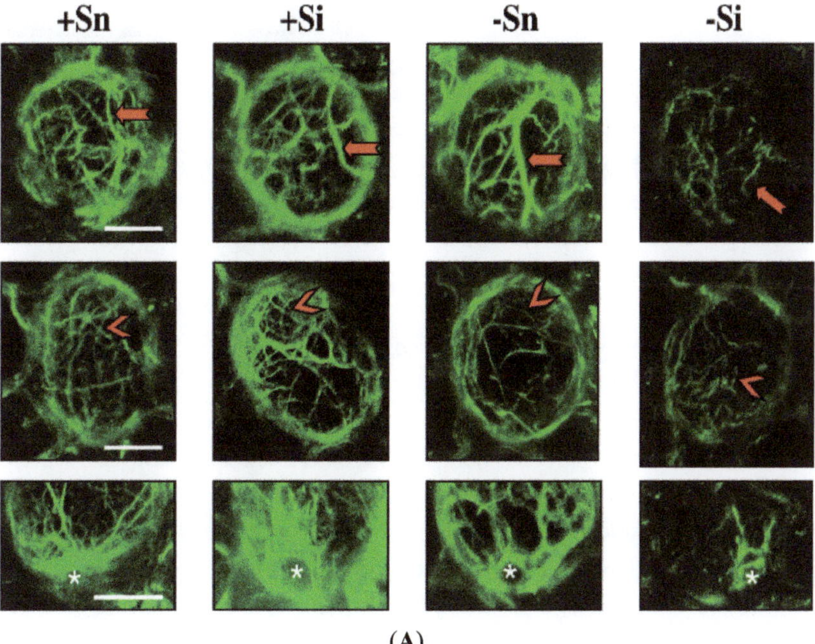

(A)

Figure 7. *Cont.*

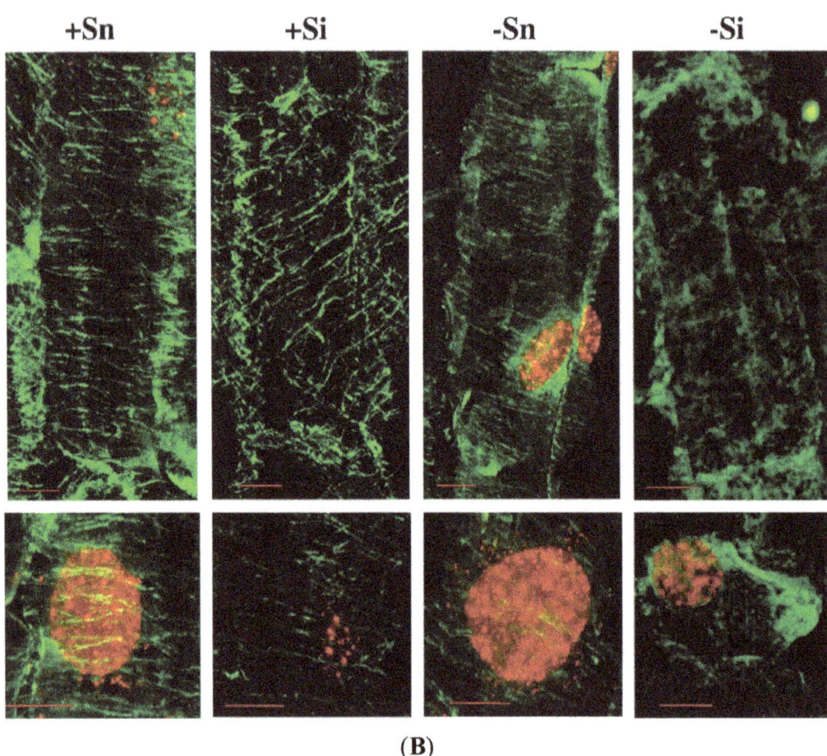

(B)

A particularly high accumulation of actin was observed in the vicinity of cell nuclei. Additionally, when cells were inoculated with *F. oxysporum* (+Si), bundles of actin were observed to thicken and fluorescence intensity increased in relation to +Sn cells. Quantitative analysis indicated a significant increase in the global density of microfilaments (MFs) in cells. This increase resulted mainly from an increase in occupancy regions of high fluorescence, which corresponded to actin cables (Figure 8A).

In non-inoculated cells of embryo axes cultured at a carbohydrate deficit (−Sn) a partial changes of the microfilaments meshwork was observed in relation to axes with an increased sugar level. Although the global density of MFs in the cell was not changed in comparison to +Sn, there was a slight reorganization of actin cytoskeleton: slightly increased occupancy of regions of high fluorescence. In turn, in inoculated cells with carbohydrate deficit (−Si) the greatest changes were observed in the actin cytoskeleton, *i.e.*, the length of all forms of microfilament bundles was reduced and the meshwork of microfilament bundles was fragmented. The global MF density was significantly decreased compared to all other variants (this may be a manifestation of depolymerization/fragmentation). This was done by a significant decrease in occupancy of regions of high intensity fluorescence corresponding to actin cables.

Figure 8. (**A**) The effect of sucrose and *F. oxysporum* on actin cytoskeleton density in cells of embryo axes of *L. luteus*. To estimate the density of microfilaments (MFs), the occupancy (%) of the pixel areas of AlexaFluor signals from segmented images (example image in legend) of cell was calculated. The red color represents areas of high fluorescence intensity corresponded to actin cables, green color-areas of low fluorescence corresponding *i.e.*, to remaining places containing actin and blue color-regions of cells without fluorescence signal. Different small letters (a,b,c,d,f) denote statistically significant differences ($p < 0.05$); (**B**) Mean length of microtubules bundles (MTs) in pixels expressed by the parameter of fiber length; (**C**) Mean width of microtubules bundles (MTs) in pixels expressed by the parameter of fiber width.

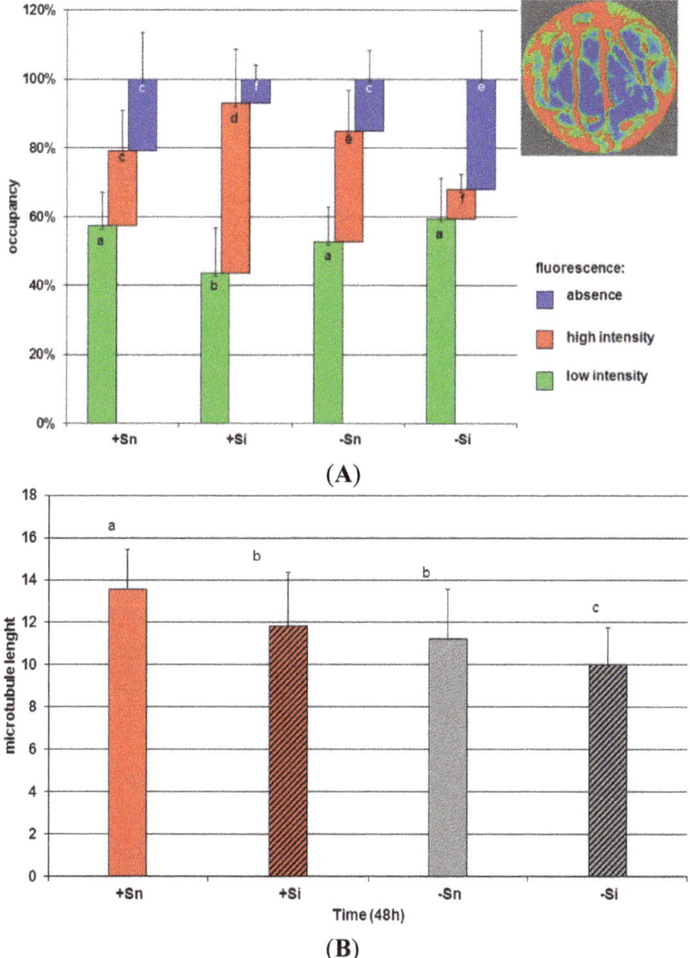

(**A**)

(**B**)

Figure 8. *Cont.*

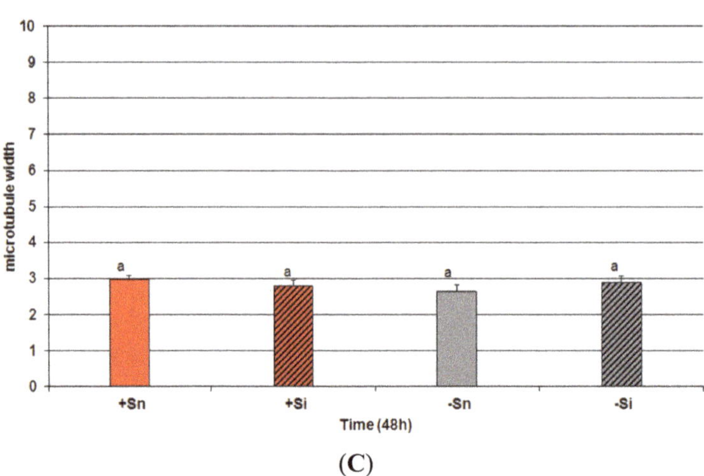

(C)

Tubulin cytoskeletons +Sn, +Si and −Sn cells of cortex in embryo axis (in horizontal section of cells) consisted of microtubules, which created bundles structure parallel to each other (Figure 7B). For the most part, these bundles were situated perpendicularly to the longer axis of the cell, and only occasionally at a small angle. In comparison to +Sn, in +Si and −Sn variants only slight changes were observed. Their structure was usually intact. Their reciprocal arrangement rarely changed. Therefore the cells, which were observed, had their microtubules bundles randomly scattered *i.e.*, within a cell there were regions, where bundles were more or less densely packed than in other regions. Moreover, the length of microtubules bundles in +Si and −Sn cells of embryo axes was a little shorter than in +Sn (Figure 8B). The width of microtubules bundles was not changed (Figure 8C). The tubulin cytoskeleton in −Si cells showed the most visible disorders *i.e.*, a decline or a considerable reduction of the number of microtubules, a significant shortening of length (fragmentation) of microtubules (Figure 8B), occurrence of big, irregularly shaped tubulin aggregates displaying strong fluorescence and the region with diffusive character of fluorescence (Figure 7B).

2.8. Discussion

This study demonstrates enhanced biosynthesis of genistein and other isoflavones, free aglycones by sucrose and monosaccharides (glucose and fructose) in *in vitro* cultured embryo axes of yellow lupine cv. Juno, resistant to fusariosis. As signal molecules, sugars strongly stimulated both the accumulation of genistein and other isoflavones (Figure 1) and the expression of the isoflavonoid biosynthetic genes (Figure 2). Infection with a hemibiotrophic fungus *F. oxysporum* significantly enhanced the synthesis of genistein, wighteone and luteone in cells of embryo axes of yellow lupine with high endogenous levels of sucrose, glucose and fructose (Figure 1A,C,D). A strong accumulation of genistein, particularly at later times post infection, was as a rule correlated with reduced levels of both sucrose and monosaccharides in tissues (Figure 3), which may indicate the involvement of these sugars as carbon skeletons in phenylpropanoid metabolism. Genistein as a free aglycone may be an inhibitor of pathogenic fungus *F. oxysporum*.

A study by Shinde *et al.* [33] showed the effectiveness of sucrose for genistein and daidzein production, and growth when compared to glucose, fructose and maltose in suspension cultures of *Psoralea corylifolia*. The maximum production of genistein and daidzein was recorded when sucrose and maltose were used as sole sources of carbon. Suspension cell cultures enriched with sucrose (3%) stimulated accumulation of isoflavones, genistein and daidzein, when compared to glucose, fructose and maltose. Successive studies confirmed that sucrose in the medium is energetically the most advantageous source of carbon for cultivation of cell cultures of *P. corylifolia* and mainly for the biosynthesis of isoflavones. The productivity of daidzein was greater in hairy roots transformed by *Agrobacterium rhizogenes* than in untransformed roots [24]. Moreover, enhanced accumulation of isoflavone [34] and anthocyanin production [29,35,36] was also provided by controlled feeding of the carbon source in the growth medium. Increased accumulation of isoflavones by sucrose was also found in embryo axes of yellow lupine cv. Polo, being sensitive to fusariosis [9].

In turn, results recorded in this study are the first which show the effects of glucose and fructose on the mechanism regulating synthesis and accumulation of isoflavone free aglycones, including genistein, in the case of infection by *F. oxysporum*. Our other investigations revealed a strong accumulation of isoflavonoids, particularly free aglycones, in *F. oxysporum*-inoculated embryo axes of yellow lupine pretreated with nitric oxide (NO) and cultured with a high level of sucrose, to result from the amplification of the signal coming from sucrose, the nitric oxide donor and the pathogenic fungus [37]. The role of sugars as regulatory molecules is now generally recognized, as well as the dynamic nature of primary carbon metabolism (with efficient invertase-mediated conversion of sucrose to glucose and fructose) and complex interactions with hormone signaling pathways [30]. Moreover, results reported by Li *et al.* [29] suggest that the DELLA proteins can act as key positive regulators in the sucrose signaling pathway controlling anthocyanin biosynthesis and a point of integration of diverse metabolic and hormonal signals and growth. Moreover, enhanced post-infection accumulation of genistein, 2'hydroxygenistein, wighteone and luteone (prenylated isoflavones) was observed in leaves of lupine plants infected with *Colletotrichum lupini* spores [38]. A similar relationship was demonstrated by Muth *et al.* [39]. Synthesis of luteone and 2'hydroxygenistein was enhanced in the youngest leaves *L. angustifolius*, while that of wighteone in older leaves infected with a pathogenic fungus *C. lupini*, causing anthracnose.

As it was reported by Jeandet *et al.* [23], genetic manipulation of phytoalexins is investigated to increase disease resistance of plants. The first example of disease resistance resulting from foreign phytoalexin expression in a novel plant concerned a phytoalexin from grapevine, which was transferred to tobacco. Secondary metabolites characterized by antifungal activity can be an important line of defence against phytopathogenic fungi [8]. Plant antifungal metabolites present constitutively in healthy plants called phytoanticipins may be inhibitors of pathogenic fungi, or they may be synthesized *de novo* in response to their attack or other stress conditions (phytoalexins) [40]. These molecules may be used directly or be considered as precursors for the development of better fungicidal molecules. In addition, the same molecule may be a phytoalexin or a phytoanticipin in different organs of the same plant [41].

Results of investigations recorded within this and previous studies [9,42] showed a post-infection increase in the activity of β-glycosidase (Figure 4); however, it is not dependent on the level of carbohydrates in tissues. In embryo axes with high endogenous levels of sucrose, glucose and fructose (Figure 3), showing a considerable accumulation of isoflavones, including genistein (Figure 1), and a high level of expression of genes encoding enzymes of genistein biosynthesis (Figure 2), also a very strong generation of $O_2^{\bullet -}$ was observed in their tissues (Figure 5). As early as 24 h and at the further time points post infection in embryo axes infected with *F. oxysporum* a very strong generation of this radical was found in cultures run on the medium with glucose, fructose and sucrose (Figure 5). Obviously, generation of $O_2^{\bullet -}$ in these tissues is connected with the adopted defence strategy in infected embryo axes with a high carbohydrate level. During later stages after inoculation, lower generation of semiquinone radicals in +Si, +Gi and +Fi tissues in relation to −Si is probably connected with the strengthening and sealing of cell walls (Figure 6).

In cells of embryo axes inoculated with *F. oxysporum* with a high endogenous level of sucrose (+Si), exhibiting high accumulation of genistein, actin bundles were observed to thicken and fluorescence intensity increased in relation to +Sn cells and other variants (Figure 7A). In turn, the length of microtubules bundles in +Si cells was a little shorter than in +Sn (Figure 8B).

Interestingly, the architecture of microfilaments and microtubules was preserved despite the inoculation, which may indicate that sucrose provides stabilization of the cytoskeleton. In turn, in inoculated cells with a carbohydrate deficit (−Si) the length of all forms of the actin cytoskeleton were reduced and fragmented. The global MF density was significantly decreased compared to all other variants which might suggest depolymerization/fragmentation [43]. Observations of the tubulin cytoskeleton in −Si also showed a decline or a considerable reduction of the number of microtubules, a significant shortening of length (fragmentation) of microtubules and the diffusive character of fluorescence (Figures 7B and 8B). Hardham [44] reported that during the attack by fungal pathogens rapid morphological changes in the plant cytosol were observed that included the reorganization of the cortical microtubule array. Microtubule reorganization has been documented during both resistant and susceptible interactions. Nick [45] also reported that microtubules act as sensors and integrators for stimuli including pathogen attack. A mechanosensory function of microtubules, the cellular competence to induce defence genes in response to an elicitor correlates with microtubule stability. For example, the reduction of microtubule treadmilling to inhibit cell-to-cell transport of plant viruses. Disturbances of the microtubule structure were observed not only under the influence of biotic, but also abiotic stresses [46].

The regulation and turnover of the actin cytoskeleton requires concerted activities of hundreds of actin-binding proteins, which may respond to signals to polymerize or destroy actin filament networks. This constant formation and destruction of actin networks requires a huge expense of energy-on the order of millions of ATPs per second-and is thought to represent a surveillance mechanism to various biotic and abiotic stresses [47,48]. As it was reported by Henty-Ridilla *et al.* [49], the cytoskeleton plays a major role in organizing the structures, as well as responds rapidly to biotic stresses and supports numerous fundamental cellular processes, including vesicle trafficking, endocytosis and the spatial distribution of organelles and protein complexes. For years the actin cytoskeleton has been assumed to play a role in plant innate immunity against fungi and oomycetes (based on pharmacological studies). However, there is little evidence that the host-cell actin cytoskeleton participates in plant responses to phytopathogenic fungi. Actin microfilaments, on which targeted cytoplasm flow is dependent, may affect deposition of material outside the protoplast, which role during cell penetration by the attacking pathogen is to form papillae [50,51]. Studies of Henty-Ridilla *et al.* [49] revealed two distinct and statistically significant changes in actin filament organization following infection with a virulent pathogen, *i.e.*, an early and transient increase in actin filament density as well as a late increase in the extent of actin filament bundling. The requirement of the actin cytoskeleton for the activation of NADPH oxidase at the plasma membrane, the superoxide-generating system, as well as the Golgi complex, peroxisomes and endoplasmic reticulum trafficking toward sites of fungal and oomycete penetration, has been demonstrated mainly through pharmacological studies [52]. The reconfiguration and reorganization of the actin cytoskeleton as a result of pathogens in the human body cells is well-known [53], but in plant-host cells have more research is still required to elucidate this problem. It is important how this network is manipulated during infection by a diverse array of pathogens. Pathogenic microorganisms subvert normal-cell processes to create a specialized niche, which enhances their survival. A common and recurring target is the cytoskeleton, mainly the actin cytoskeleton, since the F-actin filaments are highly dynamic structures, which supramolecular organization is constantly modified to meet cellular needs. Actin dynamic behavior is regulated by a large number of binding proteins, which drive intracellular and extracellular signaling pathways [54].

Smertenko and Franklin-Tong [55] reported that the plant cytoskeleton is reorganised in response to programmed cell death (PCD), with remodelling of both microtubules and microfilaments taking place. In a majority of cases, the microtubule network depolymerises, remodelling of microfilaments can follow two scenarios, either being depolymerised and then forming stable foci, or forming distinct bundles and then depolymerising. Moreover, density of the filamentous actin (F-actin) network undergoes periodic alternating polymerisation and depolymerisation processes, while alterations of this periodicity may have a role in perceiving signals and generating responses. Moreover, cytoskeletal reorganization of the microtubules and microfilaments in *P. sylvestris* root cells after inoculation by *Heterobasidion annosum*, *Heterobasidion parviporum*, and *Heterobasidion abietinum* was noted also by Zadworny *et al.* [56].

Regardless of the above, in cells of non-inoculated axes of yellow lupine with a high endogenous level of sucrose (+Sn), the actin cytoskeleton was formed of long and thick actin cables surrounding the cells and their branches, creating a dense actin meshwork extending in various directions. A particularly high accumulation of actin in these cells was observed in the vicinity of cell nuclei.

Wojtaszek *et al.* [57] reported that selective disruption of wall polysaccharides or covalent interactions between wall components leads to alterations in the organization of the actin cytoskeleton as well as disturbances of the mechanical stability of the whole cell. Short osmotic stress induces a rapid increase of F-actin at cross walls of maize root apices. Banaś *et al.* [58] reported that glucose and sucrose did not influence the pattern of the actin cytoskeleton in detached *Arabidopsis thaliana* leaves. In contrast, mannose caused the disappearance of filamentous structures and generated actin foci. Hexokinase as hexose sensor has been found to be associated with actin filaments in *Arabidopsis* [59]. Also another enzyme such as sucrose synthase, involved in sugar metabolism have been found to be associated with plant cytoskeleton [60]. Additionally, the organization of the actin cytoskeleton and vesicle trafficking may be modulated via another signal molecule such as NO in maize root apices [61].

3. Experimental Section

3.1. Materials

3.1.1. Plant Material and Growth Conditions

Experiments in this study were carried out on yellow lupine (*Lupinus luteus* L. cv. Juno). Cultivar Juno is resistant to fusariosis according to the data provided by the Plant Breeding Company in Poznań, Poland. Seeds were surface-sterilized, immersed in sterile water and left in an incubator (25 °C). After 6 h of imbibition, the seeds were transferred onto filter paper (in Petri dishes) and immersed in a small amount of water to support further absorption. After a subsequent 18 h, the seed coats were removed from the imbibed seeds and the cotyledons were removed to isolate the embryo axes. At the beginning of the experiment (time 0 h), the embryo axes were either inoculated with a *Fusarium oxysporum* f.sp. *lupini* spore suspension, or they were not inoculated. These were placed, within the next 20 min after cotyledon removal, in groups of four onto Whatman filter papers, which were subsequently transferred to sterile glass test tubes (diameter 3 cm, height 13.5 cm) containing 14 mL of Heller's mineral medium [62]. There they were suspended in such a way that one end of the axis was immersed in the medium. A space was left below the paper to allow better aeration. After removal of the cotyledons, the embryo axes were dependent on the carbon source provided by the medium. Eight culture variants were applied: +S, embryo axes cultured *in vitro* on Heller's medium supplemented with 60 mM sucrose [9,28,42,63,64], +G, 120 mM glucose, +F, 120 mM fructose and −S, embryo axes cultured *in vitro* on Heller's medium without sucrose (−Sn, non-inoculated cultured without sucrose, −Si, inoculated cultured without sucrose). In addition, embryo axes before being transferred to sterile glass test tubes containing Heller's mineral medium were not inoculated (the control, *i.e.*, +Sn, +Gn, +Fn) or inoculated with a pathogenic fungus *Fusarium oxysporum* f. sp. *lupini* (+Si, +Gi, +Fi).

The applied sucrose, glucose and fructose concentration was optimal to ensure appropriate growth of embryo axes, fresh and dry weight, as well as the uptake of minerals from the medium. Embryo axes were incubated in the dark at 25 °C. The model system was an equivalent of the stage in the development of germinating lupine seeds before the developing seedling emerges above the soil surface. At this heterotrophic stage, drastic changes may occur in the level of carbohydrates in the embryo axe of the germinating seed. The application of this model system was aimed at comparing two situations during heterotrophic stages of seed germination. The first stage is the stage when embryo axes are provided with an endogenous pool of soluble sugars due to the appropriately progressing mobilization of reserve nutrients in cotyledons, and the second stage is marked by a deficit of carbohydrates, so-called sugar starvation, resulting from disturbances in the mobilization of reserve substances. The experimental system, *i.e.*, embryo axes cultured *in vitro*, is a valuable model system, resembling natural growth condition. The applied experimental system provides a unique possibility to study the direct effect of sugars on plant defense responses to fungal infection. Activation of the phenylpropanoid pathway, *i.e.*, the level of expression of genes encoding PAL, CHS, CHI and IFS in the response of lupine embryo axes to a pathogenic fungus, was analyzed when an endogenous pool of soluble carbohydrates was provided by the exogenous sucrose, glucose, and fructose administration and during sugar starvation. Samples were collected for analyses at 0 h and after 24, 48, 72, and 96 h of culture, following that they were frozen at −80 °C to determine the contents of genistein and other isoflavones, carbohydrate concentrations, the relative expression level of genes encoding PAL, CHS, CHI and IFS using real-time PCR, free radical concentration using EPR and β-glucosidase activity. In order to determine actin and tubulin cytoskeletons, and the superoxide anion radical level fresh samples were collected for analyses at 0 h and after 24, 48, 72 and 96 h of culture. However, actin and tubulin cytoskeletons were detected only in embryo axes cultured on the medium either with sucrose or without it.

3.1.2. Preparation of Spore Suspension and Inoculation

Fusarium oxysporum f. sp. *lupini* strain K-1018 (subsequently referred to as *F. oxysporum*) was obtained from the Collection of Plant Pathogenic Fungi, the Institute of Plant Protection, Poznań. The pathogen was incubated in the dark at 25 °C in Petri dishes (diameter 9 cm) on a potato dextrose agar (PDA) medium (pH 5.5, Sigma-Aldrich, Poznań, Poland). After 3 weeks of growth an *F. oxysporum* spore suspension was prepared. The spore suspension was obtained by washing the mycelium with sterile water and shaking with glass pearls. Then the number of spores was determined using a hemocytometer chamber (Bürker, Labart, Gdańsk, Poland). Embryo axes were inoculated with the spore suspension at a concentration of 5×10^6 spores per 1 mL. Inoculation was performed by injecting 10 μL of spore suspension into the upper part of the embryo axis shoot and additionally also by spraying the upper part of the embryo axis shoot with the inoculum.

3.2. Methods

3.2.1. Analysis of Isoflavonoids

3.2.1.1. Isolation of Phenolic Compounds

Prior to LC profiling of isoflavone glucosides, frozen plant tissue was homogenized in 80% methanol (6 $mL \cdot g^{-1} \cdot FW$) at 4 °C. After homogenization, 30 μL of luteolin standard (0.3 μmol) was added to each analyzed sample as an internal standard (LC retention time and UV spectral data did not interfere with those of the assayed compounds). Moreover, 2 mL of 80% methanol were also added to each sample. Samples were vortex-mixed and ultrasonic treatment was applied at room temperature for 30 min. Next, samples were centrifuged at 3000× *g* for 10 min. Supernatants obtained after centrifugation were evaporated at 45 °C. After evaporation of samples 1 mL of 80% methanol was added. Next samples were vortex-mixed, ultrasonic treatment was applied at room temperature for 15 min, and samples were centrifuged at 8000× *g* for 10 min.

3.2.1.2. Liquid Chromatography (LC/UV/MS)

Quantitative analyses were performed on a Waters Acquity UPLC system (Manchester, UK), equipped with a diode array detector and a Poroshell 120 RP-18 column (100 mm × 2.1 mm, 2.7 μm; Agilent Technologies, Waldbronn, Germany). The UPLC system was additionally connected to an MS detector (micrOTOFq, Bruker Daltonics, Bremen, Germany) to provide proper identification of particular isoflavones based on MS and fragmentation spectra. Quantification of total isoflavones was achieved by integration of UV chromatograms at 259 nm and normalization to the peak of an internal standard (luteolin). The concentrations of lupine free aglycones, including genistein, were expressed in arbitrary units. During LC/UV/MS analyses the elution protocol was carried out using a 20 min gradient of two solvents: A (95% H_2O, 4.5% acetonitrile, 0.5% formic acid, v/v/v) and B (95% acetonitrile, 4.5% H_2O, 0.5% formic acid, v/v/v). Elution steps were as follows: 0–5 min gradient from 10% B to 30% B, isocratic to 12 min, 12–13 min linear gradient 95% B, 13–15 min isocratic at 95% of B. Free isoflavones (genistein, 2'-hydroxygenistein, wighteone, luteone) were identified by comparing their retention times with data obtained for standards. Genistein, 2'-hydroxygenistein, wighteone and luteone were run under identical chromatographic conditions. The above mentioned standards were obtained and characterized during earlier studies in our laboratory [65].

3.2.2. Real-time RT-PCR

For the isolation of RNA, lupine embryo axes (500 mg) were frozen in liquid nitrogen and ground with a mortar and pestle in the presence of liquid nitrogen. Total RNA was isolated from 50 mg of tissue using the SV Total RNA Isolation System (Promega, Manheim, Germany), according to the supplier's recommendations [28,37]. This protocol is adapted to processing of small tissue samples. The RNA level in samples was assayed spectrophotometrically at 260 nm. The A260/A280 ratio varied from 1.8 to 2.0 according to the manufacturer's protocol. The transcript levels of target genes were analyzed by two-step quantitative RT-PCR (qRT-PCR). First-strand cDNA was synthesized from 1 μg of RNA using a High Capacity cDNA Reverse Transcription Kit (Applied Biosystems, Life Technologies Polska, Warszawa, Poland), according to the manufacturer's protocol. qRT-PCR was performed using a 7900HT Fast Real-Time PCR System (Applied Biosystems) apparatus and the Power SYBR Green Master Mix kit (Applied Biosystems) in a final volume of 10 μL containing 2 μL of three-fold diluted cDNA or digested plasmid standard dilution and 2.5 pmol of each primer. Primers for amplification were designed on the basis of the cDNA sequences encoding genes of PAL, CHS, CHI, IFS and actin from yellow lupine (Table 1). In order to minimize inaccuracies due to genomic DNA contamination, amplicons were located in plausible joining regions of exons. In assays of CHI and IFS gene expression the applied thermal cycling conditions consisted of an initial denaturation at 95 °C for 10 min followed by 50 cycles at 95 °C for 15 s, 57 °C for 20 s and 60 °C for 1 min. In assays of CHS and actin gene expression the used program consisted of an initial denaturation at 95 °C for 10 min, followed by 50 cycles at 95 °C for 15 s, 52 °C for 20 s and 65 °C for 45 s. For the PAL gene expression assay thermal cycling conditions consisted of an initial denaturation at 95 °C for 10 min, followed by 45 cycles at 95 °C for 15 s, 56 °C for 15 s and 60 °C for 40 s. The quantification analysis was performed using the standard curve method. In each assay for a specific cDNA target standard curves were prepared using six 10-fold dilutions of the linear form of plasmid-cloned specific amplicons (pre-amplified PCR products) from 100 to 10,000,000 copies. Standards, cDNA samples and the no-template control were analyzed in three replications in each assay. The specificity of products was validated by dissociation curve analyses. The results were analyzed using the SDS 2.3 software (Applied Biosystems). The expression level of target genes was normalized to the actin expression value as a constitutively expressed reference gene.

Table 1. Real-time PCR primers used for quantification of mRNA levels of different genes.

Protein Name	Primer Name	Sequence (5'-3')	Amplicon (bp)
chalcone synthase	CHS F	ATCCTGATTTCTACTTCAGA	160
	CHS R	GGTGCCATATAAGCACAAA	
phenylalanine ammonia-lyase	PAL F	ATTTAACTCTGTACCATTGCCG	132
	PAL R	GGAGAACCAAACAGGGCG	
chalcone isomerase	CHI F	AGAATCAGCTGAGAAATGATA	207
	CHI R	GAGAAGGTTGTTAGACTTGT	
isoflavone synthase	IFS F	TGGGTTGTTGATGAGCTCTG	162
	IFS R	GTTTTTCTTGATACTTTGCTTG	
actin	Actin F	TGGTCGTCCTCGTCACACT	72
	Actin R	TGTGCCTCATCCCCAACATA	

3.2.3. Extraction and Assay of β-glucosidase Activity

Frozen embryo axes (500 mg) were homogenized at 4 °C with a mortar and pestle in 3 mL of 0.05 M phosphate buffer (pH 7.0) containing 0.5% polyethylene glycol (PEG 6000). Polyclar AT (10 mg per 100 mg tissue) was added during extraction. Supernatants obtained after centrifugation (at 15,000× *g* for 20 min) were used to determine β-glucosidase (EC 3.2.1.21) activity. The activity was determined by the method of Nichols *et al.* [66]. The mixture containing 0.2 mL of extract, 0.2 mL 0.05 M phosphate buffer (pH 7.0) and 0.2 mL of 4-nitrophenyl-β-D-glucopyranoside as the substrate (2 mg·mL^{-1}) was incubated for 1 h at 30 °C. After that time 0.6 mL of 0.2 N Na_2CO_3 was added. The formation of p-nitrophenol (p-NP) was measured at 400 nm (Perkin–Elmer Lambda 11 spectrophotometer, Norwalk, CT, USA).

3.2.4. Electron Paramagnetic Resonance (EPR)

Samples of 800 mg fresh weight of embryo axes were frozen in liquid nitrogen and lyophilized in a Jouan LP3 freeze dryer (Saint-Herblain, France). The lyophilized material was transferred to EPR-type quartz tubes of 4 mm in diameter. Electron paramagnetic resonance measurements were performed at room temperature with a Bruker ELEXSYS X-band spectrometer (Rheinstettenstate, Germany). The EPR spectra were recorded as first derivatives of microwave absorption. A microwave power of 2 mW and a magnetic field modulation of about 2 G were used for all experiments to avoid signal saturation and deformation. EPR spectra of free radicals were recorded in the magnetic field range of 3,330–3,400 G and with 4096 data points. In order to determine the number of paramagnetic centers in the samples the spectra were double-integrated and compared with the intensity of the standard $Al_2O_3:Cr^{3+}$ single crystal with a known spin concentration [37,64,67–71]. Before and after the first integration some background corrections of the spectra were made to obtain a reliable absorption signal before the second integration. These corrections were necessary due to the presence of small amounts of paramagnetic Mn^{2+} ions in the examined samples. Finally, EPR intensity data were calculated per 1 g of dry sample.

3.2.5. Determination of Superoxide Anion Radical Content

Determination of superoxide anion radical ($O_2^{\bullet-}$) content in biological samples was based on its ability to reduce nitro blue tetrazolium (NBT) [72] modified by Mai *et al.* [71]. Embryo axes (500 mg) were immersed in 10 mM potassium phosphate buffer (pH 7.8) containing 0.05% NBT and 10 mM NaN_3 in a final volume of 3 mL and incubated for 1 h at room temperature. After incubation 2 ml of the reaction solution were heated at 85 °C for 15 min and rapidly cooled. The levels of $O_2^{\bullet-}$ were expressed as absorbance at 580 nm per 1 g of fresh materials. The measurements were carried out in the Perkin Elmer Lambda 15 UV-Vis spectrophotometer.

3.2.6. Carbohydrate Analysis

Extraction

Plant material (150 mg) was ground in liquid nitrogen using a 30 Hz laboratory ball mill (1 min, 2 balls per 2 mL Eppendorf tube) and flooded with 1.4 mL 80% cooled methanol (MeOH, HPLC). Next the samples were supplemented with 25 μL ribitol (1 mg/1 mL). Test tube contents were vortex-mixed in a thermomixer at 950 rpm for 10 min at room temperature, followed by centrifugation at 11.000× *g* for 10 min at 4 °C. The produced supernatant (250 μL) was transferred to Eppendorf tubes and evaporated in a speedvac at room temperature.

Derivatisation

After sample desiccation in a dessicator each sample was supplemented with 50 μL methoxyamine (20 mg/mL in dry pyridine) and vortex-mixed in a thermomixer for 1.5 h at 37 °C, afterwards it was centrifuged for 10 s (short spin). Following centrifugation the samples were supplemented with 80 μL MSTF, again vortex-mixed in a thermomixer (30 min, 37 °C) and centrifuged at 11.000× *g* for 10 min. Prepared samples were transferred to inserts at 200 μL.

GC-MS Analyses

Endogenous carbohydrate levels were determined by gas chromatography coupled with mass spectrometry (GC-MS) (6890N gas chromatograph by Agilent with a GCT Premier mass spectrometer by Waters) using a DB-5MS column (30 m × 0.25 mm × 0.25 μm, J&W Scientific, Agilent Technologies, Palo Alto, CA, USA). Gradient: 70 °C for 2 min, followed by 10 °C/min up to 300 °C (10 min). Injector 250 °C, interface 250 °C, source 250 °C, *m/z* range: 50–650, EI+, electron energy 70 eV. Mass spectra were recorded in the electron ionization mode at a potential of 70 eV. Carbohydrate content was expressed in μg per 100 mg fresh matter.

3.2.7. Detection of Actin and Tubulin Cytoskeleton

In order to detect the actin cytoskeleton, material was collected from the tested experimental variants, *i.e.*, 48 h embryo axes of yellow lupine, both those non-inoculated and those inoculated with *F. oxysporum* and cultured on a medium with sucrose or without it. The cytoskeleton was detected using the fluorescence labelled phalloidin with the use of glycerol, applying the permeation method ([73] with modifications). Cross-sections were obtained from fragments of embryo axe shoots (fresh material), 3 mm below the inoculation site. Excised samples were incubated immediately in darkness for 3 h at 14–16 °C in the staining solution containing 0.02 μM Alexa Fluor® 488-phalloidin (Invitrogen, **Life Technologies Polska, Warszawa, Poland**) and 3% (w/v) glycerol in microtubule-stabilizing buffer (MTBS: 50 mM PIPES, 5 mM EGTA and 5 mM $MgSO_4$; pH 6.9). Next sections were rinsed several times and placed in MTSB buffer on a slide and examined under an LSM 510 confocal microscope (Zeiss, Jena, Germany) with fluorescence excitation with a light beam from a krypton-argon laser (488 nm). Emission of the fluorescent dye was recorded with a filter at 500–550 nm.

The tubulin cytoskeleton was detected in cells of embryo axes in yellow lupine by immunofluorescence ([74] with modifications). Plant material, *i.e.*, 3 mm fragments of shoots of yellow lupine embryo axes below the inoculation site were fixed in 4% PFA (paraformaldehyde) in MTSB buffer. Next the material was rinsed once with a mixture containing MTSB and PBS (phosphate buffered saline) (1:1) for 5 min, afterwards rinsed with PBS buffer (twice for 5 min each) at room temperature. Next, the material was dehydrated in a series of ethanol solution. In the last rinsing of the material in 100% ethanol 0.1% toluidine blue was added and then the material was embedded in Steedman's wax. Samples were sectioned longitudinally into 12-μm-thick sections and mounted on glass slides coated with 0.1% poly-L-lysine solution. After the sections were dewaxed and rehydrated, they were incubated in 100 μL of diluted mouse monoclonal anti-α-tubulin (Sigma-Aldrich, Poznań, Poland) in PBS buffer (1:100) with 1% BSA and 5 mM sodium azide at 4 °C overnight in a humidity chamber. Samples were rinsed three times for 10 min each in PBS with 1% BSA and subsequently incubated with the secondary antibody, FITC-conjugated goat anti-mouse (Sigma) diluted in PBS (1:100), at 37 °C for 2 h in a dark humidity chamber. Samples were rinsed four times in PBS and mounted in Citifluor Antifadent AF1 (Citifluor Ltd., London, UK) under a glass coverslip. The materials were examined with an LSM 510 confocal microscope (Carl Zeiss). The samples were excited at 488 nm using a krypton-argon laser line and detected using a 505–550 nm bandpass filter.

To evaluate the microfilaments (MF) density, first the maximum intensity projections from the serial optical sections was obtained. The image was then colored by thresholding segmentation using. Red pixels were defined as areas of high intensity AlexaFluor fluorescence corresponded to the actin cables, green pixels represented areas of low intensity of fluorescence (corresponded, *i.e.*, to remaining places containing actin) and blue pixels regions without MF. Sum of red and green pixel areas represented the global amount of MF. Then were defined the AlexaFluor signal occupancy, which is an indicator of the proportion of pixel areas constituting the MFs of the total pixel areas constituting the cell region. The occupancy is defined by:

$$\text{Occupancy (\%)} = 100\,\frac{AreaMF}{AreaCell}$$

where *AreaMF* and *AreaCell* are the pixel areas constituting the MFs and cell region. The occupancy could evaluate the extent of MF depolymerization and fragmentation [43]. Projection feature of LSM510 software (Zeiss, Jena, Germany) was used for 3D image reconstruction of actin distribution in cells, while histogram display mode was used for the MF density in cells.

In the case of microtubules analyzes the segmented images were converted into binary images. Thus obtained images were analyzed with software package KS300 3.0 (Carl Zeiss Vision, Germany). Individual MTs bundles were marked and the program then determined the following parameters: fiber length and fiber width (to characterize the morphology of MTs bundles). Fiber length and fiber width are defined by:

$$\text{Fiber length} = \frac{PerimeterF + \sqrt{PerimeterF^2 - 16\,AreaF}}{4}$$

$$\text{Fiber width} = \frac{PerimeterF - \sqrt{PerimeterF^2 - 16\,AreaF}}{4}$$

where *PerimeterF* and *AreaF* are the perimeter and area of the filled region, respectively. These parameters were measured in pixels. Calculations were made on the basis of three replicates of three plants each and at least 3–4 cells of each plant (about 30 cells in total).

3.2.8. Statistical Analysis

All determinations were performed in three independent experiments. Data shown are means of triplicates for each treatment; standard deviation was calculated and its range is shown in figures. The analysis of variance (ANOVA) was applied and results were compared in order to verify whether means from independent experiments within a given experimental variant were significantly different. Analysis of variance between treatment means was also carried out. The effects of two factors, *i.e.*, sugar (sucrose or glucose or fructose) and the pathogenic fungus (*F. oxysporum*), were investigated in the experiments.

Statistical significance of differences between average values for each pairs of indicators of physiological and biochemical analyses were carried out as appropriate elementary contrasts. Statistically significant differences (*p*-value) was set at $p < 0.05$ (Supplementary Materials). All calculations in the range of statistical analysis was performed using statistic packet Genstat 15.

4. Conclusions

In summary, this is the first study showing a stimulating effect of monosaccharides, glucose and fructose, on the mechanism regulating synthesis and accumulation of genistein and other isoflavones free aglycones, β-glucosidase activity, generation of free radicals in the case of infection by a pathogenic fungus. Furthermore, our findings are the first, which revealed the effect of infection on the organization of actin and tubulin cytoskeletons in terms of varied sucrose levels.

Supplementary Materials

Supplementary materials can be accessed at: http://www.mdpi.com/1420-3049/19/9/13392/s1.

Statistical significance of differences between the average values of each pairs for accumulation of genistein and other isoflavones. Statistically significant differences (*p*-value) was assumed at $p < 0.05$.

Genistein

Time (h)	Contrast	Value of Contrast	*p*-Value	Time (h)	Value of Contrast	*p*-Value
0	0n:0i	−0.0473	0.033			
	+Sn:+Si	0.1033	0.101		−0.1270	0.106
	+Gn:+Gi	−0.2893	0.010		−0.2417	0.012
	+Fn:+Fi	0.0486	0.395		−0.6283	<0.001
24	−Sn:−Si	−0.0706	0.047	72	−0.2017	0.009
	+Si:−Si	0.0330	0.289		0.2020	0.026
	+Gi:−Si	0.2640	0.014		0.1110	0.150
	+Fi:−Si	0.00006	0.999		0.4007	0.002
	+Sn:+Si	−0.1833	0.031		−1.478	<0.001
	+Gn:+Gi	−0.3913	<0.001		−0.9520	<0.001
	+Fn:+Fi	−0.2333	0.029		−1.103	0.019
48	−Sn:−Si	−0.6420	<0.001	96	−0.1153	0.172
	+Si:−Si	−0.4627	0.005		1.490	<0.001
	+Gi:−Si	−0.1460	0.086		0.8577	0.002
	+Fi:−Si	−0.3877	0.014		0.9517	0.004

2'-Hydroxygenistein

Time (h)	Contrast	Value of Contrast	*p*-Value	Time (h)	Value of Contrast	*p*-Value
0	0n:0i	0.0533	0.022			
	+Sn:+Si	0.0120	0.840		0.9657	<0.001
	+Gn:+Gi	−0.2523	<0.001		1.749	<0.001
	+Fn:+Fi	0.3080	<0.001		1.282	<0.001
24	−Sn:−Si	−0.0663	0.163	72	0.0133	0.106
	+Si:−Si	0.1370	0.049		0.2887	<0.001
	+Gi:−Si	0.2147	0.006		0.1607	0.044
	+Fi:−Si	−0.0043	0.924		0.0433	0.034
	+Sn:+Si	1.165	<0.001		1.543	<0.001
	+Gn:+Gi	1.325	0.007		0.5640	<0.001
	+Fn:+Fi	1.219	0.003		0.9840	<0.001
48	−Sn:−Si	0.0833	<0.001	96	0.0986	<0.001
	+Si:−Si	0.2460	0.013		0.2210	0.013
	+Gi:−Si	0.1010	<0.001		0.2590	<0.001
	+Fi:−Si	0.0860	<0.001		0.1540	0.031

Wighteon

Time (h)	Contrast	Value of Contrast	*p*-Value	Time (h)	Value of Contrast	*p*-Value
0	0n:0i	−0.0272	0.038			
	+Sn:+Si	−0.3823	<0.001		1.121	<0.001
	+Gn:+Gi	−0.2583	<0.001		1.255	<0.001
	+Fn:+Fi	−0.4880	<0.001		−0.4727	<0.001
24	-Sn:−Si	−0.2910	<0.001	**72**	0.0103	0.691
	+Si:−Si	0.1947	0.002		0.5460	<0.001
	+Gi:−Si	−0.0736	0.004		0.5170	<0.001
	+Fi:−Si	0.2277	0.018		0.7720	<0.001
	+Sn:+Si	−0.7563	<0.001		0.7067	<0.001
	+Gn:+Gi	−1.298	<0.001		1.684	<0.001
	+Fn:+Fi	−0.0066	0.678		1.315	<0.001
48	−Sn:−Si	0.0413	0.419	**96**	0.1980	0.006
	+Si:−Si	0.7147	<0.001		0.3933	0.003
	+Gi:−Si	1.323	<0.001		0.2870	0.001
	+Fi:−Si	0.3567	<0.001		0.1707	0.002

Luteon

Time (h)	Contrast	Value of Contrast	*p*-Value	Time (h)	Value of Contrast	*p*-Value
0	0n:0i	0.0310	0.041			
	+Sn:+Si	−0.2467	<0.001		−0.4324	<0.001
	+Gn:+Gi	−0.5930	<0.001		−0.5050	0.001
	+Fn:+Fi	−0.4550	<0.001		−1.446	<0.001
24	−Sn:−Si	−0.3887	<0.001	**72**	−0.4323	<0.001
	+Si:−Si	0.0996	0.009		0.4427	0.002
	+Gi:−Si	0.1103	0.007		0.6590	<0.001
	+Fi:−Si	0.1303	0.010		1.126	<0.001
	+Sn:+Si	−0.8723	<0.001		−1.290	<0.001
	+Gn:+Gi	−1.701	<0.001		−0.6073	<0.001
	+Fn:+Fi	−0.5290	<0.001		−0.2763	0.001
48	−Sn:−Si	−0.2123	0.001	**96**	0.5250	<0.001
	+Si:−Si	0.7347	<0.001		1.794	<0.001
	+Gi:−Si	1.441	<0.001		1.139	<0.001
	+Fi:−Si	0.4437	<0.001		0.8910	<0.001

Statistical significance of differences between the average values of each pairs for expression levels of isoflavone biosynthetic pathway genes. Statistically significant differences (*p*-value) was assumed at $p < 0.05$.

PAL

Time (h)	Contrast	Value of Contrast	p-Value	Time (h)	Value of Contrast	p-Value
0	0n:0i	−0.064	0.108			
	+Sn:+Si	0.1590	0.028		−1.475	<0.001
	+Gn:+Gi	−0.3607	<0.001		−1.552	<0.001
	+Fn:+Fi	−0.1913	0.001		−0.786	<0.001
24	−Sn:−Si	−0.9157	<0.001	**72**	−2.231	<0.001
	+Si:−Si	−0.4617	<0.001		−0.694	<0.001
	+Gi:−Si	0.0106	0.757		−0.354	<0.001
	+Fi:−Si	−0.3613	<0.001		−1.249	<0.001
	+Sn:+Si	−0.6617	<0.001		−1.869	<0.001
	+Gn:+Gi	−0.6737	<0.001		−0.824	<0.001
	+Fn:+Fi	−0.3940	<0.001		−4.280	<0.001
48	−Sn:−Si	−0.6370	<0.001	**96**	0.517	0.002
	+Si:−Si	0.3623	<0.001		2.561	<0.001
	+Gi:−Si	0.4097	<0.001		2.209	0.002
	+Fi:−Si	0.2067	0.004		5.033	<0.001

CHS

Time (h)	Contrast	Value of Contrast	p-Value	Time (h)	Value of Contrast	p-Value
0	0n:0i	−1.116	<0.001			
	+Sn:+Si	−1.652	<0.001		−4.289	<0.001
	+Gn:+Gi	−2.040	<0.001		−3.911	<0.001
	+Fn:+Fi	−2.240	<0.001		−3.029	<0.001
24	−Sn:−Si	−2.549	<0.001	**72**	−8.179	<0.001
	+Si:−Si	0.068	0.699		−3.405	<0.001
	+Gi:−Si	0.339	0.168		−3.497	<0.001
	+Fi:−Si	0.325	0.198		−4.761	<0.001
	+Sn:+Si	−2.176	<0.001		−7.276	<0.001
	+Gn:+Gi	−2.202	<0.001		−4.210	<0.001
	+Fn:+Fi	−2.953	<0.001		−8.054	<0.001
48	−Sn:−Si	−2.985	<0.001	**96**	0.129	0.004
	+Si:−Si	0.367	0.027		7.798	<0.001
	+Gi:−Si	−0.150	0.241		5.29	<0.001
	+Fi:−Si	0.584	0.01		8.679	<0.001

CHI

Time (h)	Contrast	Value of Contrast	*p*-Value	Time (h)	Value of Contrast	*p*-Value
0	0n:0i	−0.188	0.003			
	+Sn:+Si	−0.580	0.04		−1.024	<0.001
	+Gn:+Gi	−0.536	0.006		−1.266	<0.001
	+Fn:+Fi	−0.574	0.002		−0.568	0.004
24	−Sn:−Si	−0.962	<0.001	**72**	−1.572	<0.001
	+Si:−Si	−0.070	0.612		−0.231	0.126
	+Gi:−Si	−0.007	0.948		0.001	0.995
	+Fi:−Si	−0.026	0.614		−0.654	0.003
	+Sn:+Si	−0.406	0.004		−1.328	<0.001
	+Gn:+Gi	−0.650	0.009		−0.717	0.001
	+Fn:+Fi	−1.124	<0.001		−2.276	<0.001
48	−Sn:−Si	−0.688	<0.001	**96**	0.629	0.011
	+Si:−Si	0.362	0.013		2.179	<0.001
	+Gi:−Si	0.306	0.008		1.889	<0.001
	+Fi:−Si	0.574	<0.001		3.328	<0.001

IFS

Time (h)	Contrast	Value of Contrast	*p*-Value	Time (h)	Value of Contrast	*p*-Value
0	0n:0i	1.916	<0.001			
	+Sn:+Si	−1.087	<0.001		−2.685	<0.001
	+Gn:+Gi	−1.706	<0.001		−3.310	<0.001
	+Fn:+Fi	−1.483	<0.001		−1.539	<0.001
24	-Sn:−Si	−1.897	<0.001	**72**	−4.512	<0.001
	+Si:−Si	−0.110	0.419		−1.314	0.004
	+Gi:−Si	0.372	0.035		−0.522	0.05
	+Fi:−Si	−0.001	0.993		−2.381	<0.001
	+Sn:+Si	−1.102	<0.001		−4.453	<0.001
	+Gn:+Gi	−1.772	<0.001		−2.625	<0.001
	+Fn:+Fi	−1.806	<0.001		−5.880	<0.001
48	-Sn:−Si	−1.984	<0.001	**96**	0.117	0.006
	+Si:−Si	0.002	0.981		5.025	<0.001
	+Gi:−Si	0.284	0.058		3.757	0.002
	+Fi:−Si	0.255	0.02		6.755	<0.001

Statistical significance of differences between the average values of each pairs for endogenous level of sugars. Statistically significant differences (*p*-value) was assumed at $p < 0.05$.

Sucrose

Time (h)	Contrast	Value of Contrast	*p*-Value	Time (h)	Value of Contrast	*p*-Value
0	0n:0i	−651.0	0.020			
	+Sn:+Si	−84.47	0.781		2010	0.010
	+Gn:+Gi	802.3	0.005		2123	<0.001
	+Fn:+Fi	90.20	0.556		1635	<0.001
24	−Sn:−Si	272.5	0.015	**72**	53.10	0.008
	+Si:−Si	1385	0.006		398.4	<0.001
	+Gi:−Si	962.4	0.002		278.7	0.002
	+Fi:−Si	1527	<0.001		457.6	<0.001
	+Sn:+Si	607.7	0.005		2141	0.009
	+Gn:+Gi	176	0.359		1255	0.015
	+Fn:+Fi	202.6	0.112		1761	<0.001
48	−Sn:−Si	216.9	0.007	**96**	10.07	0.562
	+Si:−Si	1699	0.003		242.9	0.017
	+Gi:−Si	1316	0.003		28.27	<0.001
	+Fi:−Si	1862	<0.001		12.57	0.008

Glucose

Time (h)	Contrast	Value of Contrast	*p*-Value	Time (h)	Value of Contrast	*p*-Value
0	0n:0i	−53.53	<0.001			
	+Sn:+Si	−37.73	0.208		1668	0.007
	+Gn:+Gi	428.3	0.012		842.2	0.010
	+Fn:+Fi	77.27	0.104		1284	0.023
24	-Sn:−Si	−5.833	0.468	**72**	15.70	<0.001
	+Si:−Si	200.8	0.001		309.1	0.002
	+Gi:−Si	328.0	0.014		644.9	0.016
	+Fi:−Si	203.1	<0.001		168.6	0.002
	+Sn:+Si	196.0	0.011		1380	<0.001
	+Gn:+Gi	10.27	0.845		502.8	0.021
	+Fn:+Fi	−351.5	0.029		1659	0.013
48	−Sn:−Si	−2.133	0.170	**96**	16.50	0.003
	+Si:−Si	498.7	0.003		160.7	0.014
	+Gi:−Si	542.3	0.005		912.7	<0.001
	+Fi:−Si	799.4	0.015		62.90	<0.001

Fructose

Time (h)	Contrast	Value of Contrast	*p*-Value	Time (h)	Value of Contrast	*p*-Value
0	0n:0i	−391.4	0.003			
	+Sn:+Si	−626.5	0.006		2566	0.003
	+Gn:+Gi	953.1	<0.001		2078	0.020
	+Fn:+Fi	1509	<0.001		1703	0.001
24	−Sn:−Si	164.5	0.020	72	11.43	0.046
	+Si:−Si	1092	<0.001		981.5	0.008
	+Gi:−Si	387.9	0.003		347.1	0.015
	+Fi:−Si	−14.53	0.514		1654	<0.001
	+Sn:+Si	256.8	0.002		2417	<0.001
	+Gn:+Gi	−210.4	0.218		2422	0.017
	+Fn:+Fi	−915.0	0.001		−1804	0.011
48	-Sn:−Si	20.60	0.005	96	18.07	<0.001
	+Si:−Si	1665	<0.001		247.7	0.016
	+Gi:−Si	1087	0.011		29.37	<0.001
	+Fi:−Si	2662	0.001		4688	0.003

Statistical significance of differences between the average values of each pairs for β-glucosidase activity. Statistically significant differences (*p*-value) was assumed at $p < 0.05$.

β-Glucosidase

Time (h)	Contrast	Value of Contrast	*p*-Value	Time (h)	Value of Contrast	*p*-Value
0	0n:0i	−0.1550	0.049			
	+Sn:+Si	0.2533	0.180		−1.393	0.177
	+Gn:+Gi	−0.4433	0.201		−4.530	0.003
	+Fn:+Fi	0.7667	0.090		−1.773	0.047
24	−Sn:−Si	−2.400	0.008	72	−8.940	<0.001
	+Si:−Si	0.9267	0.060		−3.770	0.016
	+Gi:−Si	−0.4800	0.315		−2.087	0.088
	+Fi:−Si	−0.7633	0.170		−7.280	<0.001
	+Sn:+Si	−1.350	0.021		−7.613	<0.001
	+Gn:+Gi	−2.113	0.008		−4.163	0.009
	+Fn:+Fi	−1.217	0.155		−2.823	0.036
48	-Sn:−Si	−2.800	0.001	96	−10.44	<0.001
	+Si:−Si	−2.140	0.006		−3.617	0.007
	+Gi:−Si	−1.577	0.017		−7.020	<0.001
	+Fi:−Si	−2.577	0.009		−9.230	<0.001

Statistical significance of differences between the average values of each pairs for superoxide anion radical generation. Statistically significant differences (*p*-value) was assumed at $p < 0.05$.

Superoxide Anion Radical

Time (h)	Contrast	Value of Contrast	*p*-Value	Time (h)	Value of Contrast	*p*-Value
0	0n:0i	1.550	<0.001			
	+Sn:+Si	−2.500	<0.001		−3.200	<0.001
	+Gn:+Gi	−4.300	<0.001		−2.133	<0.001
	+Fn:+Fi	−0.100	0.468		−4.883	<0.001
24	−Sn:−Si	−1.300	0.002	**72**	−4.950	<0.001
	+Si:−Si	2.700	<0.001		−1.450	0.002
	+Gi:−Si	3.733	<0.001		−1.250	0.003
	+Fi:−Si	0.700	0.009		0.5000	0.041
	+Sn:+Si	−3.750	<0.001		−2.967	<0.001
	+Gn:+Gi	−4.467	<0.001		−2.217	<0.001
	+Fn:+Fi	−5.100	<0.001		−2.350	<0.001
48	−Sn:−Si	−4.750	<0.001	**96**	−2.750	<0.001
	+Si:−Si	−0.5000	0.041		0.5667	0.038
	+Gi:−Si	−0.1833	0.212		−0.4333	0.050
	+Fi:−Si	0.4167	0.048		−0.2000	0.335

Statistical significance of differences between the average values of each pairs for semiquinone radical concentration. Statistically significant differences (*p*-value) was assumed at $p < 0.05$.

Concentration of Free Radicals

Time (h)	Contrast	Value of Contrast	*p*-Value	Time (h)	Value of Contrast	*p*-Wartość
0	0n:0i	0.025				
	+Sn:+Si	−0.3867	0.006		−1.310	<0.001
	+Gn:+Gi	−0.3867	<0.001		−1.660	<0.001
	+Fn:+Fi	−0.2067	0.232		−1.057	<0.001
24	−Sn:−Si	0.0866	0.427	**72**	−2.437	0.021
	+Si:−Si	0.1167	0.354		−0.5600	0.263
	+Gi:−Si	0.2767	0.051		−0.0566	0.890
	+Fi:−Si	0.1967	0.315		−0.8900	0.133
	+Sn:+Si	−0.1000	0.621		-3.507	0.012
	+Gn:+Gi	−0.0333	0.653		-2.744	<0.001
	+Fn:+Fi	−0.0200	0.895		-2.490	0.006
48	−Sn:−Si	−0.7000	0.051	**96**	-5.950	0.010
	+Si:−Si	−0.1300	0.623		-2.027	0.051
	+Gi:−Si	−0.1900	0.369		2.620	0.014
	+Fi:−Si	−0.4333	0.111		-3.023	0.017

Acknowledgments

This study was supported by the Polish Ministry of Science and Higher Education (MNiSW, grant no. N N303 414437). MF is scholarship holder founded within the project from Sub-measure 8.2.2 Human Capital Operational Program, co-financed by European Union Fund.

Author Contributions

I.M.: Conceived and designed the experiments, wrote the paper. S.S.: wrote the chapter concerning actin and tubulin cytoskeleton. M.F. and S.S. prepared figures. M.F., I.M., S.S., Ł.M., W.N., D.N., W.B., A.K.-M.: performed experiments. I.M., M.F. and S.S.: analyzed the data.

Abbreviations

+Sn; embryo axes non-inoculated and cultured *in vitro* on Heller's medium supplemented with 60 mM sucrose; +Gn; embryo axes non-inoculated and cultured *in vitro* on Heller's medium supplemented with 120 mM glucose; +Fn; embryo axes non-inoculated and cultured *in vitro* on Heller's medium supplemented with 120 mM fructose; −Sn; non-inoculated cultured without sucrose; +Si; inoculated and cultured with 60 mM sucrose; +Gi; inoculated and cultured with 120 mM glucose; +Fi; inoculated and cultured with 120 mM fructose; −Si; inoculated and cultured without sucrose; CHI; chalcone isomerase; CHS; chalcone synthase; EPR; electron paramagnetic resonance; IFS; isoflavone synthase; PAL; phenylalanine ammonia-lyase.

Conflicts of Interest

The authors declare no conflict of interest

References

1. Dixon, R.A.; Ferreira, D. Molecules of interest: Genistein. *Phytochemistry* **2002**, *60*, 205–211.
2. Rusin, A.; Krawczyk, Z.; Grynkiewicz, G.; Gogler, A.; Zawisza-Puchalka, J.; Szeja, W. Synthetic derivatives of genistein, their properties and possible applications. *Acta Biochim. Pol.* **2010**, *57*, 23–34.
3. Maćkowiak, P.; Nogowski, L.; Nowak, K.W. Effect of isoflavone genistein on insulin receptors in perfused liver of ovariectomized rats. *J. Recept. Signal. Transduct. Res.* **1999**, *19*, 283–292.
4. Szkudelski, T.; Nogowski, L.; Pruszyńska-Oszmałek, E.; Kaczmarek, P.; Szkudelska, K. Genistein restricts leptin secretion from rat adipocytes. *J. Steroid Biochem. Mol. Biol.* **2005**, *96*, 301–307.
5. Andersen, O.M.; Markham, K.R. *Flavonoids: Chemistry, Biochemistry, and Applications*; CRC Taylor & Francis: Boca Raton, FL, USA, 2006.
6. Graham, T.L.; Graham, M.Y. Defense potentiation and elicitation competency: Redox conditioning effects of salicylic acid and genistein. In *Plant Microbe Interactions*; Stacey, G., Keen, N.T., Eds.; APS Press: St. Paul, MN, USA, 2000; Volume 5, pp. 181–220.

7. Groβkinsky, D.K.; van der Graaff, E.; Roitsch, T. Phytoalexin transgenics in crop protection-Fairy tale with a happy end? *Trends Plant Sci.* **2012**, *195*, 54–70.
8. Ribera, A.E.; Zuñiga, G. Induced plant secondary metabolites for phytopatogenic fungi control: A review. *J. Soil Sci. Plant Nutr.* **2012**, *12*, 893–911.
9. Morkunas, I.; Marczak, Ł.; Stachowiak, J.; Stobiecki, M. Sucrose-stimulated accumulation of isoflavonoids as a defense response of lupine to *Fusarium oxysporum*. *Plant Physiol. Biochem.* **2005**, *43*, 363–373.
10. Kosslak, R.M.; Rookland, R.; Barkei, J.; Paaren, H.E.; Appelbaum, E.R. Induction of *Bradyrhizobium japonicum* common nod genes by isoflavones isolated from *Glycine max*. *Proc. Natl. Acad. Sci. USA* **1987**, *84*, 7428–7432.
11. Dolatabadian, A.; Sanavy, S.A.M.M.; Ghanati, F.; Gresshoff, P.M. Morphological and physiological response of soybean treated with the microsymbiont *Bradyrhizobium japonicum* pre-incubated with genistein. *S. Afr. J. Bot.* **2012**, *79*, 9–18.
12. Smith, D.A.; Banks, S. Biosynthesis, elicitation and biological activity of isoflavonoid phytoalexins (review). *Phytochemistry* **1986**, *25*, 979–995.
13. Kneer, R.; Poulev, A.A.; Olesinski, A.; Raskin, I. Characterization of the elicitor-induced biosynthesis and secretion of genistein from roots of *Lupinus luteus* L. *J. Exp. Bot.* **1999**, *50*, 1553–1559.
14. Morkunas, I.; Stobiecki, M.; Marczak, Ł.; Stachowiak, J.; Narożna, D.; Remlein-Starosta, D. Changes in carbohydrate and isoflavonoid metabolism in yellow lupine in response to infection by *Fusarium oxysporum* during the stages of seed germination and early seedling growth. *Physiol. Mol. Plant Pathol.* **2010**, *75*, 46–55.
15. Bednarek, P.; Kerhoas, L.; Einhorn, J.; Frański, R.; Wojtaszek, P.; Rybus-Zajac, M.; Stobiecki, M. Profiling of flavonoid conjugates in *Lupinus albus* and *Lupinus angustifolius* responding to biotic and abiotic stimuli. *J. Chem. Ecol.* **2003**, *29*, 1127–1142.
16. Zielonka, J.; Gębicki, J.; Grynkiewicz, G. Radical scavenging properties of genistein. *Free Radic. Biol. Med.* **2003**, *35*, 958–965.
17. McPartland, J.M.; Guy, G.W.; di Marzo, V. Care and feeding of the endocannabinoid system: A systematic review of potential clinical interventions that upregulate the endocannabinoid system. *PLoS One* **2014**, *9*, e89566.
18. Harborne, J.B. *Flavonoids: Advances in Research since 1986*; Harborne, J.B., Ed.; Chapman & Hall: London, UK, 1994; p. 152.
19. Hoffmann, D. *Medical herbalism: The Science and Practice of Herbal Medicine*; Healing Arts Press: Rochester, VT, USA; 2003; p. 106.
20. Deavours, B.E.; Dixon, R.A. Metabolic engineering of isoflavonoid biosynthesis in alfalfa. *Plant Physiol.* **2005**, *138*, 2245–2259.
21. Kessmann, H.; Choudhary, A.D.; Dixon, R.A. Stress responses in alfalfa (*Medicago sativa* L.) III. Induction of medicarpin and cytochrome P450 enzyme activities in elicitor-treated cell suspension cultures and protoplasts. *Plant Cell Rep.* **1990**, *9*, 38–41.
22. Dixon, R.A. Phytoestrogens. *Annu. Rev. Plant Biol.* **2004**, *55*, 225–261.

23. Jeandet, P.; Clément, C.; Courot, E.; Cordelier, S. Modulation of phytoalexin biosynthesis in engineered plants for disease resistance. *Int. J. Mol. Sci.* **2013**, *14*, 14136–14170.
24. Shinde, A.N.; Malpathak, N.; Fulzele D.P. Impact of nutrient components on production of the phytoestrogens daidzein and genistein by hairy roots of *Psoralea corylifolia. J. Nat. Med.* **2010**, *64*, 346–353.
25. Baque, M.A.; Elgirban, A.; Lee, E.J.; Paek, K.Y. Sucrose regulated enhanced induction of anthraquinone, phenolics, flavonoids biosynthesis and activities of antioxidant enzymes in adventitious root suspension cultures of *Morinda citrifolia* (L.). *Acta Physiol. Plant.* **2012**, *34*, 405–415.
26. Hara, M.; Oki, K.; Hoshino, K.; Kuboi, T. Enhancement of anthocyanin biosynthesis by sugar in radish (*Raphanus sativus*) hypocotyls. *Plant Sci.* **2003**, *164*, 259–265.
27. Solfanelli, C.; Poggi, A.; Loreti, E.; Alpi, A.; Perata, P. Sucrose-specific induction of the anthocyanin biosynthetic pathway in *Arabidopsis*. *Plant Physiol.* **2006**, *140*, 637–646.
28. Morkunas, I.; Narożna, D.; Nowak, W.; Samardakiewicz, S.; Remlein-Starosta, D. Cross-talk interactions of sucrose and *Fusarium oxysporum* in the phenylpropanoid pathway and the accumulation and localization of flavonoids in embryo axes of yellow lupine. *J. Plant Physiol.* **2011**, *168*, 424–433.
29. Li, Y.; van den Ende, W.; Rolland, F. Sucrose induction of anthocyanin biosynthesis is mediated by DELLA. *Mol. Plant* **2014**, *7*, 570–572.
30. Rolland, F.; Baena-Gonzalez, E.; Sheen, J. Sugar sensing and signaling in plants: Conserved and novel mechanisms. *Annu. Rev. Plant Biol.* **2006**, *57*, 675–709.
31. Bolouri-Moghaddam, M.R.; Le Roy, K.; Xiang, L.; Rolland, F.; van den Ende, W. Sugar signalling and antioxidant network connections in plant cells. *FEBS J.* **2010**, *277*, 2022–2037.
32. Bolouri Moghaddam, M.R.; van den Ende, W. Sugars and plant innate immunity. *J. Exp. Bot.* **2012**, *63*, 3989–3998.
33. Shinde, A.N.; Malpathak, N.; Fulzele, D.P. Studied enhancement strategies for phytoestrogens production in shake flasks by suspension culture of *Psoralea corylifolia. Bioresour. Technol.* **2009**, *100*, 1833–1839.
34. Fang, C.B.; Li, H.Q.; Wan, X.C.; Jiang, C.J. Effect of several physiochemical factors on cell growth and isoflavone accumulation of *Pueraria lobata* cell suspension culture. *China J. Chin. Mater. Med.* **2006**, *31*, 1580–1583.
35. Pasqua, G.; Monacelli, B.; Mulinacci, N.; Rinaldi, S.; Giaccherini, C.; Innocenti, M.; Vinceri, F.F. The effect of growth regulators and sucrose on anthocyanin production in *Camptotheca acuminatacell* cultures. *Plant Physiol. Biochem.* **2005**, *43*, 293–298.
36. Teng, S.; Keurentjes, J.; Bentsink, L.; Koornneef, M.; Smeekens, S. Sucrose-specific induction of anthocyanin biosynthesis in *Arabidopsis* requires the MYB75/PAP1 gene. *Plant Physiol.* **2005**, *139*, 1840–1852.
37. Morkunas, I.; Formela, M.; Floryszak-Wieczorek, J.; Marczak, Ł.; Narożna, D.; Nowak, W.; Bednarski, W. Cross-talk interactions of exogenous nitric oxide and sucrose modulates phenylpropanoid metabolism in yellow lupine embryo axes infected with *Fusarium oxysporum. Plant Sci.* **2013**, *211*, 102–121.

38. Wojakowska, A.; Muth, D.; Narożna, D.; Mądrzak, C.; Stobiecki, M.; Kachlicki, P. Changes of phenolic secondary metabolite profiles in the reaction of narrow leaf lupin (*Lupinus angustifolius*) plants to infections with *Colletotrichum lupini* fungus or treatment with its toxin. *Metabolomics.* **2013**, *9*, 575–589.
39. Muth, D.; Kachlicki, P.; Krajewski, P.; Przystalski, M.; Stobiecki, M. Differential metabolic response of narrow leafed lupine (*Lupinus angustifolius*) leaves to infection with *Colletotrichum lupini*. *Metabolomics* **2009**, *5*, 354–362.
40. Dixon, R.A. Natural products and plant disease resistance. *Nature* **2001**, *411*, 843–847.
41. Grayer, R.J.; Kokubun, T. Plant-fungal interactions: The search for phytoalexins and other antifungal compounds from higher plants. *Phytochemistry* **2001**, *56*, 253–263.
42. Morkunas, I.; Kozłowska, M.; Ratajczak, L.; Marczak, Ł. Role of sucrose in the development of Fusarium wilt in lupine embryo axes. *Physiol. Mol. Plant Pathol.* **2007**, *70*, 25–37.
43. Higaki, T.; Kutsuna, N.; Sano, T.; Kondo, N.; Hasezawa, S. Quantification and cluster analysis of actin cytoskeletal structures in plant cells: Role of actin bundling in stomatal movement during diurnal cycles in *Arabidopsis* guard cells. *Plant J.* **2010**, *61*, 156–165.
44. Hardham, A.R. Microtubules and biotic interactions. *Plant J.* **2013**, *75*, 278–289.
45. Nick, P. Microtubules and the tax payer. *Protoplasma* **2012**, *249*, 81–94.
46. Samardakiewicz, S.; Krzesłowska, M.; Woźny, A. Tubulin cytoskeleton in plant cell response to trace metals. In *Compartmentation of Responses to Stresses in Higher Plants, True or False*; Maksymiec, R., Ed.; Transword Research Network: Trivandrum, India, 2009; pp. 149–162.
47. Staiger, C.J.; Sheahan, M.B.; Khurana, P.; Wang, X.; McCurdy, D.W.; Blanchoin, L. Actin filament dynamics are dominated by rapid growth and severing activity in the *Arabidopsis* cortical array. *J. Cell Biol.* **2009**, *184*, 269–280.
48. Yuan, H.Y.; Yao, L.L.; Jia, Z.Q.; Li, Y.; Li, Y.Z. Verticillium dahliae toxin induced alterations of cytoskeletons and nucleoli in *Arabidopsis thaliana* suspension cells. *Protoplasma* **2006**, *229*, 75–82.
49. Henty-Ridilla, J.L.; Shimono, M.; Li, J.; Chang, J.H.; Day, B.; Staiger, C.J. The plant actin cytoskeleton responds to signals from microbe-associated molecular patterns. *PLoS Pathog.* **2013**, *9*, e1003290.
50. Wojtaszek, P. Cytoszkielet. In *Biologia Komórki Roślinnej. Struktura*; Wojtaszek, P., Woźny, A., Ratajczak, L., Eds.; Wydawnictwo Naukowe PWN: Warszawa, Poland, 2006; Volume 1, pp. 194–226.
51. Nick, P. Mechanics of the cytoskeleton. In *Mechanical Integration of Plant Cells and Plants*; Wojaszek, P., Ed.; Springer: Berlin, Germany, 2011; pp. 53–90.
52. Hardham, A.R.; Jones, D.A.; Takemoto, D. Cytoskeleton and cell wall function in penetration resistance. *Curr. Opin. Plant Biol.* **2007**, *10*, 342–348.
53. Delorme-Axford, E.; Coyne, C.B. The actin cytoskeleton as a barrier to virus infection of polarized epithelial cells. *Viruses* **2011**, *3*, 2462–2477.
54. Navarro-Garcia, F.; Serapio-Palacios, A.; Ugalde-Silva, P.; Tapia-Pastrana, G.; Chavez-Dueñas, L. Actin cytoskeleton manipulation by effector proteins secreted by diarrheagenic *Escherichia coli* pathotypes. *BioMed Res. Int.* **2013**, *2013*, doi:10.1155/2013/374395.

55. Smertenko, A.; Franklin-Tong, V.E. Organisation and regulation of the cytoskeleton in plant programmed cell death. *Cell. Death Differ.* **2011**, *18*, 1263–1270.
56. Zadworny, M.; Guzicka, M.; Łakomy, P.; Samardakiewicz, S.; Smoliński, D.J.; Mucha, J. Analysis of microtubule and microfilament distribution in *Pinus sylvestris* roots following infection by *Heterobasidion species. For. Pathol.* **2013**, *43*, 222–231.
57. Wojtaszek, P.; Baluška, F.; Kasprowicz, A.; Łuczak, M.; Volkmann, D. Domain-specific mechanosensory transmission of osmotic and enzymatic cell wall disturbances to the actin cytoskeleton. *Protoplasma* **2007**, *230*, 217–230.
58. Banaś, A.K.; Krzeszowiec, W.; Dobrucki, J.; Gabryś, H. Mannose, but not glucose or sucrose, disturbs actin cytoskeleton in *Arabidopsis thaliana* leaves. *Acta Physiol. Plant.* **2010**, *32*, 773–779.
59. Balasubramanian, R.; Karve, A; Kandasamy, M.; Meagher, R.B.; Moore, B. A role for F-actin in hexokinase-mediated glucose signaling. *Plant Physiol.* **2007**, *145*, 1423–1434.
60. Winter, H.; Huber, J.L.; Huber, S.C. Identification of sucrose synthase as an actin-binding protein. *FEBS Lett.* **1998**, *430*, 205–208.
61. Kasprowicz, A.; Szuba, A.; Volkmann, D.; Baluška, F.; Wojtaszek, P. Nitric oxide modulates dynamic actin cytoskeleton and vesicle trafficking in a cell type-specific manner in root apices. *J. Exp. Bot.* **2009**, *60*, 1605–1617.
62. Heller, R. Recherches sur la nutrition minerale des tissues vegetaux cultives *in vitro. Ann. Sci. Nat. Bot. Biol. Veg.* **1954**, *14*, 1–223.
63. Morkunas, I.; Gmerek, J. The possible involvement of peroxidase in defense of yellow lupine embryo axes against *Fusarium oxysporum. J. Plant Physiol.* **2007**, *164*, 185–194.
64. Morkunas, I.; Bednarski, W. *Fusarium oxysporum* induced oxidative stress and antioxidative defenses of yellow lupine embryo axes with different level of sugars. *J. Plant Physiol.* **2008**, *165*, 262–277.
65. Frański, R.; Bednarek, P.; Wojtaszek, P.; Stobiecki, M. Identification of flavonoid diglycosides in yellow lupin (*Lupinus luteus* L.) with mass spectrometric techniques. *J. Mass Spectrom.* **1999**, *344*, 486–495.
66. Nichols, E.J.; Beckman, J.M.; Hadwiger, L.A. Glycosidic enzyme activity in pea tissue and pea-*Fusarium solani* interaction. *Plant Physiol.* **1980**, *66*, 199–204.
67. Morkunas, I.; Garnczarska, M.; Bednarski, W.; Ratajczak, W.; Waplak, S. Metabolic and ultrastructural responses of lupine embryo axes to sugar starvation. *J. Plant Physiol.* **2003**, *160*, 311–319.
68. Morkunas, I.; Bednarski, W.; Kozłowska, M. Response of embryo axes of germinating seeds of yellow lupine to *Fusarium oxysporum. Plant Physiol. Biochem.* **2004**, *42*, 493–499.
69. Morkunas, I.; Bednarski, W.; Kopyra, M. Defense strategies of pea embryo axes with different levels of sucrose to *Fusarium oxysporum* and *Ascochyta pisi. Physiol. Mol. Plant Pathol.* **2008**, *72*, 167–178.
70. Bednarski, W.; Ostrowski, A.; Waplak, S. Low temperature short-range ordering caused by Mn^{2+}doping of $Rb_3H(SO_4)_2$. *J. Phys. Condens. Matter* **2010**, *22*, 225901.

71. Mai, V.C.; Bednarski, W.; Borowiak-Sobkowiak, B.; Wilkaniec, B.; Samardakiewicz, S.; Morkunas, I. Oxidative stress in pea seedling leaves in response to *Acyrthosiphon pisum* infestation. *Phytochemistry* **2013**, *93*, 49–62.
72. Doke, N. Generation of superoxide anion by potato tuber protoplasts during hypersensitive response to hyphal wall components of *Phytophtora infestans* and specific inhibition of the reaction with supressors of hypersensitivity. *Physiol. Plant Pathol.* **1983**, *23*, 359–367.
73. Olyslaegers, G.; Verbelen, J.P. Improved staining of F-actin and co-localization of mitochondria in plant cells. *J. Microsc.* **1998**, *192*, 73–77.
74. Vitha, S.; Baluška, F.; Mews, M.; Volkmann, D. Immunofluorescence detection of F-actin on low melting point wax sections from plant tissues. *J. Histochem. Cytochem.* **1997**, *45*, 89–95.

Sample Availability: Not available.

Regulation of Plant Immunity through Modulation of Phytoalexin Synthesis

Olga V. Zernova, Anatoli V. Lygin, Michelle L. Pawlowski, Curtis B. Hill, Glen L. Hartman, Jack M. Widholm and Vera V. Lozovaya

Abstract: Soybean hairy roots transformed with the *resveratrol synthase* and *resveratrol oxymethyl transferase* genes driven by constitutive *Arabidopsis* actin and CsVMV promoters were characterized. Transformed hairy roots accumulated glycoside conjugates of the stilbenic compound resveratrol and the related compound pterostilbene, which are normally not synthesized by soybean plants. Expression of the non-native stilbenic phytoalexin synthesis in soybean hairy roots increased their resistance to the soybean pathogen *Rhizoctonia solani*. The expression of the *AhRS3* gene resulted in 20% to 50% decreased root necrosis compared to that of untransformed hairy roots. The expression of two genes, the *AhRS3* and *ROMT*, required for pterostilbene synthesis in soybean, resulted in significantly lower root necrosis (ranging from 0% to 7%) in transgenic roots than in untransformed hairy roots that had about 84% necrosis. Overexpression of the soybean *prenyltransferase (dimethylallyltransferase) G4DT* gene in soybean hairy roots increased accumulation of the native phytoalexin glyceollin resulting in decreased root necrosis.

Reprinted from *Molecules*. Cite as: Zernova, O.V.; Lygin, A.V.; Pawlowski, M.L.; Hill, C.B.; Hartman, G.L.; Widholm, J.M.; Lozovaya, V.V. Regulation of Plant Immunity through Modulation of Phytoalexin Synthesis. *Molecules* **2014**, *19*, 7480-7496.

1. Introduction

Diseases and pests can keep soybean grain producers from achieving maximum productivity. Host plant resistance is an economical and sustainable disease and pest management option. There is a strong demand for soybean cultivars with improved pest and disease resistance. Progress achieved during the last decade in studies of plant defense mechanisms indicates that one of the important overall transgenic approaches to combat pathogens could be over-expression of genes that produce proteins (pathogenesis-related proteins), involved in the biosynthesis of compounds (phytoalexins or phytoanticipins) that are toxic to pathogens or inhibit their growth and development [1].

Synthesis of phytoalexins or phytoanticipins is considered to be an important part of the plant innate immune response to a variety of pathogens [2–9]. Induction of synthesis of the pterocarpan phytoalexin glyceollin in soybean plants during invasion by pathogens and pests, and by abiotic stresses, has been extensively reviewed by [9]. It was also reported that genetic modification of various plant species capacity to produce phytoalexins affects the transgenic plant resistance to pathogens [9–18]. Our previous studies showed that transgenic modulation of soybean plant potential to accumulate glyceollin in response to pathogen attacks increased soybean disease resistance, with higher resistance found in plants with elevated glyceollin synthesis [16,19–23].

Expression of the non-native phytoalexin synthesis in crop plants could increase plant immunity and resistance to pests and diseases because host pathogens may have low capability to detoxify non-native phytoalexins, which reduce the rate of colonization until other parts of innate plant defense are activated to maximum levels, such as the production of antimicrobial reactive oxygen species.

Stilbenic compounds with a broad spectrum of biological activities recently generated an interest as nutraceuticals with health promoting effects in humans, such as cardioprotection, anti-inflamatory, neuroprotective and anticancer effects [24,25], and also as antibiotics protecting plants from various microorganisms, nematodes, or herbivores [14,15,18]. Stilbenes occur in plants of several families, such as Cyperaceae, Dipterocarpaceae, Gnetaceae, Fabaceae, Leguminoseae, Pinaceae, Poaceae, and Vitaceae, and are synthesized from the phenylpropanoid substrates that are present in all higher plants and involves the activity of stilbene synthase or resveratrol synthase (STS) [9]. Transformation of different plant species with the *STS* genes was successfully carried out beginning with introduction of the *STS* gene from peanut into tobacco, which resulted in resveratrol synthesis following induction with UV-light [26]. Since then, the *STS* genes have been expressed in tomato [27], barley and wheat [28], alfalfa [29], and grapevine [30]. The expression of the *STS* genes in these studies led to increased STS activity and accumulation of resveratrol glycoside conjugates in transgenic plants and, importantly, resulted in increased plant resistance to fungal pathogens [31]. Thus, transgenic expression of stilbene synthase from grape into tobacco, tomato, and alfalfa resulted in accumulation of conjugates of the non-native phytoalexin resveratrol, and correspondingly, increased resistance to *Botrytis cinerea* [10], to *Phytophthora infestans* [27] and to *Phoma medicaginis* [29]. The combined expression of the *STS* and *resveratrol oxymethyl transferase (ROMT)* genes in transgenic plants can result in accumulation of conjugates of pterostilbene, a double methylated version of resveratrol, which was reported to have higher fungitoxicity compared to resveratrol [32–34]. The stable expression of these two genes in the model plants tobacco and *Arabidopsis* resulted in accumulation of pterostilbene conjugates in both species [35,36]. However, we are unaware of any report attributing induction of resveratrol and pterostilbene synthesis in soybean and their effects on plant disease resistance. We have previously found that resveratrol and pterostilbene strongly suppressed the growth of important soybean fungi when present in growth medium containing the compounds [34]. In this paper we provide molecular and biochemical characteristics of soybean hairy roots expressing *resveratrol synthase* and *resveratrol oxymethyl transferase* under control of constitutive promoters which are capable of producing different levels of stilbenic compounds that are normally not synthesized by soybean, and describe the effects of the non-native phytoalexin expression on severity of infection caused by the soybean generalist fungus, *Rhizoctonia solani*. We also describe how overexpression of the *prenyltransferase (dimethylallyltransferase) G4DT* gene, responsible for the key prenylation reaction in the glyceollin synthesis [37], affects the capacity of hairy roots to accumulate native phytoalexin glyceollin in response to fungal infection.

2. Results and Discussion

2.1. Expression of Non-Native Stilbenoid Phytoalexins in Soybean Hairy Roots

We produced over 50 independent hairy root lines as a model system for functional gene analysis which express either the peanut resveratrol synthase 3 (*AhRS3*) gene or two genes: *AhRS3* and the resveratrol o-methyltransferase (*ROMT)* gene from *Vitis vinifera*, for synthesis of resveratrol and/or pterostilbene, respectively. These genes mediate the synthesis of the non-native stilbenic compounds that are not normally present in soybean plants.

HPLC analysis showed a wide range of resveratrol concentrations (10–350 µg per gram of fresh root tissue) in transformed root tissues and we did not find resveratrol in the control untransformed hairy roots (Figures 1A and 2A,C). Resveratrol was found in transformed hairy root tissues as glucosyl (trans-piceid) and malonylglucosyl conjugates (Figure 3). The structures of these compounds were confirmed by LC-MS. Trans-piceid gave a molecular ion $[M-H]^-$ at *m/z* 389, MS-MS gave the fragment at *m/z* 227, which corresponded to the loss of glucose residue, $[M-Glu-H]^-$. Malonyl glucoside resveratrol gave low intensity ion $[M-H]^-$ at *m/z* 475 and high intensity ion at *m/z* 431, which corresponded to the loss of CO_2 from malonyl glucoside. Ms-Ms of this ion gave a fragment at *m/z* 227. Concentration of isoflavones in these lines accumulating resveratrol conjugates varied between 100–900 µg/g FW and did not correlate with resveratrol levels (Figures 1B and 2B,D). Generally, higher levels of resveratrol (75–350 µg/g FW) were found in hairy root lines transformed with the *AhRS3* gene only; resveratrol concentrations varied within 30–110 µg/g FW range in roots expressing the *AhRS3* gene under CsVMV promoter and the *ROMT* gene and less than 50 µg/g FW were present in lines expressing the *AhRS3* gene driven by the actin promoter and the *ROMT* gene (these latter hairy roots were produced from seeds of our transgenic plants previously transformed with the *AhRS3* gene under *Arabidopsis* actin promoter after treatment with *A. rhizogenes* carrying the *ROMT* gene under the CsVMV promoter) (Figures 1A and 2A,C).

Figure 1. Concentrations of resveratrol (**A**) and isoflavones (**B**) in different soybean hairy root lines derived from soybean cultivar Spencer transformed with the *AhRS3* gene driven by the CsVMV promoter compared with untransformed Spencer hairy roots (control).

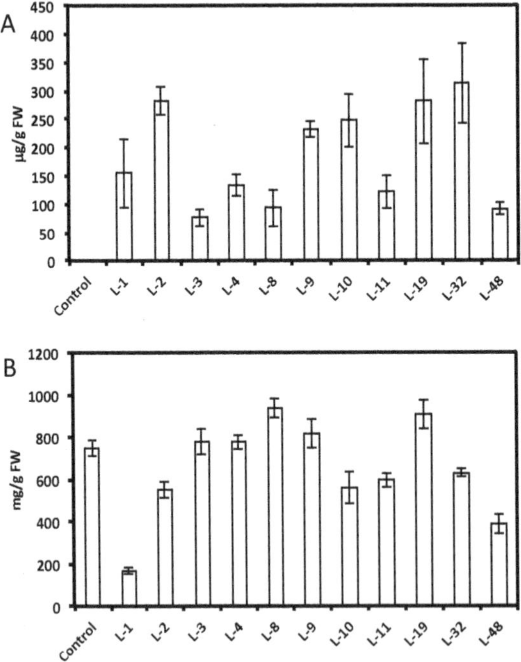

Pterostilbene was found in soybean hairy roots that were transformed with the two genes required for pterostilbene synthesis (Figure 4). LC-MS of extracts showed a peak of molecular ion at *m/z* 255 $[M-H]^-$ with fragment at *m/z* 241 in MS-MS after loss of the methyl group. It was identical to the commercial standard of pterostilbene. Pterostilbene levels in soybean hairy roots ranged from 5 to 8 μg/g FW.

HPLC analysis also showed a peak with R_T of 24.7 min in hairy root lines transformed with both the *AhRS3* and *ROMT* genes and this peak may correspond to a yet non-identified stilbene conjugate (Figure 4—unknown peak).

Several Spencer hairy root lines were selected for further biochemical and molecular analysis: lines 32 and 48 transformed with the *AhRS3* gene under CsVMV promoter; lines 14 and 45 transformed with the *AhRS3* and *ROMT* genes, both driven by CsVMV promoters; and lines 56 and 67 transformed with two genes driven by different promoters, the *AhRS3* gene—under the actin promoter and the *ROMT* gene—under the CsVMV promoter.

Figure 2. Concentrations of resveratrol (**A** and **C**) and isoflavones (**B** and **D**) in soybean hairy root lines derived from soybean cultivar Spencer transformed with either the *AhRS3* and the *ROMT* genes both driven by the CsVMV promoters (A and B) or with the *AhRS3* driven by the *Arabidopsis* actin promoter and the *ROMT* gene driven by the CsVMV promoter (C and D). The control is the untransformed hairy roots.

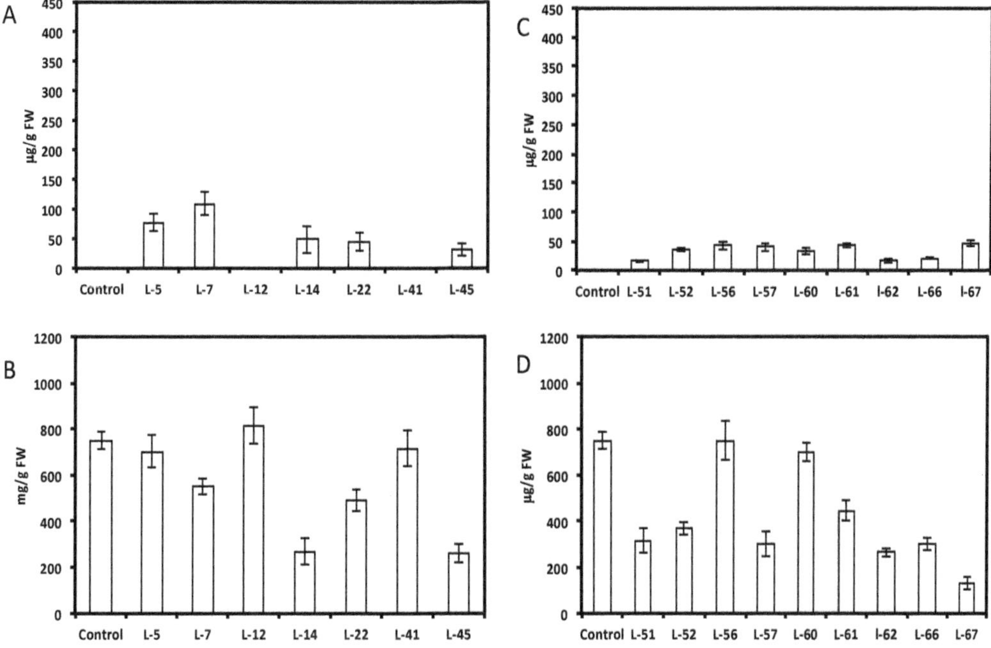

Figure 3. HPLC chromatogram of extracts of soybean hairy roots derived from soybean cultivar Spencer untransformed (**A**) and Spencer transformed with the *AhRS3* gene driven by the CsVMV promoter (**B**). Identified peaks are trans-resveratrol malonyl glucoside, (RT 17.5 min) and trans-piceid (trans-resveratrol glucoside, RT 21.1 min at UV detection (λ = 306 nm).

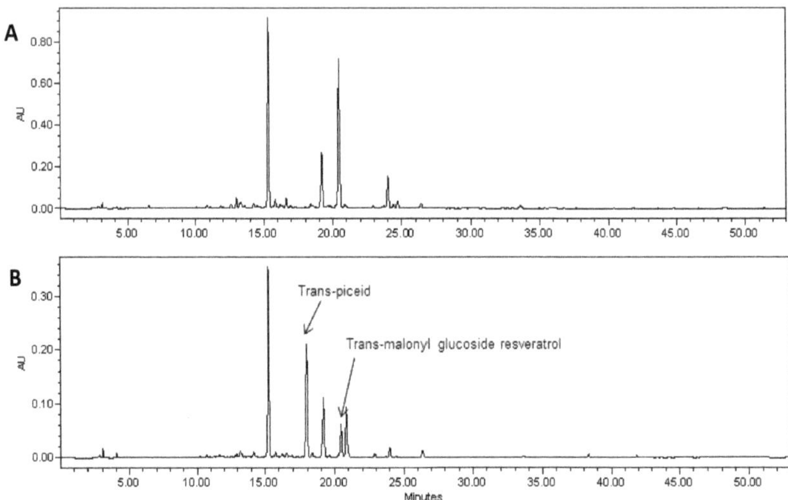

Figure 4. A portion of the HPLC chromatogram of a methanol extract of hairy roots derived from soybean cultivar Spencer (**A**) and Spencer transformed with the *AhRS3* and the *ROMT* genes, both driven by the CsVMV promoters (**B**) and treated with β-glucosidase showing the presence of resveratrol (R_T 25.9 min), pterostilbene (R_T 41.0 min) and unknown peak (R_T 24.3 min) detected only in lines transformed with both genes at UV detection (λ = 306 nm).

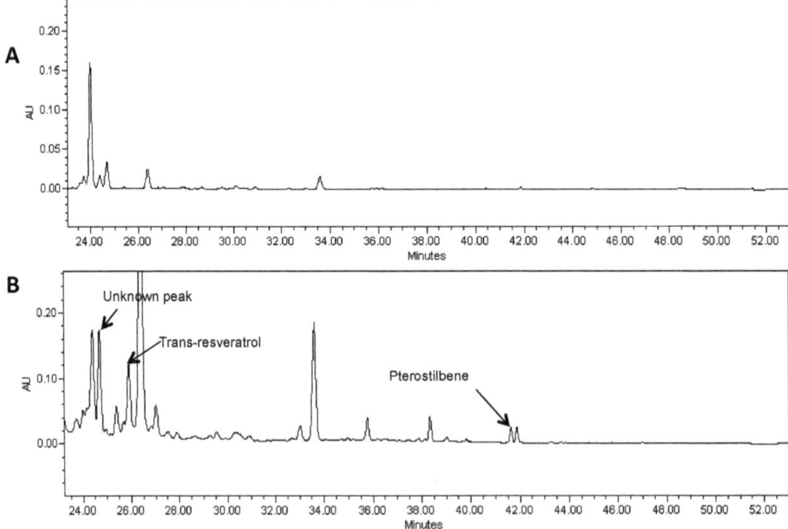

Presence of transgenes in these selected lines was confirmed by PCR (Figure 5) and the transgene expression was confirmed by RT-PCR (Figure 6) as well as by the presence of stilbenic compounds, the transgene products, which are normally not synthesized in soybean plants (Figures 1–4 and 7).

Figure 5. PCR analysis of transgenic hairy root lines derived from soybean cultivar Spencer: Lines 14 and 45 were transformed with the *AhRS3* and the *ROMT* genes both driven by the CsVMV promoters; Lines 56 and 67 were transformed with the *AhRS3* gene driven by the *Arabidopsis* actin promoter and the *ROMT* gene driven by the CsVMV promoter; Lines 32 and 48 were transformed with the *AhRS3* gene driven by the CsVMV promoter; Lines 83 and 98 were transformed with the *G4DT* gene driven by the CsVMV promoter. – negative control, Spencer hairy root, transformed with K599 strain without vector + positive control, plasmid.

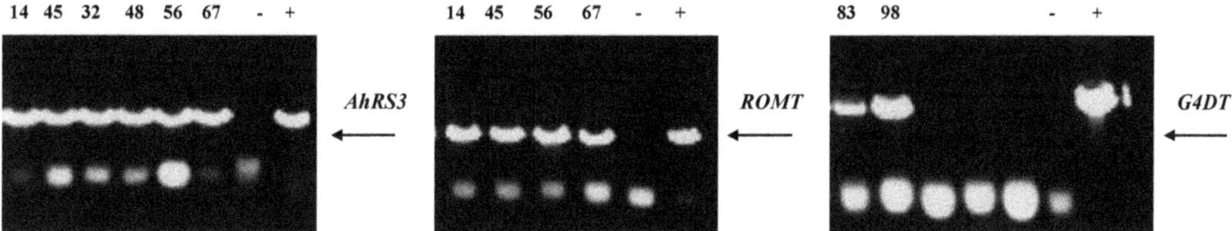

Figure 6. RT-PCR analysis of hairy roots lines derived from soybean cultivar Spencer: Lines 14 and 45 were transformed with the *AhRS3* and the *ROMT* genes, both driven by the CsVMV promoter; Lines 56 and 67 were transformed with the *AhRS3* gene driven by the *Arabidopsis* actin promoter and the *ROMT* gene driven by the CsVMV promoter; Lines 32 and 48 were transformed with the *AhRS3* gene driven by the CsVMV promoter; Lines 83 and 98 were transformed with the *G4DT* gene driven by the CsVMV promoter; – negative control, Spencer hairy root, transformed with K599 strain without vector + positive control, plasmid.

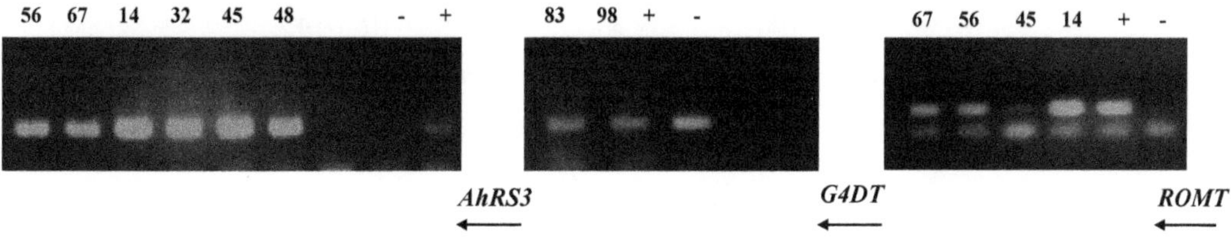

Figure 7. Resveratrol (**A**) and isoflavones and coumestrol (**B**) in selected transgenic hairy root lines derived from soybean cultivar Spencer: Lines 14 and 45 were transformed with the *AhRS3* and the *ROMT* genes, both driven by the CsVMV promoters; Lines 56, and 67 were transformed with the *AhRS3* gene driven by the *Arabidopsis* actin promoter and the *ROMT* gene driven by the CsVMV promoter; Lines 32 and 48 were transformed with the *AhRS3* gene driven by the CsVMV promoter; Lines 83 and 98 are transformed with the *G4DT* gene driven by the CsVMV promoter. The control is the untransformed hairy roots.

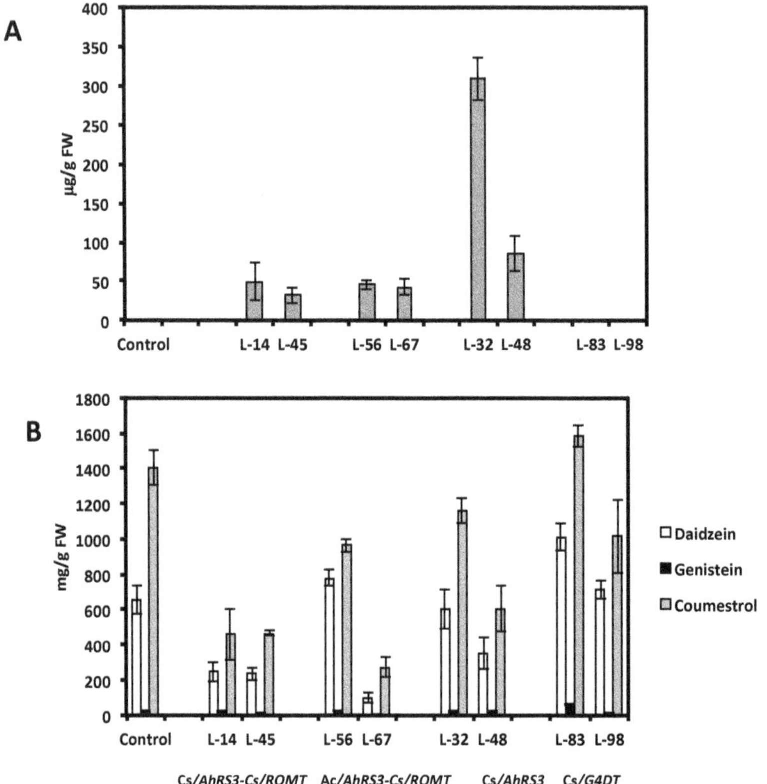

Figure 8. Soluble phenolic acids in selected transgenic hairy root lines after β-glucosidase hydrolysis: Cs—CsVMV promoter, Ac—actin promoter; VA—vanillic acid, CoumA—coumaric FerA—ferulic acid; Lines 14 and 45 were transformed with the *AhRS3* and the *ROMT* genes both driven by the CsVMV promoter; Lines 56 and 67 were transformed with the *AhRS3* gene driven by the *Arabidopsis* actin promoter and the *ROMT* gene driven by the CsVMV promoter; Lines 32 and 48 were transformed with the *AhRS3* gene driven by the CsVMV promoter; Lines 83 and 98 were transformed with the *G4DT* gene driven by the CsVMV promoter. The control is the untransformed hairy roots.

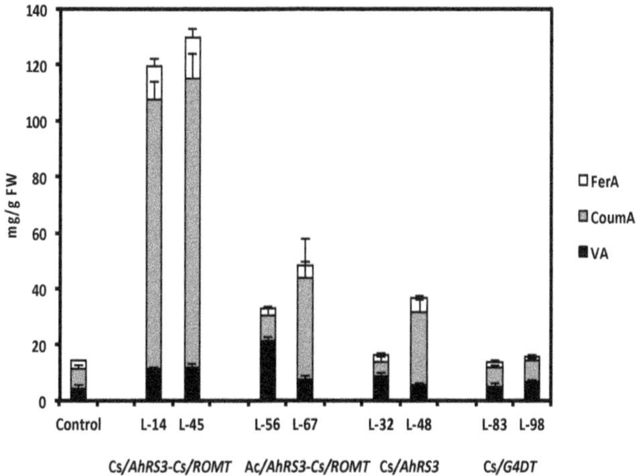

As stated above, concentrations of non-native resveratrol in lines expressing both the *AhRS3* and *ROMT* genes accumulated pterostilbene, and presumably an unidentified stilbenic conjugate, were much lower than in lines transformed with the *AhRS3* gene only. Hairy root lines transformed with both the *AhRS3* and *ROMT* genes also contained markedly reduced isoflavone concentrations compared to untransformed lines, with the exception of line 56 (Figure 7). These lines had elevated concentration of soluble phenolic acids, especially coumaric acid (up to 5 times compared to hairy root lines expressing only the *AhRS3* gene or untransformed lines, see Figure 8). Phenolic acids, detected in these samples, were present as conjugates with glucose, asparagine and aspartate. HPLC chromatograms of hairy root lines transformed with two genes had peaks which were not detected in the untransformed line or in the line transformed with the *AhRS3* gene only. A peak with R_T 11.7 min had maximum UV absorption at 308 nm and gave the pseudomolecular ion $[M-H]^-$ at *m/z* 277, which produced in MS-MS fragments at *m/z* 260 $[M-H-NH_3]$ and 145, which corresponded to the loss of asparagine $[M-H-\text{asparagine}]^-$ and this compound is assigned as coumaroyl asparagine. We previously detected the accumulation of coumaroyl asparagine in our soybean transgenic lines with suppressed chalcone synthase, which accumulated increased concentrations of phenolic acids [20]. Another additional peak (R_T 13.9 min, λ_{max} = 307.5 nm) gave a pseudomolecular ion at *m/z* 278 fragmented in MS-MS to the ions at *m/z* 260 and 163 and we assigned its structure as 4-O-aspartate [20].

We carried out a series of experiments testing the effects of the non-native phytoalexin expression in hairy roots on root necrosis caused by the soybean pathogen *R. solani*. The expression of the *AhRS3* gene resulted in significantly less necrosis in hairy roots that accumulated resveratrol conjugates; it was decreased to about 20%–50% of that of the untransformed hairy roots (Table 1). The expression of two genes, the *AhRS3* and *ROMT*, required for pterostilbene synthesis in soybean, resulted in significantly lower root necrosis (ranging from 0% to 7%) in transgenic roots than in untransformed hairy roots that had about 84% necrosis (Table 1). These results are in agreement with our previous results [34] that indicated a significant inhibition of fungal growth *in vitro* by resveratrol and pterostilbene, with pterostilbene having high fungicidal activity at a lower level (25 μg/mL) than resveratrol (100 μg/mL). Interestingly, levels of native phytoalexin glyceollin accumulated in roots in response to *R. solani* inoculation and the ratio of glyceollins I, II and III were not different between untransformed control roots and hairy roots accumulating stilbenic compounds (Figure 9A,B). This result indicated that synthesis of non-native phytoalexins did not negatively affect accumulation of glyceollin.

Table 1. Summary of percent necrosis of hairy root lines derived from soybean cultivar Spencer inoculated with *Rhizoctonia solani* AG4 4 days post-inoculation.

Construct	Line	Percent Root Necrosis [1]	
None	Control	83.7	a
CsVMV/AhRS3	32	42.8	b
CsVMV/G4DT	98	40.5	bc
CsVMV/G4DT	83	39.0	bc
CsVMV/AhRS3	48	18.6	cd
CsVMV/AhRS3& CsVMV/ROMT	22	6.7	de
CsVMV/AhRS3& CsVMV/ROMT	45	6.5	de
CsVMV/AhRS3& CsVMV/ROMT	14	2.8	e
Actin/AhRS3& CsVMV/ROMT	67	1.8	de
Actin/AhRS3& CsVMV/ROMT	56	0.0	e

[1] Least squares means were estimated from three-six repeated tests with four replications in each test. Common letters indicated means not significantly different ($P > 0.05$) from each other.

Results of these tests described above demonstrate that molecular engineering of soybean to enable plants to synthesize non-native phytoalexins has high potential to increase broad spectrum and durable innate immunity if resistance is similarly increased against other soybean pathogens as was found against *R. solani* in this study and can withstand potential pathogen evolutionary responses to the resistance over time. This approach for improving innate soybean defense against diseases and pests through genetic engineering is novel and to the best of our knowledge, is not being used in the soybean seed industry.

Figure 9. Concentrations (**A**) and proportions (**B**) of glyceollins in hairy roots derived from soybean cultivar Spencer 72 h after inoculation with *Rhizoctonia solani.* Lines 14 and 45 were transformed with the *AhRS3* and the *ROMT* genes, both driven by the CsVMV promoter; Lines 32 and 48 were transformed with the *AhRS3* gene driven by the CsVMV promoter; Lines 83 and 98 were transformed with the *G4DT* gene driven by the CsVMV promoter. The control is the untransformed hairy roots.

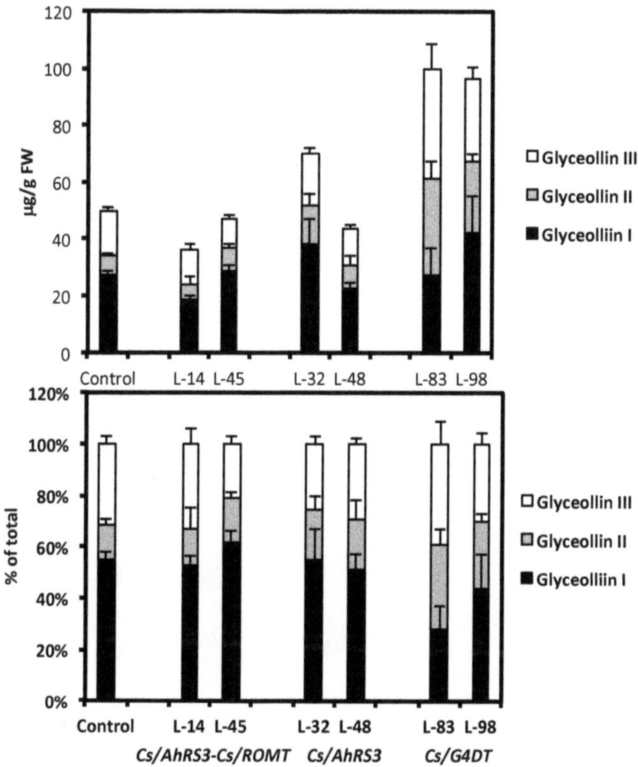

2.2. Genetic Modulation of the Native Phytoalexin Glyceollin Synthesis in Soybean Hairy Roots

With the attempt to increase the capacity of soybean tissues to produce the native soybean phytoalexin glyceollin, we generated new hairy root lines expressing the soybean pterocarpan 4-dimethylallyltransferase (*G4DT)* gene, which controls a key reaction in glyceollin biosynthesis [37]. When several independent hairy root lines derived from soybean cultivar Spencer expressing the *G4DT* gene were subjected to abiotic stress (mercuric chloride), higher levels of glyceollins (up to over 2-fold) accumulated at significantly ($p < 0.05$) higher levels in transformed lines 83 and 98 compared to the untransformed hairy root line in response to this treatment, although induction of glyceollin at some transformed lines was at a similar level (line 85) or significantly lower (line 89) than untransformed roots (Figure 10). The presence of this *G4DT* transgene in hairy roots and its expression was confirmed by PCR and PT-PCR, respectively (Figures 5 and 6, lines 83 and 98). There were no differences in daidzein, genistein and coumestrol concentrations, as well as in concentration and ratio of soluble phenolic acids between these transformed lines and untransformed lines (Figures 7 and 8). When roots of lines 83 and 98 were tested for their response to *R. solani* infection significantly less (about 50%) root necrosis caused by this pathogen was found in these lines compared with the untransformed infected roots (Table 1). Importantly, higher glyceollin accumulation occurred in root tissues of transformed lines 83 and 98 and proportion of glyceollin II was higher in these lines compared to untransformed hairy roots (Figure 9). These results are consistent with our previous results indicating that soybean innate resistance inversely correlates with plant capacity to rapidly induce higher glyceollin synthesis [16,19–23,38]. Therefore, the *G4DT* transgene appears to be a good candidate transgene to move ahead with plant transformation aimed at enhancing plant innate resistance.

Figure 10. Levels of glyceollins in soybean hairy root lines derived from the soybean cultivar Spencer transformed with the *G4DT* gene 72 h after treatment with $HgCl_2$. The control is the untransformed hairy roots. Levels not connected by same letter are significantly different.

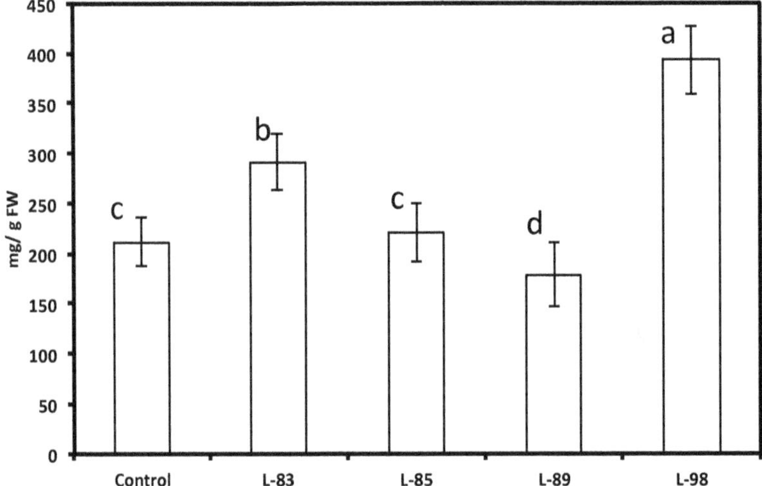

Thus, soybean resistance to fungal infection is in part a consequence of the plant's ability to develop basal defense responses involving phytoalexin production/accumulation and our data reported here reinforce the role played by native and non-native phytoalexins in plant defense. Our results indicated that transgenic modulation of soybean innate resistance, producing a greater accumulation of glyceollin in response to general pathogen invasion, as well as synthesis of stilbenic compounds could help reduce the effects of disease on soybean grain yields and help stabilize soybean production in the presence of diseases. While the pathways studied in the manuscript were constitutively expressed to determine the impact of phytoalexins on soybean pathogens, measure the capacity of soybean tissues to synthesize them, and study the biochemistry of their accumulation, inducible transgenic expression will likely be more sustainable in practice due to less overall demand on plant resources than constitutive expression.

3. Experimental

3.1. Genetic Modification of Soybean Hairy Roots

Genetic transformation of soybean hairy roots was carried out using the efficient *Agrobacterium*-mediated introduction of genes as previously described [20]. We have used the following three transgenes: 1.5 kb *AhRS3*, peanut *resveratrol synthase* [39] (gene bank accession number AF227963, Celtek Bioscience, Nashville, TN, USA), 1.074 kb *resveratrol o-methyltransferase*, *ROMT* cDNA from *Vitis vinifera* [40] (accession number FM178870.1, Celtek Bioscience), and 1.242 kb soybean cDNA *pterocarpan 4-dimethylallyltransferase*, *G4DT* [37] (accession number AB434690.1, Celtek Bioscience). These transgenes were inserted in the BamHI site in the pILTAB-357 vector for *Agrobacterium rhizogenes* (K599)-mediated transformation of hairy roots [20] and were driven by the CsVMV promoter. To generate hairy root cultures, soybean cotyledons of the cultivar Spencer were treated either with *Agrobacterium* transformed with CsVMV/*AhRS3* or with CsVMV/*G4DT* plasmids or with a mixture (in equal proportion) of *Agrobacterium* cultures transformed with either *AhRS3* or *ROMT* genes driven by the CsVMV promoter. We also treated seeds of transgenic soybean plants previously transformed in our laboratory by biolistic delivery of the *AhRS3* gene under *Arabidopsis* actin promoter in the pAPCH-7 plasmid (gift from Dr. R. Meagher, Department of Genetics, University of Georgia, Athens, GA, USA) with *A. rhizogenes* carrying the *ROMT* gene under control of the CsVMV promoter (lines 56 and 67).

3.2. Molecular Analysis

Genomic DNA was extracted from hairy roots using the method of Dellaporta [41] and was used in PCR with primers: 5'*AhRS3*-: ATG GTG TCT GTG AGT GGA ATT CGC AAT GTT, 3'RT-*AhRS3*: AGA TAT ACC CAA AGG ATC AAA to amplify the 1,100 bp fragment of *AhRS3* gene; primers from the CsVMV promoter: 5'AGG ATA CAA CTT CAG AGA and 3'*ROMT*-BamH: GGA TCC TCA AGG ATA AAC CTC AAT GAG GGA CCT CAA to amplify the 1,074 bp of *ROMT* gene; primers from CsVMV promoter (as above) and 3'*G4DT*: TCA TCT AAT TAA TGC CAT GAG AAA GAA CCC to amplify 1242 bp of *G4DT*. The presence of transgenes in root tissues was confirmed by the PCR reaction using Tag DNA polymerase (New England Biolabs, Ipswich, MA, USA) with denaturation at 95 °C-1 min, annealing at 48 °C-40 s and extension at 72 °C-90 s, 35 cycles. "RNeasy Plant Mini Kit" (Qiagen, Valencia, CA, USA) was used for extraction of total RNA from root tissues. DNA was digested with DNase I, using the Ambion@RNA Life Technologies TURBO DNA-free kit (Carlsbad, CA, USA). cDNA was synthesized by reverse transcriptase reaction using oligo (dT) primer and reverse transcriptase enzyme, Promega GoScript kit, (Madison, WI, USA) for positive control or excluding reverse transcriptase enzyme in negative control for genomic DNA contamination. To amplify *AhRS3* transcript from cDNA we used the same primers as for genomic DNA detection. To amplify the 110 bp fragment of the *ROMT* transcript, we used the following primers: RT-*ROMT*-5': TGC CTC TAG GCT CCT TCT AA and RT-*ROMT* -3': TTT GAA ACC AAG CAC TCA GA. To amplify the 570 bp of *G4DT* fragment we used the following primers: 5'CsVMV: AGG ATA CAA CTT CAG AGA and 3'-*G4DT*:GTC CAG ATG CCA TTG GAA GAT GTGG.

3.3. Analysis of Phenylpropanoids in Soybean Tissue

Soybean samples frozen in liquid nitrogen were freeze-dried and extracted with 80% methanol [16,19,20,38,42]. Methanol was removed on a rotatory evaporator and the residue was re-suspended in 0.1 M acetate buffer, pH 5.0 and treated with β-glucosidase from almonds (Sigma-Aldrich, St. Louis, MO, USA) overnight at 37 °C. Aglucones were extracted with ethyl acetate, transferred to 80% methanol and analyzed by HPLC using Waters 2690 Separation Module (Waters Corp, Milford, MA, USA) with Prevail C18 column , 250 × 46 mm, 5 μm particle size (Alltech Assoc, Deerfield, IL, USA) with PDA detector Waters 996 and authentic standards as described by [16,19–21].

Solvent A was water-acetonitrile-acetic acid (95:5:0.5) and solvent B was acetonitrile-water-acetic acid (95:5:0.5). The flow rate was 1 mL/min. Elution was done with a linear gradient from 5% to 20% B in 10 min, then to 50% for 30 min and to 95% for 7 min. After that the column was washed with 95% acetonitrile for 3 min and equilibrated at 5% B between runs for 3 min. Total sample to sample time was 58 min. Detection was done by UV absorbance at 306 nm for stilbenes and their derivatives, at 295 nm for phenolic acids, and at 285 nm for glyceollins.

LC-MS analysis of soluble phenolics was carried out in Agilent 1100 LC/MSD Trap XCT Plus (Santa Clara, CA, USA) supplied with column and using the same HPLC conditions as above except that flow rate was 500 μL/min. Ion source was ESI, negative mode, nebulizing gas N_2, 350 °C, 9 mL/min, capillary voltage 3500 V.

Glyceollins II and III were identified using published UV and LC-MS-MS spectra and quantified respectively to glyceollin I isomer. Cis-resveratrol and other stilbenic compounds were identified by UV spectrum and LC-MS. Levels of accumulated glyceollins in transformed and untransformed hairy root lines were analyzed using JMP 11 (SAS Institute, Cary, NC, USA, 2014) Fit Y by X Procedure. Means were separated by least significant difference at $\alpha = 0.05$. Resveratrol, pterostilbene, isoflavones and coumestrol and their derivatives were identified by comparing of retention times and UV spectra with authentic standards.

3.4. Evaluation of Fungal Colonization of Hairy Root Cultures

Hairy roots transformed with vectors containing the four constructs and control vector without constructs were evaluated for disease severity caused by an isolate of *Rhizoctonia solani*, anastomosis group 4 (AG4), originally isolated from infected soybean roots in Urbana, Illinois in 2000, and maintained on potato dextrose agar by serial transfer. The tests were set up as completely randomized designs with four replicates. A 2cm long section of root was cut from an actively growing hairy root culture and placed onto a 60 × 15 mm petri dish (Falcon, Franklin Lakes, NJ, USA) filled with antibiotic (100 ppm/L penicillin and 100 ppm/L streptomycin) − treated 1% water agar. A 4 mm plug, cut from the edge of an actively growing 4-day-old culture using a cork borer, was placed mycelium side down onto the middle of the excised root segment. Plates containing root segments were placed into a closed plastic container and covered with a black cloth to keep the roots in the dark. The plates were kept on a bench-top at 24 °C through the duration of the test. Mycelial plugs were removed from the root segments 24 h post-inoculation. Roots were evaluated by visually estimating the percent necrosis 4 days post-inoculation. Each test was repeated 3–6 times. The percent necrosis data was transformed prior to analysis using arc sine square root transformation to correct for non-constant variance among the treatments (hairy root lines). Transformed data were analyzed using the JMP 11 Fit Model Procedure. Variables were the constructs and lines nested within constructs. Means were separated by least significant difference at $\alpha = 0.05$.

4. Conclusions

Results of this work indicated that the expression of transges, *AhRS3* and *ROMT* controlling the biosynthesis of the non-native stilbenes resveratrol and pterostilbene significantly reduced *R. solani* root colonization in transformed soybean hairy roots. In addition, overexpression of the *G4DT* gene in soybean hairy roots resulted in higher accumulation of the native phytoalexin glyceollins in response to biotic and abiotic stresses. These encouraging results validate the approach to use genetic engineering of soybean and other important crop plants to elevate biosynthesis of native and non-native phytoalexins, with the aim to increase broad-spectrum innate host defenses against pathogens and pests.

Acknowledgments

This work was supported by funds from the United Soybean Board, USB Project #1272 and the USDA Cooperative State Research, Education and Extension Service, Hatch project number ILLH-802-309 and 802-352.

Author Contributions

VV, OZ, AL and JW designed and carried out research related to transformation and biochemical parts of project, GH, MP and CH designed and performed studies on disease resistance testing; all authors participated in analysis of the data obtained and writing the part of manuscript involving their specific expertise. All authors read and approved the final manuscript.

Conflicts of Interest

The authors declare no conflict of interest.

References

1. Islam, A. Fungus resistant transgenic plants: Strategies, progress and lessons learnt. *Plant Tissue Cult. Biotech.* **2006**, *16*, 117–138.
2. Paxton, J.D.; Groth, J. Constraints on pathogens attacking plants. *Crit. Rev. Plant Sci.* **1994**, *13*, 77–95.
3. VanEtten, H.D.; Mansfield, J.W.; Bailey, J.A.; Farmer, E.E. Two classes of plant antibiotics: Phytoalexins *versus* phytoanticipins. *Plant Cell* **1994**, *6*, 1191–1192.

4. VanEtten, H.; Temporini, E.; Wasmann, C. Phytoalexin (and phytoanticipin) tolerance as a virulence trait: Why is it not required by all pathogens? *Physiol. Mol. Plant Pathol.* **2001**, *59*, 83–93.
5. Kliebenstein, D.J. Secondary metabolites and plant/environment interactions: A view through *Arabidopsis thaliana* tinged glasses. *Plant Cell Environ.* **2004**, *27*, 675–684.
6. Pedras, M.S.C.; Zheng, Q.A.; Sarma-Mamillapalle, V.K. The phytoalexins from Brassicaceae: Structure, biological activity, synthesis and biosynthesis. *Nat. Prod. Commun.* **2007**, *2*, 319–330.
7. Bednarek, P.; Osbourn, A. Plant-microbe interactions: Chemical diversity in plant defense. *Science* **2009**, *324*, 746–748.
8. Ahuja, I.; Kissen, R.; Bones, A.M. Phytoalexins in defense against pathogens. *Trends Plant Sci.* **2012**, *17*, 73–90.
9. Grobkinsky, D.K.; van der Graaff, E.; Roitsch, T. Phytoalexin transgenics in crop protection—Fairy tale with a happy end? *Plant Sci.* **2012**, *195*, 54–70.
10. Hain, R.; Reif, H.J.; Krause, E.; Langebartels, R.; Kindl, H.; Vornam, B.; Wiese, W.; Schmelzer, E.; Schreier, P.H.; Stocker, R.H.; *et al.* Disease resistance results from foreign phytoalexin expression in a novel plant. *Nature* **1993**, *361*, 153–156.
11. Papadopoulou, K.; Melton, R.E.; Leggett, M.; Daniels, M.J.; Osbourn, A.E. Compromised disease resistance in saponin-deficient plants. *Proc. Natl. Acad. Sci. USA* **1999**, *96*, 12923–12928.
12. Essenberg, M. Prospects for strengthening plant defenses through phytoalexin engineering. *Physiol. Mol. Plant Pathol.* **2001**, *59*, 71–81.
13. Campbell, M.A.; Fitzgerald, H.A.; Ronald, P.C. Engineering pathogen resistance in crop plants. *Transgenic Res.* **2002**, *11*, 599–613.
14. Jeandet, P.; Douillet, A.C.; Debord, S.; Sbaghi, M.; Bessis, R.; Adrian, M. Phytoalexins from the *Vitaceae*: Biosynthesis, phytoalexin gene expression in transgenic plants, antifungal activity, and metabolism. *J. Agric. Food Chem.* **2002**, *50*, 2731–2741.
15. Jeandet, P.; Delaunois, B.; Conreux, A.; Donnez, D.; Nuzzo, V.; Cordelier, S.; Clement, C.; Courot, E. Biosynthesis, metabolism, molecular engineering and biological functions of stilbene phytoalexins in plants. *Biofactors* **2010**, *36*, 331–341.
16. Lozovaya, V.V.; Lygin, A.V.; Zernova, O.V.; Widholm, J.M. Genetic engineering of plant; disease resistance by modification of the phenylpropanoid pathway. *Plant Biosyst.* **2005**, *139*, 20–23.
17. Aharoni, A.; Galili, G. Metabolic engineering of the plant primary-secondary metabolism interface. *Curr. Opin. Biotechnol.* **2011**, *22*, 239–244.
18. Jeandet, P.; Clement, C.; Courot, E.; Cordelier, S. Modulation of Phytoalexin Biosynthesis in Engineered Plants for Disease Resistance. *Int. J. Mol. Sci.* **2013**, *14*, 14136–14170.
19. Lozovaya, V.V.; Lygin, A.V.; Li, S.; Hartman, G.L.; Widholm, J.M. Biochemical response of soybean roots to *Fusarium solani* f. sp. *glycines* infection. *Crop Sci.* **2004**, *44*, 819–826.
20. Lozovaya, V.V.; Lygin, A.V.; Zernova, O.V.; Ulanov, A.V.; Li, S.; Hartman, G.L.; Widholm, J.M. Modification of phenolic metabolism in soybean hairy roots through down regulation of chalcone synthase or isoflavone synthase. *Planta* **2007**, *225*, 665–679.

21. Lygin, A.V.; Li, S.; Vittal, R.; Widholm, J.M.; Hartman, G.L.; Lozovaya, V.V. The importance of phenolic metabolism to limit the growth of *Phakopsora pachyrhizi*. *Phytopathology* **2009**, *99*, 1412–1420.
22. Lygin, A.V.; Hill, C.B.; Zernova, O.V.; Crull, L.; Widholm, J.M.; Hartman, G.L.; Lozovaya, V.V. Response of Soybean Pathogens to Glyceollin. *Phytopathology* **2010**, *100*, 897–903.
23. Lygin, A.V.; Zernova, O.V.; Hill, C.B.; Kholina, N.A.; Widholm, J.M.; Hartman, G.L.; Lozovaya, V.V. Glyceollin is an important component of soybean plant defense against *Phytophthora sojae* and *Macrophomina phaseolina*. *Phytopathology* **2013**, *103*, 984–994.
24. Pervaiz, S.; Holme, A.L. Resveratrol: Its biologic targets and functional activity. *Antioxid. Redox Signal.* **2009**, *11*, 2851–2897.
25. Fulda, S. Resveratrol and derivatives for the prevention and treatment of cancer. *Drug Discov. Today* **2010**, *15*, 757–765.
26. Hain, R.; Bieseer, B.; Kindl, H.; Schroeder, G.; Stoecker, R. Expresshion of a stilbene synthase gene in Nicotiana-tabacum results in synthesis of the phytoalexin resveratrol. *Plant Mol. Biol.* **1990**, *15*, 325–336.
27. Thomzik, J.E.; Stenzel, K.; Stocker, R.; Schreier, P.H.; Hain, R.; Stahl, D.J. Synthesis of a grapevine phytoalexin in transgenic tomatoes (*Lycopersicon esculentum* Mill.) conditions resistance against *Phytophthora* infestans. *Physiol. Mol. Plant Pathol.* **1997**, *51*, 265–278.
28. Leckband, G.; Lorz, H. Transformation and expression of a stilbene synthase gene of *Vitis* V*inifera* L. in barley and wheat for increased fungal resistance. *Theor. Appl. Genet.* **1998**, *96*, 1004–1012.
29. Hipskind, J.D.; Paiva, N.L. Constitutive accumulation of a resveratrol-glucoside in transgenic alfalfa increases resistance to Phoma medicaginis. *Mol. Plant Microbe Interact.* **2000**, *13*, 551–562.
30. Coutos-Thevenot, P.; Poinssot, B.; Bonomelli, A.; Yean, H.; Breda, C.; Buffard, D.; Esnault, R.; Hain, R.; Boulay, M. *In vitro* tolerance to *Botrytis cinerea* of grapevine 41B rootstock in transgenic plants expressing the stilbene synthase *VST*1 gene under the control of a pathogen-inducible PR 10 promoter. *J. Exp. Botany* **2001**, *52*, 901–910.
31. Delaunois, B.; Cordelier, S.; Conreux, A.; Clément, C.; Jeandet, P. Molecular engineering of resveratrol in plants. *Plant Biotech. J.* **2009**, *7*, 2–12.
32. Pezet, R.; Pont, V. Ultrastractural observation of pterostilbene fungitoxity in dormant conidia of *Botrytis-cinerea* pers. *J. Phytopathol.* **1990**, *129*, 19–30.
33. Pezet, R.; Gindro, K.; Viret, O.; Spring, J.-L. Glycosylation and oxidative dimerization of resveratrol are respectively associated to sensitivity and resistance of grapevine cultivars to downy mildew. *Physiol. Mol. Plant Pathol.* **2004**, *65*, 297–303.
34. Lygin, A.V.; Hill, C.B.; Pawlowski, M.; Zernova, O.V.; Widholm, J.M.; Hartman, G.L.; Lozovaya, V.V. Inhibitory effects of stilbenes on soybean pathogen growth. *Phytopathology* **2014**, in press.
35. Rimando, A.M.; Pan, Z.; Polashok, J.J.; Dayan, F.E.; Mizuno, C.S.; Snook, M.E.; Liu, C.-J.; Baerson, S.R. In planta production of the highly potent resveratrol analogue pterostilbene via stilbene synthase and 0-methyltransferase co-expression. *Plant Biotechnol. J.* **2011**, *10*, 269–283.

36. Xu, Y.H.; Wang, J.W.; Wang, S.; Wang, J.Y.; Chen, X.Y. Characterization of GaWRKY1, a cotton transcription factor that regulates the sesquiterpene synthase gene (+)-delta-cadinene synthase-A. *Plant Physiol.* **2004**, *135*, 507–515.
37. Akashi, T.; Sasaki, K.; Aoki, T.; Ayabe, S.; Yazaki, K. Molecular cloning and characterization of a cDNA for pterocarpan 4-dimethylallyltransferase catalyzing the key prenylation step in the biosynthesis of glyceollin, a soybean phytoalexin. *Plant Physiol.* **2009**, *149*, 683–693.
38. Lozovaya, V.V.; Lygin, A.V.; Zernova, O.V.; Li, S.; Hartman, G.L.; Widholm, J.M. Isoflavonoid accumulation in soybean hairy roots upon treatment with *Fusariun solani. Plant Physiol. Biochem.* **2004**, *42*, 671–679.
39. Lim, J.; Song, J.; Chung, I.; Yu, C. Resveratrol synthase transgene expression and accumulation of resveratrol glycoside in *Rehmannia glutinosa*) ubiqiutin promoter for selection. *Mol. Breed.* **2005**, *16*, 219–233.
40. Schmidlin, L.; Poutaraud, A.; Claudel, P.; Mestre, P.; Prado, E.; Santos-Rosa, M.; Wiedemann-Merdinoglu, S.; Karst, F.; Merdinoglu, D.; Hugueney, P. A stress-inducible resveratrol o-methyltransferase involved in the biosynthesis of pterostilbene in grapevine. *Plant Physiol.* **2008**, *148*, 1630–1639.
41. Dellaporta, S.L. Plant DNA miniprep version 2.1–2.3. In *The Maize Handbook*; Freeling, M., Walbot, V., Eds.; Springer-Verlag: New York, NY, USA, 1993; pp. 522–525.
42. Zernova, O.V.; Lygin, A.V.; Widholm, J.M.; Lozovaya, V.V. Modification of isoflavones in soybean seeds via expression of multiple phenolic biosynthetic genes. *Plant Physiol. Biochem.* 2009, *47*, 769–777.

Sample Availability: Not available.

A Sorghum MYB Transcription Factor Induces 3-Deoxyanthocyanidins and Enhances Resistance against Leaf Blights in Maize

Farag Ibraheem, Iffa Gaffoor, Qixian Tan, Chi-Ren Shyu and Surinder Chopra

Abstract: Sorghum responds to the ingress of the fungal pathogen *Colletotrichum sublineolum* through the biosynthesis of 3-deoxyanthocyanidin phytoalexins at the site of primary infection. Biosynthesis of 3-deoxyanthocyanidins in sorghum requires a MYB transcription factor encoded by *yellow seed1* (*y1*), an orthologue of the maize gene *pericarp color1* (*p1*). Maize lines with a functional *p1* and flavonoid structural genes do not produce foliar 3-deoxyanthocyanidins in response to fungal ingress. To perform a comparative metabolic analysis of sorghum and maize 3-deoxyanthocyanidin biosynthetic pathways, we developed transgenic maize lines expressing the sorghum *y1* gene. In maize, the *y1* transgene phenocopied *p1*-regulated pigment accumulation in the pericarp and cob glumes. LC-MS profiling of fungus-challenged *Y1*-maize leaves showed induction of 3-deoxyanthocyanidins, specifically luteolinidin. *Y1*-maize plants also induced constitutive and higher levels of flavonoids in leaves. In response to *Colletotrichum graminicola*, *Y1*-maize showed a resistance response.

Reprinted from *Molecules*. Cite as: Ibraheem, F.; Gaffoor, I.; Tan, Q.; Shyu, C.-R.; Chopra, S. A Sorghum MYB Transcription Factor Induces 3-Deoxyanthocyanidins and Enhances Resistance against Leaf Blights in Maize. *Molecules* **2015**, *20*, 2388-2404.

1. Introduction

Maize (*Zea mays* L.) is an important cereal crop. In 2013, the total area planted under maize for all purposes in the United States amounted to 95.37 million acres, with about 87.67 million acres for grain production (U.S. Department of Agriculture, National Agricultural Statistics Service). In the field, maize plants frequently encounter a wide variety of pathogens. Anthracnose caused by *Colletotrichum graminicola* (Ces.) G. W. Wils. and southern corn leaf blight caused by *Cochliobolus heterostrophus* (Drechsler) are among the most serious fungal diseases that affect productivity.

Application of synthetic fungicides is among the strategies used to control fungal infections, but their cost and environmental impact are a concern for producers and consumers. To prevent further epidemics and reduce the need for synthetic chemicals, there is an ongoing search for crop germplasm with natural resistance [1]. Metabolic engineering of defense-related compounds has proven effective in enhancing plant performance against biotic stress [2,3]. This approach offers an opportunity to either transfer a complete defense-related metabolic pathway or activate a preexisting one by the transfer of genes between distant plant species [4,5].

In maize, the flavonoid pathway gives rise to many defense related compounds such as flavan-4-ols, 3-deoxyanthocyanidins, and *C*-glycosyl flavones (Figure 1). Flavan-4-ols are the precursors of the brick red phlobaphene pigments that accumulate in mature pericarp and cob glumes. Their biosynthesis requires a functional *pericarp color1* (*p1*) gene, which encodes an R2R3 MYB transcription factor [6–8].

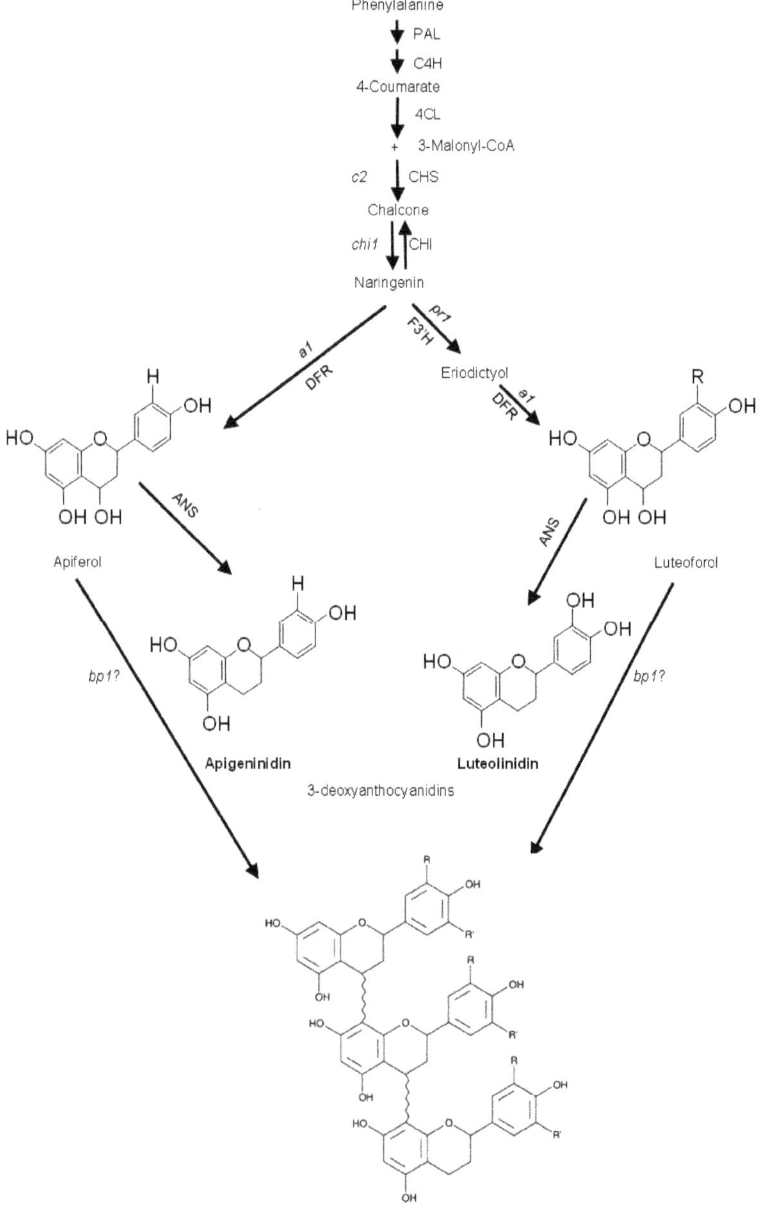

Figure 1. Schematic representation of the biosynthetic pathway of flavonoid compounds. Enzyme names are (gene names in parentheses): PAL, Phenylalanine ammonia lyase; C4H, Cinnamate-4-hydroxylase; 4CL, 4-coumarate: coenzymeA ligase; C3'H, p-coumarate 3'-hydroxlase; CHS (*c2*), Chalcone synthase; CHI (*chi1*), Chalcone isomerase, DFR (*a1*), Dihydroflavonol reductase; and F3'H, Flavonoid 3'-hydroxylase (*pr1*). Pathway modeled after [1,21–24].

We have performed a comparative characterization of the flavonoid pathway in sorghum and maize [9]. These two species are genetically related and are suggested to have diverged from a common ancestor more than 16.5 million years ago [10,11]. The two genomes have a high degree of synteny and sequence similarity [12,13]. The co-linearity between their genomes may suggest a similarity between their metabolic pathways. In fact, sorghum has also been shown to accumulate phlobaphenes in the pericarp under the control of *yellow seed1* (*y1*), an orthologue of maize *pericarp color1* [14–16]. *y1* and *p1* activate the transcription of chalcone synthase (*chs*), chalcone isomerase (*chi*), and dihydroflavonol reductase (*dfr*) during biosynthesis of flavan-4-ols in sorghum and maize [7,9,17].

Regardless of the similarities mentioned above, the flavonoid pathways in sorghum and maize exhibit a number of differences. For example, in maize, phlobaphenes are obvious in the floral tissues, husk, and leaf sheath but not in the leaf, whereas in sorghum these compounds appear in all the above mentioned tissues and in the mature leaf [9]. The presence of phlobaphenes in sorghum leaves may indicate that the *y1* promoter is active in this tissue. Another difference is the response of these two species to fungal challenges. Sorghum responds to anthracnose and other foliar fungi by the induction of red-brown 3-deoxyanthocyanidin phytoalexins [18]. However, there is no published report that leaves of maize lines carrying a similar set of functional flavonoid regulatory and structural genes synthesize detectable levels of 3-deoxyanthocyanidins either constitutively or induced in response to biotic or abiotic stresses. With the exclusion of chalcone synthase in maize, neither the flavonoid structural nor regulatory genes showed induction after fungal infection [19]. Silks of some maize lines have been reported to accumulate very low levels of luteolinidin under the control of *p1* [20].

Sorghum 3-deoxyanthocyanidins include apigeninidin, luteolinidin, and their derivatives. Upon fungal challenge, these compounds accumulate around the primary infection sites and prevent further proliferation of the fungus within sorghum tissues [18,25]. These compounds have been shown to inhibit fungal germ tube growth and distort fungal structures. Their potent antifungal activity against *Colletotrichum sublineolum*, *C. graminicola*, and *C. heterostrophus* has been demonstrated [26]. 3-deoxyanthocyanidins have a structure similar to flavan-4-ols and their biosynthesis requires the activity of *chs*, *chi*, *dfr*, and *f3'h*. The induction of these genes requires a functional *y1* gene because *y1* mutants are deficient in 3-deoxyanthocyanidins and exhibit symptoms of anthracnose susceptibility [16]. The sequences of *y1* and *p1* genes have a high level of similarity (92%) in the coding region but very poor similarity in the non-coding regions [9].

We developed transgenic plants to investigate the heterologous expression of sorghum *y1* in maize and to test if *y1* can induce anthracnose resistance in maize. Our results demonstrate that the *y1* transgenes are active in maize tissues. Biochemical analyses established that *y1* successfully drives the maize–flavonoid pathway towards production of flavan-4-ols and 3-deoxyanthocyanidins. Transgenic *Y1*-maize plants were resistant to both *C. heterostrophus* and *C. graminicola*; this interaction is the result of the induction of 3-deoxyanthocyanidins.

2. Results and Discussion

2.1. *y1* Transgenes Phenocopy *p1* Pigmentation Patterns in Maize

Transgenic maize lines expressing a sorghum *y1* gene (p*Y1::Y1*) (Figure 2A) exhibited distinct patterns of pericarp and cob glume pigmentation. Three ear pigmentation patterns from independent, representative transformation events and a negative segregant are shown (Figure 2B). The pericarp and cob glume pigmentation patterns described here are based on the nomenclature of the maize *p1* alleles [27]. Transgenic events were divided into four classes based on their ear phenotypes: *Y1-rr* (red pericarp, red cob glumes); *Y1-pr* (patterned pericarp, red cob glumes); *Y1-wr* (white pericarp, red cob glumes); and *y1-ww* (white pericarp, white cob glumes). Sibling maize plants were genotyped using *y1* gene specific primers; those lacking the transgene and showing a susceptible response against BASTA herbicide exhibited a white pericarp and white cob glume phenotype (*y1-ww*). These negative segregants (NS) represented similar genetic background to *Y1* transgenic events and thus were used as controls throughout this study.

Unlike the maize *p1* gene, the sorghum *y1* gene induced accumulation of phlobaphenes in the husk and tassel glumes of the three functional categories of transgenic events (*Y1*-maize). Apart from the accumulation of phlobaphenes in floral tissues, an orange pigment was also observed in the leaf midrib in *Y1*-maize. The mid-rib pigmentation appeared at the three-leaf stage of plant growth and persisted through the maturation of the plant. Additionally, the silk tissue of *Y1*-maize plants showed a rapid "silk-browning" phenotype at the cut ends or upon injury. The silk browning phenotype is thus under the control of the *y1* transgene and is similar to the one produced by *p1* and *p2* genes in maize [28]. In *Y1*-maize, this phenotype is more intense compared to the one observed with the endogenous *p1* alleles (see Figure 2B). These distinct phenotypes produced by *Y1*-maize were stably inherited across seven generations. To further confirm if *y1* regulated phenotypes are the result of the activation of flavonoid structural genes in transgenic maize, expression of the *y1* transcription factor and four marker genes was assayed: chalcone synthase (*c2*), chalcone isomerase (*chi*) dihydroflavonol reductase (*a1*), and flavonoid 3'-hydroxylase (*pr1*). Pericarp tissues of the *Y1-rr* and *Y1-pr* transgenic events showed induction of *c2*, *a1*, and *pr1*, and upregulation of *chi* flavonoid structural genes and the *y1* transcription factor in *Y1*-maize while tissue obtained from their respective NS plants showed no detectable expression by RT-PCR (Figure 2C). Overall our phenotypic and gene expression data demonstrated that the sorghum *y1* gene can target maize flavonoid structural genes and either induce or upregulate the flavonoid biosynthetic pathway in maize floral and vegetative tissues.

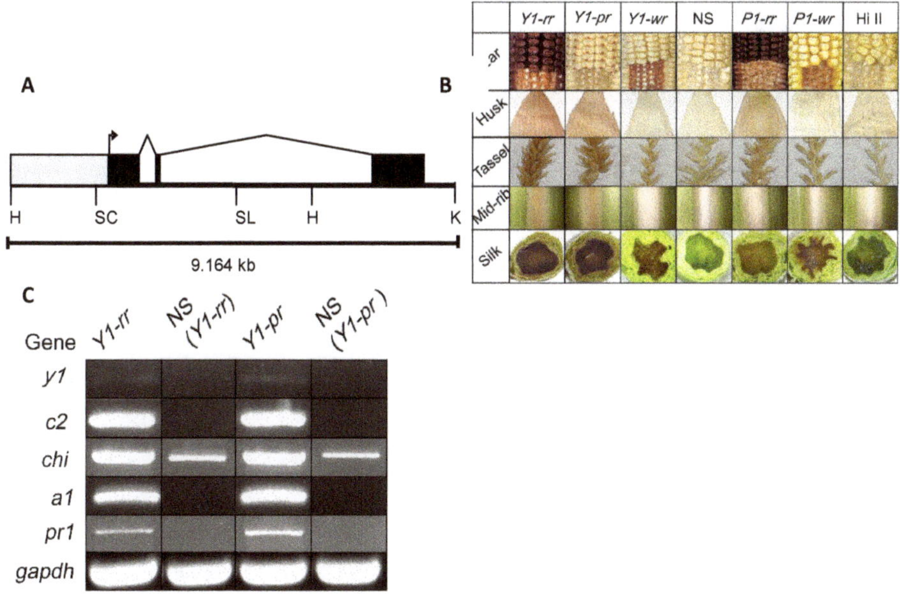

Figure 2. Characterization of *y1* transgenes. (**A**) Structural features of the sorghum *y1* gene. The gray box represents the upstream regulatory region. The bent arrow indicates the transcription start site. Solid boxes correspond to exons that are joined by angled lines representing introns. The restriction enzyme sites shown are: H, *Hin*dIII; K, *Kpn*I; SL, *Sal*I; SC, *Sca*I. Illustration not drawn to scale. (**B**) Sorghum *y1* gene-induced pigmentation phenotypes in transgenic *Y1*-maize. Three *y1* transgenic events representing *Y1-rr*, *Y1-pr* and *Y1-wr* were characterized for ear, husk, tassel glumes, leaf mid-rib, and silk browning phenotypes. Comparable controls included are: plants segregating for the absence of *y1* transgene shown as negative segregant (NS) and native *p1* expressing alleles *P1-rr* and *P1-wr* and *HII* (from A188 X B73), used for transformation. (**C**) Sorghum *y1* gene induces flavonoid structural genes in *Y1*-maize. The expression of the *y1* transgene and four flavonoid structural genes relative to the housekeeping gene glyceraldehyde phosphate dehydrogenase was assayed using RT-PCR. Expression was tested in the pericarp tissues of the *Y1-rr* and *Y1-pr* transgenes and their respective negative segregants (*Y1-rr* and *Y1-pr*). *c2:* chalcone synthase, *chi:* chalcone isomerase, *a1:* dihydroflavonol reductase, *pr1:* flavonoid 3'-hydroxylase, *gapdh:* glyceraldehyde phosphate dehydrogenase.

2.2. y1 Regulates Accumulation of 3-Deoxyflavonoids (flavan-4-ols) in Maize

In sorghum, *y1* has been shown to be required for the biosynthesis of flavan-4-ols or 3-deoxyflavonoid compounds that are precursors to the phlobaphenes [9]. To investigate the effect of *y1* on the flavonoid pathway in maize, we assayed flavan-4-ol accumulation in the pericarp, cob glumes, silks, and leaves (see Figure S1). Spectral results indicated the presence of flavan-4-ols with an absorption maximum of 564 nm. Quantitative measurement of total flavonoids in the leaf showed significantly higher accumulation in *Y1-rr* and *Y1-pr* as compared to *Y1-wr* (p = 0.0039 and p = 0.0152, respectively) and the endogenous *p1* alleles ($p \leq 0.01$, Figure 3A). The high level of flavonoid compounds in the two *y1* transgenes could be due to the accumulation of flavonoid pathway intermediates such as chalcone and naringenin or novel compounds produced by the activity of maize enzymes induced by an active *y1*.

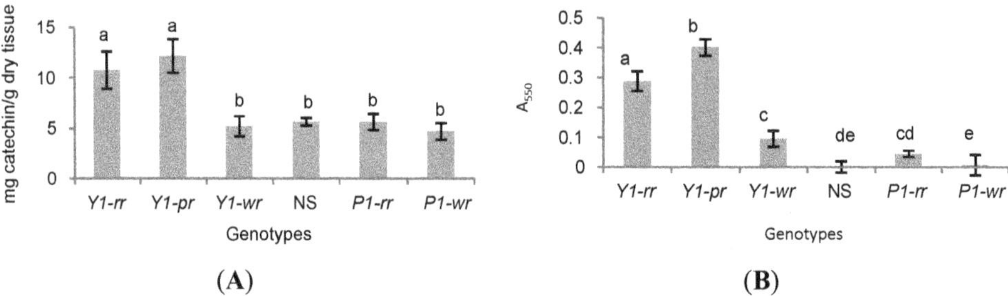

Figure 3. Sorghum *y1* gene induces accumulation of flavonoid compounds in transgenic maize leaves. *Y1-rr*, *Y1-pr*, and *Y1-wr* are independent transgenic events; NS, Negative segregant; *P1-rr* and *P1-wr*, maize lines carrying endogenous *p1* alleles. Values shown are mean ± SE. (**A**) Total flavonoids expressed as catechin equivalents; (**B**) Flavan-4-ols expressed as absorbance at 550 nm.

To further identify compounds regulated by the *y1* gene, we surveyed the leaves of *Y1*-maize plants for the presence of flavonoid precursors that give rise to either phlobaphenes or anthocyanins. The acid-butanol extracts were boiled to differentiate between flavan-4-ols and flavan-3,4-diols. In maize, flavan-4-ols (3-deoxyflavonoids) give rise to flavylium cations that have a λ_{max} of 564 nm and are heat labile, while flavylium cations obtained from the flavan-3,4-diol (3-hydroxyflavonoids)-derived compounds exhibit a λ_{max} of 533 nm and are unaffected by boiling [21,29]. Our results revealed that the *Y1*-maize extracts exhibited a major peak at 564 nm, which disappeared upon boiling, confirming the presence of flavan-4-ols (See Supplemental Figure S1). Quantification of flavan-4-ols in leaf tissue revealed that *Y1-rr* and *Y1-pr* transgenic events had significantly higher levels of these compounds as compared to the endogenous *p1* alleles and NS ($p < 0.0001$). Although *Y1-wr* transgenics accumulated significant levels of flavan-4-ols compared to the NS and *P1-wr* [B73] (p = 0.0178 and p = 0.0008 respectively), they did not significantly differ from the *P1-rr* allele (p = 0.1692) in this trait (Figure 3B).

2.3. *Y1-Maize Exhibits Enhanced Resistance to C. heterostrophus and C. graminicola*

In sorghum, a functional *y1* gene is required for resistance against *C. sublineolum* [16]. To test the response of *Y1*-maize to fungal challenges, plants were infected with either *C. heterostrophus* or *C. graminicola*. We compared the response of *Y1*-maize plants to both NS and *P1-wr*. When infected with *C. heterostrophus*, the inoculated leaves of *Y1*-maize plants produced a reduced number of chlorotic lesions compared to the control genotypes (Figure 4A). In different *Y1*- maize events, the mean values of the infected area ranged from 4% to 16% (Figure 4B). In contrast, these lesions were spread over about 40% and 62% of the leaf area in NS and *P1-wr* genotypes, respectively. These results indicate that the disease severity was significantly reduced in *Y1*-maize plants compared to the control genotypes ($p < 0.01$). Similarly, when entire plants were infected with *C. graminicola*, we found that the two transgenic maize lines (*Y1-rr* and *Y1-pr*) were more resistant, with averages of only 24%–29% of the leaf covered in lesions compared to the NS and *P1-wr*, which had 34% and 83% lesion area, respectively ($p < 0.05$) (Figure 4C,D).

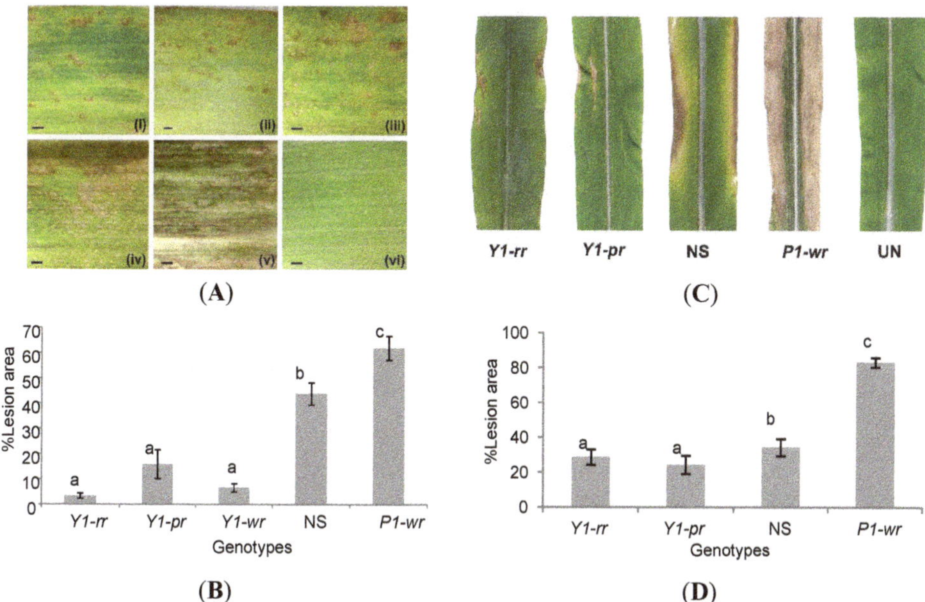

Figure 4. Sorghum *y1* gene enhances resistance against *C. heterostrophus* and *C. graminicola* in *Y1*-maize. (**A**) Detached leaf assay showing disease symptoms that developed 4 days post infection (dpi) when infected with *C. heterostrophus*. (**i**) *Y1-rr*; (**ii**) *Y1-pr*; (**iii**) *Y1-wr*; (**iv**) NS; (**v**) *P1-wr*; (**vi**) un-inoculated *Y1-pr*. Scale bar indicates 1 mm. (**B**) Quantification of the lesion area 4 dpi with *C. heterostrophus*. Values shown are the mean ± SE. (**C**) Symptoms that developed 11 dpi when whole plants were infected with *C. graminicola*. (**D**) Quantification of lesion area 11 dpi with *C. graminicola*. Values shown are the mean of 44 replicates ± SE. The x-axis in Figure 4A,C shows different genotypes used: *Y1-rr*, *Y1-pr*, *Y1-wr*; NS and *P1-wr*.

2.4. Induction of 3-Deoxyanthocyanidins during Y1-Maize–C. graminicola Interaction

Transgenic maize plants carrying the sorghum *Y1* gene were shown to be more resistant to the foliar pathogens *C. graminicola* and *C. heterostrophus* relative to the NS. In sorghum, resistance to foliar pathogens is in part due to the induced biosynthesis of 3-deoxyanthocyanidins [16]. Similarly, the resistant phenotype relative to the NS may be due to the biosynthesis of these novel compounds driven by the *Y1* gene in maize. Extracts obtained from infected leaves were analyzed using LC-MS to identify these compounds. The *m/z* ratios and elution times of peaks similar to those of apigeninidin and luteolinidin were scrutinized. Chromatograms obtained from the *Y1* transgenes *Y1-rr* and *Y1-pr* indicated novel peaks eluting with a retention time and *m/z* ratio similar to luteolinidin (271.060) compared to the NS (Figure 5), though no peaks were indicative of apigeninidin.

2.5. Discussion

In the current study, the activity of the sorghum *y1* gene was tested as a transgene in maize. First, our results established that, similar to P1, the Y1 protein is able to activate the same suite of known maize flavonoid genes, resulting in maize-like phlobaphene accumulation patterns in seed pericarp and cob glumes. In addition, the *y1* gene also induced the biosynthesis of flavan-4-ols in maize leaves, a property that has not been reported for maize lines expressing *P1-rr* or *P1-wr* [27,30]. The presence of flavan-4-ols in these leaves suggests that *y1* behaves in maize as it does in sorghum and actively interacts with the promoters of flavonoid genes to drive the pathway towards the production of these flavonoid compounds in the maize leaf. In *Y1*-maize, we observed phlobaphenes in the mature leaf tissue, which suggests that, like sorghum, the polymerization of flavan-4-ols to phlobaphenes can occur as also documented in the case of a maize mutant *Unstable factor for orange1* (*Ufo1*) [31,32]. Thus, the absence of flavan-4-ols and phlobaphenes in maize leaves containing active endogenous *p1* alleles could be due to poor activity of the *p1* or the inability of P1 to activate transcription of flavonoid structural genes in leaves [27,28,30]. The biochemical analysis of the *Y1*-maize leaves revealed that Y1 induced significant accumulation of flavan-4-ols and flavonoids to levels that are not commonly found in maize lines. This further establishes that the *y1* promoter is active in maize leaves.

The *Y1*-maize plants showed enhanced resistance against *C. graminicola* and *C. heterostrophus*. LC-MS profiling of induced flavonoids showed the presence of 3-deoxyanthocyanidin phytoalexins, specifically luteolinidin. In addition to luteolinidin, a second unknown small peak is observed in *Y1*-maize, which is absent from the NS profile. The improved disease resistance of *Y1*-maize plants is thus due to the induced 3-deoxyanthocyanidins as well as higher levels of pre-formed flavonoids, especially flavan-4-ols, which are known to contribute to plant defense [33]. Flavan-4-ols have also been suggested as putative precursors of 3-deoxyanthocyanidins [6,17,21,34–37]. Sorghum grains and leaves with higher levels of flavan-4-ols exhibited better resistance against mold compared to those that were deficient [38–40]. Flavan-4-ols include two main compounds—luteoforol and apiforol. Luteoforol has been demonstrated to have potent biocidal effects against many fungi and bacteria, including *C. graminicola* [41]. This antimicrobial activity might justify its presence in the epidermal cells of pericarp, silk, husk, and leaves. In fact, a mechanism describing the release of flavan-4-ols from their intracellular compartments to the sites of pathogen infection, similar to that of sorghum 3-deoxyanthocyanidins, has been proposed [18,25,41]. Although we were able to identify luteolinidin, we did not detect any apigeninidin, possibly due to a very active flavonoid 3′-hydroxylase, which converts apigeninidin to luteolinidin [23,42].

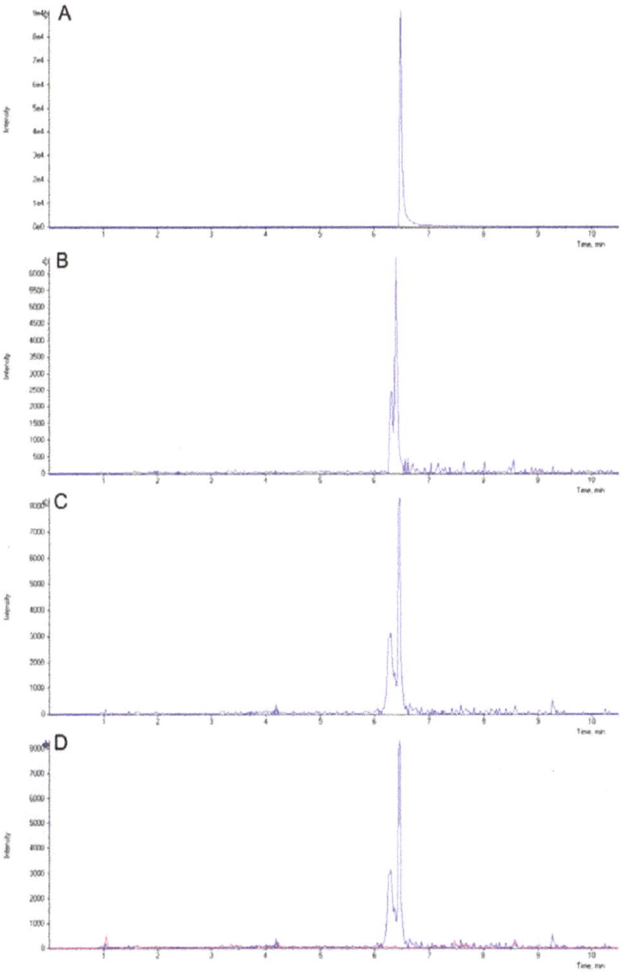

Figure 5. Induction of 3-deoxyanthocyanidins and their derivatives in *Y1*-maize. LC-MS chromatograms obtained from the luteolinidin standard (**A**), infected leaves of *Y1-rr* ((**B**,**C**) representing two biological replicates), and an overlay (**D**) of *Y1-rr* (sample B; blue trace) and NS (pink trace) are presented for comparison. The *m*/*z* values of the extracted chromatograms were similar to those of the luteolinidin standard (271.060).

Induction of phenylpropanoids in maize cell suspensions after transformation with the *p1* transgene has been reported [21,24,43]. These, along with our current results, demonstrate that *p1* and *y1* transgenes play similar regulatory roles in the phenylpropanoid pathway in maize. However, the basis of this mechanism is not yet clear. One possibility is that the R2R3 MYB protein products of *p1* and *y1* genes might interact directly with structural genes in the phenylpropanoid pathway to secure the efficient flow of intermediates between the phenylpropanoids and flavonoids. In fact, in BMS maize cells, the *p1* transgene induced the expression of *pal1*, which controls the flow of the amino acid phenylalanine into the phenylpropanoid pathway [43]. Since *y1* and *p1* are known to regulate flavonoid biosynthesis downstream of chalcone, it is also possible that these transcription factors may interfere with feedback regulation controlling the activity of enzymes working in other branches of the phenylpropanoid pathway [21,32,44–46]. Thus future exploitation of *y1*, an R2R3 MYB regulatory gene, to produce desirable biopesticides is a viable strategy [2,47–50].

3. Experimental Section

3.1. Maize Genetic Stocks

Maize genetic stocks of inbred lines 4co 63 (*p1-ww*) and B73 (*P1-wr*) were obtained from Maize Genetic Coop Center, USDA, Urbana-Champaign, IL, USA. Genetics stocks carrying *p1* alleles *P1-rr 4B2*, *P1-ww-1112* and *p-del2* were obtained from Thomas Peterson, Iowa State University, Ames, IA, USA.

3.2. Transgene Constructs

All plasmids used in this study were developed based on the pBluescript II vector (Stratagene, La Jolla, CA, USA). The plasmid p*Y1::Y1* contains 9164 bp of the *y1* gene [AY860968 [16]], which includes the 2375 bp of the 5' regulatory region, 6946 bp sequence with three exons and two introns, and 820 bp of the 3'UTR. This plasmid was prepared by ligating the *Hin*dIII-*Kpn*I DNA fragment of the *Y1-rr* gene into a pBluescript II vector.

3.3. Tissue Culture, Transformation, and Regeneration of Transgenic Maize Plants

We used maize *HiII* line as a transgene recipient because it carries a non-functional *p1* allele. *HiII* was developed from a cross between A188 × B73 [51]. Immature zygotic embryos derived from the *HiII* maize [52] were used to develop friable embryogenic type II calli. Callus induction, maintenance, and transformation were carried out according to a previously described protocol at the Plant Transformation Facility at Iowa State University [53]. A plasmid carrying the *BAR* gene for Bialphos herbicide tolerance was co-bombarded, along with the p*Y1:Y1* construct. A total of 16 p*Y1:Y1* independent transformation events were generated from calli resistant to the herbicide Bialphos (BASTA™, AgrEvo, Wilimington, DE, USA). The selection for transgenic plants in T_1 and subsequent generations was based on herbicide resistance as well as PCR analysis, using *y1* specific gene primers. The transgenic plants were maintained in a hemizygous state by out-crossing with pollen from the inbred line 4Co63 that carries a *p1-ww* allele (null *p1* allele). Progenies derived from such crosses always segregated in a 1:1 ratio, indicating stable expression patterns of the transgenic plants included in this study. All maize plants carrying p*Y1::Y1* transgenes exhibited normal growth and morphology when compared with the sibling negative segregant (NS) maize plants.

3.4. Expression Analysis of Genes Induced by y1

Total RNA was extracted from pericarp 18 days after pollination (dap) by RNAzolRT (Molecular Research Center Inc., Cincinnati, OH, USA) and used to synthesize first strand cDNA using the High-Capacity cDNA Reverse Transcription Kit (Applied Biosystems, Foster City, CA,

USA). PCR using primers specific to flavonoid pathway genes was used to determine the expression of said genes: yellow seed 1 (*y1*): RT_PWREx_2F (5'-TCCGGTGCGGCAAGAG-3') and RT_PWREx_2R (5'-GGAGCTTGATGATGATGTCTTCTTC-3'); chalcone synthase (*c2*): CHSF (5'-TCGATCGGTCTCTCTGGTACAACGTA-3') and CHSR (5'-TACATCATGAGGCGGTTCAC GGA-3'); chalcone isomerase (*chi1*): CHIF (5'-GTGCGGAATTTAACATGGCGTGC-3') and CHIR (5'-CGGCGCGAAAGTCTCTGGCTT-3'); flavonoid 3'-hydroxylase (*pr1*): 5F3H-F2 (5'-GAGCAC GTGGCGTACAACTA-3') and ZMR4 (5'-AAACGTCTCCTTGATCACCGC-3'); dihydroflavonol reductase (*a1*): A1 (5'-CAATTCGTTGAACATGGAAGTAAG-3') and A2 (5'-CAATTCGTT GAACATGGAAGTAAG-3') and glyceraldehyde 3-phosphate dehydrogenase (*gapdh*): GAP1 (5'-AGGGTGGTGCCAAGAAGGTTG-3') and GAP2 (5'-GTAGCCCCACTCGTTGTCGTA-3').

3.5. Tissue Collection for Chemical Analyses

All tissues used for chemical analyses were collected at the specified time, flash frozen in liquid nitrogen and either lyophilized or stored at −80 °C. All analyses were performed on three independent sample replicates.

3.6. Quantification of Total Flavonoids and Flavan-4-ols

All biochemical analyses were carried out on the second leaf above the primary ear of greenhouse grown plants collected at the time of pollination. For total flavonoid quantification, ground tissue (20 mg) was washed three times in ether to remove waxes and chlorophyll pigments and then extracted three times under sonication in 70% acetone supplemented with 1 mM ascorbic acid. The supernatant was collected and acetone was evaporated using a speed vacuum drier. The extract was used for determination of total flavonoids [54]. The extracts were diluted with 1 M NaOH and the absorbance was recorded at 510 nm using a SpectraMAX 190 plate reader (Molecular Devices Corp., Sunnyvale, CA, USA). Total flavonoid content was expressed as mg catechin equivalent g^{-1} dry weight.

To quantify flavan-4-ols, 30 mg ground leaf tissue was washed in ether and suspended in 500 μL of HCl:butanol (3:7). The homogenate was incubated at 37 °C for 1 h, followed by centrifugation at 20,000 *g* for 10 min. The absorbance of the supernatant was recorded at 550 nm using an UV mini-1240 spectrophotometer (Shimadzu Scientific Instruments, Inc. Columbia, MD, USA). The flavan-4-ols were expressed as the relative concentration of flavylium ions [21].

3.7. Evaluation of Y1-Transgenic Plants for Resistance to Cochliobolus heterostrophus and Colletotrichum graminicola

Resistance to southern corn leaf blight caused by *C. heterostrophus* was evaluated using the detached leaf assay [55]. *C. heterostrophus* was grown on potato dextrose agar (PDA) under continuous light at room temperature for ten days. Conidia were collected in 0.001% Tween 20. The second leaf above the primary ear was collected 15 days after pollination (dap). Leaf discs were prepared from both sides of the midrib and those on the left side were used as controls. Discs with adaxial surface facing upward were placed on water agar (1% w/v) supplemented with 2 mg·L^{-1} kinetin. Whatman filter papers soaked in either a spore suspension of *C. heterostrophus*

(10^5 spores·mL^{-1}) or 0.001% Tween 20 for the control were placed on the leaf discs. Plates were incubated under illumination at 28 °C. The filter paper discs were removed after 24 h and plates were kept under the same conditions until collection. The disease phenotypes of the p*Y1:Y1* transgenic plants and control genotypes were recorded using a dissection microscope (Nikon SMZ1000) connected to a Nikon digital camera (DXM1200F). Disease severity was quantified as described below.

To test the role of *y1* in resistance to anthracnose leaf blight, four- to six-week-old greenhouse-grown plants were inoculated and disease was quantified using *C. graminicola* as described previously [16].

3.8. Image Analysis for Evaluation of Disease Response

For quantitative analysis, images were processed by Automated Lesion Extraction using algorithms (PhenoPhyte) developed for a visual phenotype database [42]. This technique depends on differentiating the lesion pixels (foreground) from the healthy ones (background) and measuring the area of lesions. The percentage of the infected area was used to evaluate disease severity.

3.9. LC-MS Analysis of 3-Deoxyanthocyanidins

Flag leaves of field-grown plants were used to identify compounds induced in response to *C. graminicola* infection, using the detached leaf assay as described above and harvested for analysis 3 dpi. Infections were carried out in triplicate for each genotype, each of which consisted of three individual leaves. Tissue samples (~100 mg) were extracted in 2 mL of 2 N HCl by boiling for 40 min, then centrifuged at 20,000 *g* for 15 min. The resulting supernatant was extracted twice in 1 mL of isoamyl alcohol, which was evaporated to dryness and re-suspended in 250 μL of methanol supplemented with 0.1% HCl [56]. Extracts (5 μL) were separated by reverse phase HPLC using a Prominence 20 UFLCXR system (Shimadzu, Columbia, MD, USA) with a Waters BEH C18 column (100 × 2.1 mm, 1.7 μm particle size) and a 20 min aqueous/acetonitrile gradient, at a flow rate of 250 μL/min. Solvent A was water with 0.1% formic acid, and Solvent B was acetonitrile with 0.1% formic acid. The initial conditions were 97% A and 3% B, increasing to 45% B at 10 min and 75% B at 12 min, then held at 75% B until 17.5 min before returning to the initial conditions at 18 min. The eluate was delivered into the 5600 (QTOF) TripleTOF using a Duospray™ ion source (all AB Sciex, Framingham, MA, USA), samples were analyzed in positive ion mode, and the mass spectrometer was operated in IDA (Information Dependent Acquisition) mode with a 100 m survey scan from 50 to 1250 *m*/*z*, and up to 10 MS/MS product ion scans per duty cycle. The survey scan data was used to generate the extracted ion chromatographs with PeakView software package (AB Sciex, Framingham, MA, USA).

4. Conclusions

We engineered 3-deoxyanthocyanidin phytoalexins in maize by transforming it with a sorghum transcription factor, *yellow seed 1* (*y1*). In maize, *Y1* expression drives the biosynthesis of foliar flavonoid compounds, especially flavan-4-ols. Furthermore, fungal infection resulted in the induction of luteolinidin, which is a potent antifungal compound. We believe the preformed flavan-4-ols, in addition to the induced luteolinidin, contributed to increased resistance of the transgenic maize compared to the near-isogenic non-transgenic lines. The introduction of *y1* is thus a viable strategy for introducing anthracnose resistance to maize lines carrying the downstream flavonoid pathway structural genes.

Supplementary Materials

Supplementary materials can be accessed at: http://www.mdpi.com/1420-3049/20/02/2388/s1.

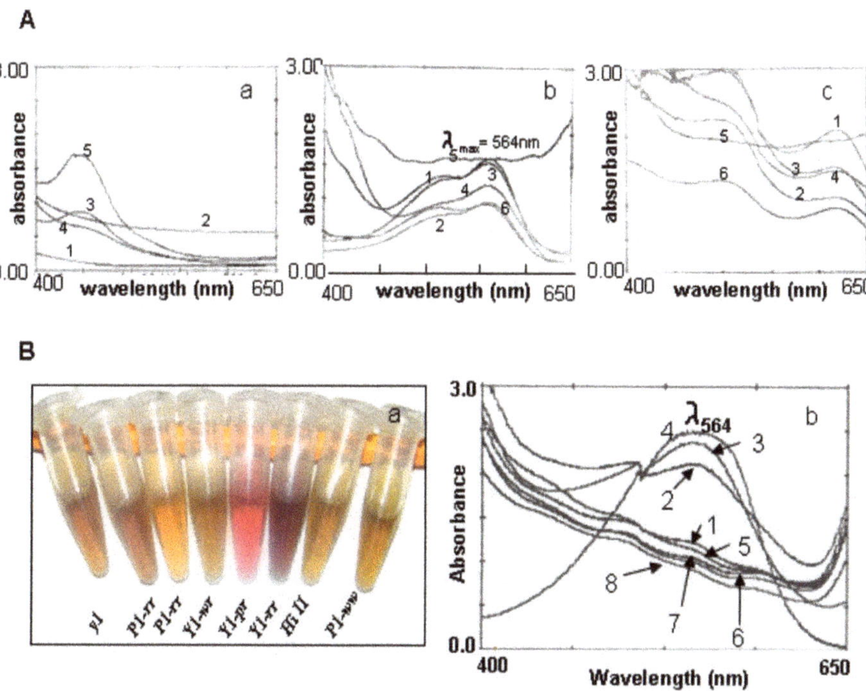

Figure S1. Spectral analysis of flavonoids from floral and vegetative tissues of transgenic and non-transgenic maize plants. (**A**) Spectral analysis of acidic-butanol extracts from sibling maize plants segregating for either the absence (a) or the presence (b) of *y1* transgene. The absorption maxima at 564 nm were eliminated by boiling confirming the presence of flavan-4-ols (c). Numbers on curves represent extracts prepared from: pericarp (1), cob glumes (2), tassel glumes (3), anthers (4), and mid-rib (5). *P1-rr* Pericarp extract (curve 6) was used as a positive control for flavan-4-ols. (**B**) Spectral analysis of acidic-butanol extracts from mature leaves of transgenic and non transgenic maize plants. Extracts from leaves of standard maize lines expressing native *P1-rr* and *P1-wr* alleles were included for comparison. Leaves were deprived of chlorophyll followed by acidic-butanol assay. The clear extracts were photographed (a) and screened for the presence of flavan-4-ols (b). Curves are for leaf extracts from: 1, *Y1-wr*; 2, *Y1-rr*; 3, *Y1-pr*; 4, *P1-rr* pericarp (positive control for flavan-4-ols); 5, NS; 6, HiII; 7, *P1-rr*; and 8, *P1-wr*.

Acknowledgments

We thank Scott Harkcom and Penn State Agronomy farm staff for assistance with field preparation and tending the summer crops, Scott Diloreto for greenhouse maintenance, German Sandoya for his advice for statistical analyses, Nur Suhada Abu Bakar for plant disease assays, Maurice Snook for HPLC analyses, and Kameron Wittmeyer for his critical review of the manuscript. Farag Ibraheem received a graduate fellowship from the Egyptian Government, and Qixian Tan was supported by a graduate assistantship from the Plant Science Department, Penn State University. This work was supported by a NIFA-AFRI competitive grant award 2011-67009-30017 and Hatch projects 4452 and 4430 to SC. LC-MS was performed with the support of the Metabolomics Core Facility at Penn State using 5600 (QTOF) TripleTOF instrumentation funded by a grant from NSF MRI 1126373.

Author Contributions

S.C., F.I., and I.G. designed the experiments; F.I., I.G., and Q.T. performed the experiments; C-R.S. analyzed the disease lesion data; F.I., I.G., C-R.S. and S.C. wrote the paper. All authors discussed, edited and approved the final manuscript.

Conflicts of Interest

The authors declare no conflicts of interest.

References

1. Winkel-Shirley, B. Flavonoid biosynthesis. A colorful model for genetics, biochemistry, cell biology, and biotechnology. *Plant Physiol.* **2001**, *126*, 485–493.
2. Dixon, R.A.; Liu, C.; Jun, J.H. Metabolic engineering of anthocyanins and condensed tannins in plants. *Curr. Opin. Biotechnol.* **2013**, *24*, 329–335.
3. Verpoorte, R.; Memelink, J. Engineering secondary metabolite production in plants. *Curr. Opin. Biotechnol.* **2002**, *13*, 181–187.
4. Kristensen, C.; Morant, M.; Olsen, C.E.; Ekstrøm, C.T.; Galbraith, D.W.; Møller, B.L.; Bak, S. Metabolic engineering of dhurrin in transgenic arabidopsis plants with marginal inadvertent effects on the metabolome and transcriptome. *Proc. Natl. Acad. Sci. USA* **2005**, *102*, 1779–1784.
5. Tattersall, D.B.; Bak, S.; Jones, P.R.; Olsen, C.E.; Nielsen, J.K.; Hansen, M.L.; Høj, P.B.; Møller, B.L. Resistance to an herbivore through engineered cyanogenic glucoside synthesis. *Science* **2001**, *293*, 1826–1828.
6. Styles, E.D.; Ceska, O. Pericarp flavonoids in genetic strains of *zea mays*. *Maydica* **1989**, *34*, 227–237.
7. Grotewold, E.; Drummond, B.J.; Bowen, B.; Peterson, T. The *myb*-homologous *p* gene controls phlobaphene pigmentation in maize floral organs by directly activating a flavonoid biosynthetic gene subset. *Cell* **1994**, *76*, 543–553.

8. Morohashi, K.; Casas, M.I.; Ferreyra, L.F.; Mejía-Guerra, M.K.; Pourcel, L.; Yilmaz, A.; Feller, A.; Carvalho, B.; Emiliani, J.; Rodriguez, E.; *et al.* A genome-wide regulatory framework identifies maize *pericarp color1* controlled genes. *Plant Cell* **2012**, *24*, 2745–2764.
9. Boddu, J.; Jiang, C.H.; Sangar, V.; Olson, T.; Peterson, T.; Chopra, S. Comparative structural and functional characterization of sorghum and maize duplications containing orthologous myb transcription regulators of 3-deoxyflavonoid biosynthesis. *Plant Mol. Biol.* **2006**, *60*, 185–199.
10. Bennetzen, J.L.; Freeling, M. The unified grass genome: Synergy in synteny. *Genome Res.* **1997**, *7*, 301–306.
11. Gaut, B.S.; le Thierry d'Ennequin, M.; Peek, A.S.; Sawkins, M.C. Maize as a model for the evolution of plant nuclear genomes. *Proc. Natl. Acad. Sci. USA* **2000**, *97*, 7008–7015.
12. Devos, K.M.; Gale, M. Genome relationships: The grass model in current research. *Plant Cell* **2000**, *12*, 637–646.
13. Melake-Berhan, A.; Hurber, S.H.; Butler, L.G.; Bennetzen, J.L. Structure and evolution of the genome of sorghum bicolor and zea mays. *Theor. Appl. Genet.* **1993**, *86*, 598–604.
14. Zanta, C.A.; Yang, X.; Axtell, J.D.; Bennetzen, J.L. The candystripe locus, *y-cs*, determines mutable pigmentation of the sorghum leaf, flower, and pericarp. *J. Hered.* **1994**, *85*, 23–29.
15. Chopra, S.; Brendel, V.; Zhang, J.B.; Axtell, J.D.; Peterson, T. Molecular characterization of a mutable pigmentation phenotype and isolation of the first active transposable element from *sorghum bicolor*. *Proc. Natl. Acad. Sci. USA* **1999**, *96*, 15330–15335.
16. Ibraheem, F.; Gaffoor, I.; Chopra, S. Flavonoid phytoalexin dependent resistance to anthracnose leaf blight requires a functional *yellow seed1* in *sorghum bicolor*. *Genetics* **2010**, *184*, 915–926.
17. Chopra, S.; Gevens, A.; Svabek, C.; Wood, K.V.; Peterson, T.; Nicholson, R.L. Excision of the *candystripe1* transposon from a hyper-mutable *y1-cs* allele shows that the sorghum *y1* gene controls the biosynthesis of both 3-deoxyanthocyanidin phytoalexins and phlobaphene pigments. *Physiol. Mol. Plant Pathol.* **2002**, *60*, 321–330.
18. Snyder, B.A.; Nicholson, R.L. Synthesis of phytoalexins in sorghum as a site-specific response to fungal ingress. *Science* **1990**, *248*, 1637–1639.
19. Hipskind, J.D.; Nicholson, R.L.; Goldsbrough, P.B. Isolation of a cdna encoding a novel leucine-rich repeat motif from *Sorghum bicolor* inoculated with fungi. *Mol. Plant-Microbe Interact.* **1996**, *9*, 819–825.
20. McMullen, M.D.; Snook, M.; Lee, E.A.; Byrne, P.F.; Kross, H.; Musket, T.A.; Houchins, K.; Coe, E.H., Jr. The biological basis of epistasis between quantitative trait loci for flavone and 3-deoxyanthocyanin synthesis in maize (*Zea mays L.*). *Genome* **2001**, *44*, 667–676.
21. Grotewold, E.; Chamberlin, M.; Snook, M.; Siame, B.; Butler, L.; Swenson, J.; Maddock, S.; Clair, G.S.; Bowen, B. Engineering secondary metabolism in maize cells by ectopic expression of transcription factors. *Plant Cell* **1998**, *10*, 721–740.
22. McMullen, M.D.; Kross, H.; Snook, M.E.; Cortes-Cruz, M.; Houchins, K.E.; Musket, T.A.; Coe, E.H. *Salmon silk* genes contribute to the elucidation of the flavone pathway in maize (*Zea mays* L.). *J. Hered.* **2004**, *95*, 225–233.

23. Sharma, M.; Chai, C.; Morohashi, K.; Grotewold, E.; Snook, M.E.; Chopra, S. Expression of flavonoid 3′-hydroxylase is controlled by p1, the regulator of 3-deoxyflavonoid biosynthesis in maize. *BMC Plant Biol.* **2012**, *12*, 196.
24. Zhang, P.; Wang, Y.; Zhang, J.; Maddock, S.; Snook, M.; Peterson, T. A maize qtl for silk maysin levels contains duplicated *myb*-homologous genes which jointly regulate flavone biosynthesis. *Plant Mol. Biol.* **2003**, *52*, 1–15.
25. Nicholson, R.L.; Kollipara, S.S.; Vincent, J.R.; Lyons, P.C.; Cadena-Gomez, G. Phytoalexin synthesis by the sorghum mesocotyl in response to infection by pathogenic and nonpathogenic fungi. *Proc. Natl. Acad. Sci. USA* **1987**, *84*, 5520–5524.
26. Bate-Smith, E.C. Luteoforol (3',4,4',5,7-pentahydroxyflavan) in *Sorghum vulgare* L. *Phytochemistry* **1969**, *8*, 1803–1810.
27. Chopra, S.; Athma, P.; Peterson, T. Alleles of the maize p gene with distinct tissue specificities encode myb-homologous proteins with c-terminal replacements. *Plant Cell* **1996**, *8*, 1149–1158.
28. Zhang, P.; Chopra, S.; Peterson, T. A segmental gene duplication generated differentially expressed *myb*-homologous genes in maize. *Plant Cell* **2000**, *12*, 2311–2322.
29. Boddu, J.; Svabek, C.; Ibraheem, F.; Jones, A.D.; Chopra, S. Characterization of a deletion allele of a sorghum myb gene, *yellow seed1* showing loss of 3-deoxyflavonoids. *Plant Sci.* **2005**, *169*, 542–552.
30. Cocciolone, S.M.; Nettleton, D.; Snook, M.; Peterson, T. Transformation of maize with the *p1* transcription factor directs production of silk maysin, a corn earworm resistance factor, in concordance with a hierarchy of floral organ pigmentation. *Plant Biotechnol.* **2005**, *3*, 225–235.
31. Chopra, S.; Cocciolone, S.M.; Bushman, S.; Sangar, V.; McMullen, M.D.; Peterson, T. The maize unstable factor for orange1 is a dominant epigenetic modifier of a tissue specifically silent allele of pericarp color1. *Genetics* **2003**, *163*, 1135–1146.
32. Robbins, M.L.; Roy, A.; Wang, P.-H.; Gaffoor, I.; Sekhon, R.S.; de O. Buanafina, M.M.; Rohila, J.S.; Chopra, S. Comparative proteomics analysis by DIGE and iTRAQ provides insight into the regulation of phenylpropanoids in maize. *J. Proteomics* **2013**, *93*, 254–275.
33. Hammerschmidt, R. Phenols and plant-pathogen interactions: The saga continues. *Physiol. Mol. Plant Pathol.* **2005**, *66*, 77–78.
34. Kambal, A.E.; Bate-Smith, E.C. A genetic and biochemical study on pericarp pigmentation between two cultivars of grain sorghum, *sorghum bicolor*. *Heredity* **1976**, *37*, 417–421.
35. Schutt, C.; Netzly, D. Effect of apiforol and apigeninidin on growth of selected fungi. *J. Chem. Ecol.* **1991**, *17*, 2261–2266.
36. Stich, K.; Forkmann, G. Biosynthesis of 3-deoxyanthocyanins with flower extracts from *Sinningia cardinalis*. *Phytochemistry* **1988**, *27*, 785–789.
37. Winefield, C.S.; Lewis, D.H.; Swinny, E.E.; Zhang, H.B.; Arathoon, H.S.; Fischer, T.C.; Halbwirth, H.; Stich, K.; Gosch, C.; Forkmann, G.; *et al.* Investigation of the biosynthesis of 3-deoxyanthocyanins in *Sinningia cardinalis*. *Physiol. Plant.* **2005**, *124*, 419–430.

38. Jambunathan, R.; Butler, L.G.; Bandyopadhyay, R.; Lewisk, N. Polyphenol concentration in grain leaf and callus of mold susceptible and mold resistant sorghum cultivars. *J. Agric. Food Chem.* **1986**, *34*, 425–420.
39. Jambunathan, R.; Kherdekar, M.S.; Bandyopadhyay, R. Flavan-4-ols concentration in mold-susceptible and mold resistant sorghum at different stages of grain development. *J. Agric. Food Chem.* **1990**, *38*, 545–548.
40. Menkir, A.; Ejeta, G.; Butler, L.; Melakeberhan, A. Physical and chemical kernel properties associated with resistance to grain mold in sorghum. *Cereal Chem.* **1996**, *73*, 613–617.
41. Spinelli, F.; Speakman, J.B.; Rademacher, W.; Halbwirth, H.; Stich, K.; Costa, G. Luteoforol, a flavan 4-ol, is induced in pome fruits by prohexadione-calciumand shows phytoalexin-like properties against *erwinia amylovora* and other plant pathogens *Eur. J. Plant Pathol.* **2005**, *112*, 133–142.
42. Boddu, J.; Svabek, C.; Sekhon, R.; Gevens, A.; Nicholson, R.L.; Jones, A.D.; Pedersen, J.F.; Gustine, D.L.; Chopra, S. Expression of a putative flavonoid 3'-hydroxylase in sorghum mesocotyls synthesizing 3-deoxyanthocyanidin phytoalexins. *Physiol. Mol. Plant Pathol.* **2004**, *65*, 101–113.
43. Bruce, W.; Folkerts, O.; Garnaat, C.; Crasta, O.; Roth, B.; Bowen, B. Expression profiling of the maize flavonoid pathway genes controlled by estradiol-inducible transcription factors crc and p. *Plant Cell* **2000**, *12*, 65–80.
44. Bushman, B.S.; Snookc, M.E.; Gerkeb, J.B.; Szalmaa, S.J.; Berhowd, M.A.; Houchinse, K.E.; McMullen, M.D. Two loci exert major effects on chlorogenic acid synthesis in maize *Crop Sci.* **2002**, *42*, 1669–1678.
45. Dixon, R.A.; Achnine, L.; Kota, P.; Liu, C.; Reddy, M.S.; Wang, L. The phenylpropanoid pathway and plant defence—A genomics perspective. *Mol. Plant Pathol.* **2002**, *3*, 371–390.
46. Blount, J.W.; Korth, K.L.; Masoud, S.A.; Rasmussen, S.; Lamb, C.; Dixon, R.A. Altering expression of cinnamic acid 4-hydroxylase in transgenic plants provides evidence for a feedback loop at the entry point into the phenylpropanoid pathway. *Plant Physiol.* **2000**, *122*, 107–116.
47. Jin, H.; Martin, C. Multifunctionality and diversity within the plant myb-gene family. *Plant Mol. Biol.* **1999**, *41*, 577–585.
48. Jirschitzka, J.; Mattern, D.J.; Gershenzon, J.; D'Auria, J.C. Learning from nature: New approaches to the metabolic engineering of plant defense pathways. *Curr. Opin. Biotechnol.* **2013**, *24*, 320–328.
49. Mengiste, T.; Chen, X.; Salmeron, J.; Dietrich, R. The *Botrytis susceptible1* gene encodes an r2r3myb transcription factor protein that is required for biotic and abiotic stress responses in arabidopsis. *Plant Cell* **2003**, *15*, 2551–2565.
50. Stracke, R.; Werber, M.; Weisshaar, B. The r2r3-myb gene family in arabidopsis thaliana. *Curr. Opin. Plant Biol.* **2001**, *4*, 447–456.
51. Armstrong, C.L.; Romero-Severson, J.; Hodges, T.K. Improved tissue culture response of an elite maize inbred through backcross breeding, and identification of chromosomal regions important for regeneration by rflp analysis. *Theor. Appl. Genet.* **1992**, *84*, 755–762.

52. Armstrong, C.L.; Green, C.E. Establishment and maintenance of friable, embryogenic maize callus and the involvement of l-proline. *Planta* **1985**, *164*, 207–214.
53. Frame, B.R.; Zhang, H.; Cocciolone, S.M.; Sidorenko, L.V.; Dietrich, C.R.; Pegg, S.E.; Zhen, S.; Schnable, P.S.; Wang, K. Production of transgenic maize from bombarded type ii callus: Effect of gold particle size and callus morphology on transformation efficiency. *Vitr. Cell. Dev. Biol.* **2000**, *36*, 21–29.
54. Wolfe, K.; Wu, X.Z.; Liu, R.H. Antioxidant activity of apple peels. *J. Agric. Food Chem.* **2003**, *51*, 609–614.
55. Coca, M.; Bortolotti, C.; Rufat, M.; Penas, G.; Eritja, R.; Tharreau, D.; del Pozo, A.M.; Messeguer, J.; San Segundo, B. Transgenic rice plants expressing the antifungal afp protein from *Aspergillus giganteus* show enhanced resistance to the rice blast fungus *Magnaporthe grisea*. *Plant Mol. Biol.* **2004**, *54*, 245–259.
56. Harborne, J.B. *Phytochemical Methods. A Guide to Modern Techniques of Plant Analysis*, 3rd ed.; Champman and Hall: New York, NY, USA, 1998.

Sample Availability: Not available.

EDTA a Novel Inducer of Pisatin, a Phytoalexin Indicator of the Non-Host Resistance in Peas

Lee A. Hadwiger and Kiwamu Tanaka

Abstract: Pea pod endocarp suppresses the growth of an inappropriate fungus or non-pathogen by generating a "non-host resistance response" that completely suppresses growth of the challenging fungus within 6 h. Most of the components of this resistance response including pisatin production can be elicited by an extensive number of both biotic and abiotic inducers. Thus this phytoalexin serves as an indicator to be used in evaluating the chemical properties of inducers that can initiate the resistance response. Many of the pisatin inducers are reported to interact with DNA and potentially cause DNA damage. Here we propose that EDTA (ethylenediaminetetraacetic acid) is an elicitor to evoke non-host resistance in plants. EDTA is manufactured as a chelating agent, however at low concentration it is a strong elicitor, inducing the phytoalexin pisatin, cellular DNA damage and defense-responsive genes. It is capable of activating complete resistance in peas against a pea pathogen. Since there is also an accompanying fragmentation of pea DNA and alteration in the size of pea nuclei, the potential biochemical insult as a metal chelator may not be its primary action. The potential effects of EDTA on the structure of DNA within pea chromatin may assist the transcription of plant defense genes.

Reprinted from *Molecules*. Cite as: Hadwiger, L.A.; Tanaka, K. EDTA a Novel Inducer of Pisatin, a Phytoalexin Indicator of the Non-Host Resistance in Peas. *Molecules* **2015**, *20*, 24-34.

1. Introduction

The pea endocarp system [1] has been employed to follow and understand the transcription initiation of the nonhost resistance response in plants at chromatin sites targeted by DNA-specific gene activators, one of which is a fungal DNase [2,3]. Some of the components that elicit phytoalexin production may also act without directly targeting DNA [4]. The compound EDTA (Figure 1), used traditionally in biochemistry as a chelating agent [5], was not expected to target cellular DNA or induce the phytoalexin pisatin.

$$(HOOC{-}CH_2)_2N{-}CH_2{-}CH_2{-}N(CH_2{-}COOH)_2$$

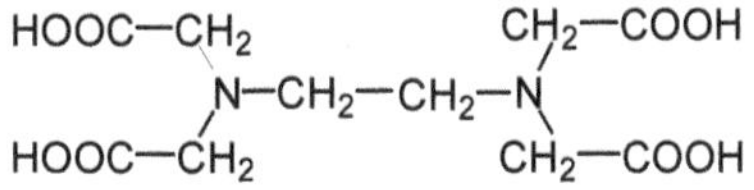

Figure 1. Structural formula of EDTA (ethylenediaminetetraacetic acid).

FsphDNase is a natural biotic elicitor of both pisatin and the nonhost disease resistance response [3]. It is released from *Fusarium solani* f. sp. *phaseoli* (Fsph), a pathogen of bean, and causes single strand nicks in DNA in temporal association with the induction of pisatin [3]. This fungal enzyme's catalytic action is dependent on Mn^{2+} [3]. EDTA was employed in the pea/Fsph interaction with the intention to chelate metal cofactors from, and negate the *in vivo* function of, FsphDNase in the fungal/host interaction and therein to block phytoalexin production. However, EDTA applied at certain concentrations, can enhance the production of phytoalexin. An assay was developed [6] using agarose gel electrophoresis for DNA damage analysis that detects DNA alterations resulting from a component's direct action on the DNA molecule. This technique offers an assay for assessing EDTA-caused DNA fragmentation. DNA fragmentation is temporally associated with early changes in the host chromatin [7]. Some of the chromatin changes have been shown to be the result of the ubiquitination of proteins (histones H2A/H2B and the transcription factor, HMG A) affecting regions containing PR genes. In addition to pathogen-derived-elicitors, the pea tissue responds to an extensive array of abiotic defense gene inducers [4] that possess defined actions within many biological systems. EDTA is now one that is of interest as a cellular DNA targeting compound.

Molecular and Cytological Observations of the Pea Immune Response

The use of the pea pod endocarp model system allows an examination of synchronous molecular events as the elicitor applications rapidly contact all of the cells within the epidermal surface layer. The action of successful elicitors culminates in resistance that within 6 h completely suppresses further growth of a true, compatible pea pathogen [1] such as *Fusarium solani* f. sp. *pisi* (Fspi). If a functional induction of resistance is unsuccessful, the endocarp tissue hosts a susceptibility reaction that develops within 24 h. The mapping of regions within pea chromosomes revealed that some QTLs associated with disease resistance encompass map sites of pathogenesis-response (PR) genes [8,9].

The PR genes constitute the major components of disease resistance [10]. Thus, there are sensitive regions within pea chromatin associated with the pea disease resistance response. The induction and biosynthesis of the anti-fungal phytoalexin pisatin, as an isoflavonoid, requires the participation of multiple enzymes in this secondary pathway [11].

This suggests that the sensitive regions of pea chromatin itself may often serve as the initial target of the elicitors. Such a general action would support the prospect of a target being able simultaneously to activate genes located at multiple sites throughout the genome. This report on EDTA is designed to examine if its effect on pisatin accumulation, DNA damage, defense gene induction and cytological changes observed in relationship to the expression of disease resistance, may also be associated with a DNA target.

2. Results and Discussion

2.1. Phytoalexin Induction by EDTA

Table 1 indicates the concentrates of EDTA that affect the induction of the isoflavonoid phytoalexin, pisatin within 24 h after treatment. Pisatin as a phytoalexin possesses some antifungal properties. Concentrations as low as 7.8 mM EDTA independently induce pisatin accumulations. This concentration synergistically increases pisatin accumulation above that by the Fsph spores alone, e.g., {7.8 mM EDTA + spores treatment = 474 µg − (spores only treatment = 221 µg) = 85 µg/g fresh wt.}.

Table 1. The effect of a broad range of EDTA concentrations on the induction of pisatin accumulations in pea endocarp tissue in the presence and absence of *Fusarium solani* f. sp. *phaseoli* (Fsph) a pathogen of bean.

Treatment [a]	Conc. EDTA mM	Pisatin µg/g frs.wt.
Water	0	0.0
EDTA	250	0.0
EDTA	125	12.4 ± 9.7
EDTA	62.5	62.2 ± 22.2
EDTA	31.2	194.2 ± 53.7
EDTA	15.6	136.2 ± 5.5
EDTA	7.8	85.8 ± 19.3
EDTA	3.9	3.3 ± 2.9
EDTA	1.9	0.6 ± 0.5
EDTA	0.9	0.0 ± 0.0
Water + Fsph spores	0.0	221.5 ± 67.9
EDTA + Fsph spores	250	1.7 ± 1.7
EDTA + Fsph spores	125	15.1 ± 2.1
EDTA + Fsph spores	62.5	86.3 ± 48.3
EDTA + Fsph spores	31.2	292.2 ± 115.2
EDTA + Fsph spores	15.6	428.9 ± 29.5
EDTA + Fsph spores	7.8	474.8 ± 16.3
EDTA + Fsph spores	3.9	353.6 ± 3.1
EDTA + Fsph spores	1.9	376.9 ± 39.7
EDTA + Fsph spores	0.9	258.0 ± 53.0

Notes: [a] Twenty µL of the indicated treatments were applied per pea pod half (~180 mg fresh weight) followed by 5 µL suspension of *F. solani* f. sp. *phaseoli* (Fsph) macroconidia 1.7×10^7 spores/mL within 20 min (where indicated). Following a 24 h incubation period at 22 °C the pisatin was extracted overnight in 5 mL hexanes. Pisatin was quantified at 309 nm in ethanol.

2.2. Effect of EDTA on the Defense Response of Pea; Defense Gene Activation

In addition to the induced accumulation of pisatin, defense gene induction is often correlated with the actual suppression of the growth of a pathogen on peas [1]. However, resistance reportedly occurs as a result of the induction of multiple defense responses [10]. The gene products of the defense genes often referred to as "pathogenesis related" or PR proteins. Some of the genes code for enzymes on the pathway to pisatin production.

Primers were constructed (Table 2) to detect the activation of early expressed PR genes [12] coding for an array of functions: *DRR206* codes for an enzyme associated with a secondary pathway toward lignin (lignan) production. The pea gene *DRR230* coding for a defensin [13] has been established as resistance-conferring trait with defined antimicrobial activity [14]; *DRR49* (*PR-10*) codes for a product that enters the nucleus [15] and is putative RNase. *DRR49* trans-genetically confers resistance in potato to early blight [16]. The *PR1b* gene in *Arabidopsis* has a PR-1 function and is a "non-expressor" of *NPR1* which reportedly is a master, positive regulator of plant immunity in *Arabidopsis* [17]. NPR1 binds directly to salicylic acid (SA) and works as a SA receptor. In the presence and accumulation of SA [18], the NPR1 is reduced, monomerized and translocated to the nucleus. In the nucleus NPR1 interacts with the TGACG motif binding factor that binds to elements of the PR1 promoter. As a result, NPR1 is proposed to up-regulate a set of disease resistance genes via this route.

Table 2. Primers selected for real-time PCR analysis of pea PR genes.

Target	Genbank #	Real-time F primer	Real-time R primer
Pea Ubiquitin	L881142	GGCTAAGATACAGGACAAGGAG	AACGAAGGACAAGATGAAGGG
Pea Actin (Pea-ACT)	U81046	CACAATTGGCGCTGAAAGATT	GATCATCGATGGCTGGAACA
DRR206	U11716	CTTGGCTTAGTTTCACATTTGTTCTT	GGGTCAGCTCCAGCAAAAGTAA
DRR230 (defensin)	L01579	TGTGGTGACAGAGGCAAACAC	TCGTGAAGCATACTCCCCTGTA
PR10 (AKA *DRR49*)	U31669	GATCTCATTCGAGGCTAAACTGTCT	CACACTCAGCTTTGCAATGGA
PR1b	AJ586324.1	AACTCATGTGCTGCTGGTTATCA	AACCGAATTGCGCCAAAC

The data in Figure 2 indicate that EDTA concentrations of 3.9 and 15 mM effectively activate the PR genes: *DRR206*, the defensing, DRR230, *DRR49* (*PR10*) and *PR1b*, to levels above the water control. The 5 h activation level represents a stage 1 h prior to 6 h, a time point when nonhost resistance is maximal. The additive or synergistic enhancement of the induction of EDTA/Fsph spores treatments observed with pisatin accumulations was not consistently obtained in the induction of PR genes.

2.3. Direct Effect of EDTA on Fungal DNase

As expected 50 mM EDTA directly inhibits *in vitro in vitro* activity of a fungal DNase expectantly by the chelation of Mn^{2+}, a metal cofactor required for activity [2] (Figure 3). These results suggest that the inhibition of a fungal DNase enzyme by EDTA does occur *in vitro*. However, this suppression is apparently not sufficient *in vivo* to block the phytoalexin-induction-potential of the fungus. Rather it appears that *in vivo* EDTA acted both independently and synergistically with the total eliciting potential of the fungus.

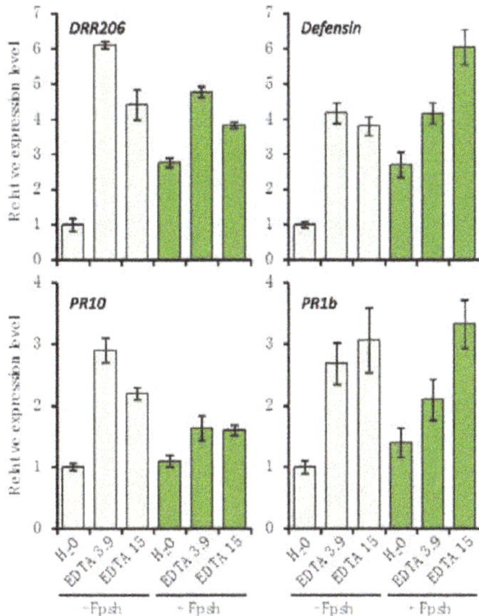

Figure 2. Effect of EDTA treatment on gene expression levels of pea pathogenesis-related genes. Pea endocarp tissues (0.4 g) were treated (20 μL/pod half) for 5 h with water (H_2O), EDTA 3.9 mM, and EDTA 15 mM and after 20 min with (or without) 5 μL fungal spores (1.3×10^6 spores/mL) of, *Fusarium solani* f. sp. *phaseoli* (Fsph). The tissues were then subjected to the qRT-PCR analysis. Histograms show the expression levels of pea pathogenesis-related genes. Data were normalized by the reference gene *Ubiquitin* and converted into a value relative to that of the water treatment control (-Fsph). Error bars represent standard error.

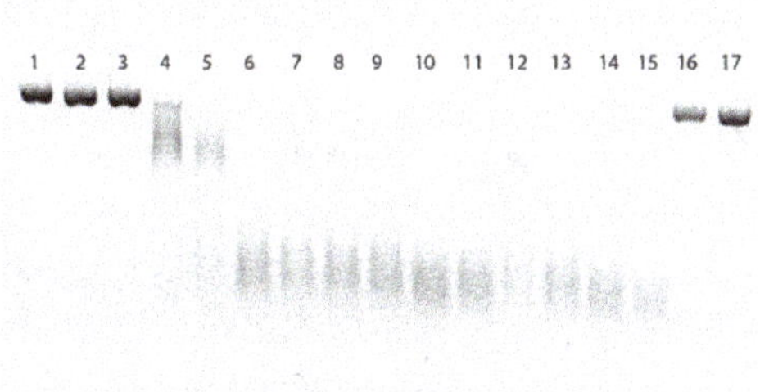

Figure 3. Effects of varying concentrations of EDTA on the enzymatic degradation of DNA. Each of the reactions loaded (1–15) contained 3 μL of fungal (*Verticillium dahlia*) DNase (2 Units), 2.2 μg plasmid DNA (in 10 mM MES pH 6.0) and 1 μL of the EDTA concentrations from 50 mM EDTA through half-fold dilutions to 0.003 mM EDTA. Reaction time was 10 min at 22 °C. Lanes 1, 2 and 3 with 50, 25 and 12 mM concentrations of EDTA, respectively, completely blocked DNA degradation. Treatments of 6.25 mM through 0.003 mM (Lanes 4–15) demonstrate the diminishing ability of EDTA to inhibit the DNase enzyme. Lanes 16 and 17 represent DNA controls without enzyme or EDTA.

2.4. DNA Damage by EDTA

It was of interest to determine if direct DNA damage is a part of the EDTA action. The pea endocarp tissue was treated with EDTA only, EDTA with fungal spores, and fungal spores only (Figure 4). The extraction of pea genomic DNA from treated tissue represents primarily high molecular weight molecules. Therefore the detection of subtle single strand nicks requires that the resultant fragments in gel separations must occur under alkaline conditions that will allow the nicked DNA fragment to dis-engage as a single-stranded entity (Figure 4A). To further dis-engage the small fragments from the predominance of genomic DNA, the total DNA was trapped in a CHEF gel-type agar under alkaline conditions. The smaller single stranded fragments are allowed to diffuse out into an alkaline buffer over 48 h with slow stirring (Figure 4B). Gel separations of the total DNA aliquots reveal marginal effects of fragmentation associated with the EDTA treatments. Further separations of alkaline processed-and-released-DNA fragments reveal greater contrasts between EDTA-induced damage and control tissue. There are at least two confounding actions likely to be associated with the 2 and 6 h sampling times. First DNA repair is a rapid process in eukaryotic tissue and secondly some natural DNA fragmentation is present in excised water treated tissue. However the greater intensity of EDTA-induced DNA fragment is repeatedly observed.

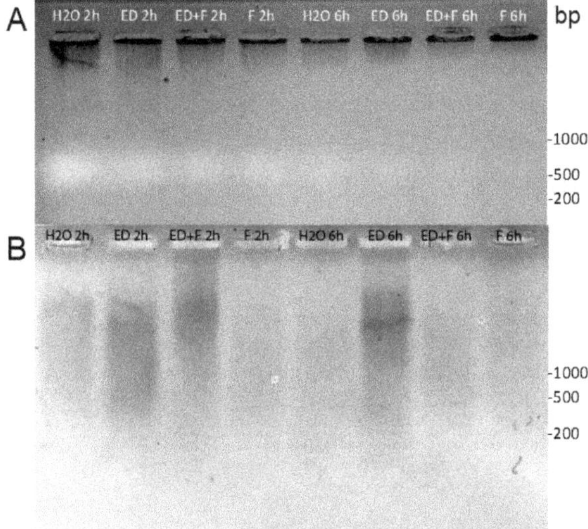

Figure 4. EDTA applications to pea endocarp tissue result in fragmentation of DNA in the period disease resistance is initiated (1–6 h). DNA was extracted from pea endocarp tissue (**A**) treated for 2 h and 6 h with (10 µL) water (H_2O), 6 mM EDTA (ED), 6 mM EDTA + *Fusarium solani* f. sp. *phaseoli* spores (ED + F) or *F. solani* spores only (F). DNA (2 µg of each treatment was loaded per well) was separated on 1% agarose gels. The fragmented DNA was further separated from the bulk of the high molecular pea DNA (25 µg) by agarose retention disks under alkaline conditions (**B**) and the diffused fragmented DNA from the agarose disk recovered and separated on standard agarose gels. (Photos are inverted images of ethidium bromide stained DNA).

2.5. Nonhost Resistance Induced by EDTA

The effect of EDTA applied 20 min prior to the inoculation with spores of the true pea pathogen, *Fusarium solani* f. sp. *pisi* (Fspi) is concentration dependent. EDTA at 6 mM effectively broke resistance whereas EDTA applied at 3–0.3 mM effectively promoted complete resistance against the pathogen (Figure 5).

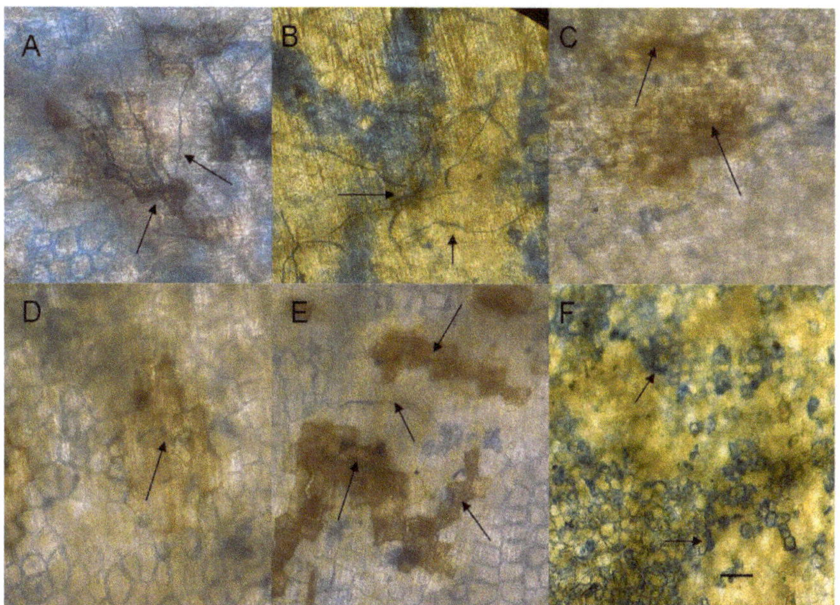

Figure 5. Low concentrations of EDTA induce resistance in pea endocarp tissue against the pea pathogen, *Fusarium solani* f. sp. *pisi* (Fspi). Twenty min of pretreatments (10 μL) of (**A**) water and (**B**) 6 mM EDTA; (**C**) 3 mM EDTA; (**D**) 1.5 mM EDTA; (**E**) 0.7 mM; and (**F**) 0.3 mM EDTA concentrations were applied to pea tissue 20 min prior to 5 μL of a 1.3×10^6 spore suspension of Fspi. Photos represent the extent of growth typical from the lesion observations of 30 spores for each concentration of EDTA applied. Arrows indicate the location of the spores or mycelia distorted or suppressed by the treatments. (Note: Bar = 50 microns.)

2.6. Cytological Effect of EDTA on Pea Nuclear Condition

The exposed endocarp surface cells of a split pea pod are without a cuticle layer and thus can be DNA-stained with DAPI without fixing and the condition of a large number of the nuclei in the surface cell layer can be determined via a fluorescent microscope. The photographs of Figure 6 indicate nuclear changes and the diminishing detection of DAPI staining. The sizing of the visible nuclei in tissue treated 3 h with 12, 6 and 3 mM EDTA indicates slight but significant reductions in size from those in water treated tissue. The effects that caused the diminished detection of the DAPI stain are unknown.

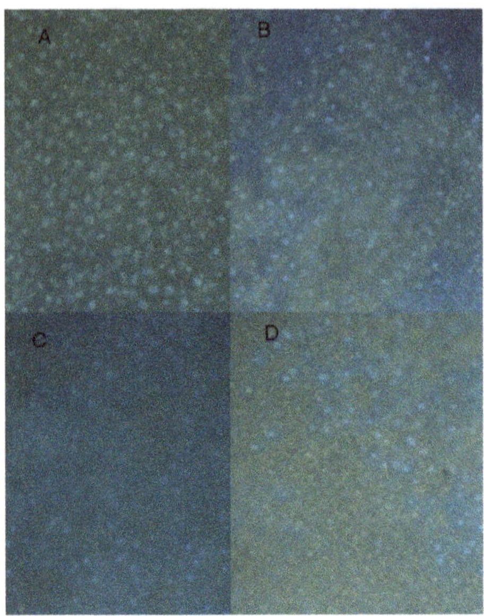

Figure 6. Effect of EDTA on the condition of nuclei within pea endocarp tissue. Pea endocarp tissue was treated 3 h with (**A**) water, (**B**) 12 mM EDTA, (**C**) 6mM EDTA or (**D**) 3 mM EDTA. Surface cell sections were treated with DAPI and photographed via a fluorescence microscope. Nuclei (average of 40 visible nuclei remaining intact) measured 10.0, 7.94, 9.74 and 8.7 microns, respectively, adjusted proportionally to the normal average of 10 microns previously determined by electron microscope analysis in control tissue.

3. Experimental Section

3.1. Plant and Pathogens

Immature pea pods (2 cm in length) were harvested from *Pisum sativum* cv. Lance plants grown in sand in the greenhouse (12 h light). All fungi were cultured on potato dextrose agar (PDA) (Difco) supplemented with pea pods (5 g/L). Fungal cultures used in this study were *Fusarium solani* f. sp. *pisi* Snyder & Hansen (ATCC No. 38136) (Fspi) and *F. solani* f. sp. *phaseoli* Snyder & Hansen (ATCC No. 38135) (Fsph) from pea and bean respectively. Cell death was assayed by staining plant cells with 1 mg/mL trypan blue.

3.2. EDTA Treatment and Pisatin Quantization

Immature pea pods (2 cm) were separated into halves with a smooth spatula. Treatments (20 μL/pod half) were applied and distributed to the exposed endocarp surface. Following 24 h at 100% humidity and 22 °C the pod halves were submerged into 5 mL hexanes for 24 h. The hexanes were volatilized away and the residual material containing pisatin was dissolved in 95% ethanol and quantitated at UV 309 nm in a spectrophotometer.

3.3. Activation of PR Genes

Procedures for total RNA isolation and purification were briefly described as follows: pea endocarp tissues (pod halves) were treated the designated treatment and after 5 ½ h the tissues were ground in liquid nitrogen and solubilized in the extraction buffer, (sodium perchlorate, 5 M; Tris base, 0.5 M; sodium dodecyl sulfate, 2.5%; NaCl, 0.05%; and disodium-EGTA, 0.05 M). The nucleic acids were precipitated from the aqueous phase of a chloroform/phenol extraction and then dissolved in water. The nucleic acid solution was adjusted to 2.0 M lithium chloride to precipitate the RNA. The total RNA from each sampled was subjected to quantitative real-time reverse transcription-polymerase chain reaction (qRT-PCR) using the CFX96 system (Bio-Rad Laboratories, Inc., Hercules, CA, USA). The primers used are described in Table 2.

4. Conclusions

The defense response of pea endocarp tissue to pathogen challenge or to chemical elicitors appears to depend on both the chemistry of the eliciting compound and of the cellular target. Although EDTA is most known as a chelating agent it has the potential to alter the DNA of pea nuclei. This potential was realized in the absence of EDTA-related cell death (data not shown). Further the concentrations of EDTA active in inducing resistance in pea to the pea pathogen, Fsph, were not effective in directly inhibiting fungal growth in liquid culture (data not shown). Interestingly a structurally similar chelator [19], EGTA, was not effective in inducing pisatin accumulations (data not shown) Previous research on the pea/Fsph model for studying non-host disease resistance has indicated that chromatin can be a primary target for activating the defense genes [7,12]. These PR (pathogenesis-related) genes are distributed in various regions within the pea genome [8–10]. Other eukaryotic research has revealed the numerous ways that RNA polymerase complexes, that are stalled upstream of some genes, can be released for transcription by helical changes in the DNA or the substitution or removal of nuclear proteins [20,21]. In the defense response of pea tissue the nuclear proteins HMG A and histones H2A and H2B are reduced by ubiquitination [7]. EDTA at low concentrations appears capable of causing some of these changes through its action of subtly fragmenting pea DNA. Applications of EDTA to corneal epithelial cells also increases DNA single and double stranded breaks [22] and were proposed to be caused indirectly by way of the reactive oxygen species of oxidative stress. The EDTA-related mechanism for activating defense genes in peas appears to be directly on the DNA, since an extensive screening of DNA-specific components with known actions, e.g., intercalation, minor groove insertion, helical alteration, DNase action and thymidine dimerization are also capable of activating PR genes and/or phytoalexin accumulations [4,23–25].

Pathologists are in a continual search for components that can activate defense responses in the absence of major detrimental side effects and major costs. Thus there appear to be future practical applications of these EDTA properties in agriculture.

Acknowledgments

I thank Haley McDonel for assistance in manuscript preparation and Pat Okubara and Lei Zhang for manuscript review. PPNS No. 0667, Department of Plant Pathology, College of Agricultural Human and Natural Resource Sciences, Agriculture Research Center, Project No. WNPO1844, Washington State University Pullman, WA 99164 6430.

Author Contributions

L.H. designed research; and L.H. and K.T. performed research, analyzed data, and wrote the paper.

Conflicts of Interest

The authors declare no conflict of interest.

References

1. Hadwiger, L.A. Pea-*Fusarium solani* interactions: Contributions of a system toward understanding disease resistance. *Phytopathology* **2008**, *98*, 372–379.
2. Klosterman, J.; Chen, J.; Choi, J.J.; Chinn, E.E.; Hadwiger, L.A. Characterization of a 20 kDa DNase elicitor from *Fusarium solani* f. sp. *phaseoli* and its expression at the onset of induced resistance in *Pisum sativum* . *Mol. Plant Pathol.* **2001**, *2*, 147–158.
3. Gerhold, D.L.; Pettinger, A.J.; Hadwiger, L.A. Characterization of a plant stimulated nuclease from *Fusarium solani*. *Physiol. Mol. Plant Pathol.* **1933**, *43*, 33–46.
4. Hartney, S.; Carson, J.; Hadwiger, L.A. The use of chemical genomics to detect functional systems affecting the non-host disease resistance of pea to *Fusarium solani* f. sp. *phaseoli*. *Plant Sci.* **2007**, *172*, 45–56.
5. Flora, S.J.S.; Pachauri, V. Chelation in metal intoxication. *Int. J. Environ. Res. Public Health* **2010**, *7*, 2745–2788.
6. Drouin, R.; Gao, S.; Holmquist, G.P. *Technologies for Detection of DNA Damage and Mutations*; Pfeifer, G.P., Ed.; Plenum Press: New York, NY, USA, 1996; pp. 37–43.
7. Isaac, J.; Hartney, S.L.; Druffel, K.; Hadwiger, L.A. The non-host disease resistance response in peas; alterations in phosphorylation and ubiquitination of HMG A and histones. H2A/H2B. *Plant Sci.* **2009**, *177*, 439–449.
8. Pilet-Nayel, M.L.; Muehlbauer F.J.; McGee, R.J.; Kraft, J.M.; Baranger, A.; Coyne, C.J. Quantitative trait loci for partial resistance to Aphanomyces root rot in pea. *Theor. Appl. Genet.* **2002**, *106*, 28–39.
9. Prioul-Gervais, S.; Deniot, G.; Recevur, E.-M.; Frankewitz, A.; Fourmann, M.; Rameau, C.; Pilet-Nayel, M.-L.; Baranger, A. Candidate genes for quantitative resistance to *Mycosphaerella pinodes* in peas (*Pisum sativum* L.). *Theor. Appl. Genet.* **2007**, *114*, 971–984.
10. Hadwiger, L.A. Localization predictions for gene products involved in non-host resistance responses in a model plant/fungal pathogen interaction. *Plant Sci.* **2009**, *177*, 257–265.

11. DiCenzo, G.L.; VanEtten, H.D. Studies on the late steps of (+) pisatin biosynthesis: Evidence for (−) enatiomeric intermediates. *Phytochemistry* **2006**, *67*, 675–683.
12. Hadwiger, L.A.; Polashock, J. Fungal mitochondrial DNases: Effectors with the potential to activate plant defenses in nonhost resistance. *Phytopathology* **2013**, *103*, 81–90.
13. Chiang, C.C.; Hadwiger, L.A. The *Fusarium solani*-induced expression of a pea gene family encoding cysteine content proteins. *MPMI* **1991**, *4*, 324–331.
14. Almeida, M.S.; Cabral, K.M.; Zingali, R.B.; Kurtenbach, E. Characterization of two novel defense peptides from pea (*Pisum sativum*) seeds. *Arch. Biochem. Biophys.* **2006**, *378*, 278–286.
15. Allaire, B.S.; Hadwiger, L.A. Immunogold localization of a disease resistance response protein in *Pisum sativum* endocarp cells. *Physiol. Mol. Plant Pathol.* **1994**, *44*, 9–17.
16. Chang, M.M.; Chiang, C.C.; Martin, M.W.; Hadwiger, L.A. Expression of a pea disease resistance response protein in potatoes. *Am. Potato J.* **1993**, *70*, 635–647.
17. Yu, D.; Chen, C.; Zhixiang, C. Evidence for an important role of WRKY DNA binding proteins in the regulation of *NPR*1 gene expression. *Plant Cell* **2001**, *13*, 1527–1539.
18. Maier, F.; Zwicker, S.; Huckelhoven, A.; Meissner, M.; Funk, J.; Pfitznener, S.J.; Pfitzner, U.M. Nonexpressor of Pathogenesis-Related Proteins 1 (NPR1) and some NPR1-related proteins are sensitive to salicylic acid. *Mol. Plant Pathol.* **2011**, *12*, 73–91.
19. Barr, R.; Troxel, K.S.; Crane, F.L. EGTA, a calcium chelator, inhibits electron transport in photosystem II of spinach chloroplasts at two different sites. *Biochem. Biophys. Res. Commun.* **1980**, *92*, 206–212.
20. Weake, V.M.; Workman, J.L. Histone ubiquitination triggering gene activity. *Mol. Cell* **2008**, *29*, 653–663.
21. Adkins, M.W.; Tyler, J.K. Transcriptional activators are dispensable for transcription in the absence of Spt6-mediated chromatin reassembly of promoter regions. *Mol. Cell* **2006**, *21*, 405–416.
22. Ye, J.; Wu, H.; Wu, Y.; Wang, C.; Zhang, H.; Shi, X.; Yang, J. High molecular weight Hyaluronan decreases oxidative DNA damage induced by EDTA in human coryneal epithelial cells. *Eye* **2012**, *26*, 1012–1020.
23. Schochau, M.E.; Hadwiger, L.A. Regulation of gene expression by actinomycin D and other compounds which change the conformation of DNA. *Arch. Biochem. Biophys.* **1969**, *134*, 34–41.
24. Hadwiger, L.A.; Jafri, A.; von Broembsen, S.; Eddy, R. Mode of pisatin induction. Increased template activity and dye-binding capacity of chromatin isolated from polypeptide-treated pea pods. *Plant Physiol.* **1974**, *53*, 52–63.
25. Choi, J.J.; Klosterman, S.J.; Hadwiger, L.A. A comparison of the effects of DNA-damaging agents and biotic elicitors on the induction of plant defense genes, nuclear distortion and cell death. *Plant Physiol.* **2001**, *125*, 752–762.

Sample Availability: Samples described in this publication are available upon request in a timely manner for noncommercial research purposes.

Effectiveness of Phenolic Compounds against Citrus Green Mould

Simona M. Sanzani, Leonardo Schena and Antonio Ippolito

Abstract: Stored citrus fruit suffer huge losses because of the development of green mould caused by *Penicillium digitatum*. Usually synthetic fungicides are employed to control this disease, but their use is facing some obstacles, such public concern about possible adverse effects on human and environmental health and the development of resistant pathogen populations. In the present study quercetin, scopoletin and scoparone—phenolic compounds present in several agricultural commodities and associated with response to stresses—were firstly tested *in vitro* against *P. digitatum* and then applied *in vivo* on oranges cv. Navelina. Fruits were wound-treated (100 μg), pathogen-inoculated, stored and surveyed for disease incidence and severity. Although only a minor (≤13%) control effect on *P. digitatum* growth was recorded *in vitro*, the *in vivo* trial results were encouraging. In fact, on phenolic-treated oranges, symptoms appeared at 6 days post-inoculation (DPI), *i.e.*, with a 2 day-delay as compared to the untreated control. Moreover, at 8 DPI, quercetin, scopoletin, and scoparone significantly reduced disease incidence and severity by 69%–40% and 85%–70%, respectively, as compared to the control. At 14 DPI, scoparone was the most active molecule. Based on the results, these compounds might represent an interesting alternative to synthetic fungicides.

Reprinted from *Molecules*. Cite as: Sanzani, S.M.; Schena, L.; Ippolito, A. Effectiveness of Phenolic Compounds against Citrus Green Mould. *Molecules* **2014**, *19*, 12500-12508.

1. Introduction

Penicillium digitatum [Pers.: Fr.] Sacc. is the causal agent of green mould, one of the most common postharvest diseases of citrus fruit. This wound-obligate pathogen has a relatively short disease cycle (3 to 5 days at 25 °C) and, on a single fruit, can produce 1 to 2 billion conidia that efficiently disperse through the air [1]. *P. digitatum* may attack the fruit on the tree, in the packinghouse, in transit, in storage and in the market. However, handling and storage under ambient conditions particularly favours its growth. During this stage, green mould reaches 60%–80% of decay caused by *Penicillium* genera [2]. Youssef *et al.* [3] evaluated the presence and abundance of *Penicillium* spp. conidia in packinghouses, reporting significantly higher values in "bin emptying" area, with a density exceeding 400 CFU/g fw on fruit surface and 66 CFU on semi-selective PDA plates left open in the atmosphere for 10 min. Moreover, the incidence of penicillium rots showed an increasing trend, with values ranging from 23% (bin emptying) to 40% (calibration).

When permitted, synthetic fungicides are the primary means to control green mould. However, the public growing concern for consequences on human and environmental health of toxic residues [4] and the development of fungicide-resistant strains in pathogen populations [1] have motivated the search for alternative approaches.

Among unconventional control strategies, the induction of fruit resistance, the use of plant or animal-derived products with fungicidal activity and the application of antagonistic microorganisms or physical means can be considered, either alone or as part of an integrated pest management policy [5]. Within plant product category, the role of phenolic compounds in the active expression of resistance has been reported [6]. Some of them occur constitutively in the plant (phytoanticipins), whereas others form in response to biotic or abiotic stresses (phytoalexins), such as injuries [7,8].

For example, the exposure of citrus fruit to salt application [9], heat [10], gamma radiation [11] or ultraviolet (UV) light [12] induced the accumulation in the fruit peel of compounds, as the coumarins scopoletin and scoparone, associated with the development of resistance against fungal pathogens. Moreover, following pathogen infection, tissues of *Morinda tomentosa* Roth. and *Cassia fistula* L. proved to contain several flavonoids including quercetin [13]. Similarly, Mayr *et al.* [14] reported that quercetin glycosides are released from apple cell vacuoles as aglycones, developing their toxic activity after pathogen attack. In a previous study [15], we tested the efficacy of several phenolic compounds including quercetin, scopoletin and scoparone against *Penicillium expansum*, causal agent of the blue mould of apple, and the production of its mycotoxin patulin.

The selective accumulation of quercetin, scopoletin and scoparone in plants, as well as their antifungal character and antioxidant properties [15–18], make them good "natural pesticide" candidates to improve plant resistance to fungal infections. Although several investigations on the accumulation of phenolic compounds following the induction of host resistance were carried out [7,12,19], to the best of our knowledge, there are no bibliographic records on their direct exogenous application on citrus to maintain their postharvest quality.

The aim of the present investigation was to evaluate *in vivo* the activity of the phenolic compounds quercetin, scopoletin and scoparone against green mould on "Navelina" oranges. Moreover, the putative control effect on *P. digitatum* growth was tested by *in vitro* trials.

2. Results and Discussion

The effect of different concentrations of quercetin, scopoletin, and scoparone on *in vitro* radial growth of *P. digitatum* is shown in Table 1. At 3 days post inoculation (DPI), only scoparone at the highest tested concentration (100 μg/mL, 1000 μg/plate), significantly, although slightly (up to 13%) reduced fungal growth. On the contrary, a significant enhancement of *P. digitatum* growth was observed in presence of quercetin at the same concentration. When colony diameters were measured at 6 DPI, all treatments proved to reduce significantly fungal growth at 100 μg/mL, being quercetin the best one with a 14% reduction. These findings are in agreement with previous experiments [15] in which quercetin only slightly (13%) reduced *in vitro P. expansum* growth. On the contrary, scopoletin and scoparone did not have any significant effect. Since quercetin is associated to apples, target host of *P. expansum*, and scopoletin/scoparone to citrus, target host of *P. digitatum*, these results seem to confirm the existence of a specificity in the host-pathogen interaction, as reported by Sanzani *et al.* [20]. Afek and Sztejnberg [21] already proved the *in vitro* activity of scoparone against *P. digitatum*. The ED_{50} for spore germination was 64 μg/mL, whereas, as far as we know, no data on radial growth are available in literature. Similarly, Garcia *et al.* [22] reported the *in vitro* fungitoxic effect of 2 mM scopoletin on germ tube elongation and conidia germination of *Microcyclus ulei*.

The better results obtained at a higher concentration seem to suggest that the efficacy of phenolics is dose-dependent, thus, in the future higher concentrations will be tested.

Table 1. Effect of quercetin, scopoletin and scoparone at 10 and 100 µg/mL (100 and 1000 µg/plate, respectively) on *Penicillium digitatum* colony diameter (mm) after 3 and 6 days post-inoculation (DPI) at 24 °C in the dark.

Treatment	Colony Diameter (mm)	
	3 DPI	6 DPI
Control [no phenolics]	14.7 ± 1.4 ^{b,c}	54.2 ± 0.5 ^{a,b}
Quercetin 10 µg/mL	16.3 ± 0.6 ^{a,b}	56.5 ± 0.4 ^{a}
Quercetin 100 µg/mL	17.8 ± 0.2 ^{a}	47.7 ± 1.2 ^{d}
Scopoletin 10 µg/mL	14.0 ± 0.4 ^{c,d}	52.5 ± 1.1 ^{b,c}
Scopoletin 100 µg/mL	13.0 ± 0.8 ^{c,d}	51.5 ± 1.8 ^{c}
Scoparone 10 µg/mL	14.7 ± 0.5 ^{b,c}	54.7 ± 1.2 ^{a,b}
Scoparone 100 µg/mL	12.8 ± 0.6 ^{d}	50.3 ± 0.9 ^{c}

Each value corresponds to the mean of three replicates ± standard error of the mean (SEM). For each assessment time, values with the same letter are not significantly different according to Duncan's Multiple Range Test (DMRT, $p \leq 0.05$).

Considering the absence of a relevant effect on fungal growth and the results of previous investigations [15], we decided to test quercetin, scopoletin and scoparone at 100 µg/wound against green mould incidence and severity on "Navelina" oranges. Results are reported in Figures 1 and 2, respectively. Concerning incidence of decay, infections started at 6 DPI, *i.e.*, with a 2 day-delay on treated oranges, as compared to the control, and quercetin, scopoletin and scoparone significantly ($p \leq 0.05$) reduced them by 60%, 40% and 69%, respectively, at 8 DPI (Figure 1). Quercetin and scopoletin maintain their significant effect up to 10 DPI, whereas scoparone was effective for all the incubation period, being the best treatment at 14 DPI (27% reduction). The control activity was confirmed also as far as disease severity concerns (Figure 2). Indeed, *Penicillium* lesion diameters were significantly ($p \leq 0.05$) reduced by all treatments up to the end of the incubation period. In particular, the three tested phenolic compounds equally reduced disease severity up to 12 DPI (36%–47%), whereas, at 14 DPI, scoparone proved to be the most effective treatment (37% reduction). These results seem to suggest that tested phenolics, rather than completely blocking infections, exert a fungistatic effect. Moreover, the significant efficacy demonstrated by scoparone throughout the incubation period is particularly interesting, considering the average shelf life of oranges in markets and supermarkets.

Figure 1. Incidence of decay (infected wounds, %) on "Navelina" oranges treated with quercetin, scopoletin or scoparone (100 µg/wound), inoculated with *Penicillium digitatum*, and incubated at 24 ± 1 °C for 14 days. Untreated fruits served as a control. Each value corresponds to the mean of three replicates ± standard error of the mean (SEM). Means separation according to Fisher's Least Significant Difference (LSD).

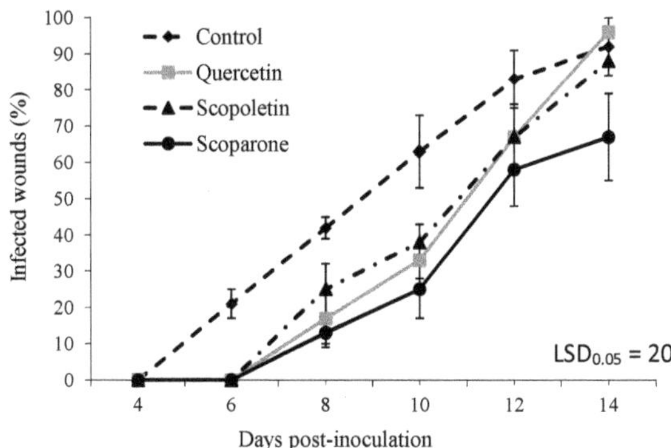

Figure 2. Disease severity (lesion diameter, mm) on "Navelina" oranges treated with quercetin, scopoletin or scoparone (100 µg/wound), inoculated with *Penicillium digitatum* and incubated at 24 ± 1 °C for 14 days. Untreated fruits served as a control. Each value corresponds to the mean of three replicates ± standard error of the mean (SEM). Means separation according to Fisher's Least Significant Difference (LSD).

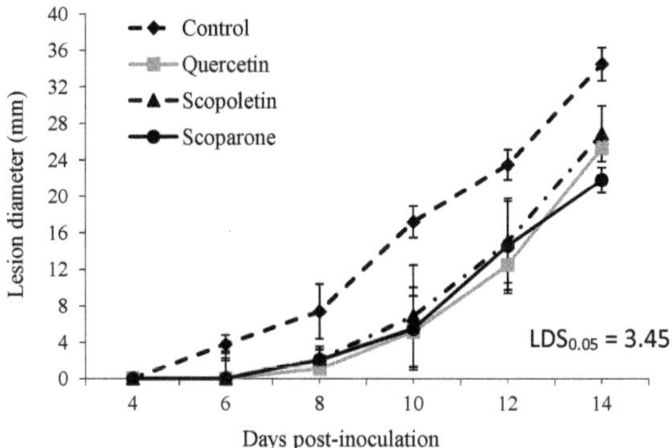

The role of phenolic compounds in defense mechanisms against pathogens is well known. In particular, quercetin efficacy in reducing apple blue mould incidence and severity has already been reported [15], and results were similar to those recorded in the present study on oranges cv. Navelina. Scopoletin and scoparone induction as alternative control means against blue and green mould of oranges has been also extensively studied [12,23–25], however, to date there are no reports on their exogenous application on oranges. Tested phenolics seem to be more effective *in vivo* than *in vitro*. This behavior might be a consequence of: (i) the activation of orange defensive genes; (ii) the addition to phenolics already present in the host, thus reaching a concentration toxic to the fungus; (iii) the lack of detoxification mechanisms in the pathogen for the unknown added compounds; (iv) the interaction with one or more of the pre-existing compounds in fruit tissues, thus forming a new toxic compound.

As well known, during ripening the phenolic profile of fruit skin changes markedly, with a consistent content reduction in mature fruit [26], which becomes at the same time more susceptible to infections. Considering that the inner white layer of citrus peel (albedo) produces quercetin, scoparone, and scopoletin at concentrations even of 80–140 μg/g DW [9,27], it could be conceivable that, with the addition of the tested compounds, a phenolic concentration toxic to the pathogen was restored.

The ability of quercetin to induce resistance to *P. expansum* in apples has been demonstrated [28]. Furthermore, it proved to reduce pathogen toxigenic ability by acting on the biosynthetic pathway of patulin [29], which has not only a health significance, but also was recently classified as a pathogenicity/virulence factor [30].

P. digitatum usually needs a wound to initiate the infection process. Harvest and postharvest fruit handling frequently produces wounds on fruit, especially when the peduncle is present. In addition, the environment in the degreening room (warm temperature and high humidity) is ideal for the proliferation of microorganisms and the ineffective sanitizing of packinghouse dump tanks, flumes and hydrocoolers might further promote infection by pathogens. Thus, careful handling of fruits and hygiene of the packing line are of crucial importance in preventing rot development. For instance, water should be treated (either chemically or physically) to prevent unintentional contamination of clean produce [31]. In this context, the use of alternative compounds, such as quercetin, scopoletin or scoparone, might be an interesting alternative to the chemicals currently used, such as chlorine. In fact, our substances showed a promising activity on "Navelina" oranges at an inoculum concentration (5×10^4 conidia/mL) superior to the one commonly present in the floating water (10^3 conidia/mL) of numerous packinghouses [32]. Moreover, in recent years plant-derived compounds have grown in popularity among consumers [33], especially for their presumed anti-inflammatory and antioxidant properties [34], although toxicity of most plant products for their use as food components has not been proved yet.

Finally, quercetin was successfully applied on apples by dipping in large-scale experiments at a concentration (1.25 g/L) comparable to those of thiabendazole (0.5–1.15 g/L) and imazalil (2 g/L), two fungicides used as postharvest antifungal treatments [1]. Therefore, based on a survey on costs, the application of phenolics might be comparable to the most common fungicides used in packinghouses [15].

3. Experimental Section

3.1. Chemicals

Sodium dihydrogen phosphate, sodium hydrogen diphosphate, sodium hydroxide, quercetin (3,3',4',5,7-pentahydroxyflavone), scopoletin (7-hydroxy-5-methoxycoumarin) and scoparone (6,7-dimethoxycoumarin) with a purity of >95% were purchased from Sigma-Aldrich (Milan, Italy).

3.2. Preparation of Phenolic Compounds Solutions

Stock solutions of quercetin, scopoletin and scoparone were prepared at concentration 10 mg/mL by dissolving pure standards into a mixture of phosphate buffer (50.0 mmol/L, pH 7.4)/NaOH (1.0 mol/L, pH 13) (9:1, v/v). In preliminary trials, this solving buffer did not show any significant antifungal activity (data not shown). Solutions of quercetin, scopoletin or scoparone at concentration 1 mg/mL were obtained by appropriate dilutions of the respective stock solutions.

3.3. Penicillium Conidial Suspension

To produce inoculum, a strain of *P. digitatum* molecularly and morphologically identified and deposited in the "Fungal Culture Collection" of the Department of Soil, Plant and Food Sciences (University of Bari Aldo Moro, Bari, Italy) was cultured on PDA dishes for 8 days at 24 ± 1 °C. The surface of the colony was washed with 6 mL of sterile distilled water containing 0.05% (v/v) Tween 80. The resulting suspension was filtered through two layers of sterile gauze and spores were counted by a Thoma chamber (HGB Henneberg-Sander GmbH, Lutzellinden, Germany). A suspension with a concentration of 5×10^4 conidia/mL was used for all *in vitro* and *in vivo* trials.

3.4. In Vitro Trials

Aliquots of quercetin, scopoletin and scoparone stock solutions (10 and 1 mg/mL) were incorporated into molten PDA before pouring into Petri dishes (90 mm diameter, 10 mL/dish) so to reach 100 and 10 μg/mL as final concentrations (corresponding to 1000 and 100 μg/dish). Plates were centrally inoculated with 10 μL of *P. digitatum* suspension (500 conidia), and incubated at 24 ± 1 °C in the dark for 6 days. For each compound and concentration, three dishes were prepared. Non amended PDA plates were used as a control. Colony growth (average of the two orthogonal diameters) was recorded at 3 and 6 days post-inoculation (DPI). The experiment was performed twice.

3.5. In Vivo Trials

Forty-eight oranges cv. Navelina were surface-sterilised with Na-hypochlorite (2 min in a 2% solution), rinsed under running tap water for 1 min and allowed to dry before wounding with a sterile nail (3 × 3 mm). Then, 10 μL of phenolic buffer solution (10 mg/mL, 100 μg/wound) were pipetted into each wound and, after drying (approximately 30 min later), wounds were inoculated with 10 μL of pathogen conidial suspension (500 conidia). Wounds treated with solving buffer served as a control. Each treatment was replicated three times and each replicate consisted of a tray containing four oranges with three wounds each. Replicates were individually wrapped into a plastic bag, avoiding contact with wounds, and incubated at 24 ± 1 °C for 14 days. Disease incidence (infected wounds, %) and severity (lesion diameter, mm) were recorded every 2 days for 14 days. The experiment was performed twice.

3.6. Statistical Analysis

Data were subjected to ANOVA (one-way analysis of variance). Significant differences ($p \leq 0.05$) were tested by the General Linear Model (GLM) procedure using the Duncan's Multiple Range Test (DMRT) for colony growth and the Fisher's Least Significant Difference (LSD) for disease severity and incidence. Percentage data were arcsine-square-root transformed before ANOVA analysis. Data were processed using the statistical software package Statistics for Windows (StatSoft, Tulsa, OK, USA).

3.7. Control Index Calculation

The effect of phenolics was expressed by a control index [CI] calculated with the following formula:

$$\text{CI } (\%) = [(\text{A} - \text{B})/\text{A}] \times 100 \quad (1)$$

where A and B correspond to the mean colony diameter or percentage of infected wounds or lesion diameter measured in control (not amended) dishes/oranges and treated dishes/oranges, respectively.

4. Conclusions

In conclusion, quercetin, scopoletin and scoparone proved to be effective in reducing green mould severity and incidence in "Navelina" oranges. Since we recorded no consistent effect on fungal growth, further trials are in progress to confirm their activity on a larger scale and to elucidate their possible mode(s) of action.

Author Contributions

Sanzani, S.M., Schena, L., and Ippolito, A. conceived and designed the experiments; Sanzani, S.M. performed the experiments; Sanzani, S.M., Schena, L., and Ippolito, A. analyzed the data and wrote the manuscript.

Conflicts of Interest

The authors declare no conflict of interest.

References

1. Holmes, G.J.; Eckert, J.W. Sensitivity of *Penicillium digitatum* and *P. italicum* to postharvest citrus fungicides in California. *Phytopathology* **2001**, *89*, 716–721.
2. Plaza, P.; Usall, J.; Teixidó, N.; Viñas, I. Effect of water activity and temperature on germination and growth of *Penicillium digitatum, P. italicum* and *Geotrichum candidum. J. Appl. Microbiol.* **2003**, *94*, 549–554.
3. Youssef, K.; Ligorio, A.; Sanzani, S.M.; Nigro, F.; Ippolito, A. Investigations of *Penicillium* spp. population dynamic in citrus packinghouse. *J. Plant Pathol.* **2011**, *93* (Suppl. 4), 61–62.

4. Daferera, D.J.; Ziogas, B.N.; Polissiou, M.G. GC-MS analysis of essential oils from some Greek aromatic plants and their fungitoxicity on *Penicillium digitatum*. *J. Agric. Food Chem.* **2000**, *48*, 2576–2581.
5. Ippolito, A.; Sanzani, S.M. Control of postharvest decay by the integration of pre and postharvest application of nonchemical compounds. *Acta Hortic.* **2011**, *905*, 135–143.
6. Nicholson, R.L.; Hammerschmidt, R. Phenolic compounds and their role in disease resistance. *Annu. Rev. Phytopathol.* **1992**, *30*, 369–389.
7. Ismail, M.A.; Rouself, R.L.; Brown, G.E. Wound healing in citrus: Isolation and identification of 7-hydroxycoumarin [umbelliferone] from grapefruit flavedo and its effects on *Penicillium digitatum* Sacc. *Hortic. Sci.* **1978**, *13*, 358.
8. Jeandet, P.; Clément, C.; Courot, E.; Cordelier, S. Modulation of phytoalexin biosynthesis in engineered plants for disease resistance. *Int. J. Mol. Sci.* **2013**, *14*, 14136–14170.
9. Youssef, K.; Sanzani, S.M.; Ligorio, A.; Ippolito, A.; Terry, L.A. Sodium carbonate and bicarbonate treatments induce resistance to postharvest green mould on citrus fruit. *Postharvest Biol. Technol.* **2014**, *87*, 61–69.
10. Ben-Yehoshoua, S.; Kim, J.J.; Shapiro, B. Elicitation of resistance to the development of decay in citrus fruit by curing of sealed fruit. *Acta Hortic.* **1989**, *258*, 623–630.
11. Dubery, I.A.; Holzapfel, C.W.; Kruger, G.J.; Schabor, J.C.; Dyk, M.V. Characterization of a γ-radiation-induced antifungal stress metabolite in citrus peel. *Phytochemistry* **1988**, *27*, 2769–2772.
12. Kim, J.J.; Ben-Yehoshua, S.; Shapiro, B.; Henis, Y.; Carmeli, S. Accumulation of scoparone in heat-treated lemon fruit inoculated with *Penicillium digitatum* Sacc. *Plant Physiol.* **1991**, *97*, 880–885.
13. Gottstein, D.; Gross, D. Phytoalexins of woody plants. *Trees* **1992**, *6*, 55–68.
14. Mayr, U.; Treutter, D.; Santos-Buelga, C.; Bauer, H.; Feucht, W. Developmental changes in the phenol concentrations of Golden Delicious apple fruits and leaves. *Phytochemistry* **1995**, *38*, 1151–1155.
15. Sanzani, S.M.; de Girolamo, A.; Schena, L.; Solfrizzo, M.; Ippolito, A.; Visconti, A. Control of *Penicillium expansum* and patulin accumulation on apples by quercetin and umbelliferone. *Eur. Food Res. Technol.* **2009**, *228*, 381–389.
16. Nijveldt, R.J.; van Nood, E.; van Hoorn, D.E.C.; Boelens, P.G.; van Norren, K.; van Leeuwen, P.A.M. Flavonoids: A review of probable mechanisms of action and potential applications. *Am. J. Clin. Nutr.* **2001**, *74*, 418–425.
17. Shaw, C.Y.; Chen, C.H.; Hsu, C.C.; Chen, C.C.; Tsai, Y.C. Antioxidant properties of scopoletin isolated from *Sinomonium acutum*. *Phytother. Res.* **2003**, *17*, 823–825.
18. Sourivong, P.; Schronerová, K.; Babincova, M. Scoparone inhibits ultraviolet radiation-induced lipid peroxidation. *Z. Naturforschung* **2007**, *62*, 61–64.
19. Rodov, V.; Ben-Yehoshua, S.; D'hallewin, G.; Castina, T.; Farg, D. Accumulation of the phytoalexins, scoparone and scopoletin in citrus fruit subjected to various postharvest treatments. *Acta Hortic.* **1994**, *381*, 517–524.
20. Sanzani, S.M.; Montemurro, C.; di Rienzo, V.; Solfrizzo, M.; Ippolito, A. Genetic structure and natural variation associated with host of origin in *Penicillium expansum* strains causing blue mould. *Int. J. Food Microbiol.* **2013**, *165*, 111–120.

21. Afek, U.; Sztejnberg, A. Accumulation of scoparone, a phytoalexin associated with resistance of citrus to *Phytophthora citrophthora*. *Phytopathology* **1988**, *78*, 1678–1682.
22. Garcia, D.; Sanier, C.; Macheix, J.J.; D'Auzac, J. Accumulation of scopoletin in *Hevea brasiliensis* infected by *Microcyclus ulei* [P. Henn.] V. ARX and evaluation of its fungitoxcity for three leaf pathogens of rubber tree. *Physiol. Mol. Plant Pathol.* **1995**, *47*, 213–223.
23. Rodov, V.; Ben-Yehoshua, S.; Kim, J.J.; Shapiro, B.; Ittah, Y. Ultraviolet illumination induces scoparone production in kumquat and orange fruit and improve decay resistance. *J. Am. Soc. Hortic. Sci.* **1992**, *117*, 788–792.
24. Rodov, V.; Burns, P.; Ben-Shalom, N.; Fluhr, R.; Ben-Yehoshua, S. Induced local disease resistance in Citrus mesocarp [albedo]: Accumulation of phytoalexins and PR-proteins. *Proc. Int. Soc. Citric.* **1996**, *2*, 1101–1104.
25. Ortuño, A.; Díaz, L.; Alvarez, N.; Porras, I.; García-Lidón, A.; del Río, J.A. Comparative study of flavonoid and scoparone accumulation in different Citrus species and their susceptibility to *Penicillium digitatum*. *Food Chem.* **2011**, *125*, 232–239.
26. McRae, K.B.; Lidster, P.D.; Demarco, A.C.; Dick, A.J. Comparison of the polyphenol profiles of apple fruit cultivars by correspondence analysis. *J. Sci. Food Agric.* **1990**, *50*, 329–342.
27. Wang, Y.C.; Chuang, Y.C.; Hsu, H.W. The flavonoid, carotenoid and pectin content in peels of citrus cultivated in Taiwan. *Food Chem.* **2008**, *106*, 277–284.
28. Sanzani, S.M.; Schena, L.; de Girolamo, A.; Ippolito, A.; González-Candelas, L. Characterization of genes associated to induced resistance against *Penicillium expansum* in apple fruit treated with quercetin. *Postharvest Biol. Technol.* **2010**, *56*, 1–11.
29. Sanzani, S.M.; Schena, L.; Nigro, F.; de Girolamo, A.; Ippolito, A. Effect of quercetin and umbelliferone on the transcript level of *Penicillium expansum* genes involved in patulin biosynthesis. *Eur. J. Plant Pathol.* **2009**, *125*, 223–233.
30. Sanzani, S.M.; Reverberi, M.; Punelli, M.; Ippolito, A.; Fanelli, C. Study on the role of patulin on pathogenicity and virulence of *Penicillium expansum*. *Int. J. Food Microbiol.* **2012**, *153*, 323–331.
31. Fallanaj, F.; Sanzani, S.M.; Zavanella, C.; Ippolito, A. Salt addition improves the control of citrus postharvest diseases using electrolysis with conductive diamond electrodes. *J. Plant Pathol.* **2013**, *95*, 373–383.
32. Mari, M.; Leoni, O.; Iori, R.; Cembali, T. Antifungal vapour-phase activity of ally-isothiocyanate against *Penicillium expansum* on pears. *Plant Pathol.* **2002**, *51*, 231–236.
33. Feliziani, E.; Santini, M.; Landi, L.; Romanazzi, G. Pre-and postharvest treatment with alternatives to synthetic fungicides to control postharvest decay of sweet cherry. *Postharvest Biol. Technol.* **2013**, *78*, 133–138.
34. Gatto, M.A.; Sanzani, S.M.; Tardia, P.; Linsalata, V.; Pieralice, M.; Sergio, L.; di Venere, D. Antifungal activity of total and fractionated phenolic extracts from two wild edible herbs. *Nat. Sci.* **2013**, *5*, 895–902.

Sample Availability: Not available.

Analysis on Blast Fungus-Responsive Characters of a Flavonoid Phytoalexin Sakuranetin; Accumulation in Infected Rice Leaves, Antifungal Activity and Detoxification by Fungus

Morifumi Hasegawa, Ichiro Mitsuhara, Shigemi Seo, Kazunori Okada, Hisakazu Yamane, Takayoshi Iwai and Yuko Ohashi

Abstract: To understand the role of the rice flavonoid phytoalexin (PA) sakuranetin for blast resistance, the fungus-responsive characteristics were studied. Young rice leaves in a resistant line exhibited hypersensitive reaction (HR) within 3 days post inoculation (dpi) of a spore suspension, and an increase in sakuranetin was detected at 3 dpi, increasing to 4-fold at 4 dpi. In the susceptible line, increased sakuranetin was detected at 4 dpi, but not at 3 dpi, by which a large fungus mass has accumulated without HR. Induced expression of a PA biosynthesis gene *OsNOMT* for naringenin 7-*O*-methyltransferase was found before accumulation of sakuranetin in both cultivars. The antifungal activity of sakuranetin was considerably higher than that of the major rice diterpenoid PA momilactone A *in vitro* and *in vivo* under similar experimental conditions. The decrease and detoxification of sakuranetin were detected in both solid and liquid mycelium cultures, and they took place slower than those of momilactone A. Estimated local concentration of sakuranetin at HR lesions was thought to be effective for fungus restriction, while that at enlarged lesions in susceptible rice was insufficient. These results indicate possible involvement of sakuranetin in blast resistance and its specific relation to blast fungus.

Reprinted from *Molecules*. Cite as: Hasegawa, M.; Mitsuhara, I.; Seo, S.; Okada, K.; Yamane, H.; Iwai, T.; Ohashi, Y. Analysis on Blast Fungus-Responsive Characters of a Flavonoid Phytoalexin Sakuranetin; Accumulation in Infected Rice Leaves, Antifungal Activity and Detoxification by Fungus. *Molecules* **2014**, *19*, 11404-11418.

1. Introduction

Phytoalexins (PAs) are induced by stresses, including microbial infection, in many plant-microbe interactions, and are defined as low molecular weight antifungal compounds biosynthesized *de novo* by host plants [1–3]. PAs have diverse structures, including flavonoid, isoflavonoid, diterpenoid, sesquiterpenoid, polyacetylenes, indoles and stilbenoid, and the importance as general defense compounds was demonstrated in dicot plants such as tobacco, tomato and alfalfa by ectopic expression of a PA biosynthesis gene [4]. At the same time, metabolism and detoxification of PAs by the pathogenic fungus, which are thought to be a way to overcome the defense of plants, were reported in dicot plants [5]. In monocot plants, characteristic natures and roles of diterpenoid PAs for defense were indicated [6], however many subjects remain to be studied. In blast fungus-infected rice leaves, accumulation of sixteen PAs, including fifteen diterpenes such as momilactone A and one flavonoid PA, sakuranetin, was reported [7,8]. A previous study on the major rice diterpenoid PA momilactone A indicated the importance for blast resistance via experiments on its time-dependent accumulation profiles in resistant and susceptible rice lines, antifungal activity of the PA *in vitro* and *in vivo*, and detoxification of the PA by the fungus in mycelium cultures [9].

Manuscripts in the 1990s on the rice flavonoid PA sakuranetin described that blast fungus infection induced higher levels of sakuranetin in resistant rice lines than in susceptible rice lines [10,11]. However, the experimental systems used for the sakuranetin quantification contained a wound process, which itself can induce PA increases. In the former study [10], a spore suspension was applied on the finger-rubbed leaf surface of detached leaves. After an appropriate time, droplets on the leaves were collected and combined with excised leaf discs which contained fungus-induced spots, and subjected to sakuranetin quantification. In the later paper [11], a spore suspension was applied after crushing the leaf surfaces with a pair of headless pliers and the leaf was used for the quantification. These procedures were accompanied by severe wounding. Actually, treatment with the defense signal compound jasmonic acid, whose increase is generally induced after wounding, increased sakuranetin accumulation [12]. To reveal the net increase in sakuranetin caused by blast fungus infection, experiments should be conducted under natural infection conditions eliminating the wound effect. At the same time, experiments to reveal its antifungal activity and detoxification by blast fungus are necessary to provide information about the defensive characteristics of the PA against fungi.

Thus, we here studied these using resistant and susceptible rice lines and a corresponding blast fungus race. The results indicate that: (1) sakuranetin has stronger antifungal activity to blast fungus than the diterpenoid PA momilactone A *in vitro* and *in vivo*; (2) the concentration of sakuranetin accumulated in infected regions in the resistant rice line was estimated to be effective enough to restrict the fungus, while that in susceptible rice was too low for it; (3) blast fungus converted sakuranetin to less toxic compounds as did momilactone A, while the rate of conversion was slower than that of momilactone A. These results indicate the fungus-responsive characteristics of sakuranetin, and the defensive role of the PA in the fight of host rice plants *vs.* invaded blast fungus.

2. Results and Discussion

2.1. The Local Concentration of Sakuranetin is Estimated to be Superior in Infected Regions in a Resistant Rice Line than in a Susceptible Line

Sakuranetin is a rice flavonoid PA with a flavanone structure, and its increase by blast fungus infection has been reported by Kodama *et al.* [10] and Dillon *et al.* [11]. They used pre-wounded rice leaves for fungal inoculation. To evaluate the net increase in the PA levels caused by fungus infection, experiments should be done under natural infection conditions without any wound effect. Then, we spray-inoculated intact rice seedlings grown in soil with a spore suspension of blast fungus race 003, and the 4th leaves were subjected to sakuranetin quantification. In this experimental system, typical susceptible and resistant responses of host rice plants to the fungal type were observed [9,13]; blast fungus grows vigorously in the susceptible rice line Nipponbare (N), and susceptible type enlarged light brown lesions (ELs) become visible at 4 days post inoculation (dpi). On the other hand, in resistant rice line IL7, which was generated by introduction of a rice resistance gene *Pii* to the fungus into N and back-crossings [14], fungal growth is severely restricted during the early infection period [9,13] accompanying hypersensitive reaction (HR), which is a typical resistance response of host plants to restrict invasive pathogens to infected regions. HR generally accompanies formation of HR lesions (HRLs), which are very small size and dark brown and become visible at 2 dpi, increasing the number at 3 dpi [9].

After spray-inoculation with a spore suspension, only a small amount of sakuranetin at a background level was found at 0–2 dpi in both susceptible N and resistant IL7 rice leaves (Figure 1A). An increase in sakuranetin was first found in IL7 at 3 dpi, and was estimated to be 23 ng per g fresh leaf (closed circles), and the level increased to 92 ng at 4 dpi. In the N leaves, the increase was not detected at 3 dpi, and 108 ng was found at 4 dpi (open circles). The majority of PAs have been reported to accumulate locally at fungus-induced lesions [15,16], and the area occupied by HRLs at 3 dpi was roughly estimated to about 0.1% of the whole IL7 leaf area; around 50 HRLs with about 0.01 mm^2 each were formed in the leaf of 500 mm^2 [9,17] indicating that about 1,000-fold higher sakuranetin than the mean level is localized at HRLs. As the mean sakuranetin content in IL7 at 3 dpi was 23 ng per g of fresh leaf, the local concentration at HRLs was 23 μg per g leaf corresponding to near 0.1 mM (Figure 1). At 4 dpi, the mean PA content was 92 ng per g of IL7 leaf, so the local concentration of HRLs was estimated to be about 0.3 mM. In susceptible N leaves, the area occupied by susceptible type enlarged lesions (ELs) was about 20% of the whole leaf at 4 dpi with accumulation of a large fungus mass [9,17]. Mean sakuranetin content at 4 dpi was 108 ng per g of N leaf, and accordingly the local concentration at ELs was estimated to 540 ng per g leaf corresponding to 2 μM (Figure 1B, N: open circles). These results suggest that the net sakuranetin concentration at fungus-induced lesions was estimated to be about 160-fold higher in IL7 than N at 4 dpi. The sakuranetin contents quantified here were converted into relative values based on fungal DNA contents in infected leaves [9], demonstrating that those were much higher in IL7 than in N (Figure 1C).

The sakuranetin contents in the resistant rice line detected here were considerably lower than the levels mentioned in previous reports; 100 μg per g leaf at 2 dpi by Kodama *et al.* [10] and 15.8 μg per g leaf at 3 dpi by Dillon *et al.* [11]. As indicated by Tamogami *et al.* [12], treatment with the wound hormone jasmonic acid elicited 40-fold more sakuranetin in rice leaves in comparison with water-treated control. Then, one of the possible reasons for the different results could come from the wound-inducible nature of sakuranetin. Actually, in our experimental system, sakuranetin was detected in a very small amount in healthy rice leaves, and increased about 30-fold by cutting leaves at 48 h. We spray-inoculated a fungal spore suspension and inoculated rice seedlings were carefully grown without wounding before sakuranetin quantification. However, in the previous papers, the inoculation methods involved severe wounding procedures. Another reason for the different results could come from the different experimental system used, such as rice lines, fungus race, incubation conditions and quantification methods.

Figure 1. Increase in sakuranetin and transcript of *NOMT* in resistant IL7 and susceptible N rice lines after inoculation with blast fungus. (**A**) Sakuranetin content in leaves; (**B**) Estimated sakuranetin concentration at lesions; (**C**) Relative content of sakuranetin/ fungal DNA; (**D**) RT-PCR analysis on a sakuranetin biosynthesis gene *OsNOMT* for naringenin 7-*O*-methyltransferase and HR-inducible control genes *OsACS2* and *OsACO7* in rice leaves after mock- or blast-inoculation according to Iwai *et al.* [13]. Arrow indicates the time of HR lesion visualization. Asterisks indicate a significant difference between inoculated N and IL7 leaves at 3 dpi (Student's paired *t* test: $p < 0.01$). gFW: g fresh leaf weight. dpi: Day post inoculation. hpi: Hour post inoculation. Data are means ± standard deviation (SD) from independent three samples for Figure 1A–C.

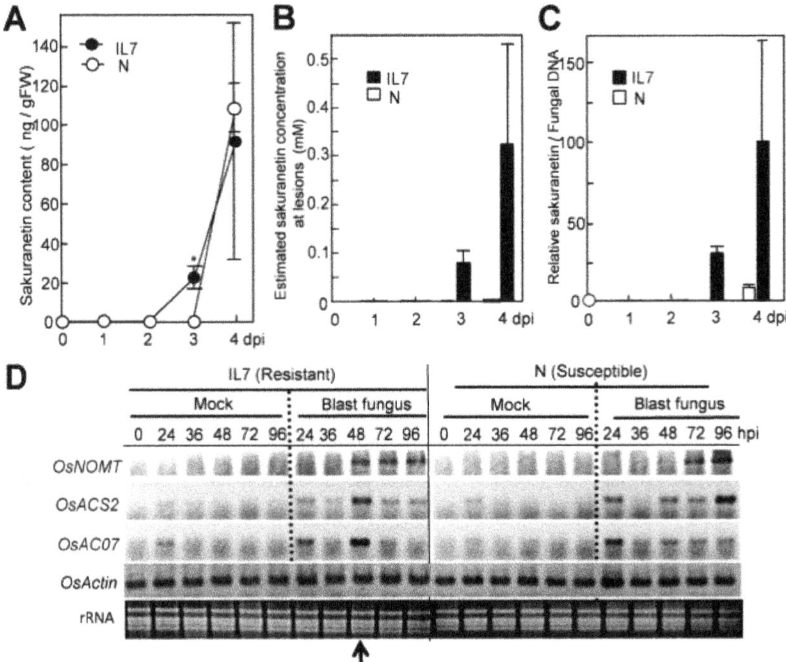

Next, induced expression of a sakuranetin biosynthesis gene upon HR was studied by RT-PCR according to Iwai *et al.* [13]. Expression of the *OsNOMT* gene for naringenin 7-*O*-methyltransferase which is a key enzyme in the biosynthesis of sakuranetin (Shimizu *et al.* [18]) was analyzed with that of HR inducible positive control genes, *OsACS2* and *OsACO7* for ethylene biosynthesis [13]. The transcripts for *OsACS2* and *OsACO7* were transiently accumulated in blast inoculated IL7 at 48 h post inoculation (hpi) at which HRLs became to be visible but not in inoculated N (Figure 1D). Transcript of *OsNOMT* was found in IL7 but not in N at 48 hpi. Expression time of *OsNOMT* in IL7 was 48–96 hpi in IL7, while it was at 72–96 hpi in N, suggesting induction of *OsNOMT* preceded sakuranetin accumulation in both lines.

2.2. Inhibition of Fungal Growth by Sakuranetin in Vitro and in Planta

The inhibitory activity of sakuranetin to mycelium growth was studied *in vitro* using a solid medium. Four blast mycelium plugs (about 10 mm^2 each) were inoculated on potato dextrose agar (PDA) containing sakuranetin at various concentrations, and fungal growth was analyzed measuring the diameter of the mycelium colony after appropriate incubation.

As shown in Figure 2A, sakuranetin inhibited fungal mycelium growth in a concentration-dependent manner on PDA. On the right side, the phenotypes of mycelium colonies at 5 dpi are shown. A time-course experiment indicates that growth of mycelium colony in diameter at 2 dpi was inhibited about 50% by sakuranetin at both 0.1 and 0.3 mM (Figure 2B). The inhibition rate by 0.3 mM sakuranetin became slightly higher than by 0.1 mM thereafter, and 51 and 36% of the growth was inhibited by 0.3 and 0.1 mM at 5 dpi, respectively. The level of antifungal activity on PDA was similar to that on potato sucrose agar (PSA), which contains sucrose instead of glucose as the naturally occurring sugar source in plants. The phenotype of the mycelium colony was altered in the presence of sakuranetin. The colony looked like more condensed and risen on both PDA and PSA containing 0.3 mM sakuranetin, and an example on PSA at 5 dpi is shown in Figure 2C.

Referring to the report by Hasegawa *et al.* on the diterpenoid PA momilactone A [9], the antifungal activity of sakuranetin on blast fungus was higher than that of momilactone A under similar experimental conditions. For example, momilactone A inhibited 12% and 17% of the mycelium growth on PDA at 0.1 and 0.3 mM at 4 dpi respectively [9], while sakuranetin inhibited 40% and 55% of it at 0.1 and 0.3 mM, respectively (Figure 2B).

Figure 2. Inhibition of blast mycelium growth by sakuranetin on solid medium. (**A**) Size of blast mycelium colony on PDA containing sakuranetin at 5 dpi; (**B**) Time course analysis on the spread of blast mycelium colony on PDA containing sakuranetin at 0.1 or 0.3 Mm; (**C**) Phenotype of the mycelium colony with or without 0.3 mM sakuranetin at 5 dpi. Data are means ± SD from four independent samples. Scale bar in Figure 2A indicates 50 mm, and that in Figure 2C 10 mm.

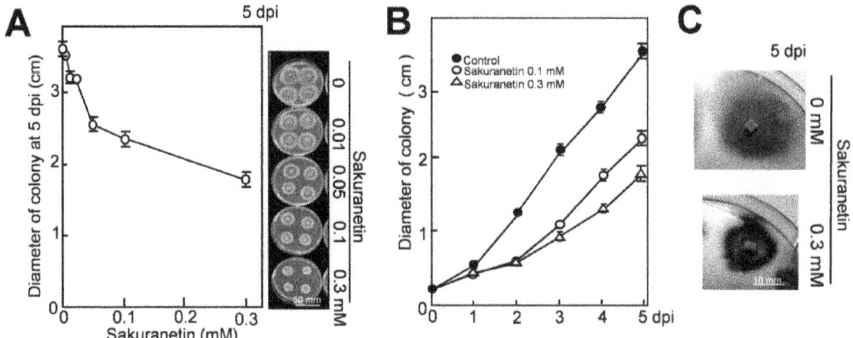

Next, the antifungal activity of sakuranetin was analyzed *in planta*. After spray inoculation with a blast spore suspension, ten 4th leaves from intact susceptible N seedlings were detached at 1 dpi, and the leaf bases were put into a glass tube containing 10 mL of chemical solution, and incubated according to the illustrated method in Figure 3A and in Seo *et al.* [17]. The visible phenotype of inoculated 4th leaves at 4 dpi is shown in the leftmost column of Figure 3B. Compared with the visible phenotype, spread of fungus was more clearly observed after lactophenol-trypan blue staining of mycelium (second column). Sakuranetin treatment at 0.1 and 0.2 mM reduced the number of susceptible type ELs, whose area was more than 0.5 mm^2, to 50% and 25% respectively (third column) and fungal DNA content to 7% and 5% respectively (most right column) at 4 dpi. When the antifungal activity of sakuranetin *in planta* was compared with that of momilactone A described by Hasegawa *et al.* [9], sakuranetin exhibited higher antifungal activity than momilactone A. For example, the number of ELs in blast-inoculated N leaves was decreased to the 25% of control by 0.2 mM sakuranetin (Figure 3B), and that was decreased to the 43% by 0.2 mM momilactone A [9]. Fungal DNA content in infected leaves was decreased to the 4% by 0.2 mM sakuranetin (Figure 3B), and that was decreased to the 39% by 0.2 mM momilactone A [9]. These results indicate sakuranetin has a considerable antifungal activity *in vitro* and *in vivo*, which is higher than that of momilactone A. The calculated concentration of sakuranetin at HRLs in IL7 seems to be effective to restrict fungal growth, while that at ELs was insufficient in susceptible N (Figure 1B), indicating a possible contribution of sakuranetin for fungal resistance upon HR in IL7.

Figure 3. Inhibition of blast fungus growth by sakuranetin in rice leaves at 4 dpi. (**A**) Experimental protocol; (**B**) Leftmost column: visible lesion phenotype on blast-inoculated Nipponbare leaves treated with sakuranetin; second column: phenotype after staining mycelium in inoculated leaves with lactophenol-trypan blue; third column: number of ELs larger than 0.5 mm^2 in a leaf. Rightmost column: fungal DNA content/transcript of *OsActin*. Data are means ± SE from ten independent samples for ELs and three samples for fungal DNA, respectively.

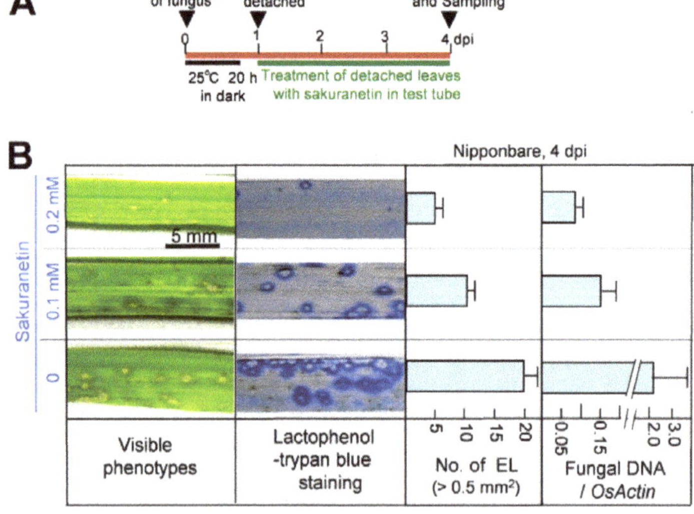

2.3. Decrease of Sakuranetin in Blast Mycelium Cultures

Detoxification of PA would be a way for the fungus to survive in the invaded host plant. Possible detoxification of sakuranetin by the fungus was examined using both solid and liquid mycelium cultures. Discs 20 mm in diameter of 15-day-old blast mycelium layer on PDA were prepared, and put upside down on new PDA containing 0.3 mM sakuranetin according to the method described by Hasegawa *et al.* [9]. After incubation for the indicated time periods, the agar discs beneath the mycelium discs were prepared and subjected to sakuranetin quantification after treatment with 70% methanol for sakuranetin extraction.

Sakuranetin in the agar medium decreased with time after being overlaid by the mycelium layer. Recovered sakuranetin levels at 1, 2, 3 and 5 dpi decreased to 65%, 45%, 20% and 17% of the initial level, respectively (Figure 4A, closed circles), while no clear decrease was seen after mock-inoculation (open circles). Compared with the data of momilactone A under similar experimental conditions, the rate of decrease of sakuranetin was considerably slower than that of momilactone A; recovered momilactone levels at 1, 2 and 3 dpi were 53%, 3% and 0% of the initial level, respectively [9].

Next, decrease and possible conversion of sakuranetin were analyzed in liquid mycelium culture. The mycelium culture of blast fungus was prepared as described by Hasegawa *et al.* [9]. After addition of sakuranetin at 0.1 mM, 1 mL of culture containing 2 mg equivalent of fungal protein was collected at appropriate time intervals, and fractionated into the supernatant and the precipitated hyphae, designated as medium fraction and fungal mass fraction, respectively. Sakuranetin was then extracted from each fraction with 70% methanol and subjected to quantification. After addition of sakuranetin, about the 60% was transferred to the fungal mass fraction and about the 40% was retained in the medium fraction at 4 h (Figure 4B). The sakuranetin content gradually decreased thereafter in both fractions, and the contents in fungal mass and the medium became 25% and 40% of the initial level, respectively, at 8 h, retaining 65% as total content. At 24 h, it was decreased to 10% in fungal mass and 2% in the medium (Figure 4B).

The decrease in momilactone A using a similar mycelium culture has been reported [9]. However, the rate of decrease was higher than that of sakuranetin. At 4 h after PA application, 35% of the initially added momilactone A remained [9], while the 100% of the sakuranetin remained (Figure 4B). At 8 h, 20% of the momilactone A remained [9], while 65% of sakuranetin was still detected in the culture. These results indicate that sakuranetin is metabolized more slowly by the fungus than momilactone A in both solid and liquid mycelium media.

Figure 4. Decrease of sakuranetin in the blast mycelium culture. (**A**) Decrease of sakuranetin content in solid medium PDA after being overlaid by blast mycelium layer (+Fungus) or mock-layer (−Fungus). At time 0, 0.3 mM sakuranetin was present in the agar medium (100%), and sakuranetin in PDA was quantified thereafter; (**B**) Decrease in sakuranetin in liquid mycelium culture. At time 0, 0.1 mM sakuranetin was present in the PD liquid mycelium culture medium (100%). After incubation for the indicated time period, sakuranetin in the medium fraction and fungus mass fraction was quantified separately. Data are means ± SD from three independent samples.

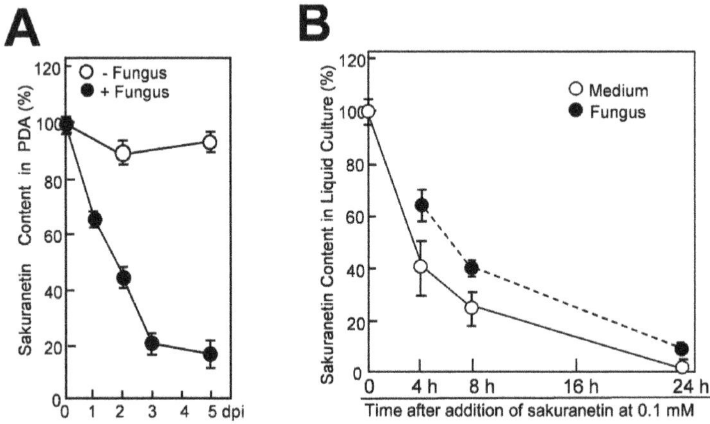

2.4. Detoxification of Sakuranetin in the Culture of Blast Mycelium

Next, we evaluated whether the decrease in sakuranetin from the liquid mycelium culture accompanies the decrease in antifungal activity or not. As the first experiment, a standard antifungal activity test was conducted using authentic sakuranetin according to the method by Hasegawa *et al.* [9]. In the presence of sakuranetin at 34, 67 and 100 μM, germ tube growth from blast spore was inhibited by 72%, 89% and 100% at 24 h, respectively (Figure 5A). In this system, the sakuranetin concentration required for 50% inhibition of germ tube elongation was estimated to be about 20 μM, which is similar to the concentration reported by Kodama *et al.* [10]. On the other hand, in the presence of momilactone A at 100, 200 and 300 μM, germ tube growth was inhibited 45%, 78% and 96%, respectively [9]. These results indicate that sakuranetin more effectively inhibits fungal growth than momilactone A in liquid culture.

Figure 5. Decrease in antifungal activity in the liquid blast mycelium culture after addition of sakuranetin. (**A**) Standard experiment on germ tube growth inhibition by sakuranetin. Length of germ tubes grown from blast spore was determined at 24 h after incubation with sakuranetin; (**B**) The same lots of samples used in Figure 4B were employed for the antifungal assay. Data are means ± SD from three independent samples each of which contains 10–20 spores for Figure 5A, and from 10 to 20 independent samples each of which contains 5–20 spores for Figure 5B, respectively. The precise method was according to Hasegawa *et al.* [9].

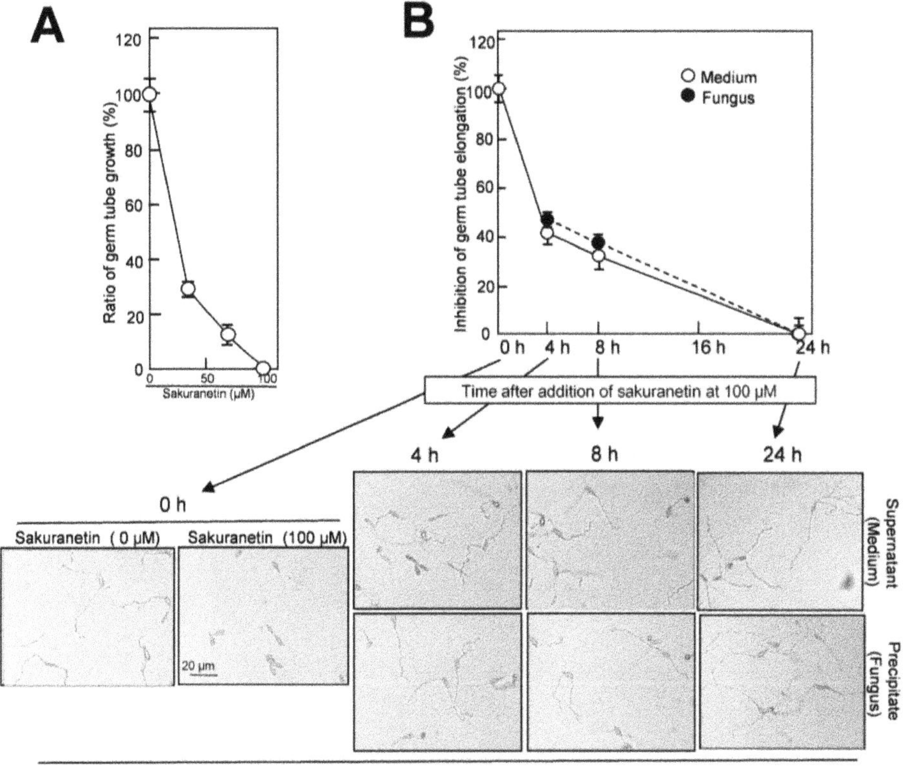

The same lots of samples which were used for the fungus-induced sakuranetin decrease experiment in liquid culture (Figure 4B), were employed for the antifungal activity analysis. Each sample from the medium and fungal mass fractions was treated with 70% methanol for extraction, and subjected to the assay after concentration. The antifungal activity in both fractions decreased with time, and almost all was lost at 24 h (Figure 5B). A representative photograph used for length determination of elongated germ tube from blast spores is shown at the bottom of Figure 5B. When the antifungal activity found in the medium fraction at time 0 was designated as 100%, the antifungal activity in the medium was decreased to 40% at 4 h, 32% at 8 h and was nearly zero at 24 h, and the antifungal activity in fungal mass fraction decreased to 46% at 4 h, 38% at 8 h and zero at 24 h

(Figure 5B). The time-course profile of the loss of antifungal activity was similar to that of the decrease in sakuranetin content shown in Figure 4B, indicating the sakuranetin content in the culture fractions reflects the level of antifungal activity determined by inhibition of germ tube elongation. Thus, the conversion of sakuranetin would accompany its detoxification by blast fungus. The detoxification of sakuranetin may be the second example on fungus-induced conversion of PA from the family Graminaceae. As the first example, Hasegawa *et al.* [9] reported detoxification of momilactone A by blast fungus *in vitro* [9], and Imai *et al.* [19] reported a degradation intermediate of momilactone A. In the mycelium culture after addition of sakuranetin, a small amount of naringenin, which is the precursor of sakuranetin biosynthesis in rice plant, was detected. Quantification and evaluation of naringenin as the metabolite of sakuranetin should be studied in future. Including this subject, isolation and identification of possible metabolites of sakuranetin are in progress.

2.5. Possible Mechanism of Sakuranetin to Restrict Fungus

The results above indicate that superior accumulation of sakuranetin at HRLs in resistant rice contributes to blast resistance. The precise mechanism of how the flavonoid PA sakuranetin inhibits fungal growth is not obvious now. Mizutani *et al.* [20] suggested that a widely used fungicide metominostrobin (SSF-126) is similar in mode of action to CN, which inhibits CN-sensitive respiration in the mitochondrial respiration chain of blast fungus. Although blast fungus then induced CN-resistant respiration to survive, the CN-resistant fungal respiration was inhibited in the presence of synthetic flavone inducing death of the fungus. Seo *et al.* [17] demonstrated that an exogenously supplied synthetic flavone induced blast resistance as well as CN, and further that treatment of CN together with flavone enhanced the resistance *in vitro* and *in vivo*. If the flavonoid PA sakuranetin could inhibit CN-resistant fungal respiration as the flavone did, the PA and CN may co-operationally contribute to the resistance by inhibition of both respiration mechanisms in the fungus. A study to reveal this possibility is in progress.

Shimizu *et al.* [18] identified the gene for naringenin 7-*O*-methyltransferase (NOMT), which confers the last step of sakuranetin biosynthesis from naringenin. To elucidate the importance of sakuranetin for the resistance, experiments on *NOMT* gene for gain of function in susceptible rice and loss of function in resistant rice would be valuable.

3. Experimental Section

3.1. Plant Materials and Blast Fungus Inoculation

As susceptible- and resistant-type rice (*Oryza sativa*) lines, Nipponbare (N) and IL7 [14] were used respectively. IL7 is a near isogenic line of N containing the *R* gene *Pii* against blast fungus (*M. oryzae*) race 003 (isolate, Kyu89-241) [21]. Rice plants were grown in cultivation soil (Bonsol No.1, Sumitomo Chem. Ltd., Chuo-ku, Tokyo, Japan) in a growth chamber for 16 h at 450 μmol m^{-2} s^{-1} at 28 °C and 8 h of dark at 25 °C. The 2 week-old seedlings were spray-inoculated with a spore suspension of blast fungus (1×10^5 conidia mL^{-1}) containing 0.05% Tween 20, and incubated under dark at high humidity for 20 h at 25 °C for effective infection. The seedlings were moved and incubated in a chamber at 25 °C under 16 h light and 8 h dark cycle. After indicated time periods of incubation, the 4th leaves were used for sakuranetin quantification, fungal DNA quantification and analysis of the number and size of lesions developed. Infected leaves were cut into 20 mm lengths, and subjected to lactophenol-trypan blue staining of mycelium according to Iwai *et al.* [13] after treatment with ethanol.

3.2. Quantification of Sakuranetin

For quantification of sakuranetin in blast fungus-inoculated leaves, 0.15 g portions of each of the 4th leaves was dipped in 6 mL of 70% methanol, and heated in a screw cap glass vial for 5 min. At this step, [methyl-$^2H_3$]sakuranetin (40 ng g^{-1} fresh leaf), which was synthesized from naringenin and deuterio-diazomethane according to the method reported by Aida *et al.* [22], was added to the extract to estimate the recovery rate of sakuranetin. Deuteriodiazomethane was prepared by using sodium deuteroxide according to the previously reported method [23,24]. The extract was transferred to a new tube, and the residue was re-extracted twice with 3 mL of 70% methanol. Extracts from three extractions were combined and concentrated to dryness. The residue was dissolved in 0.5 mL of methanol. The solution was centrifuged at 10,000 rpm for 10 min. The supernatant was subjected to the LC/MS/MS to quantify sakuranetin and [methyl-$^2H_3$]sakuranetin by the method described by Shimizu *et al.* [18]. Sakuranetin and [methyl-$^2H_3$]sakuranetin levels were determined with combinations of the precursor and product ions of *m/z* 287/167 for sakuranetin and *m/z* 290/170 for [methyl-$^2H_3$]sakuranetin in the selected reaction monitoring mode.

For quantification of sakuranetin in agar medium culture, each fraction containing sakuranetin was treated with 9 volume of methanol. Sakuranetin was extracted by shaking for 120 min, and 10 μL of the supernatant after centrifugation for 10 min at 10,000 ×*g* was subjected to an HPLC column (Inertsil ODS-3V, 4.6 × 250 mm, GL Science, Shinjuku-ku, Tokyo, Japan) equilibrated with 75% methanol. Sakuranetin was detected by the UV detector at 285 nm after eluting at 1.0 mL/min. For quantification of sakuranetin in liquid mycelium culture, 0.4 mL of methanol was added to 0.1 mL of the solution, and quantified by using LC/MS/MS as described by Inoue *et al.* [8].

3.3. Determination of Antifungal Activity of Sakuranetin in Vitro

Blast fungus was cultured on a Petri dish containing PDA at 25 °C in the dark. Fifteen days after incubation, four plugs (10 mm^2 in area) were prepared from extended front area of mycelium, and inoculated onto PDA containing sakuranetin in a Petri dish 9 cm in diameter. The diameter of mycelium colony was measured after incubation at 25 °C in the dark.

In liquid culture system, antifungal activity was determined by the inhibition of germ tube elongation from blast spores. Freshly prepared blast spores (4×10^3) were suspended in 100 μL of PD medium containing authentic chemical for standard experiment or the extract from the culture fraction to be tested, and the mixture was put on the center hole (10 mm in diameter) of a glass-bottom dish (35 mm in diameter). After incubation for indicated time period at 25 °C in the dark, photographs of germ tube elongation from blast spores were taken, and they were analyzed by measuring the length of germ tube. The liquid mycelium culture was used according to Hasegawa *et al.* [9].

One mL of the culture containing 2 mg equivalent of fungal protein was prepared at 4, 8 and 24 h, and levels of sakuranetin and antifungal activity in the precipitate and the medium fractions were separately quantified.

3.4. Determination of Antifungal Activity of Sakuranetin in Rice Leaves

Ten fungus-inoculated 4th leaves from 2 week-old seedlings were cut at the base of whole plants at 1 dpi, put into an open 70 mL glass test tube containing 10 mL of chemical solution, and incubated at 25 °C under 16 h light/8 h dark condition. Stock solution of 100 mM sakuranetin in methanol was diluted with water to suitable concentrations, and the same amount of methanol was used for the control. The infected leaves were cut into pieces (2 cm long) before subjecting to lactophenol-trypan blue staining according to Iwai *et al.* [13]. The number of susceptible type enlarged lesion (EL) larger than 0.5 mm^2 was determined using close-up photographs after staining leaves with lactophenol-trypan blue.

3.5. Determination of Blast Fungus DNA in Rice Leaves

Quantification of *M. oryzae* DNA in inoculated leaves was conducted by real-time PCR [25] after extraction with ISO PLANT (Nippon Gene, Chiyoda-ku, Tokyo, Japan). Two specific primer pairs, which were designed based on the 3' non-coding region of a *tubulin* gene in *M. oryzae* (forward, 5'-GGGATGATGGTGGTGGAGGAC-3'; and reverse, 5'-GCCAGGTGCTTAGGACGAAAC-3'). These data were normalized with the DNA amount of a rice *actin* gene (AK060893), which was quantified using following primers (forward, 5'-GAGTATGATGAGTCGGGTCCAG-3'; and reverse, 5'-ACACCAACAATCCCAAACAGAG-3').

3.6. Analysis of OsNOMT Expression in Mock- and Blast-Inoculated Rice Leaves by RT-PCR

One-step RT-PCR was conducted using the primer sets for the probe of *OsActin* (forward, 5'-GAGAAGAGCTATGAGCTGCCTGATGG-3'; and reverse, 5'-AGGGCAGTGATCTCCTTGCT CAT-3'), and for the probe of *OsNOMT* (forward, 5'-CGGGAGCAGCAGCGGCGAA-3'; and reverse; 5'-GGCGAGCGGTGATCATCCGCA-3'). The precise method and the primer sets for HR-inducible *OsACS2* and *OsACO7* genes were according to Iwai *et al.* [13].

4. Conclusions

Blast fungus-responsive characters of the rice flavonoid PA sakuranetin were studied. Increase in sakuranetin was found earlier in resistant IL7 than a susceptible N line after spray-inoculation with a suspension of blast spores. Although the mean sakuranetin concentrations detected in whole IL7 leaves were not so much, the local concentration at HRLs at 4 dpi was estimated to about 0.3 mM, which is an effective concentration to restrict blast fungus. Accumulation of sakuranetin in susceptible rice leaves was not found at 3 dpi. At 4 dpi, the local concentration at the susceptible type ELs was estimated to be about 2 μM, which is insufficient to inhibit fungal growth. The antifungal activity of sakuranetin was determined and compared with that of a major rice diterpenoid PA, momilactone A. Results indicate that the antifungal activity of sakuranetin *in vitro* and also *in vivo* was considerably higher than that of momilactone A under similar experimental conditions. Detoxification of sakuranetin by the fungus in mycelium cultures was found, however it took place more slowly than that of momilactone A. These results reveal the characteristic nature of the flavonoid PA sakuranetin against blast fungus in rice. The level of antifungal activity and the mode of detoxification by blast fungus were different from those of momilactone A, indicating the flavonoid PA sakuranetin shares the roles and collaborates with diterpenoid PAs to restrict blast fungus. The different characteristics of individual rice PA may guarantee the diversity of the resistance to blast fungus.

Acknowledgments

We thank the National Agriculture and Food Research Organization for providing IL7 rice seeds and Dr. N. Hayashi in National Institute of Agrobiological Sciences (NIAS) for providing *M. oryzae* race 003. We also thank Y. Goto, M. Teruse and T. Kimoto for their technical assistance in NIAS.

Author Contributions

I.M., M.H. and Y.O. designed the research; M.H, S.S., K.O. H.Y., T.I. and Y.O. performed the experimental work; I.M., S.S., M.H. and Y.O wrote the manuscript. All authors discussed, edited and approved the final version.

Conflicts of Interest

The authors declare no conflict of interest.

References

1. Kuć, J. Phytoalexins, stress metabolism, and disease resistance in plants. *Annu. Rev. Phytopathol.* **1995**, *33*, 275–297.
2. Müller, K.O.; Börger, H. Experimentelle Untersuchungen über die *Phytophthora*-Resistenz der Kartoffel. *Arb. Biol. Reichsanst. Land Forstwirtsch*. **1940**, *23*, 189–231.
3. Jeandet, P.; Clément, C.; Courot, E.; Cordelier, S. Modulation of phytoalexin biosynthesis in engineered plants for disease resistance. *Int. J. Mol. Sci.* **2013**, *14*, 14136–14170.
4. Großkinsky, D.K.; van der Graaff, E.; Roitsch, T. Phytoalexin transgenics in crop protection-Fairy tale with a happy end? *Plant Sci.* **2012**, *195*, 54–70.
5. Pedras, M.S.C.; Ahiahonu, P.W.K. Metabolism and detoxification of phytoalexins and analogs by phytopathogenic fungi. *Phytochemistry* **2005**, *66*, 391–411.
6. Schmelz, E.A.; Huffaker, A.; Sims, J.W.; Christensen, S.A.; Lu, X.; Okada, K.; Peters, R.J. Biosynthesis, elicitation and roles of monocot terpenoid phytoalexins. *Plant J.* **2014**, doi:10.1111/tpj.12436.
7. Grayer, R.J.; Kokubun, T. Plant-fungal interaction: The search for phytoalexins and other antifungal compounds from higher plants. *Phytochemistry* **2001**, *56*, 253–263.

8. Inoue, Y.; Sakai, M.; Yao, Q.; Tanimoto, Y.; Toshima, H.; Hasegawa, M. Identification of a novel casbane-type diterpene phytoalexin, *ent*-10-oxodepressin, from rice leaves. *Biosci. Biotechnol. Biochem.* **2013**, *77*, 760–765.
9. Hasegawa, M.; Mitsuhara, I.; Seo, S.; Imai, T.; Koga, J.; Okada, K.; Yamane, H.; Ohashi, Y. Phytoalexin accumulation in the interaction between rice and the blast fungus. *Mol. Plant Microbe Interact.* **2010**, *23*, 1000 - 1011.
10. Kodama, O.; Miyakawa, J.; Akatsuka, T.; Kiyosawa, S. Sakuranetin, a flavanone phytoalexin from ultraviolet-irradiated rice leaves. *Phytochemistry* **1992**, *31*, 3807 - 3809.
11. Dillon, V.M.; Overton, L.; Grayer, R.J.; Harborne, J.E. Differences in phytoalexin response among rice cultivars of different resistance to blast. *Phytochemistry* **1997**, *44*, 599–603.
12. Tamogami, S.; Rakwal, R; Kodama, O. Phytoalexin production elicited by exogenously applied jasmonic acid in rice leaves (*Oryza sativa* L.) is under the control of cytokinins and ascorbic acid. *FEBS Lett.* **2000**, *412*, 61–64.
13. Iwai, T.; Miyasaka, A.; Seo, S.; Ohashi, Y. Contribution of ethylene biosynthesis for resistance to blast fungus infection in young rice plants. *Plant Physiol.* **2006**, *142*, 1202–1215.
14. Ise, K.; Horisue, N. Characteristics of several near-isogenic lines of rice for blast resistance gene. *Breed Sci.* **1988**, *38* (Suppl.), 404–405.
15. Snyder, B.A.; Nicholson, R.L. Synthesis of phytoalexins in sorghum as a site-specific response to fungal ingress. *Science* **1990**, *248*, 1637–1639.
16. Umemura, K.; Ogawa, N.; Shimura, M.; Koga, J.; Usami, H.; Kono, T. Possible role of phytocassane, rice phytoalexin, in disease resistance of rice against the blast fungus *Magnaporthe grisea. Biosci. Biotechnol. Biochem.* **2003**, *67*, 899–902.
17. Seo, S.; Mitsuhara, I.; Feng, J.; Iwai, T.; Hasegawa, M.; Ohashi, Y. Cyanide, a coproduct of plant hormone ethylene biosynthesis, contributes to the resistance of rice to blast fungus. *Plant Physiol.* **2011**, *155*, 502–514.
18. Shimizu, T.; Lin, F.; Hasegawa, M.; Okada, K.; Nojiri, H.; Yamane, H. Purification and identification of naringenin 7-*O*-methyltransferase, a key enzyme in the biosynthesis of the flavonoid phytoalexin sakuranetin in rice. *J. Biol. Chem.* **2012**, *287*, 19315–19325.
19. Imai, T.; Ohashi, Y.; Mitsuhara, I.; Seo, S.; Toshima, H.; Hasegawa, M. Identification of a degradation intermediate of the momilactone A rice phytoalexin by the rice blast fungus. *Biosci. Biotechnol. Biochem.* **2012**, *76*, 414–416.
20. Mizutani, A.; Miki, N.; Yukioka, H.; Tamura, H.; Masuko, M. A possible mechanism of control of rice blast disease by a novel alkoxyiminoacetamide fungicide, SSF126. *Phytopathology* **1996**, *86*, 295–300.
21. Yamada, M.; Kiyosawa, S.; Yamaguchi, T.; Hirano, T.; Kobayashi, T.; Kushibuchi, K.; Watanabe, S. Proposal of a new method for differentiating races of *Pyricularia oryzae* Cavara in Japan. *Ann. Pathytopath. Soc. Japan* **1976**, *42*, 216–219.
22. Aida, Y.; Tamogami, S.; Kodama, O.; Tsukiboshi, T. Synthesis of 7-methoxyapigeninidin and its fungicidal activity against *Gloeocercospora sorghi. Biosci. Biotechnol. Biochem.* **1996**, *60*, 1495–1496.

23. Fales, H.M.; Jaouni, T.M.; Babashak, J.F. Simple device for preparing ethereal diazomethane without resorting to codistillation. *Anal. Chem.* **1973**, *45*, 2302–2303.
24. Ngan, F.; Toofan, M. Modification of preparation of diazomethane for methyl esterification of environmental samples analysis by gas chromatography. *J. Chromatogr. Sci.* **1991**, *29*, 8–10.
25. Qi, M.; Young, Y. Quantification of *Magnaporthe grisea* during infection of rice plants using real-time polymerase chain reaction and northern blot/phosphoimaging analysis. *Phytopathology* **2002**, *92*, 870–876.

Sample Availability: Not available.

Antimicrobial Activity of Resveratrol Analogues

Malik Chalal, Agnès Klinguer, Abdelwahad Echairi, Philippe Meunier, Dominique Vervandier-Fasseur and Marielle Adrian

Abstract: Stilbenes, especially resveratrol and its derivatives, have become famous for their positive effects on a wide range of medical disorders, as indicated by a huge number of published studies. A less investigated area of research is their antimicrobial properties. A series of 13 *trans*-resveratrol analogues was synthesized via Wittig or Heck reactions, and their antimicrobial activity assessed on two different grapevine pathogens responsible for severe diseases in the vineyard. The entire series, together with resveratrol, was first evaluated on the zoospore mobility and sporulation level of *Plasmopara viticola* (the oomycete responsible for downy mildew). Stilbenes displayed a spectrum of activity ranging from low to high. Six of them, including the most active ones, were subsequently tested on the development of *Botrytis cinerea* (fungus responsible for grey mold). The results obtained allowed us to identify the most active stilbenes against both grapevine pathogens, to compare the antimicrobial activity of the evaluated series of stilbenes, and to discuss the relationship between their chemical structure (number and position of methoxy and hydroxy groups) and antimicrobial activity.

Reprinted from *Molecules*. Cite as: Chalal, M.; Klinguer, A.; Echairi, A.; Meunier, P.; Vervandier-Fasseur, D.; Adrian, M. Antimicrobial Activity of Resveratrol Analogues. *Molecules* **2014**, *19*, 7679-7688.

1. Introduction

Plants possess an innate immune system that prevents their infection by most of microorganisms such as oomycetes and fungi [1]. This self-defense potential includes the production of the secondary metabolites phytoalexins, antimicrobial compounds synthesized and accumulated in response to biotic or abiotic stresses [2,3]. In grapevines, they are stilbenes synthesized via the phenylalanine/ polymalonate pathway [4]. The key compound resveratrol (3,5,4'-trihydroxystilbene) is formed by condensation of one molecule of *p*-coumaroyl-CoA and three molecules of malonyl-CoA by stilbene synthase (EC 2.3.1.95). Subsequent glycosylation, methoxylation or dimerization reactions provides a spectrum of resveratrol derivatives [5]. Such modifications are essential for the biological activity of the so- formed compounds [6–8].

A huge number of works has reported the role of stilbenes, especially resveratrol, in human health. They have attracted attention for their high preventive or curative effects on a wide range of medical disorders and are known as cardioprotective, antitumor, neuroprotective and antioxidant agents (for reviews, see [9–13]). In comparison, the antimicrobial properties of stilbenes have been less investigated. In grapevines, stilbenes are constitutively accumulated at high concentrations in the heartwood where they act as phytoanticipins and can prevent the development of wood decay [14,15]. In other tissues, they are accumulated in response to various microorganisms including pathogens: *Plasmopara viticola, Erysiphe necator, Botrytis cinerea, Phaeomoniella chlamydospora*, *Fusarium solani*, *Cladosporium cucumerinum*, *Pyricularia oryzae*, *Aspergilli*, *Rhizopus stolonifer* (for review, see [5,16,17]). Whereas resveratrol generally shows a moderate antimicrobial activity, it is the precursor of more active derivatives such as pterostilbene and viniferins (for reviews, see [5,16,17]).

Recently, interest in the bioproduction and chemical synthesis of stilbenes has emerged to identify highly active molecules that could be used for medical applications and/or plant disease control [18–22]. Previous studies have shown that resveratrol is not the most active stilbene regarding antimicrobial activity [5,6,16,17]. In this study, 13 *trans*-resveratrol analogues were synthesized as previously described [20] to identify better candidates than *trans*-resveratrol to control two harmful grapevine pathogens: *P. viticola* (downy mildew) and *B. cinerea* (grey mold). This allowed us to compare their antimicrobial activity against both pathogens and to discuss the chemical structure/antimicrobial activity relationships of these compounds.

2. Results and Discussion

2.1. Activity of Stilbenes on *P. viticola*

A series of 13 *trans*-resveratrol analogues was synthesized via Wittig or Heck reactions and the structures were confirmed by ^{1}H, ^{13}C-NMR, HRMS, and IR after purification [20]. The compounds differed by the number and position of hydroxy and/or methoxy groups (Figure 1).

Figure 1. Structure of the stilbenes used for bioassays. (**a**) 4-OH stilbenes bearing substituents on cycle B; (**b**) 4-OH stilbenes bearing substituents on cycle A and/or cycle B; (**c**) Structure of 2-OH and 3-OH stilbenes; (**d**) Structure of stilbenes without phenolic function.

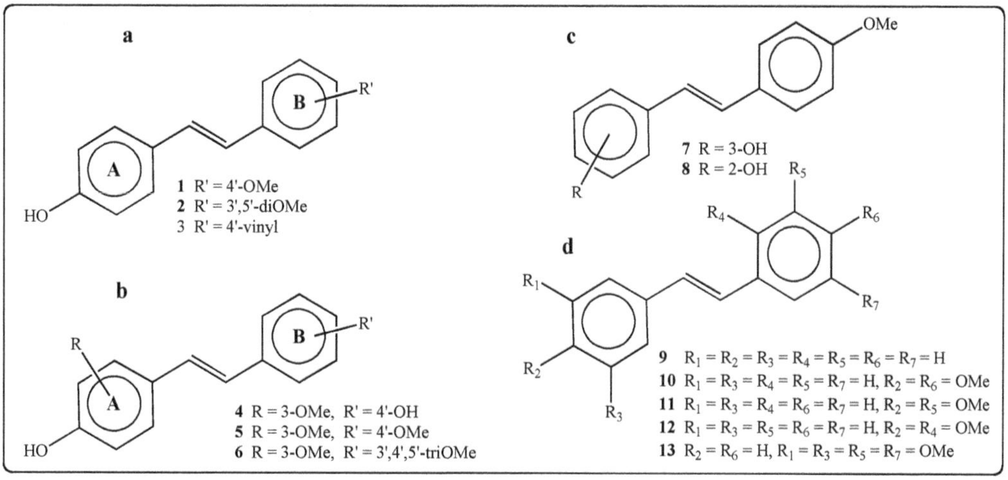

a
1 R' = 4'-OMe
2 R' = 3',5'-diOMe
3 R' = 4'-vinyl

b
4 R = 3-OMe, R' = 4'-OH
5 R = 3-OMe, R' = 4'-OMe
6 R = 3-OMe, R' = 3',4',5'-triOMe

c
7 R = 3-OH
8 R = 2-OH

d
9 $R_1 = R_2 = R_3 = R_4 = R_5 = R_6 = R_7 = H$
10 $R_1 = R_3 = R_4 = R_5 = R_7 = H$, $R_2 = R_6 = OMe$
11 $R_1 = R_3 = R_4 = R_6 = R_7 = H$, $R_2 = R_5 = OMe$
12 $R_1 = R_3 = R_5 = R_6 = R_7 = H$, $R_2 = R_4 = OMe$
13 $R_2 = R_6 = H$, $R_1 = R_3 = R_5 = R_7 = OMe$

Their activity was first assessed in parallel with resveratrol against *P. viticola*, the oomycete responsible for grapevine downy mildew. This pathogen is obligatory and biotrophic *i.e.*, it develops in leaving grapevine tissues and cannot be grown on artificial media. The assays were therefore performed by adding the compounds to a sporangia suspension (0.25, 0.5 and 0.75 mM) prior to inoculation of leaf disks. The level of sporulation was determined by visual scoring at 6 days post inoculation as previously described [23]. Stilbenes were prepared in DMSO to ensure their dissolution and the final DMSO concentration in suspensions was 2% (v/v). As there was interest in identifying stilbenes more active than resveratrol, the three concentrations assessed for this bioassay were chosen on the basis of resveratrol ID_{100} values towards *P. viticola* and/or *B. cinerea* previously reported in the literature [24–26]. The results obtained are presented in Figure 2.

DMSO alone only slightly reduced the level of sporulation in comparison with water control. All resveratrol analogues showed activity compared to both water and DMSO controls. The most active ones were **2**, **4**, **8**, and resveratrol (**RSV**); especially **2** and **RSV** that totally inhibited the sporulation at all concentrations assessed. Compounds **6** and **12** were also highly active at 0.5 and 0.75 mM and, to a lesser extent, compounds **3** and **7** at 0.75 mM. Compound **2** corresponds to pterostilbene, previously reported as having the highest antimicrobial activity among natural resveratrol derivatives [6,25,26]. In most cases, the presence of lateral groups increases the activity against *P. viticola.* For disubstituted stilbenes, the nature of the groups (-OH and -OMe *vs.* -OMe and -OMe) is important for activity although their position looks essential (comparisons **1** *vs.* **10**, **7** *vs.* **11**, **8** *vs.* **12**). The presence of a methoxy group in position 4' together with a methoxy or hydroxy group in position 2 confer high activity. Resveratrol and tri-substituted derivatives were the most active compounds, suggesting the importance of the hydroxy group at position 4'. The presence of more than three groups does not increase activity.

Figure 2. Effect of stilbenes (0.25, 0.5 and 0.75 mM) on *P. viticola* sporulation. Leaf disks (10/condition) were inoculated with a *P. viticola* sporangia suspension added by the stilbenes, water or DMSO (2% v/v final concentration) as controls. The index of sporulation was scored at 6 days post-inoculation on a scale of 0 to 4, where 0 = no visible sporulation, 1 = 1% to 25%, 2 = 26% to 50%, 3 = 51% to 75%, and 4 = 76% to 100% of the disk area covered. Values represent the mean index from three independent experiments.

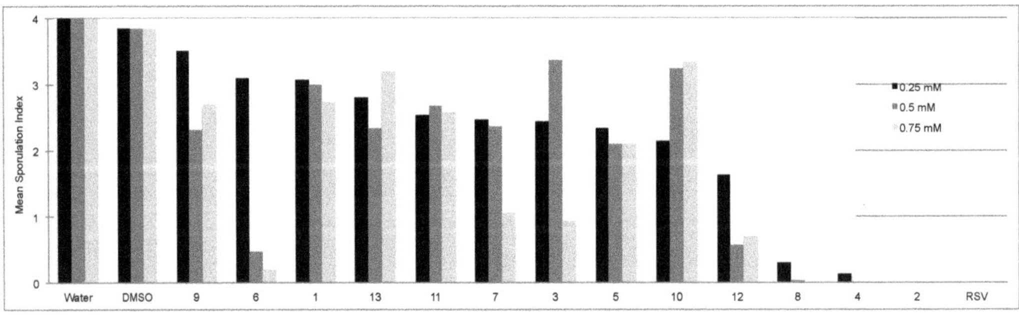

A prerequisite for successful downy mildew infection is zoospore mobility. Hence, once released by sporangia, bi-flagellated zoospores have to swim into water droplets present at the leaf surface to reach stomata where they encyst and form a germ tube that uses these natural pores to enter the leaf. The effects of the most active stilbenes were therefore measured on zoospore mobility (Table 1).

Table 1. Effects of stilbenes (**RSV** and compounds **2**, **4**, **6** and **8**) on the mobility of *P. viticola* zoospores. Values correspond to the percentage of mobile zoospores released by sporangia in a suspension added by stilbenes (0.25, 0.5, and 0.75 mM). The value of 100 was attributed to the number of mobile zoospores determined in the water control.

Compound	0.25 mM	0.5 mM	0.75 mM
2	0	0	0
8	0	0	0
RSV	38	0	0
4	38	23	0
6	46	15	0

For these bioassays, DMSO slightly reduced the mobility of zoospores (90% of mobile zoospores towards the water control) and the most active compounds were **2** and **8**. Curiously, no mobile zoospores could be observed in the presence of compound **8** whereas sporulation occured at 0.25 mM and, to a lesser extent, at 0.5 mM. The mobility of zoospores in these conditions is maybe too low to be measured but sufficient to allow few zoospores to reach stomata and initiate infection. The opposite was observed for **RSV** at 0.25 mM for which mobile zoospores were observed whereas no sporulation could be detected. Mobile zoospores might reach stomata but they could not germinate. The effects of natural stilbenes on the mobility of zoospores have been previously reported [26]. Resveratrol was less active than pterostilbene, as in our conditions, and also than viniferins.

2.2. Activity of Stilbenes on B. cinerea

We were interested in identifying stilbenes with high antimicrobial activity against both grapevine pathogens. On the basis of the results obtained with *P. viticola*, compounds **2**, **4**, **6**, and **8** were evaluated against *B. cinerea*, the necrotroph fungus responsible for grey mold. Compounds **1** and **7** were included to allow comparison with **8**. Conidia suspensions were prepared in the culture medium and added by stilbenes or DMSO (2% v/v final concentration) as control. Bioassays were performed in microplates and the mycelium development was automatically recorded at regular time intervals by spectrophotometry for 60 h. Stilbenes were tested at concentrations ranging from 0.01 to 0.75 mM to allow the determination of their IC_{50} value (Table 2). As previous studies have clearly shown that resveratrol is less active than pterostilbene (**2**) against *B. cinerea* [5,6,16,17,25] it was not included in these assays at different concentrations but only at 0.5 mM as positive control. The IC_{50} of pterostilbene was therefore used for comparison. Resveratrol (0.5 mM) was systematically included in each bioassay as reference.

Table 2. Effects of stilbenes on *B. cinerea* development. Values correspond to the concentration that inhibits 50% of the mycelial growth (IC_{50}). SE: Standard error.

Compound	IC_{50} ± SE (μM)
7	28 ± 3
8	30 ± 5
2	52 ± 4
4	55 ± 11
1	>100
6	>100

DMSO alone did not inhibit the mycelial development (data not shown). Compound **8** appeared highly interesting since it showed high antimicrobial activity against both pathogens. It was more efficient than pterostilbene (**2**), generally described as the most active resveratrol analogue, on *B. cinerea*. This highlights the interest of this disubstituted stilbene. According to the results obtained with *P. viticola*, compound **4** also showed a high activity whereas compounds **1** and **6** were less active. Interestingly, compound **7** showed a high antimicrobial activity on *B. cinerea* but was less effective on *P. viticola*. Altogether, these results confirm that the chemical structure of stilbenes is essential for their biological activity. As example, resveratrol methylation confers a higher antifungal activity [6,19,25]. Monoglucosides also exhibit an antioxidant activity that depends on the location of the hydroxy groups and the type of the sugar residue [7]. Albert *et al.* [19] also reported this structure/biological activity relationship. However, in some cases, it depends on microorganisms. As example, they observed that the activity of methoxylated stilbenes was significant for fungi but low for bacteria. It would be interesting to determine what make the specificity of the activity of such compounds.

At the end of the experiments, the fungus was collected from the microplates and observed by microscopy. Characteristic images are presented in Figure 3. In control conditions, *B. cinerea* developed a thick network of long hyphae (Figure 3a). Conversely, in presence of active compounds (**2**, **4**, **7** and **8**), the mycelium development was restricted (Figure 3b). One could observe short gem tubes and thinned hyphal tips, indicating that the fungal development was stopped. In some cases, dead conidia released their intercellular content (Figure 3c, arrow). These observations are typical of the toxic effects of stilbenes [25,27].

Figure 3. Representative photographs showing *B. cinerea* mycelial development (**a**) in the culture medium alone as control or (**b**,**c**) added by compounds **2**, **4**, **7** or **8**. Observations were made using a Leitz DMRB microscope.

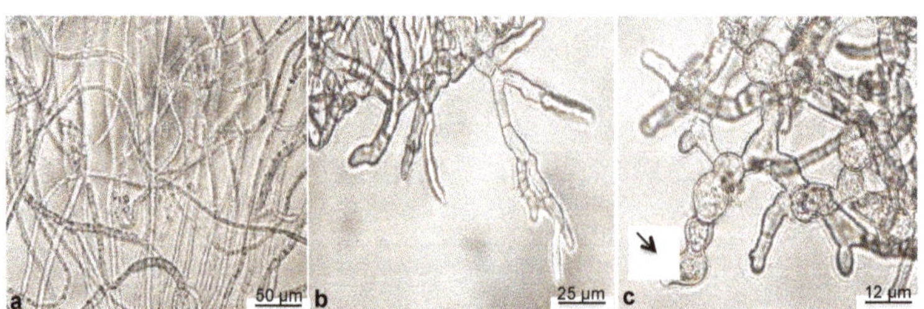

3. Experimental

3.1. Synthesis

The syntheses and characterization of the 13 stilbenes assessed were previously reported [20]. Briefly, preparation of compounds **4–8**, and **10–12** was achieved using palladium catalysis Heck coupling reactions. Stilbenes **1–3**, **10** and **13** were prepared through classic Wittig reactions.

3.2. Plant Material

Grapevine (*V. vinifera* L. cv. Marselan) herbaceous cuttings were grown in a glasshouse at a temperature of 24 and 18 °C (day and night, respectively), with a photoperiod of 16 h light, and at a relative humidity (RH) of 70% ± 10%, as previously described [28]. Once they developed six leaves, the second and third leaves below the apex were excised and used for *P. viticola* inoculation.

3.3. *P. viticola* Bioassays

3.3.1. Assessment of Stilbene Effects on Sporulation

The *P. viticola* strain was maintained on Marselan plants as previously described [28]. For bioassays, leaf disks (1 cm diameter) were punched out and placed in a Petri dish on wet filter paper. They were inoculated with 30 μL of a 10^5 sporangia/mL suspension prepared in distilled water and added by stilbenes dissolved in DMSO (0.25, 0.5 and 0.75 mM, 2% v/v final concentration of DMSO). The suspension was also added by water and DMSO alone (2% v/v final concentration) as controls. The antimicrobial activity of compounds was determined 6 days post inoculation (dpi) by visual scoring of the index of sporulation on a scale of 0 to 4, where 0 = no visible sporulation, 1 = 1% to 25%, 2 = 26% to 50%, 3 = 51% to 75%, and 4 = 76% to 100% of the disk area covered. Ten leaf disks were prepared per condition and the experiment was repeated three times.

3.3.2. Assessment of Stilbene Effects on Zoospore Mobility

Assays were performed using a modified method previously described [26]. A sporangia suspension of *P. viticola* was prepared at 5×10^4 sporangia·mL^{-1} in distilled water and added by stilbene dissolved in DMSO (0.25, 0.5 and 0.75 mM, 2% v/v final concentration of DMSO). The suspension was also added by water and DMSO (2% v/v final concentration) alone as controls. After 45 min at room temperature with frequent and gentle handly stirring to allow the release of zoospores, 30 μL of the suspension was deposited on a Malassez cell. The number of zoospores passing through an area of one rectangle of the cell was determined during 1 min using a microscope (magnification ×40).

3.4. *B. cinerea* Bioassays

A conidial suspension of the *B. cinerea* strain BMM was prepared at 2×10^5 conidia/mL of PDB (Potato Dextrose Broth 1/4) medium. Stilbenes prepared in DMSO were added to the suspension (at 0.01, 0.02, 0.05, 0.1, 0.25 and 0.5 mM, 2% v/v final concentration of DMSO). The suspension was also added by DMSO alone (2% v/v final concentration) as control. One hundred microliters of each suspension were distributed in microplates for automatic spectrophotometry recording (Bioscreen C, Thermoelectron Led, St Herblain, France). The absorbance at 492 nm was measured and recorded every 2 h for 60 h.

4. Conclusions

This study allowed us to compare the antimicrobial activity of 13 *E*-stilbenes with resveratrol on two grapevine pathogens. Altogether, the results confirm the importance of the chemical structure of stilbenes regarding their biological activity. However, they did not allow us to draw a clear and direct relationship between the structure of a compound (number/position of OH- and OCH_3- groups) and its antimicrobial activity, making it difficult to predict the level of antimicrobial activity of a stilbene. The results also highlight that the activity of resveratrol analogues is microorganism dependent. Among the 13 *trans*-resveratrol derivatives assessed, only three of them were highly active against both pathogens. One of them is the already known pterostilbene. The other ones are 2-hydroxy, 4-methoxystilbene (**8**) and 3-methoxy-4,4'-hydroxystilbene (**4**). Compound **7** (3-hydroxy-4'-methoxystilbene) also looks interesting, but only against *B. cinerea*. Interestingly, compounds 7 and **8** showed a higher antimicrobial activity against *B. cinerea*, compared to the highly active pterostilbene. It would be interesting to investigate the biological activity of these new compounds against a larger spectrum of microorganisms.

Acknowledgments

We thank the Conseil Régional de Bourgogne and Bureau Interprofessionnel des Vins de Bourgogne for financial support, Adrian Variot for technical assistance, and Arnaud Haudrechy for helpful comments concerning the manuscript. Malik Chalal benefited of a PhD grant from Algerian government (Ministère de la Recherche et de l'Enseignement), which is sincerely acknowledged.

Author Contributions

MA, DVF and PM designed research; MC, AK and AE performed research, MA, MC and AE analyzed the data; MA, MC and DVF wrote the paper. All authors read and approved the final manuscript.

Conflicts of Interest

The authors declare no conflict of interest.

References

1. Nürnberger, T.; Brunner, F.; Kemmerling, B.; Piater, L. Innate immunity in plants and animals: Striking similarities and obvious differences. *Immunol. Rev.* **2004**, *198*, 249–266.

2. Kuc, J. Phytoalexins, stress metabolism, and disease resistance in plants. *Annu. Rev. Phytopathol.* **1995**, *33*, 275–297.
3. Jeandet, P.; Clément, C.; Courot, E.; Cordelier, S. Modulation of phytoalexin biosynthesis in engineered plants for disease resistance. *Int. J. Mol. Sci.* **2013**, *14*, 14136–14170.
4. Langcake, P.; Pryce, R.J. The production of resveratrol by *Vitis vinifera* and other members of the *Vitaceae* as a response to infection or injury. *Physiol. Mol. Plant Pathol.* **1976**, *9*, 77–86.
5. Chong, J.L.; Poutaraud, A.; Hugueney, P. Metabolism and roles of stilbenes in plants. *Plant Sci.* **2009**, *177*, 143–155.
6. Pont, V.; Pezet, R. Relation between the chemical structure and the biological activity of hydroxystilbenes against *Botrytis cinerea. J. Phytopathol.* **1990**, *130*, 1–8.
7. Orsini, F.; Pellizzoni, F.; Verotta, L.; Aburjai, T.; Rogers, C.B. Isolation, synthesis, and antiplatelet aggregation activity of resveratrol 3-o-*beta*-d-glucopyranoside and related compounds. *J. Nat. Prod.* **1997**, *60*, 1082–1087.
8. Regev-Shoshani, G.; Shoseyov, O.; Bilkis, I.; Kerem, Z. Glycosylation of resveratrol protects it from enzymic oxidation. *Biochem. J.* **2003**, *374*, 157–163.
9. Aggarwal, B.B.; Bhardwaj, A.; Aggarwal, R.S.; Seeram, N.P.; Shishodia, S.; Takada, Y. Role of resveratrol in prevention and therapy of cancer: Preclinical and clinical studies. *Anticancer Res.* **2004**, *24*, 2783–2840.
10. Baur, J.A.; Sinclair, D.A. Therapeutic potential of resveratrol: The *in vivo* evidence. *Nat. Rev. Drug Discov.* **2006**, *5*, 493–506.
11. Roupe, K.A.; Remsberg, C.M.; Yanez, J.A.; Davies, N.M. Pharmacometrics of stilbenes: Seguing towards the clinic. *Curr. Clin. Pharmacol.* **2006**, *1*, 81–101.
12. Patel, K.R.; Scott, E.; Brown, V.A.; Gescher, A.J.; Steward, W.P.; Brown, K.; Ann, N.Y. Clinical trials of resveratrol. *Acad. Sci.* **2011**, *1215*, 161–169.
13. Pezzuto, J.M. The phenomenon of resveratrol: Redefining the virtues of promiscuity. *Ann. N. Y. Acad. Sci.* **2011**, *1215*, 123–130.
14. Hart, J.H.; Shrimpton, D.M. Role of stilbenes in resistance of wood to decay. *Phytopathology* **1979**, *69*, 1138–1143.
15. Hart, H. Role of phytostilbenes in decay and disease resistance. *Annu. Rev. Phytopathol.* **1981**, *19*, 437–458.
16. Jeandet, P.; Delaunois, B.; Conreux, A.; Donnez, D.; Nuzzo, V.; Cordelier, S.; Clément, C.; Courot, E. Biosynthesis, metabolism, molecular engineering, and biological functions of stilbene phytoalexins in plants. *Biofactors* **2010**, *36*, 331–341.
17. Adrian, M.; de Rosso, M.; Bavaresco, L.; Poinssot, B.; Héloir, M.C. Resveratrol from vine to wine. In *Resveratrol Sources, Production and Health benefits*; Delmas, D., Ed.; Nova Science Publishers Inc.: New York, NY, USA, 2013; pp. 3–19.
18. Mazue, F.; Colin, D.; Gobbo, J.; Wegner, M.; Rescifina, A.; Spatafora, C.; Fasseur, D.; Delmas, D.; Meunier, P.; Tringali, C.; *et al.* Structural determinants of resveratrol for cell proliferation inhibition potency: Experimental and docking studies of new analogs. *Eur. J. Med. Chem.* **2010**, *45*, 2972–2980.

19. Albert, S.; Horbach, R.; Deising, H.B.; Siewert, B.; Csuk, R. Synthesis and antimicrobial activity of (E) stilbene derivatives. *Bioorg. Med. Chem.* **2011**, *19*, 5155–5166.
20. Chalal, M.; Vervandier-Fasseur, D.; Meunier, P.; Cattey, H.; Hierso, J.C. Syntheses of polyfunctionalized resveratrol derivatives using Wittig and Heck protocols. *Tetrahedron Lett.* **2012**, *68*, 3899–3907.
21. Bhusainahalli, V.M.; Spatafora, C.; Chalal, M.; Vervandier-Fasseur, D.; Meunier, P.; Latruffe, N.; Tringali, C. Resveratrol-related dehydrofimers: Laccase mediated biomimetic synthesis and antiproliferative activity. *Eur. J. Org. Chem.* **2012**, *27*, 5217–5224.
22. Jeandet, P.; Vasserot, Y.; Chastang, T.; Courot, E. Engineering microbial cells for the biosynthesis of natural compounds of pharmaceutical significance. *BioMed Res. Int.* **2013**, *2013*, 780145:1–780145:13.
23. Trouvelot, S.; Varnier, A.L.; Allègre, M.; Mercier, L.; Baillieul, F.; Arnould, C.; Gianinazzi-Pearson, V.; Klarzynski, O.; Joubert, J.M.; Pugin, A.; *et al.* A *beta*-1,3 glucan sulfate induces resistance in grapevine against *Plasmopara viticola* through priming of defense responses, including HR-like cell death. *Mol. Plant Microbe Int.* **2008**, *21*, 232–243.
24. Langcake, P. Disease resistance of *Vitis* spp. and the production of the stress metabolites resveratrol, ε-viniferin, α-viniferin and pterostilbene. *Physiol. Plant Pathol.* **1981**, *18*, 213–226.
25. Adrian, M.; Jeandet, P.; Veneau, J.; Weston, L.A.; Bessis, R. Biological activity of resveratrol, a stilbenic compound from grapevines, against *Botrytis cinerea*, the causal agent for gray mold. *J. Chem. Ecol.* **1997**, *23*, 1689–1702.
26. Pezet, R.; Gindro, K.; Viret, O.; Richter, H. Effect of resveratrol, viniferins and pterostilbene on *Plasmopara viticola* zoospore mobility and disease development. *Vitis* **2004**, *43*, 145–148.
27. Adrian, M.; Jeandet, P. Effects of resveratrol on the ultrastructure of Botrytis cinerea conidia and biological significance in plant/pathogen interactions. *Fitoterapia* **2012**, *83*, 1345–1350.
28. Allègre, M.; Héloir, M.C.; Trouvelot, S.; Daire, X.; Pugin, A.; Wendehenne, D.; Adrian, M. Are grapevine stomata involved in the elicitor-induced protection against downy mildew? *Mol. Plant Microbe Int.* **2009**, *22*, 977–986.

Sample Availability: Samples of the compounds are not available.

Transcriptional Responses of the *Bdtf1*-Deletion Mutant to the Phytoalexin Brassinin in the Necrotrophic Fungus *Alternaria brassicicola*

Yangrae Cho, Robin A. Ohm, Rakshit Devappa, Hyang Burm Lee, Igor V. Grigoriev, Bo Yeon Kim and Jong Seog Ahn

Abstract: *Brassica* species produce the antifungal indolyl compounds brassinin and its derivatives, during microbial infection. The fungal pathogen *Alternaria brassicicola* detoxifies brassinin and possibly its derivatives. This ability is an important property for the successful infection of brassicaceous plants. Previously, we identified a transcription factor, *Bdtf1*, essential for the detoxification of brassinin and full virulence. To discover genes that encode putative brassinin-digesting enzymes, we compared gene expression profiles between a mutant strain of the transcription factor and wild-type *A. brassicicola* under two different experimental conditions. A total of 170 and 388 genes were expressed at higher levels in the mutants than the wild type during the infection of host plants and saprophytic growth in the presence of brassinin, respectively. In contrast, 93 and 560 genes were expressed, respectively, at lower levels in the mutant than the wild type under the two conditions. Fifteen of these genes were expressed at lower levels in the mutant than in the wild type under both conditions. These genes were assumed to be important for the detoxification of brassinin and included *Bdtf1* and 10 putative enzymes. This list of genes provides a resource for the discovery of enzyme-coding genes important in the chemical modification of brassinin.

Reprinted from *Molecules*. Cite as: Cho, Y.; Ohm, R.A.; Devappa, R.; Lee, H.B.; Grigoriev, I.V.; Kim, B.Y.; Ahn, J.S. Transcriptional Responses of the *Bdtf1*-Deletion Mutant to the Phytoalexin Brassinin in the Necrotrophic Fungus *Alternaria brassicicola*. *Molecules* **2014**, *19*, 10717-10732.

1. Introduction

Plants protect themselves from the attack of potential pathogens. Plant resistance mechanisms include the production of antimicrobial compounds called phytoalexins. Brassinin and its derivatives are phytoalexins produced by plants of the genus *Brassica*. They are induced during the infection process by microbes, including pathogenic fungi [1–3]. Brassinin also has antimicrobial activity *in vitro* [4]. Mutant strains of *A. brassicicola* with cell wall integrity defects are more sensitive to brassinin [5,6]. It is possible that brassinin affects the cell integrity of pathogens similar to camalexin, a phytoalexin that disrupts the cell membrane of the bacterium *Pseudomonas syringae* [7].

Brassinin probably contributes to the resistance of plants against pathogenic fungi because of its antifungal activity [4]. In spite of the induction of brassinin, however, several fungi establish parasitic growth in *Brassica* species. Their success as parasites might be partly due to their ability to detoxify brassinin. The stem rot fungus *Sclerotinia sclerotiorum* metabolizes brassinin into its corresponding glucosyl derivatives, which have no detectable antifungal activity [8]. In comparison, the blackleg fungus *Leptosphaeria maculans* detoxifies brassinin by the unusual oxidative transformation of a dithiocarbamate to an aldehyde [9]. *Alternaria brassicicola* detoxifies brassinin by converting it into the intermediate metabolites N'-indolylmethanamine and N'-acetyl-3-indolylmethanamine [10].

We recently produced direct molecular evidence that brassinin is important in plant resistance to *A. brassicicola*. Wild-type *A. brassicicola* detoxified brassinin by transforming it into non-indolyl products during mycelial growth in glucose yeast extract broth (GYEB). The transcription factor, brassinin detoxification factor 1 (*Bdtf1*), is an important regulator of unknown enzymes that detoxify brassinin and possibly its derivatives [11]. Mutants of the *Bdtf1* gene failed to detoxify brassinin and showed a 70% reduction in virulence on *Brassica* species but no measurable effects on *Arabidopsis thaliana*, which is in the brassica family, but produces camalexin instead of brassinin. Under test conditions, wild-type *A. brassicicola* completely degraded the indolyl compound brassinin, but did not produce intermediate products, such as N'-indolylmethanamine and N'-acetyl-3-indolylmethanamine. Brassinin hydrolase in *A. brassicicola* (BHAb) produces these intermediates [12]. However, expression of the *BHAb* gene is not regulated by the transcription factor *Bdtf1* and its expression level is very low in both the Δ*bdtf1* mutant and wild-type *A. brassicicola*. The data suggest that *A. brassicicola* produces additional enzymes important for digestion of brassinin during saprophytic growth in a nutrient-rich medium containing brassinin. The aim of this study was to investigate the gene expression profiles of wild-type *A. brassicicola* and the Δ*bdtf1* mutant in a search for novel genes that encode brassinin-detoxification enzymes.

2. Results

2.1. Effects of Brassinin on Δbdtf1 Mycelium

Each strain of the Δ*bdtf1* mutants was indistinguishable from wild-type *A. brassicicola* in mycelial growth on nutrient-rich potato dextrose agar (PDA) (Figure 1A) or glucose-yeast-extract-broth medium (GYEB). They were also identical to wild-type *A. brassicicola* in conidium development and response to stress-inducing chemicals. A major difference between the mutants and the wild type was their response to brassinin [11]. Brassinin in 0.2 mM concentration caused a slight delay in germination of the wild type, but the mutants were unable to germinate. Active mycelial growth of the mutants also stopped when transferred to PDA or GYEB containing 0.2 mM brassinin (Figure 1B). In the presence of 0.1 mM brassinin the mutants germinated and grew, but their growth rate was significantly reduced ($p < 0.01$) compared to wild-type *A. brassicicola* (Figure 1C). In GYEB with 0.1 mM brassinin, pre-grown mycelia of the wild type digested about 50% and 100% of the brassinin respectively during 4 and 8 h of incubation (Figure 1D). The mutant mycelia, however, did not degrade a measurable amount of brassinin during 8 h of growth and over 80% of the brassinin still remained after 24 h. We investigated the effects of brassinin on gene expression in nutrient-rich GYEB after actively growing mycelium had been exposed to brassinin for 4 h. This specific time was selected because the amount of brassinin was reduced by 50% in the medium with wild type but it was not reduced in the medium with Δ*bdtf1*mutants (Figure 1D).

Figure 1. Effects of brassinin on colony growth of Δ*bdtf1* mutants of *Alternaria brassicicola* on potato dextrose agar calculated as the slope of the linear regression line with four data points. Y-axes show colony diameter in millimeters and their inability to digest brassinin. (**A**). Similar growth rates in the absence of brassinin. (**B**). No growth of Δ*bdtf1* mutants in the presence of 0.2 mM brassinin. (**C**). Reduced growth of Δ*bdtf1* mutants in the presence of 0.1 mM brassinin. (**D**). Reduced degradation of brassinin by Δ*bdtf1* during mycelial growth in a liquid medium. Y-axis shows relative amounts of brassinin compared to the input amount. Bars represent standard deviations.

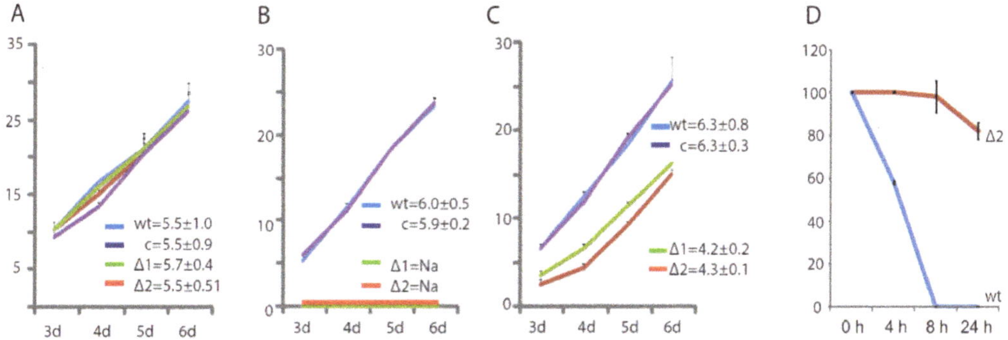

Abbreviations: wt = wild-type *Alternaria brassicicola*; c = Δ*bdtf1-5* mutants complemented with a native allele of the *Bdtf1* gene (Δ*bdtf1*:*Bdtf1*); Δ*1* = (Δ*bdtf1*-5); Δ*2* = *Bdtf1* deletion mutant (Δ*bdtf1*-9).

2.2. Gene Expression during Plant Infection

We used green cabbage (*B. oleracea*) to compare gene expression profiles between the mutants and wild-type *A. brassicicola* during the infection of host plants. Infection protocol using green cabbage was well established in our laboratory and most previous experiments were performed with this plant [11,13]. In addition, significant differences in the diameters of lesions on green cabbage between the wild type and mutants became noticeable at about 44 hpi. We speculated that the smaller lesions caused by the mutants were due to the induction of brassinin by the plant around 44 hpi. Notably, four phytoalexins, including brassinin and its derivatives, were induced and detected 44 h post-inoculation (hpi) in *Brassica junceae* [2]. Thus, we also compared gene expression profiles during the infection of green cabbage at 44 hpi as a complementary experiment.

2.3. Statistics of Gene Expression Profiles

Infection samples of mixed tissue from host plants and fungal hyphae at 44 hpi produced a total of 90.2 and 66.8 million reads of sequence tags for the wild type and Δ*bdtf1* mutants, respectively. Of these, respective 5.42 × 10^7 (60%) and 4.01 × 10^7 (60%) were mapped to the genome of *A. brassicicola*. Among 10,688 predicted genes in the *A. brassicicola* genome [14], 93 genes in the Δ*bdtf1* mutant were expressed at levels over twofold lower and 170 genes were expressed at levels over twofold higher ($p < 0.05$) than the wild type (Table S1).

Fungal tissue samples from GYEB with 0.1 mM brassinin produced a total of 61.9 and 77.2 million reads for the wild type and Δ*bdtf1* mutant, respectively. Of these, 4.60 × 10^7 (74%) and 6.83 × 10^7 (88%) were mapped to the genome of *A. brassicicola*. Among 10,688 predicted genes in the *A. brassicicola* genome, 560 genes from the Δ*bdtf1* mutants were expressed at levels over twofold lower and 388 genes expressed at levels over twofold higher ($p < 0.05$) than the wild type (Table S2). No sequence tags of the *Bdtf1* gene were expressed by the Δ*bdtf1* mutants in either set of data (Table S1 and S2). We examined the reliability of the RNA-seq data with semi-quantitative real time polymerase chain reaction (qRT-PCR) using four differentially expressed genes: *Bdtf1*, BHAb, AB02263.1, and AB08641.1. The qRT-PCR results were similar to the RNA-seq data (Tables S1 and S2).

2.4. Brassinin Effects: Genes at Lower Levels in the Mutant

During mycelial growth the Δ*bdtf1* mutant expressed fewer transcripts of 560 genes than the wild type in the presence of brassinin. Among the 560 genes, 286 had no homologs with functional annotations in public databases and the other 274 had similar genes either in predicted protein sequences or well-studied functional domains. The 274 genes belonged to the categories of helicases, nucleotide binding proteins, DNA-dependent ATPases, and ATP binding proteins (Table 1). Many of these genes were associated with ribosome biogenesis. These genes included eight RNA-binding proteins, three RNA polymerases, three ribosomal proteins, two ribonucleases, pre-rRNA processing complex protein, rRNA processing proteins, and 20 helicases. Other genes included several proteins associated with tRNA processing and translation initiation (Table S2). In contrast, there were few or no genes directly associated with DNA replication, transcription, recombination, or DNA damage-repair. The expression of HSP70 (AB02816.1) was reduced more than twofold ($p < 0.05$) in the mutant during mycelial growth in the presence of brassinin but was not affected during plant infection. The expression level of HSP90, which is activated by brassinin [15], was similar between the mutant and wild type. Expression levels of *AbSlt2* and *AbHog1* activated by brassinin [6], possibly via phosphorylation, were also similar between the wild type and the mutant.

Table 1. Functional groups of proteins over-represented among 560 genes that were expressed at lower levels in the mutant compared to wild-type *Alternaria brassicicola* during saprophytic growth in the presence of brassinin.

Annotation	Description	*p*-value
GO:0043141	ATP-dependent 5'-3' DNA helicase activity	0.000191
GO:0008026	ATP-dependent helicase activity	0.000191
GO:0008758	UDP-2,3-diacylglucosamine hydrolase activity	0.000191
GO:0047429	nucleoside-triphosphate diphosphatase activity	0.000191
GO:0004787	thiamin-pyrophosphatase activity	0.000191
GO:0008413	8-oxo-7,8-dihydroguanine triphosphatase activity	0.000191
GO:0004386	helicase activity	0.000191
GO:0019176	dihydroneopterin monophosphate phosphatase activity	0.000191
GO:0019177	dihydroneopterin triphosphate pyrophosphohydrolase activity	0.000191
GO:0008828	dATP pyrophosphohydrolase activity	0.000191
GO:0000810	diacylglycerol pyrophosphate phosphatase activity	0.000191
GO:0043139	5'-3' DNA helicase activity	0.000191
GO:0005488	binding	0.000213
GO:0030554	adenyl nucleotide binding	0.000213
GO:0017110	nucleoside-diphosphatase activity	0.000213
GO:0008796	bis(5'-nucleosyl)-tetraphosphatase activity	0.000312
GO:0004551	nucleotide diphosphatase activity	0.000312
GO:0000166	nucleotide binding	0.000587
GO:0003678	DNA helicase activity	0.000629
GO:0004003	ATP-dependent DNA helicase activity	0.000629
GO:0003676	nucleic acid binding	0.000772
GO:0017076	purine nucleotide binding	0.001009
GO:0008094	DNA-dependent ATPase activity	0.001162
GO:0003824	catalytic activity	0.002675
GO:0017171	serine hydrolase activity	0.004091
GO:0004086	carbamoyl-phosphate synthase activity	0.004091
GO:0008236	serine-type peptidase activity	0.004091
GO:0016887	ATPase activity	0.01031
GO:0005730	nucleolus	0.010587
GO:0050660	FAD binding	0.010587
GO:0048037	cofactor binding	0.012798
GO:0032040	small subunit processome	0.013819
GO:0031177	phosphopantetheine binding	0.020373
GO:0017111	nucleoside-triphosphatase activity	0.029526
GO:0016638	oxidoreductase activity, acting on the CH-NH2 group of donors	0.031008
GO:0005524	ATP binding	0.037275
GO:0032559	adenyl ribonucleotide binding	0.037275
GO:0016874	ligase activity	0.042934
GO:0006537	glutamate biosynthetic process	0.043581
GO:0042623	ATPase activity, coupled	0.049164

2.5. Brassinin Effects: Genes at Higher Levels in the Mutant

During mycelial growth in the presence of brassinin, the mutant expressed more transcripts of 388 genes than the wild type. Of the 388 differentially expressed genes, 238 did not have similar genes in the public databases. The other 150 genes had similar sequences with functional annotations.

Representative molecular functions of these 150 genes included catalytic activity, biological processes, metabolism, and hyrolase activity. Genes associated with cellular processes (GO:0009987) or cellular metabolic processes (GO:0044207) were less common ($p < 0.05$) among the differentially expressed genes than those randomly selected in the genome. In contrast, drug-resistance proteins and ABC transporters were over-represented (Table 2). Lipid metabolic process (GO:0006629), membrane transport (GO:0016021) and oxidative stress response (GO:0006979) genes were among the genes expressed in greater amounts by the mutants than by the wild type.

Table 2. Functional groups of proteins under- or over-represented among 388 genes that were expressed at higher levels in the mutant compared to wild-type *Alternaria brassicicola* during saprophytic growth in the presence of brassinin.

Annotation	Description	Representation	*p*-value
GO:0009987	cellular process	Under	0.001924
GO:0044237	cellular metabolic process	Under	0.013208
KOG0065	Pleiotropic drug resistance proteins (PDR1-15), ABC superfamily	Over	0.041251
PF06422	CDR ABC transporter	Over	0.044766

2.6. Gene Expression Patterns During Plant Infection

There were 93 genes expressed at lower levels in the mutant than in the wild type during plant infection. These genes included a few putative hydrolytic enzyme-coding genes, such as alcohol dehydrogenases, isochorismatase hydrolases, and mannose dehydrogenase. None of these genes formed a coherent functional group that was over-represented with statistical significance. However, AB02597.1, AB03046.1, AB07427.1, and AB10411.1 of 90 Na+/Pi symporter (KOG2493) were expressed at lower levels in the mutant than the wild type. Interestingly, the 170 genes that were expressed at twofold higher levels in the mutant compared to the wild type included many genes with hydrolyase activity (Table 3). These genes encode putative cell-wall degrading enzymes, such as a cutinase (AB1674.1), five pectate lyases (AB0565.1, AB0904.1, AB01332.1, AB04736.1, and AB04813.1) and 22 glycoside hydrolases (Table S1).

Table 3. Functional groups of proteins over-represented among 170 genes that were expressed at higher levels in the mutant than in wild-type *Alternaria brassicicola* during infection of the host plant, *Brassica oleracea*.

Annotation	Description	*p*-value
GO:0016798	hydrolase activity acting on glycosyl bonds	7.38×10^{-9}
GO:0004553	hydrolase activity, hydrolyzing O-glycosyl compounds	7.38×10^{-9}
GO:0005975	carbohydrate metabolic process	5.80×10^{-8}
GO:0005622	intracellular	9.67×10^{-4}
GO:0044424	intracellular part	0.025823
GO:0006139	nucleobase, nucleoside, nucleotide and nucleic acid metabolic process	0.025823

2.7. Fungal Genes Affected Under Both Conditions

Among several hundred genes affected by the *Bdtf1* loss-of-function mutation, we identified 52 genes that were differentially expressed under both experimental conditions (Table S3). Thirty of them were expressed consistently at either higher or lower levels in the mutant than in the wild type. We speculated that the brassinin digestion enzymes were induced in the presence of brassinin and that the *Bdtf1* gene was essential for the induction of those genes. Furthermore, the mutant was less virulent mainly due to its inability to detoxify brassinin. Thus, we were interested in genes that were expressed at lower levels in the mutant than the wild type under both experimental conditions. There were 15 such genes among the 52 differentially expressed genes (Figure 2). As suspected, the *Bdtf1* gene was not expressed in the mutant under either condition but was expressed in wild-type *A. brassicicola* (Table 4). One gene (AB08641.1) showed a 41-fold difference in its expression during mycelial growth in the presence of brassinin. The ratio was over 10-fold during plant infection. The predicted amino acid sequence of the gene showed a low sequence similarity to glutathione S-transferease. Except for the *Bdtf1* and AB08641.1 genes, expression ratios between the wild type and mutant were modest for the other 13 of 15 genes. These 13 remaining genes encoded diverse enzymes and 2 transporters (Figure 2).

3. Discussion

3.1. Effects of Brassinin on Protein Synthesis

Wild-type *A. brassicicola* promptly detoxified brassinin, thus overcoming the inhibition effects on mycelial growth, unlike Δ*bdtf1* mutant (Figure 1). Over 40 of the 274 genes with annotation that were expressed at lower levels in the mutant than in the wild type were putatively associated with protein synthesis (Table S2). We speculate that the reduced expression of these 40 genes was caused by the brassinin rather than by deletion of the *Bdtf1* gene. It is possible that cellular metabolism was significantly slowed in the mutant during mycelial growth *in vitro* with brassinin and transcription and translation were slowed accordingly.

It is notable that camalexin appears to inhibit protein synthesis [16]. We propose a possibility that brassinin also negatively affect protein synthesis in a concentration-dependent manner. The concentration of brassinin in the culture medium of the mutant was higher than the medium of the wild type after 4 h incubation although equal amounts were initially added to the medium. This higher concentration of brassinin might have caused a strong suppression of these 40 genes associated with protein synthesis. The possibility that both camalexin and brassinin inhibit protein synthesis warrants further investigation.

Figure 2. Hierarchical clustering of fungal RNA-seq data showing the number of overlapping genes among four groups of differentially expressed genes. (**A**) Number of genes expressed at higher levels in the mutant than in the wild type during host infection (44 hpi) and mycelial growth in the presence of brassinin. (**B**) Number of genes expressed at lower levels in the mutant than in the wild type. (**C**) Set of 52 genes showing differential expression patterns between the mutant and the wild type. The color key represents the log2 ratio of fragments per kilobase of exon model per million. Red indicates higher expression levels and green indicates lower expression levels in the mutant than in wild-type *A. brassicicola*. Abbreviations: number of genes differentially expressed in the mutant during mycelial culture in the presence of 0.1 mM brassinin; 44 hpi = number of genes differentially expressed in the mutant during plant infection

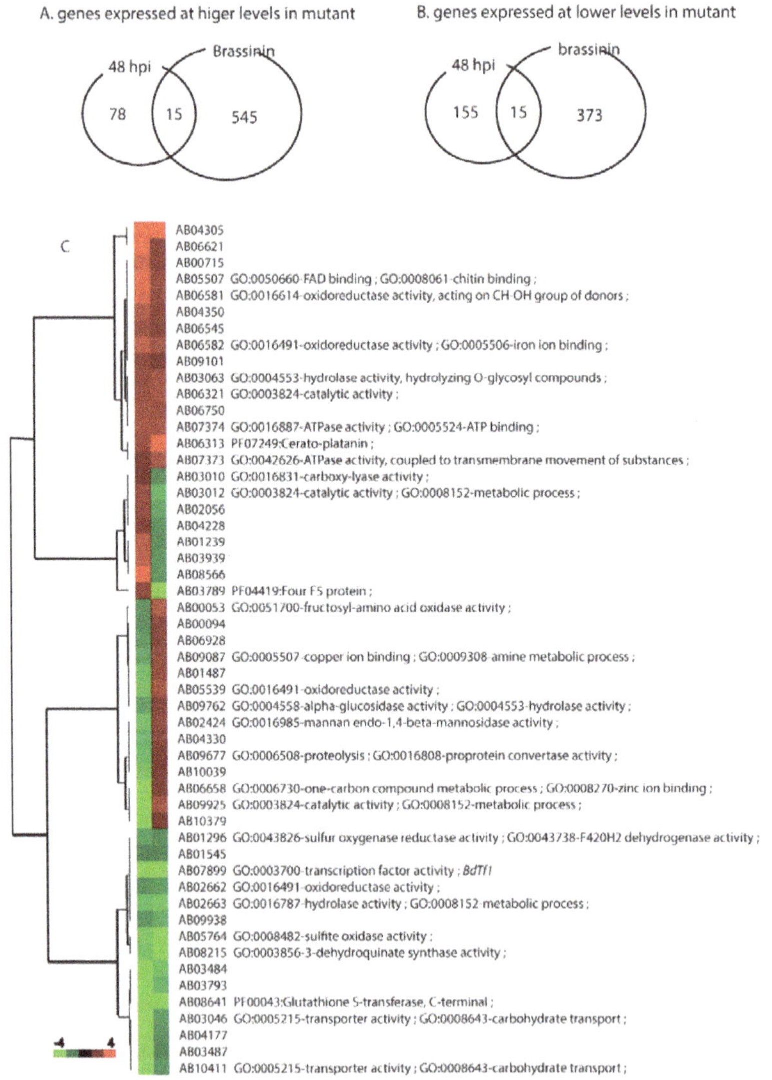

Table 4. Genes expressed at lower levels in the mutant compared to wild-type *Alternaria brassicicola* during saprophytic growth in the presence of brassinin and during the infection of host plants.

Protein ID	HMM-Secretion	[2] Mycelial Growth with Brassinin		[3] Plant Infection		Go Annotation	Manual Blast Results
		Wild type	Δ*bdtf1*	Wild type	Δ*bdtf1*		
AB01296.1		10.3	4.5	3.1	1.4	GO:0043826-sulfur oxygenase reductase activity;	Aldo/keto reductase are major group of enzymes involved in detoxification
AB01545.1		199.8	99.0	195.2	95.9	NA	xanthine phosphoribosyltransferase or purine salvage enzyme
AB02662.1		199.2	93.9	95.3	41.9	GO:0016491-oxidoreductase activity;	Aldo/keto reductase involved in detoxification
AB02663.1	[1] S	100.7	37.0	33.0	10.4	GO:0016787-hydrolase activity;	HAD-superfamily subfamily IIA hydrolase
AB03046.1		94.4	17.6	177.8	80.5	GO:0005215-transporter activity;	Sugar transporter STL1 induced when cells are subjected to osmotic shock
AB03484.1		75.9	13.3	75.6	25.6	NA	Similar to glutathione-dependent formaldehyde-activating GFA
AB03487.1		3.9	0.7	20.9	10.3	NA	Glutathione S-transferase omega-like
AB03793.1		4.6	1.1	54.3	21.8	NA	Cupin domain, salicylate hydroxylase
AB04177.1		5.3	1.1	73.6	35.1	NA	Methyltransferase involved in epigenetic regulation.
AB05764.1		3.0	0.9	1.4	0.3	GO:0008482-sulfite oxidase activity;	nitrate reductase
AB07899.1		11.0	0.0	29.7	0.1	GO:0003700-transcription factor activity;	*Bdtf1*
AB08215.1		5.1	1.7	56.7	8.5	GO:0003856-3-dehydroquinate synthase activity;	3-dehydroquinate synthase
AB08641.1		18.7	0.4	185.1	15.2	NA	glutathione S-transferase
AB09938.1		129.7	59.7	61.0	24.8	NA	arginine N-methyltransferase
AB10411.1		72.5	12.4	110.7	50.9	GO:0005215-transporter activity;	Sugar transporter STL1

1. "S" indicates a putative secretion protein predicted by hidden Markov models and signal P. 2. Numbers in these columns indicate normalized expression levels of each gene represented by fragments per kilobase of exon model per million (FPKM). 3. Numbers in these columns indicate normalized expression levels of each gene represented by FPKM during host plant infection.

3.2. Genes Important for Cell Protection

Brassinin affected germination and mycelial growth of the mutant strains (Δ*bdtf1*) more than it affected wild-type *A. brassicicola*. Neither was killed by the brassinin, however, even when exposed for several days to the concentration that totally inhibited germination and mycelial growth of the mutant [11]. The phytoalexin camalexin elicits genes involved in the biosynthesis of sterol, sphingolipid, and melanin [5]. These three compounds are probably important for cell protection against oxidative damage. Likewise, Δ*bdtf1* mutants unable to detoxify brassinin may induce other genes to mitigate its effects during exposure. We inquired whether *Bdtf1* was important for the maintenance of cell wall integrity by comparing the gene expression profiles between the wild type and the Δ*bdtf1* mutant. Differentially expressed genes did not include those associated with biosynthesis of sterol, sphingolipid, or melanin. It is possible that the expression level of these genes was elevated in both the wild type and mutant when exposed to brassinin. In addition, the expression level of the *AbHog1* or *AbSlt2* genes of *A. brassicicola* was similar in the wild type and the mutant after exposure to the brassinin. Loss-of-function mutants of either gene are hypersensitive to brassinin [6]. The Δ*bdtf1* mutant was also hypersensitive to brassinin, but there was no significant difference in the expression levels of *AbHog1* and *AbSlt2* genes in the Δ*bdtf1* mutant or wild-type *A. brassicicola*. Furthermore, the Δ*bdtf1* mutant did not have defects in osmoregulation, cell wall integrity, or oxygen stress response during mycelial growth in the absence of brassinin [11]. The gene-expression pattern and the cell-wall related phenotypes of the mutant suggest that the *Bdtf1* gene is not a downstream gene of either the *AbHog1* or *AbSlt2* gene. It also suggests that the *Bdtf1* gene is not important for the maintenance of cell wall integrity or membrane biogenesis.

3.3. Compensatory Genes

Another phytoalexin, camalexin, induces drug-efflux genes in *A. brassicicola* [5] and activates brassinin-detoxifying enzymes in mycelial cultures of *A. brassicicola* [2]. It is of note that the Δ*bdtf1* mutant was not killed in the presence of brassinin [11] and the brassinin was reduced by 20% after 24 h of incubation (Figure 1D). Survival of the mutant might have been possible by the genes expressed at higher levels in the mutants than in the wild type in response to brassinin. They included 12 and 15 genes encoding transporters and putative detoxifying enzymes, respectively. Expression of one of the transporter-coding genes (AB04925.1), for example, was five times greater in the mutant. It is possible that this higher level of expression was caused by the greater concentration of brassinin and was involved in limiting the intracellular accumulation of brassinin by pumping it out. Putative toxin digestion enzymes included three oxidoreductases, three carboxylesterases, two cyanide hydratases, two heat shock proteins, cyanate lyase, amidase, hydroperoxide reductase, latamase, monooxygenase, oxidase, multicoper oxidase, and peroxidase. Some of these genes might have been responsible for the slight reduction of brassinin.

3.4. Enzymatic Modification of Brassinin

The ability of plant-pathogenic fungi to promptly detoxify plant defense compounds is an important determinant of their virulence [17,18]. Brassinin hydrolase (BHAb), a detoxifying enzyme in *A. brassicicola* [12], however, was not included in the list of differentially expressed genes. This result indicates that the *Bdtf1* gene does not regulate *BHAb*, as we found in a previous study [11]. We doubt that other genes encoding putative enzymes described above as compensatory genes are the brassinin digestion enzymes regulated by the *Bdtf1* gene. If they were regulated by *Bdtf1* gene, their expression should have been reduced in the Δ*bdtf1* mutant. However, they were expressed at a higher level and the mutant was still unable to detoxify brassinin. Thus, unknown brassinin detoxifying enzymes among those expressed at lower levels in the mutant than the wild type in the presence of brassinin are yet to be discovered.

3.5. Candidates for Brassinin-Detoxifying Enzymes

The brassinin derivatives brassilexin and cyclobrassinin have stronger antifungal activities than brassinin [19]. Wild-type *A. brassicicola* must detoxify brassinin for successful pathogenesis. It is unclear whether the fungus also processes other brassinin-derived phytoalexins or exhaustively digests brassinin before it is converted to its derivatives. Either way, comparisons of the gene expression profiles between wild-type *A. brassicicola* and the mutant Δ*bdtf1* provided data leading to the discovery of enzymes involved in the detoxification of brassinin. Brassinin is detoxified through oxidation or hydrolysis as implied by chemical analysis [2,20]. The 15 genes expressed at lower levels in the mutant included a hydrolase (AB02663.1), two GST-like proteins (Ab08641.1 and AB03487.1), and three reductases (Table 4). Expression ratios were over fivefold lower in the mutant than the wild type for two GST-like protein genes. Other genes showed small differences. It is yet to be verified experimentally if any of these genes encode enzymes that detoxify brassinin. A loss-of-function mutation of the genes that detoxify brassinin would reduce virulence in the mutant compared to wild-type *A. brassicicola.* The severity of the reduction would depend on several factors. First, functional redundancy among the detoxifying enzyme genes will influence the effects of the loss-of-function mutation of each gene. A lack of redundancy would cause the greatest reduction in virulence. Second, an enzyme cascade would also affect virulence. Intermediate products produced by one enzyme can accumulate and slow further conversion of brassinin. The brassinin derivatives make the outcome even more complicated. Loss-of-function mutants of each gene need to be developed and their effects on virulence clarified. The enzyme activities of the proteins encoded by each gene also need to be verified.

4. Experimental Section

4.1. Fungal Strains and their Maintenance

We used the facultative plant pathogen *Alternaria brassicicola* (Schweinitz, Wiltshire, UK) (ATCC96836) in this study. Fungal strains of the wild type and its mutants, Δ*bdtf1-5* and Δ*bdtf1-9*, were purified by two rounds of single-spore isolation. To restore their vigor, each strain was inoculated on host plants and the conidia produced were transferred to potato dextrose agar. Newly formed conidia were harvested from the agar after 5 days of growth. The conidia were suspended in 20% glycerol and maintained as culture stock in separate tubes, with one tube used for each assay as described previously [13,21].

4.2. Assays for Brassinin Digestion and Preparation of Mycelium for RNA-Seq

Fungal mycelia were grown for 2 days in 1% glucose and 0.5% yeast extract broth (GYEB). The medium was refreshed 16 h before harvest. Mycelia were harvested and semi-dried by blotting with sterile paper. Subsequently, 0.15 g of semi-dried mycelium (equivalent to 0.025 g dry weight) was transferred to GYEB containing 0.1 mM brassinin. The mycelia were cultured at 25 °C in a shaker-incubator for 24 h with continuous agitation at 100 rpm. During the 24 h, 2 mL of GYEB was recovered from each culture flask at 4, 8, and 24 h. After removal of the mycelia, GYEB from each culture was transferred to a clean tube and extracted twice using 0.8 mL of chloroform for each extraction. The relative concentration and integrity of the brassinin were evaluated using a HPLC system as previously described [11]. This experiment, from the growth of fungal strains to brassinin quantification, was performed three times. For gene expression profiles, mycelia were harvested 4 h after the transfer to brassinin-containing medium. They were semi-dried by blotting with paper towels, immediately frozen by plunging them into liquid nitrogen, and then stored at −70 °C. A total of three sets of tissue were separately prepared as three biological replicates.

4.3. Preparation of Fungal Tissues from Infected Host Plants

We performed pathogenicity assays as described previously, with a slight modification [11,22]. Two healthy leaves were harvested from each of nine host plants of *B. oleracea* and placed in mini-moist chambers. Each leaf was inoculated with six droplets of wild-type inoculum on the left side and six droplets of a Δ*bdtf1* mutant strain on the right side of the central vein. The inoculum contained 1,500 conidia of either the Δ*bdtf1* mutant or wild-type *A. brassicicola* in 10 µL of water. The mini-moist chambers were sealed with plastic wrap after leaf inoculation to keep the relative humidity close to 100%. Host plant tissue and fungal hyphae were harvested at 44 hpi from six leaves (three plants) for each sample. The tissues were frozen in liquid nitrogen immediately after harvest to fix gene expression profiles. Three sets of tissues were harvested for each strain as three replicates.

4.4. Generation of RNA-Seq Data

We extracted total RNA from the frozen tissues using an RNeasy kit and residual DNA was digested in columns following the manufacturer's protocol (Qiagen, Palo Alto, CA, USA). With the DNA-free RNA, we constructed strand-specific sequencing libraries using the TruSeq™ RNA Sample Prep Kit following the manufacturer's protocol (Illumina, San Diego, CA, USA). Each library representing a replicate sample was constructed with a unique index primer. A total of six index primers were used to construct six libraries. All six libraries were mixed and 100 nucleotide-long sequence tags were determined using Illumina Hiseq2000 (Illumina). Image analysis, base-calling, and quality checks were performed with the Illumina data analysis pipeline CASAVA v1.8.0. The data were deposited in NCBI's Gene Expression Omnibus and are accessible through GEO Series Accession No. GSE59195.

The sequenced reads were mapped to the genome sequence of *A. brassicicola* in the interactive JGI fungal portal MycoCosm [23,24] using the programs Tophat 2.0.0 [25] and Bowtie 2.0.0 [26]. Default settings were used, except in the case of *A. brassicicola* the intron length was designated as a minimum of 10 nucleotides and a maximum of 500 nucleotides. The program Cuffdiff version 1.3.0, part of Cufflinks [27], was used to identify reads overlapping with previously predicted genes. The mapping bias correction method was used while running Cuffdiff [28]. The expression levels of each predicted gene were determined and normalized to Mapped Fragments Per Kilobase of exon model per Million (FPKM). Differentially expressed genes between the wild type and the mutant were determined by comparing FPKM from three biological replicates for both the wild type and the mutant using the default-allowed false discovery rate (FDR) of 0.05. In addition to this we applied a cutoff of at least a twofold change in expression value for differential expression. Custom scripts were written in Python for data analysis.

Representation Analysis of Functional Annotation Terms

Custom scripts were developed in Python and R to analyze over- and under-representation of functional annotation terms in sets of differentially regulated genes using the Fisher Exact test. The Benjamini-Hochberg correction was used to correct for multiple testing using a p-value of 0.05.

4.5. *qRT-PCR*

We generated a cDNA pool using Superscript II from 2 μg of total RNA for each sample following manufacturer's protocol (Invitrogen, Carlsbad, CA, USA). Subsequent semi-quantitative PCR was performed as described previously [22]. Relative amounts of transcripts for each gene were calculated compared to the housekeeping gene, elongation factor 1-α (*Ef1-α*) by [(number of transcripts of a gene) / (number of transcripts of *Ef1-α*)] × 100. The *Ef1-α* gene showed consistent expression patterns in all tissue samples studied previously [22,29,30]. Thus, we used it as a representing house-keeping gene to calculate relative expression levels of four genes, including the transcription factor *Bdtf1* (AHU86567.1). Brassinin hydrolase in *A. brassicicola* (BHAb, AB00197.1) was included to confirm previous study results [11,12]. We also included two genes (AB02663.1, AB08641.1) that showed differential expression under both experimental conditions. Pairs of primers were designed and used for the qRT-PCR with *BdTf1*, *BHAb*, AB02663.1, AB08641.1 and *Ef1-α* genes. Primers are Bdtf1rtF (GTCAGAGCATAGCCGACACA) Bdtf1rtR (TGAAGCTTCGGAGGAAAGAG), BHAbrtF (TTCT GGTGGAGAGGGAGCTA), BHAbrtR (GGATCCTGATAGAGCCACCA), AB02663RtF (CCCGAA CTGGCTACCTACAA) AB02663RtR (GAAGCAGGGTTGTCACCAAT), AB8641RtF (AACCCCA AAGGCAGAATACC) AB8641RtR (ATTTCTTTTCGGGGACGAGT) and Ef1αrtF (GGGTCCTC GACAAGTTGAA), and Ef1αrtR (GGGAGCGTCAATAACTGTGA).

5. Conclusions

Brassinin is detoxified through oxidation or hydrolysis as implied by chemical analysis and the enzymes responsible for the modification are yet to be identified. Previously we identified a transcription factor that is essential for efficient detoxification of brassinin *in vitro*. To discover enzyme-coding genes involved in the brassinin detoxification, we compared gene expression profiles between wild-type *A. brassicicola* and the mutant Δ*bdtf1* under two different experimental conditions. In this study, we discovered six candidate genes, including a hydrolase (AB02663.1), two GST-like proteins (AB08641.1 and AB03487.1), and three reductases. We have generated loss-of-function mutants of genes encoding either a hydrolase or one of the two GST-like proteins. We are in the process of verifying their functions in brassinin detoxification *in vitro*.

Supplementary Materials

Supplementary materials can be accessed at: http://www.mdpi.com/1420-3049/19/8/10717/s1.

Table S1. Differentially expressed genes during plant infection.

Protein Id	44hpiWt	44hpiFtf1	PFAM annotation	GO annotation	44hpiWt -> 44hpiFtf1: Up/Down	44hpiWt -> 44hpiFtf1: q value	Ration (m/w)	44hpiWt -> 44hpiFtf1: >2-fold and significant difference?	log ratio
53	84.4937	234.775	PF01619:Proline dehydrogenase;	GO:0051700-fructosyl-amino acid oxidase activity; GO:0006562-proline catabolic process; GO:0006537-glutamate biosynthetic process; GO:0004657-proline dehydrogenase activity; GO:0004154-dihydropterin oxidase activity; GO:0019116-hydroxy-nicotine oxidase activity; GO:0051699-proline oxidase activity;	Up	0	2.7786095	Yes	1.474363
94	11.5852	25.7617			Up	1.10E-11	2.2236733	Yes	1.152945
128	768.923	1792.39	PF00686:Glycoside hydrolase, starch-binding;	GO:0003824-catalytic activity; GO:0005975-carbohydrate metabolic process;	Up	0	2.3310396	Yes	1.220974
143	87.5762	227.39	PF00150:Glycoside hydrolase, family 5;	GO:0004553-hydrolase activity, hydrolyzing O-glycosyl compounds; GO:0005975-carbohydrate metabolic process; GO:0004338-glucan 1,3-beta-glucosidase activity;	Up	0	2.5964817	Yes	1.376558
164	312.506	809.009	PF00135:Carboxylesterase, type B;	GO:0004091-carboxylesterase activity;	Up	0	2.5887791	Yes	1.372272
328	15.6726	32.3234			Up	8.49E-10	2.0624147	Yes	1.044334
417	210.838	519.247			Up	0	2.4627771	Yes	1.300286
503	29.7275	63.5814		GO:0016020-membrane; GO:0006118-electron transport; GO:0016491-oxidoreductase activity;	Up	0	2.1388075	Yes	1.096807

Table S1. *Cont.*

Protein Id	44hpiWt	44hpiFtf1	PFAM annotation	GO annotation	44hpiWt -> 44hpiFtf1: Up/Down	44hpiWt -> 44hpiFtf1: q value	Ration (m/w)	44hpiWt -> 44hpiFtf1: >2-fold and significant difference?	log ratio
565	464.37	1068.77	PF00544:Pectate lyase/Amb allergen;		Up	0	2.3015483	Yes	1.202605
615	164.44	400.467	PF00457:Glycoside hydrolase, family 11;	GO:0004553-hydrolase activity, hydrolyzing O-glycosyl compounds; GO:0005975-carbohydrate metabolic process;	Up	0	2.4353381	Yes	1.284122
715	7.73127	16.7603			Up	1.03E-10	2.1678586	Yes	1.116271
732	1.24879	7.3449			Up	8.35E-05	5.8816134	Yes	2.556212
748	8.01814	18.32			Up	0	2.2848192	Yes	1.19208
904	225.249	737.856	PF00544:Pectate lyase/Amb allergen;		Up	0	3.2757349	Yes	1.711819
926	2.86878	8.44118			Up	7.12E-05	2.9424285	Yes	1.557007
952	57.1391	127.271	PF07646:Kelch; PF01344:Kelch repeat; PF00024:N/apple PAN;		Up	0	2.227389	Yes	1.155354
1151	830.565	1854.59	PF01965:ThiJ/PfpI;		Up	0	2.2329258	Yes	1.158935
1210	352.314	719.545			Up	0	2.0423401	Yes	1.030223
1332	1067.26	2492.07	PF03211: Pectate lyase;	GO:0005576-extracellular region; GO:0030570-pectate lyase activity;	Up	0	2.3350168	Yes	1.223433

Table S1. *Cont.*

Protein Id	44hpiWt	44hpiFtf1	PFAM annotation	GO annotation	44hpiWt -> 44hpiFtf1: Up/Down	44hpiWt -> 44hpiFtf1: q value	Ration (m/w)	44hpiWt -> 44hpiFtf1: >2-fold and significant difference?	log ratio
1335	9.17412	28.8727	PF04488:Glycosyltransferase sugar-binding region containing DXD motif;	GO:0033164-glycolipid 6-alpha-mannosyltransferase activity;	Up	0	3.1471901	Yes	1.654064
1336	8.94556	32.2733		GO:0005840-ribosome; GO:0003735-structural constituent of ribosome; GO:0003676-nucleic acid binding; GO:0005622-intracellular; GO:0006412-translation;	Up	0	3.6077451	Yes	1.851097
1487	11.8081	28.551			Up	8.92E-13	2.4179165	Yes	1.273764
1540	48.4789	320.148			Up	0	6.6038627	Yes	2.72331
1674	5.53953	52.7419	PF01083:Cutinase;	GO:0005576-extracellular region; GO:0016787-hydrolase activity; GO:0050525-cutinase activity; GO:0008152-metabolic process;	Up	0	9.5210063	Yes	3.251114
1760	39.9371	80.5833		GO:0046872-metal ion binding; GO:0030001-metal ion transport;	Up	1.05E-12	2.0177554	Yes	1.012751
1821	1.02698	7.20374			Up	1.18E-10	7.0144891	Yes	2.810338
1871	32.7018	66.1033	PF00445:Ribonuclease T2;	GO:0004521-endoribonuclease activity; GO:0003723-RNA binding; GO:0033897-ribonuclease T2 activity;	Up	0	2.0213964	Yes	1.015352
1940	53.1545	111.686		GO:0004308-exo-alpha-sialidase activity;	Up	0	2.1011579	Yes	1.071185
2150	6.70825	16.3836			Up	0.0459447	2.4423061	Yes	1.288244
2403	3.52354	9.28525			Up	3.51E-06	2.6352049	Yes	1.397915

Table S1. *Cont.*

Protein Id	44hpiWt	44hpiFtf1	PFAM annotation	GO annotation	44hpiWt -> 44hpiFtf1: Up/Down	44hpiWt -> 44hpiFtf1: q value	Ration (m/w)	44hpiWt -> 44hpiFtf1: >2-fold and significant difference?	log ratio
2424	60.4396	139.818	PF02156:Glycoside hydrolase, family 26;	GO:0016985-mannan endo-1, 4-beta-mannosidase activity; GO:0006080-substituted mannan metabolic process;	Up	0	2.3133508	Yes	1.209984
2447	10.3934	42.3522			Up	3.04E-10	4.0749129	Yes	2.026769
2489	28.9171	63.9221			Up	1.15E-12	2.2105294	Yes	1.144392
2531	4.16673	8.74082	PF00098:Zn-finger, CCHC type;	GO:0003676-nucleic acid binding; GO:0008270-zinc ion binding;	Up	0.00026073	2.0977649	Yes	1.068853
2607	19.6811	57.426			Up	0	2.9178247	Yes	1.544893
2713	13.9411	43.494			Up	0.0433032	3.1198399	Yes	1.641472
2942	15.3572	52.4761			Up	0	3.4170357	Yes	1.772745
2998	0.664181	2.08232		GO:0008471-laccase activity;	Up	0.00021201	3.1351695	Yes	1.648543
3016	25.2516	90.6533		GO:0003900-DNA-directed RNA polymerase I activity; GO:0003902-DNA-directed RNA polymerase III activity; GO:0003899-DNA-directed RNA polymerase activity; GO:0034062-RNA polymerase activity; GO:0003901-DNA-directed RNA polymerase II activity;	Up	0	3.5900022	Yes	1.843985
3017	37.5596	125.044			Up	0	3.3292154	Yes	1.735182
3063	334.714	870.402	PF01055:Glycoside hydrolase, family 31;	GO:0004553-hydrolase activity, hydrolyzing O-glycosyl compounds; GO:0005975-carbohydrate metabolic process;	Up	0	2.600435	Yes	1.378753

Table S1. *Cont.*

Protein Id	44hpiWt	44hpiFtf1	PFAM annotation	GO annotation	44hpiWt -> 44hpiFtf1: Up/Down	44hpiWt -> 44hpiFtf1: q value	Ration (m/w)	44hpiWt -> 44hpiFtf1: >2-fold and significant difference?	log ratio
3295	21.028	105.099			Up	0	4.9980502	Yes	2.321365
3437	376.213	845.297			Up	0	2.2468575	Yes	1.167909
3542	7.81097	32.2902	PF00150:Glycoside hydrolase, family 5;	GO:0004553-hydrolase activity, hydrolyzing O-glycosyl compounds; GO:0005975-carbohydrate metabolic process; GO:0004338-glucan 1,3-beta-glucosidase activity;	Up	0	4.1339552	Yes	2.047523
3746	189.965	382.666			Up	0	2.0144027	Yes	1.010352
3753	197.168	532.32	PF00150:Glycoside hydrolase, family 5;	GO:0004553-hydrolase activity, hydrolyzing O-glycosyl compounds; GO:0005975-carbohydrate metabolic process;	Up	0	2.6998296	Yes	1.432868
3757	182.268	409.571	PF00067:Cytochrome P450;	GO:0006118-electron transport; GO:0003676-nucleic acid binding; GO:0004497-monooxygenase activity; GO:0020037-heme binding; GO:0005506-iron ion binding;	Up	0	2.2470812	Yes	1.168052
4077	9.12494	38.2624	PF01532:Glycoside hydrolase, family 47;	GO:0004571-mannosyl-oligosaccharide 1,2-alpha-mannosidase activity; GO:0016020-membrane; GO:0005509-calcium ion binding;	Up	0	4.1931673	Yes	2.06804
4118	37.9606	82.7029	PF00704:Glycoside hydrolase, family 18;	GO:0006032-chitin catabolic process; GO:0004553-hydrolase activity, hydrolyzing O-glycosyl compounds; GO:0005975-carbohydrate metabolic process; GO:0008843-endochitinase activity; GO:0004568-chitinase activity;	Up	0	2.178651	Yes	1.123435
4154	2.33522	6.4225		GO:0003676-nucleic acid binding;	Up	4.95E-12	2.7502762	Yes	1.459577

Table S1. *Cont.*

Protein Id	44hpiWt	44hpiFtf1	PFAM annotation	GO annotation	44hpiWt -> 44hpiFtf1: Up/Down	44hpiWt -> 44hpiFtf1: q value	Ration (m/w)	44hpiWt -> 44hpiFtf1: >2-fold and significant difference?	log ratio
4192	8.19405	20.2969		GO:0016021-integral to membrane; GO:0006810-transport; GO:0005215-transporter activity;	Up	0	2.4770291	Yes	1.308611
4219	10.9793	25.9667			Up	0	2.3650597	Yes	1.241877
4265	0.0689903	1.24106			Up	0.00029256	17.988906	Yes	4.169036
4305	12.5902	71.9322			Up	0	5.7133485	Yes	2.514337
4330	6.52885	14.0642			Up	4.68E-09	2.1541619	Yes	1.107127
4331	9.3264	20.1934			Up	2.04E-09	2.165187	Yes	1.114492
4350	46.9425	105.121			Up	0	2.2393567	Yes	1.163084
4369	1.73173	3.75665			Up	9.71E-06	2.1693047	Yes	1.117233
4415	4.59435	18.2933			Up	1.02E-07	3.9816949	Yes	1.993383
4421	11.8119	47.7395	PF00355:Rieske [2Fe-2S] region;	GO:0016491-oxidoreductase activity; GO:0006725-aromatic compound metabolic process; GO:0005506-iron ion binding; GO:0018619-benzene 1,2-dioxygenase activity; GO:0006118-electron transport; GO:0051537-2 iron, 2 sulfur cluster binding;	Up	0	4.0416444	Yes	2.014942
4439	1.30184	5.85432			Up	1.75E-13	4.4969582	Yes	2.168949
4485	11.4791	31.5329			Up	0	2.7469836	Yes	1.457848

Table S1. *Cont.*

Protein Id	44hpiWt	44hpiFtf1	PFAM annotation	GO annotation	44hpiWt -> 44hpiFtf1: Up/Down	44hpiWt -> 44hpiFtf1: q value	Ration (m/w)	44hpiWt -> 44hpiFtf1: >2-fold and significant difference?	log ratio
4486	396.115	846.819	PF00083:General substrate transporter;	GO:0016021-integral to membrane; GO:0016020-membrane; GO:0005215-transporter activity; GO:0008643-carbohydrate transport; GO:0006810-transport; GO:0005351-sugar:hydrogen symporter activity; GO:0008733-L-arabinose isomerase activity;	Up	0	2.137811	Yes	1.096134
4545	511.62	1253.31	PF01055:Glycoside hydrolase, family 31;	GO:0004553-hydrolase activity, hydrolyzing O-glycosyl compounds; GO:0005975-carbohydrate metabolic process;	Up	7.43E-09	2.4496892	Yes	1.292599
4619	1.84247	6.58948	PF05270:Alpha-L-arabinofuranosidase B;	GO:0046373-L-arabinose metabolic process; GO:0046556-alpha-N-arabinofuranosidase activity;	Up	1.24E-09	3.5764382	Yes	1.838523
4634	17.2053	34.7289			Up	1.79E-14	2.0185001	Yes	1.013284
4635	4.99125	10.6726			Up	1.22E-06	2.138262	Yes	1.096439
4706	83.9617	250.379	PF01975:Survival protein SurE;	GO:0016787-hydrolase activity;	Up	0	2.9820621	Yes	1.57631
4707	3.24603	6.91629			Up	0.00363911	2.130692	Yes	1.091322
4786	2.89787	5.91024			Up	0.0383569	2.0395118	Yes	1.028224
4791	6.50859	17.3106		GO:0016021-integral to membrane; GO:0006810-transport; GO:0005215-transporter activity;	Up	4.60E-15	2.6596544	Yes	1.411239

Table S1. *Cont.*

Protein Id	44hpiWt	44hpiFtf1	PFAM annotation	GO annotation	44hpiWt -> 44hpiFtf1: Up/Down	44hpiWt -> 44hpiFtf1: q value	Ration (m/w)	44hpiWt -> 44hpiFtf1: >2-fold and significant difference?	log ratio
4813	111.699	375.878	PF03211:Pectate lyase;	GO:0005576-extracellular region; GO:0030570-pectate lyase activity;	Up	0	3.3650973	Yes	1.750648
5054	14.4124	31.2476	PF01116:Ketose-bisphosphate aldolase, class-II;	GO:0006096-glycolysis; GO:0043801-hexulose-6-phosphate synthase activity; GO:0018802-2,4-dihydroxyhept-2-ene-1,7-dioate aldolase activity; GO:0043876-D-threonine aldolase activity; GO:0008732-L-allo-threonine aldolase activity; GO:0004332-fructose-bisphosphate aldolase activity; GO:0008270-zinc ion binding; GO:0004553-hydrolase activity, hydrolyzing O-glycosyl compounds; GO:0005975-carbohydrate metabolic process; GO:0043863-4-hydroxy-2-ketopimelate aldolase activity;	Up	1.95E-12	2.1681052	Yes	1.116435
5058	3.71185	8.14927		GO:0016021-integral to membrane; GO:0006810-transport; GO:0003677-DNA binding; GO:0005215-transporter activity;	Up	3.00E-06	2.195474	Yes	1.134532
5116	21.2632	47.2766			Up	0	2.2234001	Yes	1.152768
5390	671.558	1592.19	PF01670:Glycoside hydrolase, family 12;	GO:0004190-aspartic-type endopeptidase activity; GO:0000272-polysaccharide catabolic process; GO:0008810-cellulase activity; GO:0006508-proteolysis;	Up	0	2.3708898	Yes	1.245429

Table S1. *Cont.*

Protein Id	44hpiWt	44hpiFtf1	PFAM annotation	GO annotation	44hpiWt -> 44hpiFtf1: Up/Down	44hpiWt -> 44hpiFtf1: q value	Ration (m/w)	44hpiWt -> 44hpiFtf1: >2-fold and significant difference?	log ratio
5436	4.55239	10.4336		GO:0004289-subtilase activity; GO:0006508-proteolysis;	Up	2.54E-13	2.291895	Yes	1.196541
5484	89.3087	183.964			Up	0	2.0598665	Yes	1.042551
5485	232.749	559.214		GO:0003868-4-hydroxyphenylpyruvate dioxygenase activity;	Up	0	2.4026483	Yes	1.264626
5497	58.5391	146.685	PF04433:SWIRM;		Up	0	2.5057611	Yes	1.325249
5507	11.4831	27.2864	PF01565:FAD linked oxidase, N-terminal; PF00187:Chitin-binding, type 1;	GO:0050660-FAD binding; GO:0008061-chitin binding; GO:0016491-oxidoreductase activity;	Up	0	2.3762224	Yes	1.24867
5539	2.04444	5.3948		GO:0016491-oxidoreductase activity;	Up	2.37E-05	2.6387666	Yes	1.399864
5652	367.729	756.157			Up	0	2.0562887	Yes	1.040043
5658	0.60122	1.67005		GO:0006810-transport; GO:0016020-membrane; GO:0005215-transporter activity;	Up	0.00730564	2.7777685	Yes	1.473926
5961	24.0113	136.884			Up	0	5.7008159	Yes	2.511168
6080	1122.9	2652.85	PF00128:Alpha amylase, catalytic region; PF00686:Glycoside hydrolase, starch-binding;	GO:0004556-alpha-amylase activity; GO:0003824-catalytic activity; GO:0005975-carbohydrate metabolic process; GO:0043169-cation binding;	Up	5.13E-05	2.3624989	Yes	1.240314

Table S1. *Cont.*

Protein Id	44hpiWt	44hpiFtf1	PFAM annotation	GO annotation	44hpiWt -> 44hpiFtf1: Up/Down	44hpiWt -> 44hpiFtf1: q value	Ration (m/w)	44hpiWt -> 44hpiFtf1: >2-fold and significant difference?	log ratio
6189	6.55314	33.1261	PF00891:O-methyltransferase, family 2;	GO:0003824-catalytic activity; GO:0008171-O-methyltransferase activity;	Up	0	5.0549965	Yes	2.33771
6252	49.4218	178.623	PF01341:Glycoside hydrolase, family 6;	GO:0004553-hydrolase activity, hydrolyzing O-glycosyl compounds; GO:0005975-carbohydrate metabolic process;	Up	0	3.6142552	Yes	1.853698
6313	487.098	2251.56	PF07249:Cerato-platanin;		Up	0	4.6223963	Yes	2.208641
6321	1.19397	3.16667	PF00561:Alpha/beta hydrolase fold;	GO:0003824-catalytic activity;	Up	0.00707452	2.6522191	Yes	1.4072
6461	12.7121	30.4312	PF00083:General substrate transporter;	GO:0016021-integral to membrane; GO:0016020-membrane; GO:0005215-transporter activity; GO:0008643-carbohydrate transport; GO:0006810-transport; GO:0005351-sugar:hydrogen symporter activity; GO:0008733-L-arabinose isomerase activity;	Up	0	2.3938767	Yes	1.259349
6538	18.7708	44.9731		GO:0005267-potassium channel activity; GO:0016020-membrane; GO:0006813-potassium ion transport;	Up	0	2.3959075	Yes	1.260572
6545	8.31416	18.0478			Up	0.00081845	2.1707304	Yes	1.118181
6561	4.10073	19.4747			Up	0	4.7490813	Yes	2.247648

Table S1. *Cont.*

Protein Id	44hpiWt	44hpiFtf1	PFAM annotation	GO annotation	44hpiWt -> 44hpiFtf1: Up/Down	44hpiWt -> 44hpiFtf1: q value	Ration (m/w)	44hpiWt -> 44hpiFtf1: >2-fold and significant difference?	log ratio
6581	20.173	48.1217	PF00732:Glucose-methanol-choline oxidoreductase;	GO:0016614-oxidoreductase activity, acting on CH-OH group of donors; GO:0050660-FAD binding;	Up	0	2.3854509	Yes	1.254262
6582	40.4905	103.963		GO:0016491-oxidoreductase activity; GO:0005506-iron ion binding; GO:0009058-biosynthetic process;	Up	0	2.5675899	Yes	1.360415
6621	2.21486	4.68855			Up	0.0151501	2.1168607	Yes	1.081926
6626	0.855633	1.88644			Up	0.0296639	2.2047303	Yes	1.140602
6658	217.589	453.082	PF00194:Carbonic anhydrase, eukaryotic;	GO:0006730-one-carbon compound metabolic process; GO:0008270-zinc ion binding; GO:0004089-carbonate dehydratase activity;	Up	0	2.0822836	Yes	1.058167
6720	2.1178	8.28738			Up	8.40E-05	3.9132024	Yes	1.96835
6744	17.1326	57.6079	PF00083:General substrate transporter;	GO:0016021-integral to membrane; GO:0016020-membrane; GO:0005215-transporter activity; GO:0008643-carbohydrate transport; GO:0006810-transport; GO:0005351-sugar:hydrogen symporter activity; GO:0008733-L-arabinose isomerase activity;	Up	0	3.3624727	Yes	1.749523
6750	5.22185	12.8148			Up	4.23E-08	2.4540728	Yes	1.295178
6751	109.548	303.085			Up	0	2.7666867	Yes	1.468159
6885	7.74066	20.8547			Up	1.71E-13	2.6941759	Yes	1.429844

Table S1. *Cont.*

Protein Id	44hpiWt	44hpiFtf1	PFAM annotation	GO annotation	44hpiWt -> 44hpiFtf1: Up/Down	44hpiWt -> 44hpiFtf1: q value	Ration (m/w)	44hpiWt -> 44hpiFtf1: >2-fold and significant difference?	log ratio
6906	214.69	608.809	PF00150:Glycoside hydrolase, family 5;	GO:0004553-hydrolase activity, hydrolyzing O-glycosyl compounds; GO:0005975-carbohydrate metabolic process;	Up	0	2.8357585	Yes	1.503735
6925	32.207	74.952		GO:0016021-integral to membrane; GO:0006810-transport; GO:0005215-transporter activity;	Up	0	2.327196	Yes	1.218593
6928	3.62193	8.01854			Up	0.00434041	2.213886	Yes	1.146581
7114	9.61178	20.4379		GO:0005576-extracellular region; GO:0003824-catalytic activity;	Up	1.88E-08	2.1263387	Yes	1.088371
7151	8.4286	18.6361	PF00069:Protein kinase;	GO:0005524-ATP binding; GO:0006468-protein amino acid phosphorylation; GO:0004674-protein serine/threonine kinase activity; GO:0004672-protein kinase activity;	Up	0	2.2110552	Yes	1.144735
7152	64.1946	203.106	PF00703:Glycoside hydrolase family 2, immunoglobulin-like beta-sandwich domain; PF02837:Glycoside hydrolase, family 2, sugar binding;	GO:0004566-beta-glucuronidase activity; GO:0004553-hydrolase activity, hydrolyzing O-glycosyl compounds; GO:0005975-carbohydrate metabolic process;	Up	0	3.163911	Yes	1.661709

Table S1. *Cont.*

Protein Id	44hpiWt	44hpiFtf1	PFAM annotation	GO annotation	44hpiWt -> 44hpiFtf1: Up/Down	44hpiWt -> 44hpiFtf1: q value	Ration (m/w)	44hpiWt -> 44hpiFtf1: >2-fold and significant difference?	log ratio
7290	5.66532	14.6685			Up	3.45E-06	2.5891741	Yes	1.372492
7296	25.5682	52.6269			Up	0.00041145	2.0582951	Yes	1.04145
7373	49.8568	134.009	PF06422:CDR ABC transporter; PF00005:ABC transporter;	GO:0016021-integral to membrane; GO:0042626-ATPase activity, coupled to transmembrane movement of substances; GO:0017111-nucleoside-triphosphatase activity; GO:0006810-transport; GO:0000166-nucleotide binding; GO:0016887-ATPase activity; GO:0005524-ATP binding;	Up	0	2.6878781	Yes	1.426468
7374	28.5063	75.5726	PF00005:ABC transporter;	GO:0016887-ATPase activity; GO:0005524-ATP binding;	Up	0	2.6510841	Yes	1.406582
7399	10.4364	47.9016	PF00106:Short-chain dehydrogenase/reductase SDR;	GO:0008678-2-deoxy-D-gluconate 3-dehydrogenase activity; GO:0016491-oxidoreductase activity; GO:0008152-metabolic process;	Up	0	4.5898586	Yes	2.19845
7401	3.00246	63.547		GO:0045551-cinnamyl-alcohol dehydrogenase activity;	Up	0	21.164978	Yes	4.403607
7402	6.01136	105.501	PF02668:Taurine catabolism dioxygenase TauD/TfdA;	GO:0016491-oxidoreductase activity; GO:0043826-sulfur oxygenase reductase activity; GO:0051341-regulation of oxidoreductase activity; GO:0043738-F420H2 dehydrogenase activity; GO:0006118-electron transport; GO:0043883-malolactic enzyme activity;	Up	0	17.550271	Yes	4.133421

Table S1. *Cont.*

Protein Id	44hpiWt	44hpiFtf1	PFAM annotation	GO annotation	44hpiWt -> 44hpiFtf1: Up/Down	44hpiWt -> 44hpiFtf1: q value	Ration (m/w)	44hpiWt -> 44hpiFtf1: >2-fold and significant difference?	log ratio
7403	3.05993	79.3092			Up	0	25.918632	Yes	4.695918
7405	3.95668	15.7573	PF00891:O-methyltransferase, family 2;	GO:0008171-O-methyltransferase activity;	Up	0	3.982455	Yes	1.993658
7406	4.78541	114.143			Up	0	23.852293	Yes	4.576056
7407	5.21179	48.2111	PF01040:UbiA prenyltransferase;	GO:0016021-integral to membrane; GO:0043766-Sep-tRNA:Cys-tRNA synthase activity; GO:0008318-protein prenyltransferase activity; GO:0004659-prenyltransferase activity; GO:0046428-1,4-dihydroxy-2-naphthoate octaprenyltransferase activity; GO:0008495-protoheme IX farnesyltransferase activity; GO:0004661-protein geranylgeranyltransferase activity; GO:0008412-4-hydroxybenzoate octaprenyltransferase activity; GO:0043888-(S)-2,3-di-O-geranylgeranylglyceryl phosphate synthase activity;	Up	0	9.2503919	Yes	3.209514
7409	5.42995	29.1257			Up	0	5.3638984	Yes	2.423282
7436	12.6683	32.427	PF00544:Pectate lyase/Amb allergen;		Up	0	2.5596962	Yes	1.355973

Table S1. *Cont.*

Protein Id	44hpiWt	44hpiFtf1	PFAM annotation	GO annotation	44hpiWt -> 44hpiFtf1: Up/Down	44hpiWt -> 44hpiFtf1: q value	Ration (m/w)	44hpiWt -> 44hpiFtf1: >2-fold and significant difference?	log ratio
7438	8.5369	17.1406	PF00724:NADH:flavin oxidoreductase/NADH oxidase;	GO:0003959-NADPH dehydrogenase activity; GO:0010181-FMN binding; GO:0016491-oxidoreductase activity;	Up	3.08E-08	2.0078249	Yes	1.005633
7683	10.0245	20.4262			Up	0.0240016	2.0376278	Yes	1.026891
7785	5.32964	11.3296			Up	1.63E-10	2.1257721	Yes	1.087987
8011	43.0003	98.1774			Up	0	2.2831794	Yes	1.191044
8013	208.584	495.297		GO:0008270-zinc ion binding; GO:0006508-proteolysis; GO:0008237-metallopeptidase activity;	Up	0	2.3745685	Yes	1.247665
8086	85.0454	224.34	PF00083:General substrate transporter;	GO:0016021-integral to membrane; GO:0016020-membrane; GO:0005215-transporter activity; GO:0008643-carbohydrate transport; GO:0006810-transport; GO:0005351-sugar:hydrogen symporter activity; GO:0008733-L-arabinose isomerase activity;	Up	0	2.6378852	Yes	1.399382
8145	3.68069	23.6045			Up	0	6.4130639	Yes	2.681014
8158	46.1222	106.99	PF00291:Pyridoxal-5'-phosphate-dependent enzyme, beta subunit;	GO:0003824-catalytic activity; GO:0006520-amino acid metabolic process; GO:0008152-metabolic process; GO:0030170-pyridoxal phosphate binding;	Up	0	2.3197072	Yes	1.213943
8218	0.479981	0.980489			Up	0.020824	2.0427663	Yes	1.030524

Table S1. *Cont.*

Protein Id	44hpiWt	44hpiFtf1	PFAM annotation	GO annotation	44hpiWt -> 44hpiFtf1: Up/Down	44hpiWt -> 44hpiFtf1: q value	Ration (m/w)	44hpiWt -> 44hpiFtf1: >2-fold and significant difference?	log ratio
8273	2.96515	6.93799	PF00646:Cyclin-like F-box;		Up	2.67E-08	2.3398445	Yes	1.226413
8283	3.59585	10.4771			Up	4.69E-05	2.9136644	Yes	1.542835
8359	3.17546	14.286			Up	1.23E-05	4.4988758	Yes	2.169565
8533	0.268449	1.60746			Up	0.00499088	5.987953	Yes	2.582063
8534	40.7921	160.515			Up	0	3.9349531	Yes	1.976346
8616	5.37754	26.2337	PF00187:Chitin-binding, type 1; PF00024:N/apple PAN;	GO:0008061-chitin binding;	Up	0	4.8783831	Yes	2.286403
8751	22.236	47.4245	PF00933:Glycoside hydrolase, family 3, N-terminal; PF01915:Glycoside hydrolase, family 3, C-terminal;	GO:0004553-hydrolase activity, hydrolyzing O-glycosyl compounds; GO:0005975-carbohydrate metabolic process; GO:0008422-beta-glucosidase activity;	Up	0	2.1327802	Yes	1.092735
8801	8.26696	29.0812			Up	0	3.5177623	Yes	1.814658
8874	22.9072	50.6273			Up	0	2.2101042	Yes	1.144114
9087	81.3754	192.723	PF01179:Copper amine oxidase;	GO:0005507-copper ion binding; GO:0009308-amine metabolic process; GO:0008131-amine oxidase activity; GO:0048038-quinone binding;	Up	0	2.3683202	Yes	1.243864
9101	2216.15	4575.67			Up	0	2.0646933	Yes	1.045927

Table S1. *Cont.*

Protein Id	44hpiWt	44hpiFtf1	PFAM annotation	GO annotation	44hpiWt -> 44hpiFtf1: Up/Down	44hpiWt -> 44hpiFtf1: q value	Ration (m/w)	44hpiWt -> 44hpiFtf1: >2-fold and significant difference?	log ratio
9287	9.418	19.6781			Up	4.55E-12	2.0894139	Yes	1.063098
9372	109.654	301.917			Up	0	2.7533606	Yes	1.461194
9460	77.653	217.624	PF03055:Retinal pigment epithelial membrane protein;	GO:0016215-CoA desaturase activity; GO:0018688-DDT 2,3-dioxygenase activity; GO:0042389-omega-3 fatty acid desaturase activity; GO:0018689-naphthalene disulfonate 1,2-dioxygenase activity;	Up	0	2.8025189	Yes	1.486724
9476	19.6211	39.5765			Up	0	2.0170378	Yes	1.012238
9503	13.5152	68.9334			Up	0	5.1004351	Yes	2.35062
9513	298.546	666.445			Up	0	2.2323026	Yes	1.158533
9526	1.31633	5.18107	PF03639:Glycoside hydrolase, family 81;		Up	0	3.9359963	Yes	1.976729
9637	0.8683	2.51783	PF05199:GMC oxidoreductase; PF00732:Glucose-methanol-choline oxidoreductase;	GO:0008812-choline dehydrogenase activity; GO:0016614-oxidoreductase activity, acting on CH-OH group of donors; GO:0050660-FAD binding;	Up	0.00013135	2.8997236	Yes	1.535915

Table S1. *Cont.*

Protein Id	44hpiWt	44hpiFtf1	PFAM annotation	GO annotation	44hpiWt -> 44hpiFtf1: Up/Down	44hpiWt -> 44hpiFtf1: q value	Ration (m/w)	44hpiWt -> 44hpiFtf1: >2-fold and significant difference?	log ratio
9677	255.226	516.645	PF05922:Proteinase inhibitor I9, subtilisin propeptide; PF00082:Peptidase S8 and S53, subtilisin, kexin, sedolisin;	GO:0006508-proteolysis; GO:0016808-proprotein convertase activity; GO:0008132-pancreatic elastase activity; GO:0008243-plasminogen activator activity; GO:0008991-serine-type signal peptidase activity; GO:0042802-identical protein binding; GO:0043086-negative regulation of catalytic activity; GO:0004289-subtilase activity;	Up	0	2.0242648	Yes	1.017398
9751	1.29133	2.65866	PF01565:FAD linked oxidase, N-terminal;	GO:0050660-FAD binding; GO:0016491-oxidoreductase activity;	Up	0.0128984	2.058854	Yes	1.041842
9762	68.8826	148.87	PF01055:Glycoside hydrolase, family 31;	GO:0004558-alpha-glucosidase activity; GO:0004553-hydrolase activity, hydrolyzing O-glycosyl compounds; GO:0005975-carbohydrate metabolic process;	Up	0	2.1612134	Yes	1.111842
9919	681.861	1395.77	PF00840:Glycoside hydrolase, family 7;	GO:0004553-hydrolase activity, hydrolyzing O-glycosyl compounds; GO:0005975-carbohydrate metabolic process;	Up	1.02E-13	2.0470008	Yes	1.033512
9920	642.99	1318.65	PF00840:Glycoside hydrolase, family 7;	GO:0004553-hydrolase activity, hydrolyzing O-glycosyl compounds; GO:0005975-carbohydrate metabolic process;	Up	0	2.0508095	Yes	1.036193

Table S1. *Cont.*

Protein Id	44hpiWt	44hpiFtf1	PFAM annotation	GO annotation	44hpiWt -> 44hpiFtf1: Up/Down	44hpiWt -> 44hpiFtf1: q value	Ration (m/w)	44hpiWt -> 44hpiFtf1: >2-fold and significant difference?	log ratio
9925	3.29561	8.847	PF00291:Pyridoxal-5'-phosphate-dependent enzyme, beta subunit;	GO:0003824-catalytic activity; GO:0008152-metabolic process; GO:0030170-pyridoxal phosphate binding;	Up	1.61E-10	2.6844803	Yes	1.424643
9952	292.485	1607.74	PF04758:Ribosomal protein S30;	GO:0005840-ribosome; GO:0003735-structural constituent of ribosome; GO:0005622-intracellular; GO:0006412-translation;	Up	0	5.4968289	Yes	2.4586
10039	866.144	1769.7			Up	0	2.0431937	Yes	1.030826
10079	1.06928	3.0199	PF06985:Heterokaryon incompatibility;		Up	1.67E-06	2.8242369	Yes	1.497861
10154	35.1817	104.637	PF01120:Glycoside hydrolase, family 29 (alpha-L-fucosidase);	GO:0004560-alpha-L-fucosidase activity; GO:0005975-carbohydrate metabolic process;	Up	0	2.9741883	Yes	1.572496
10170	6.69338	19.6336			Up	0	2.9332863	Yes	1.552518
10182	3.84763	10.1353	PF01341:Glycoside hydrolase, family 6;	GO:0004553-hydrolase activity, hydrolyzing O-glycosyl compounds; GO:0005975-carbohydrate metabolic process;	Up	2.34E-08	2.634167	Yes	1.397347
10283	0.91522	10.1001	PF00171:Aldehyde dehydrogenase;	GO:0004029-aldehyde dehydrogenase (NAD) activity;	Up	0	11.035707	Yes	3.464107
10379	1434.86	2903.09			Up	0	2.0232566	Yes	1.016679
10420	11.1056	33.8581			Up	0	3.0487412	Yes	1.608214
10533	3.75779	12.4494			Up	2.87E-09	3.3129579	Yes	1.72812

Table S1. *Cont.*

Protein Id	44hpiWt	44hpiFtf1	PFAM annotation	GO annotation	44hpiWt -> 44hpiFtf1: Up/Down	44hpiWt -> 44hpiFtf1: q value	Ration (m/w)	44hpiWt -> 44hpiFtf1: >2-fold and significant difference?	log ratio
10569	61.1712	161.329			Up	0	2.6373359	Yes	1.399081
10576	6.68023	20.0071	PF05730:CFEM;		Up	3.66E-07	2.9949717	Yes	1.582542
10588	13.3306	82.2857			Up	0	6.1726929	Yes	2.6259
76	1067.67	510.191	PF06355:Aegerolysin;	GO:0019836-hemolysis by symbiont of host red blood cells; GO:0030582-fruiting body development;	Down	0	0.4778546	Yes	−1.06536
77	82.3222	37.704			Down	0	0.4580053	Yes	−1.12656
232	656.432	326.135	PF01564:Spermine synthase;	GO:0003824-catalytic activity; GO:0003677-DNA binding; GO:0004766-spermidine synthase activity;	Down	0	0.4968298	Yes	−1.00918
240	94.9032	41.9752		GO:0016787-hydrolase activity; GO:0008152-metabolic process;	Down	0	0.4422949	Yes	−1.17692
875	55.6063	23.8679	PF00295:Glycoside hydrolase, family 28;	GO:0005975-carbohydrate metabolic process; GO:0047911-galacturan 1,4-alpha-galacturonidase activity; GO:0004650-polygalacturonase activity;	Down	0	0.4292301	Yes	−1.22018
890	8.26011	4.08968			Down	0.0007104	0.4951121	Yes	−1.01417
1231	32.5074	9.86123	PF01384:Phosphate transporter;	GO:0006817-phosphate transport; GO:0016020-membrane; GO:0005315-inorganic phosphate transmembrane transporter activity;	Down	0	0.3033534	Yes	−1.72093

Table S1. *Cont.*

Protein Id	44hpiWt	44hpiFtf1	PFAM annotation	GO annotation	44hpiWt -> 44hpiFtf1: Up/Down	44hpiWt -> 44hpiFtf1: q value	Ration (m/w)	44hpiWt -> 44hpiFtf1: >2-fold and significant difference?	log ratio
1239	24.3141	11.4756			Down	2.81E-06	0.4719731	Yes	−1.08322
1296	3.10941	1.44208	PF00248:Aldo/keto reductase;	GO:0043826-sulfur oxygenase reductase activity; GO:0043738-F420H2 dehydrogenase activity; GO:0043883-malolactic enzyme activity; GO:0051341-regulation of oxidoreductase activity; GO:0016491-oxidoreductase activity;	Down	0.0496335	0.4637793	Yes	−1.10849
1299	40.6385	17.9288			Down	0.0265342	0.4411777	Yes	−1.18057
1527	8.30484	3.85224			Down	0.00233966	0.4638548	Yes	−1.10825
1545	195.165	95.9314			Down	0	0.49154	Yes	−1.02462
1575	22.0231	7.60928	PF00724:NADH:flavin oxidoreductase/NADH oxidase;	GO:0003959-NADPH dehydrogenase activity; GO:0010181-FMN binding; GO:0016491-oxidoreductase activity;	Down	0	0.3455136	Yes	−1.53319
1735	3.23303	1.44114			Down	0.00420021	0.4457552	Yes	−1.16568
1745	5.86542	2.35695			Down	0.0230871	0.4018382	Yes	−1.31531
1746	12.777	6.35071	PF00984:UDP-glucose/GDP-mannose dehydrogenase; PF03721:UDP-glucose/GDP-mannose dehydrogenase; PF03720:UDP-glucose/GDP-mannose dehydrogenase;	GO:0016616-oxidoreductase activity, acting on the CH-OH group of donors, NAD or NADP as acceptor; GO:0051287-NAD binding; GO:0006118-electron transport; GO:0003979-UDP-glucose 6-dehydrogenase activity;	Down	1.70E-07	0.4970423	Yes	−1.00856

Table S1. *Cont.*

Protein Id	44hpiWt	44hpiFtf1	PFAM annotation	GO annotation	44hpiWt -> 44hpiFtf1: Up/Down	44hpiWt -> 44hpiFtf1: q value	Ration (m/w)	44hpiWt -> 44hpiFtf1: >2-fold and significant difference?	log ratio
2056	17.1054	7.17962			Down	0.0365403	0.4197283	Yes	−1.25247
2315	8.73039	4.01926		GO:0003824-catalytic activity; GO:0008115-sarcosine oxidase activity;	Down	1.18E-05	0.4603758	Yes	−1.11912
2321	3.8494	1.86035			Down	0.0125501	0.4832831	Yes	−1.04906
2412	1.29381	0.460885		GO:0016021-integral to membrane; GO:0006810-transport; GO:0005215-transporter activity;	Down	0.0403966	0.3562231	Yes	−1.48915
2442	11.64	4.2077			Down	1.99E-05	0.3614863	Yes	−1.46799
2443	19.2043	8.1329	PF01384:Phosphate transporter;	GO:0006817-phosphate transport; GO:0016020-membrane; GO:0005315-inorganic phosphate transmembrane transporter activity;	Down	0.0012659	0.4234937	Yes	−1.23959
2597	0.821571	0.148281	PF00083:General substrate transporter;	GO:0016021-integral to membrane; GO:0016020-membrane; GO:0005215-transporter activity; GO:0008643-carbohydrate transport; GO:0006810-transport; GO:0005351-sugar:hydrogen symporter activity; GO:0008733-L-arabinose isomerase activity;	Down	0.0448168	0.1804847	Yes	−2.47005
2661	82.0245	34.1762		GO:0016491-oxidoreductase activity;	Down	3.19E-05	0.4166584	Yes	−1.26306
2662	95.2627	41.9408		GO:0016491-oxidoreductase activity;	Down	0.0105089	0.4402647	Yes	−1.18356

Table S1. *Cont.*

Protein Id	44hpiWt	44hpiFtf1	PFAM annotation	GO annotation	44hpiWt -> 44hpiFtf1: Up/Down	44hpiWt -> 44hpiFtf1: q value	Ration (m/w)	44hpiWt -> 44hpiFtf1: >2-fold and significant difference?	log ratio
2663	32.9565	10.4341		GO:0016787-hydrolase activity; GO:0008152-metabolic process;	Down	0	0.3166022	Yes	−1.65926
2709	49.9174	24.0404			Down	0.00657156	0.4816036	Yes	−1.05408
2745	168.503	73.3385			Down	0	0.4352356	Yes	−1.20013
3010	99.4296	45.998		GO:0016831-carboxy-lyase activity;	Down	0	0.4626188	Yes	−1.1121
3012	4961.57	1928.23	PF00857:Isochorismatase hydrolase;	GO:0003824-catalytic activity; GO:0008152-metabolic process;	Down	2.19E-05	0.388633	Yes	−1.36352
3046	177.802	80.5266		GO:0016021-integral to membrane; GO:0016020-membrane; GO:0005215-transporter activity; GO:0008643-carbohydrate transport; GO:0006810-transport; GO:0005351-sugar:hydrogen symporter activity;	Down	0	0.4529004	Yes	−1.14273
3052	215.486	95.44	PF00397:WW/Rsp5/WWP ;	GO:0007186-G-protein coupled receptor protein signaling pathway; GO:0016021-integral to membrane; GO:0001584-rhodopsin-like receptor activity; GO:0005515-protein binding;	Down	0	0.4429058	Yes	−1.17493
3400	273.517	79.2498	PF00107:Zinc-containing alcohol dehydrogenase superfamily;	GO:0008270-zinc ion binding; GO:0032440-2-alkenal reductase activity; GO:0016491-oxidoreductase activity;	Down	0	0.2897436	Yes	−1.78715

Table S1. *Cont.*

Protein Id	44hpiWt	44hpiFtf1	PFAM annotation	GO annotation	44hpiWt -> 44hpiFtf1: Up/Down	44hpiWt -> 44hpiFtf1: q value	Ration (m/w)	44hpiWt -> 44hpiFtf1: >2-fold and significant difference?	log ratio
3458	11.5608	2.77459	PF02515:L-carnitine dehydratase/bile acid-inducible protein F;	GO:0003824-catalytic activity; GO:0008111-alpha-methylacyl-CoA racemase activity; GO:0008152-metabolic process;	Down	2.42E-09	0.2399998	Yes	−2.05889
3479	13.8135	2.02356	PF00107:Zinc-containing alcohol dehydrogenase superfamily;	GO:0008106-alcohol dehydrogenase (NADP+) activity; GO:0008270-zinc ion binding; GO:0016491-oxidoreductase activity;	Down	4.62E-14	0.1464915	Yes	−2.77111
3484	75.6363	25.5877			Down	0.00027488	0.3382992	Yes	−1.56363
3487	20.8725	10.3036			Down	4.01E-09	0.4936447	Yes	−1.01845
3780	4.88384	2.14944	PF01699:Sodium/calcium exchanger membrane region;	GO:0016021-integral to membrane;	Down	2.96E-06	0.4401127	Yes	−1.18406
3789	499.034	83.5403	PF04419:Four F5 protein;		Down	9.64E-11	0.167404	Yes	−2.57859
3793	54.3066	21.757			Down	9.05E-15	0.4006327	Yes	−1.31965
3939	9.84771	4.50872			Down	0.0124835	0.4578445	Yes	−1.12707
4177	73.5569	35.1142			Down	0	0.4773747	Yes	−1.06681
4228	154.874	70.7431			Down	0	0.4567784	Yes	−1.13043
4656	24.7543	11.0298		GO:0003824-catalytic activity; GO:0008152-metabolic process;	Down	1.41E-11	0.4455711	Yes	−1.16627
4763	63.1347	28.3504			Down	0	0.4490462	Yes	−1.15506
4783	31.9059	12.1898	PF01490:Amino acid/polyamine transporter II;		Down	0	0.3820547	Yes	−1.38815

Table S1. *Cont.*

Protein Id	44hpiWt	44hpiFtf1	PFAM annotation	GO annotation	44hpiWt -> 44hpiFtf1: Up/Down	44hpiWt -> 44hpiFtf1: q value	Ration (m/w)	44hpiWt -> 44hpiFtf1: >2-fold and significant difference?	log ratio
5050	59.1046	27.6288	PF02566:OsmC-like protein;	GO:0006950-response to stress; GO:0008152-metabolic process; GO:0016740-transferase activity;	Down	5.69E-10	0.467456	Yes	−1.0971
5152	33.6151	14.2817	PF00175:Oxidoreductase FAD/NAD(P)-binding; PF01558:Pyruvate ferredoxin/flavodoxin oxidoreductase; PF00667:FAD-binding;	GO:0006118-electron transport; GO:0016491-oxidoreductase activity;	Down	0	0.4248597	Yes	−1.23494
5202	75.7638	7.76772	PF01042:Endoribonuclease L-PSP;		Down	0	0.1025255	Yes	−3.28595
5203	10.4512	3.96549			Down	1.52E-07	0.3794292	Yes	−1.3981
5204	11.5003	3.85311			Down	5.13E-05	0.3350443	Yes	−1.57758
5764	1.41148	0.263551	PF00174:Oxidoreductase, molybdopterin binding; PF03404:Mo-co oxidoreductase dimerisation domain;	GO:0008482-sulfite oxidase activity; GO:0030151-molybdenum ion binding; GO:0006118-electron transport; GO:0016491-oxidoreductase activity;	Down	0.00565777	0.1867196	Yes	−2.42105
5842	9.46343	4.40724	PF00230:Major intrinsic protein;	GO:0006810-transport; GO:0016020-membrane; GO:0005215-transporter activity;	Down	0.00077761	0.4657127	Yes	−1.10249
6226	158.834	77.0639	PF00098:Zn-finger, CCHC type;	GO:0003676-nucleic acid binding; GO:0008270-zinc ion binding;	Down	0	0.4851852	Yes	−1.04339

Table S1. *Cont.*

Protein Id	44hpiWt	44hpiFtf1	PFAM annotation	GO annotation	44hpiWt -> 44hpiFtf1: Up/Down	44hpiWt -> 44hpiFtf1: q value	Ration (m/w)	44hpiWt -> 44hpiFtf1: >2-fold and significant difference?	log ratio
6589	13.8355	6.47784			Down	3.86E-11	0.4682043	Yes	−1.09479
6865	8.99376	4.44137			Down	0.00070478	0.4938279	Yes	−1.01792
6958	24.8588	11.6892		GO:0008270-zinc ion binding; GO:0016491-oxidoreductase activity;	Down	0.00033338	0.4702238	Yes	−1.08858
6959	27.8287	13.5988			Down	0.0036182	0.488661	Yes	−1.03309
7085	2.29267	0.790011			Down	0.00477918	0.3445812	Yes	−1.53708
7108	32.8618	16.1663	PF00397:WW/Rsp5/WWP ;	GO:0005515-protein binding;	Down	0.0177987	0.4919481	Yes	−1.02342
7145	53.4775	25.195	PF01753:Zn-finger, MYND type;	GO:0008270-zinc ion binding;	Down	4.98E-05	0.4711327	Yes	−1.08579
7427	22.7845	11.2367	PF00083:General substrate transporter;	GO:0016021-integral to membrane; GO:0016020-membrane; GO:0005215-transporter activity; GO:0008643-carbohydrate transport; GO:0006810-transport; GO:0005351-sugar:hydrogen symporter activity; GO:0008733-L-arabinose isomerase activity;	Down	6.26E-14	0.493173	Yes	−1.01983
7568	4.47445	2.16868		GO:0006364-rRNA processing;	Down	0.00340353	0.4846808	Yes	−1.04489
7899	29.7094	0.053622	PF00172:Fungal transcriptional regulatory protein, N-terminal; PF04082:Fungal specific transcription factor;	GO:0003700-transcription factor activity; GO:0006350-transcription; GO:0008270-zinc ion binding; GO:0003677-DNA binding; GO:0006355-regulation of transcription, DNA-dependent; GO:0005634-nucleus;	Down	5.38E-12	0.0018049	Yes	−9.11388

Table S1. *Cont.*

Protein Id	44hpiWt	44hpiFtf1	PFAM annotation	GO annotation	44hpiWt -> 44hpiFtf1: Up/Down	44hpiWt -> 44hpiFtf1: q value	Ration (m/w)	44hpiWt -> 44hpiFtf1: >2-fold and significant difference?	log ratio
8009	2.48162	0.907694	PF00230:Major intrinsic protein;	GO:0006810-transport; GO:0016020-membrane; GO:0031072-heat shock protein binding; GO:0005215-transporter activity;	Down	0.0074569	0.3657667	Yes	−1.451
8095	13.1387	6.25542			Down	1.83E-07	0.4761065	Yes	−1.07064
8175	1.53638	0.675297	PF06422:CDR ABC transporter; PF00005:ABC transporter;	GO:0016021-integral to membrane; GO:0006810-transport; GO:0042626-ATPase activity, coupled to transmembrane movement of substances; GO:0016887-ATPase activity; GO:0005524-ATP binding;	Down	0.00702396	0.4395377	Yes	−1.18594
8215	56.6568	8.47794	PF01761:3-dehydroquinate synthase;	GO:0003856-3-dehydroquinate synthase activity; GO:0009073-aromatic amino acid family biosynthetic process;	Down	0	0.1496368	Yes	−2.74046
8216	33.9643	10.4063		GO:0016206-catechol O-methyltransferase activity;	Down	0	0.3063894	Yes	−1.70656
8477	8.64419	4.27477	PF05042:Caleosin related;		Down	0.0226862	0.4945252	Yes	−1.01588
8527	16.1838	4.75247	PF00474:Na+/solute symporter;	GO:0006810-transport; GO:0016020-membrane; GO:0005215-transporter activity;	Down	0	0.293656	Yes	−1.7678
8566	40.8082	18.5993			Down	2.66E-11	0.4557736	Yes	−1.13361

Table S1. *Cont.*

Protein Id	44hpiWt	44hpiFtf1	PFAM annotation	GO annotation	44hpiWt -> 44hpiFtf1: Up/Down	44hpiWt -> 44hpiFtf1: q value	Ration (m/w)	44hpiWt -> 44hpiFtf1: >2-fold and significant difference?	log ratio
8641	185.05	15.1959	PF00043:Glutathione S-transferase, C-terminal; PF02798:Glutathione S-transferase, N-terminal;		Down	0	0.0821178	Yes	−3.60616
8648	5.38218	2.43502	PF00704:Glycoside hydrolase, family 18;	GO:0004812-aminoacyl-tRNA ligase activity; GO:0004568-chitinase activity; GO:0016998-cell wall catabolic process; GO:0005622-intracellular; GO:0005737-cytoplasm; GO:0003676-nucleic acid binding; GO:0006412-translation; GO:0000166-nucleotide binding; GO:0008270-zinc ion binding; GO:0006032-chitin catabolic process; GO:0004553-hydrolase activity, hydrolyzing O-glycosyl compounds; GO:0005975-carbohydrate metabolic process; GO:0003677-DNA binding; GO:0008061-chitin binding; GO:0006418-tRNA aminoacylation for protein translation; GO:0005524-ATP binding;	Down	3.33E-12	0.4524226	Yes	−1.14426
8650	17.3771	8.11315			Down	0.00015894	0.4668875	Yes	−1.09885

Table S1. *Cont.*

Protein Id	44hpiWt	44hpiFtf1	PFAM annotation	GO annotation	44hpiWt -> 44hpiFtf1: Up/Down	44hpiWt -> 44hpiFtf1: q value	Ration (m/w)	44hpiWt -> 44hpiFtf1: >2-fold and significant difference?	log ratio
8663	405.477	187.053	PF01124:Membrane-associated proteins in eicosanoid and glutathione metabolism (MAPEG);	GO:0004364-glutathione transferase activity;	Down	0	0.4613159	Yes	−1.11617
8720	10.8592	5.22441	PF01184:GPR1/FUN34/yaaH;	GO:0016020-membrane;	Down	0.00041391	0.4811045	Yes	−1.05558
8996	82.2895	36.8543	PF00245:Alkaline phosphatase;	GO:0004035-alkaline phosphatase activity; GO:0008152-metabolic process;	Down	0	0.4478615	Yes	−1.15888
9034	160.683	78.6459	PF06041:Bacterial protein of unknown function DUF924;		Down	0	0.4894475	Yes	−1.03077
9099	0.993669	0.293812	PF00135:Carboxylesterase, type B;		Down	0.0393599	0.295684	Yes	−1.75787
9181	143.948	38.5798			Down	6.96E-13	0.2680121	Yes	−1.89963
9226	250.636	72.6618			Down	0	0.2899097	Yes	−1.78632
9447	65.7008	31.0029	PF00909:Rh-like protein/ammonium transporter;	GO:0006810-transport; GO:0016020-membrane; GO:0008519-ammonium transmembrane transporter activity;	Down	0	0.4718801	Yes	−1.08351
9632	209.412	32.8422	PF01979:Amidohydrolase;	GO:0008270-zinc ion binding; GO:0006508-proteolysis; GO:0008237-metallopeptidase activity; GO:0016787-hydrolase activity;	Down	0	0.1568306	Yes	−2.67272

Table S1. *Cont.*

Protein Id	44hpiWt	44hpiFtf1	PFAM annotation	GO annotation	44hpiWt -> 44hpiFtf1: Up/Down	44hpiWt -> 44hpiFtf1: q value	Ration (m/w)	44hpiWt -> 44hpiFtf1: >2-fold and significant difference?	log ratio
9633	41.5102	13.7828	PF01979:Amidohydrolase;	GO:0016787-hydrolase activity;	Down	0	0.3320341	Yes	−1.5906
9938	60.958	24.8438			Down	0	0.407556	Yes	−1.29493
10076	6.06441	2.85031		GO:0004871-signal transducer activity; GO:0007165-signal transduction;	Down	0.0177948	0.4700062	Yes	−1.08925
10093	4.98931	1.57611	PF02714:Protein of unknown function DUF221;	GO:0016020-membrane; GO:0006629-lipid metabolic process; GO:0016298-lipase activity;	Down	3.78E-10	0.3158974	Yes	−1.66247
10094	105.965	44.3312			Down	0	0.418357	Yes	−1.25719
10399	15.1302	5.24824			Down	5.50E-06	0.3468718	Yes	−1.52753
10411	110.674	50.8975	PF00083:General substrate transporter;	GO:0016021-integral to membrane; GO:0016020-membrane; GO:0005215-transporter activity; GO:0008643-carbohydrate transport; GO:0006810-transport; GO:0005351-sugar:hydrogen symporter activity;	Down	0	0.4598867	Yes	−1.12065
10492	5.14998	1.52437	PF01184:GPR1/FUN34/yaaH;	GO:0016020-membrane;	Down	0.00432319	0.2959953	Yes	−1.75635
10571	31.076	13.6255	PF00107:Zinc-containing alcohol dehydrogenase superfamily;	GO:0008270-zinc ion binding; GO:0016491-oxidoreductase activity;	Down	1.54E-08	0.4384573	Yes	−1.18949

qRT-PCR results were calculated by delta delta Ct = [(Ct (gene i)-C t(Ef1-alpha) in Bdtf1 mutant -(Ct (gene i) - Ct (Ef1-alpha)) in wild type.

Table S2. Differentially expressed genes during mycelial growth in the presence of brassinin.

Protein Id	BrasWt	BrasFtf1	PFAM annotation	GO annotation	BrasWt -> BrasFtf1: Up/Down	BrasWt -> BrasFtf1: q value	ratio (m/w)	BrasWt -> BrasFtf1: >2-fold and significant difference?	log2 ratio
15	12.094	52.3012		GO:0005634-nucleus;	Up	2.76E-10	4.324557632	Yes	2.112552564
135	0.351741	2.39611	PF00004:AAA ATPase, central region;	GO:0000166-nucleotide binding; GO:0005524-ATP binding; GO:0017111-nucleoside-triphosphatase activity;	Up	5.00E-05	6.812143026	Yes	2.768108726
195	50.4079	131.07			Up	0.000530383	2.600187669	Yes	1.378615754
264	146.315	483.98			Up	1.57E-05	3.307794826	Yes	1.725869751
274	47.4078	101.34			Up	0.00408024	2.137622923	Yes	1.096007384
284	426.355	856.531	PF02560:Cyanate lyase, C-terminal;	GO:0009439-cyanate metabolic process; GO:0016836-hydro-lyase activity; GO:0008152-metabolic process; GO:0008824-cyanate hydratase activity;	Up	0.0164674	2.008962015	Yes	1.006450286
314	571.981	1247.99	PF01408:Oxidoreductase, N-terminal;	GO:0047115-trans-1,2-dihydrobenzene-1,2-diol dehydrogenase activity; GO:0016491-oxidoreductase activity;	Up	0.0305352	2.181873174	Yes	1.125567244
335	6.0015	13.5254	PF00248:Aldo/keto reductase;	GO:0004024-alcohol dehydrogenase activity, zinc-dependent; GO:0004025-alcohol dehydrogenase activity, iron-dependent; GO:0004023-alcohol dehydrogenase activity, metal ion-independent; GO:0016491-oxidoreductase activity; GO:0004022-alcohol dehydrogenase activity;	Up	0.0129281	2.253669916	Yes	1.172276226
366	7.55093	100.204			Up	0	13.27041835	Yes	3.730141947

Table S2. *Cont.*

Protein Id	BrasWt	BrasFtf1	PFAM annotation	GO annotation	BrasWt -> BrasFtf1: Up/Down	BrasWt -> BrasFtf1: q value	ratio (m/w)	BrasWt -> BrasFtf1: >2-fold and significant difference?	log2 ratio
367	9.81324	38.1957	PF00135:Carboxylesterase, type B;	GO:0004091-carboxylesterase activity;	Up	1.87E-08	3.892261883	Yes	1.960608782
392	33.4009	67.071			Up	0.0208415	2.008059663	Yes	1.005802135
431	110.658	249.312			Up	0.00442658	2.252995717	Yes	1.171844571
440	1107.28	9924.17			Up	3.52E-09	8.962656239	Yes	3.163926363
441	19.1576	76.9988			Up	3.54E-06	4.019229966	Yes	2.006919126
446	183.292	795.211			Up	1.55E-05	4.338492678	Yes	2.117193894
447	6.40141	23.1806			Up	6.91E-07	3.621170961	Yes	1.85645629
505	8.04796	18.4785	PF00106:Short-chain dehydrogenase/reductase SDR;	GO:0004316-3-oxoacyl-[acyl-carrier-protein] reductase activity; GO:0016491-oxidoreductase activity; GO:0008152-metabolic process;	Up	0.00835258	2.296047694	Yes	1.19915261
594	174.609	470.441	PF01679:Protein of unknown function UPF0057;	GO:0016021-integral to membrane;	Up	4.53E-05	2.694254019	Yes	1.429885877
632	151.078	339.81	PF06966:Protein of unknown function DUF1295;	GO:0016021-integral to membrane; GO:0005737-cytoplasm; GO:0006629-lipid metabolic process; GO:0016627-oxidoreductase activity, acting on the CH-CH group of donors;	Up	0.00250008	2.249235494	Yes	1.169434719
633	169.233	424.168	PF01042:Endoribonuclease L-PSP;		Up	0.000906845	2.506414234	Yes	1.325624868

Table S2. *Cont.*

Protein Id	BrasWt	BrasFtf1	PFAM annotation	GO annotation	BrasWt -> BrasFtf1: Up/Down	BrasWt -> BrasFtf1: q value	ratio (m/w)	BrasWt -> BrasFtf1: >2-fold and significant difference?	log2 ratio
638	20.4636	46.2233			Up	0.0181979	2.25880588	Yes	1.175560292
652	44.8961	163.731			Up	7.60E-08	3.646886923	Yes	1.866665468
712	235.411	502.036	PF00702:Haloacid dehalogenase-like hydrolase;	GO:0003824-catalytic activity; GO:0016787-hydrolase activity; GO:0008152-metabolic process;	Up	0.00568349	2.132593634	Yes	1.092609086
715	6.60696	21.109			Up	1.55E-05	3.194964098	Yes	1.67579972
729	476.328	1199.41			Up	0.000331888	2.518033792	Yes	1.332297644
745	781.724	2323.9			Up	0.000230055	2.97278835	Yes	1.571816753
757	224.951	809.768			Up	3.03E-07	3.599752835	Yes	1.847897852
765	21.2397	44.0455			Up	0.0191179	2.073734563	Yes	1.052231242
766	1.46866	3.75902	PF01565:FAD linked oxidase, N-terminal;	GO:0050660-FAD binding; GO:0016491-oxidoreductase activity;	Up	0.0154571	2.559489603	Yes	1.355856146
771	16.0046	43.98		GO:0000701-purine-specific mismatch base pair DNA N-glycosylase activity; GO:0017065-single-strand selective uracil DNA N-glycosylase activity; GO:0019104-DNA N-glycosylase activity; GO:0043733-DNA-3-methyladenine glycosylase III activity; GO:0004844-uracil DNA N-glycosylase activity;	Up	0.000171575	2.747959962	Yes	1.458360984
772	30.5444	67.4824			Up	0.00353437	2.209321512	Yes	1.143603383

Table S2. *Cont.*

Protein Id	BrasWt	BrasFtf1	PFAM annotation	GO annotation	BrasWt -> BrasFtf1: Up/Down	BrasWt -> BrasFtf1: q value	ratio (m/w)	BrasWt -> BrasFtf1: >2-fold and significant difference?	log2 ratio
861	21.5646	43.3826	PF00172:Fungal transcriptional regulatory protein, N-terminal; PF04082:Fungal specific transcription factor;	GO:0003700-transcription factor activity; GO:0006350-transcription; GO:0008270-zinc ion binding; GO:0003677-DNA binding; GO:0006355-regulation of transcription, DNA-dependent; GO:0005634-nucleus;	Up	0.0103715	2.01175074	Yes	1.008451563
868	17.4632	38.1893	PF04616:Glycoside hydrolase, family 43;	GO:0004553-hydrolase activity, hydrolyzing O-glycosyl compounds; GO:0005975-carbohydrate metabolic process;	Up	0.00364066	2.186844336	Yes	1.12885053
896	12.8846	26.771		GO:0008270-zinc ion binding; GO:0005515-protein binding;	Up	0.00528027	2.077751735	Yes	1.055023281
902	841.8	2149.13			Up	0.000243877	2.553017344	Yes	1.352203339
923	28.581	62.5783			Up	0.00516674	2.189507015	Yes	1.130606072
973	22.0139	44.8371			Up	0.0307163	2.036763136	Yes	1.026278213
1089	29.6134	81.2108	PF00097:Zn-finger, RING;	GO:0008270-zinc ion binding; GO:0005515-protein binding;	Up	5.38E-05	2.742366631	Yes	1.45542146
1114	51.1062	116.114	PF00722:Glycoside hydrolase, family 16;	GO:0004553-hydrolase activity, hydrolyzing O-glycosyl compounds; GO:0005975-carbohydrate metabolic process; GO:0042972-licheninase activity;	Up	0.00126508	2.272013963	Yes	1.183971701
1161	14.3178	68.2185			Up	7.65E-05	4.764593722	Yes	2.252353201
1165	71.3622	147.408			Up	0.00938551	2.065631385	Yes	1.046582826

Table S2. *Cont.*

Protein Id	BrasWt	BrasFtf1	PFAM annotation	GO annotation	BrasWt -> BrasFtf1: Up/Down	BrasWt -> BrasFtf1: q value	ratio (m/w)	BrasWt -> BrasFtf1: >2-fold and significant difference?	log2 ratio
1209	135.516	416.537	PF00043:Glutathione S-transferase, C-terminal;		Up	3.30E-06	3.073710853	Yes	1.619981456
1239	13.0686	39.2635			Up	0.000102458	3.004415163	Yes	1.587084184
1284	102.138	247.775			Up	0.00252755	2.425884588	Yes	1.278510915
1311	3.74827	10.2144		GO:0016021-integral to membrane; GO:0006810-transport; GO:0005215-transporter activity;	Up	0.000713793	2.725097178	Yes	1.446307678
1324	291.795	905.896			Up	2.04E-06	3.104563135	Yes	1.634390271
1330	7.6441	15.3889	PF06792:Protein of unknown function UPF0261;		Up	0.0252732	2.013173559	Yes	1.009471555
1349	2.79299	6.10332			Up	0.0338394	2.185228017	Yes	1.127783825
1405	6.30707	61.2663	PF06422:CDR ABC transporter; PF00005:ABC transporter;	GO:0016021-integral to membrane; GO:0042626-ATPase activity, coupled to transmembrane movement of substances; GO:0017111-nucleoside-triphosphatase activity; GO:0006810-transport; GO:0000166-nucleotide binding; GO:0016887-ATPase activity; GO:0005524-ATP binding;	Up	0	9.71390836	Yes	3.280051876
1448	90.2639	203.324			Up	0.00248111	2.252550577	Yes	1.171559499

Table S2. *Cont.*

Protein Id	BrasWt	BrasFtf1	PFAM annotation	GO annotation	BrasWt -> BrasFtf1: Up/Down	BrasWt -> BrasFtf1: q value	ratio (m/w)	BrasWt -> BrasFtf1: >2-fold and significant difference?	log2 ratio
1462	138.35	299.944	PF00071:Ras GTPase;	GO:0007264-small GTPase mediated signal transduction; GO:0006886-intracellular protein transport; GO:0005515-protein binding; GO:0003924-GTPase activity; GO:0007165-signal transduction; GO:0006913-nucleocytoplasmic transport; GO:0005525-GTP binding; GO:0005622-intracellular; GO:0015031-protein transport;	Up	0.00408024	2.168008674	Yes	1.116370529
1595	25.323	52.7597			Up	0.00612839	2.083469573	Yes	1.058988031
1631	567.303	1560.42			Up	0.000198564	2.750593598	Yes	1.459742996
1635	87.6759	186.064	PF00135:Carboxylesterase, type B;	GO:0004091-carboxylesterase activity;	Up	0.00794663	2.122179527	Yes	1.085546707
1653	17.5518	36.2207	PF00249:Myb, DNA-binding; PF00569:Zn-finger, ZZ type; PF04433:SWIRM;	GO:0003677-DNA binding; GO:0008270-zinc ion binding;	Up	0.00742758	2.063645894	Yes	1.045195436
1656	2.69771	8.17959	PF01204:Glycoside hydrolase, family 37;	GO:0004555-alpha,alpha-trehalase activity; GO:0005991-trehalose metabolic process;	Up	0.000113726	3.032049405	Yes	1.600293261
1660	6.23029	18.918		GO:0003824-catalytic activity; GO:0008152-metabolic process;	Up	1.88E-05	3.036455767	Yes	1.602388353
1662	7.97255	21.4455			Up	0.00234716	2.689917279	Yes	1.427561807

Table S2. *Cont.*

Protein Id	BrasWt	BrasFtf1	PFAM annotation	GO annotation	BrasWt -> BrasFtf1: Up/Down	BrasWt -> BrasFtf1: q value	ratio (m/w)	BrasWt -> BrasFtf1: >2-fold and significant difference?	log2 ratio
1687	3.05478	7.56126	PF00106:Short-chain dehydrogenase/reductase SDR;	GO:0050356-tropine dehydrogenase activity; GO:0016491-oxidoreductase activity; GO:0008152-metabolic process;	Up	0.0228509	2.475222438	Yes	1.30755818
1688	3.96885	9.7357			Up	0.013078	2.453027955	Yes	1.294563675
1695	25.4776	58.8566	PF02716:Isoflavone reductase;		Up	0.00201747	2.310131253	Yes	1.207974822
1699	45.3571	95.2548			Up	0.0106179	2.100107811	Yes	1.070463392
1708	14.2215	68.4692			Up	6.19E-08	4.814485111	Yes	2.267381516
1709	0.651441	3.61962		GO:0003700-transcription factor activity; GO:0046983-protein dimerization activity; GO:0043565-sequence-specific DNA binding; GO:0006355-regulation of transcription, DNA-dependent; GO:0005634-nucleus;	Up	0.0218045	5.556328202	Yes	2.474131819
1711	10.1384	22.5846			Up	0.0121918	2.227629606	Yes	1.155509372
1733	12.1598	278.182			Up	0	22.87718548	Yes	4.515837667
1743	50.3164	127.772			Up	0.000236944	2.539370861	Yes	1.344471108
1749	143.619	294.877	PF00957: Synaptobrevin;	GO:0016021-integral to membrane; GO:0016192-vesicle-mediated transport;	Up	0.00942353	2.053189341	Yes	1.037866676
1829	1447.95	3089.03	PF00887:Acyl-coA-binding protein, ACBP;	GO:0000062-acyl-CoA binding;	Up	0.0148726	2.133381678	Yes	1.093142097
1931	154.326	543.422			Up	4.71E-07	3.521260189	Yes	1.816091833
1933	71.5751	204.782			Up	5.02E-05	2.861078783	Yes	1.516559224

Table S2. *Cont.*

Protein Id	BrasWt	BrasFtfl	PFAM annotation	GO annotation	BrasWt -> BrasFtfl: Up/Down	BrasWt -> BrasFtfl: q value	ratio (m/w)	BrasWt -> BrasFtfl: >2-fold and significant difference?	log2 ratio
1960	26.8434	55.3769	PF03109:ABC-1;		Up	0.00569847	2.062961473	Yes	1.044716878
1961	36.3573	151.307	PF07646:Kelch; PF01344:Kelch repeat;		Up	4.01E-09	4.161667671	Yes	2.057161763
1985	31.2002	64.1461			Up	0.0369402	2.055951564	Yes	1.039806277
1995	5.06197	15.0832			Up	4.35E-05	2.979709481	Yes	1.575171676
2014	10.4528	31.0114	PF01243:Pyridoxamine 5'-phosphate oxidase-related; PF00612:IQ calmodulin-binding region;	GO:0010181-FMN binding;	Up	6.59E-06	2.966803153	Yes	1.568909207
2019	17.2082	52.8331	PF00251:Glycoside hydrolase, family 32;	GO:0004564-beta-fructofuranosidase activity; GO:0004553-hydrolase activity, hydrolyzing O-glycosyl compounds; GO:0005975-carbohydrate metabolic process;	Up	3.58E-06	3.070228147	Yes	1.618345865
2056	35.8073	86.9563			Up	0.00902654	2.42845174	Yes	1.280036817
2145	0.472719	66.8781	PF01965:ThiJ/PfpI;		Up	0	141.4753797	Yes	7.144407199
2153	37.3701	80.3268		GO:0008115-sarcosine oxidase activity;	Up	0.00327486	2.149493847	Yes	1.10399698
2155	21.1593	48.734			Up	0.00225944	2.303195285	Yes	1.203636741
2212	134.181	308.465			Up	0.00227939	2.298872419	Yes	1.200926402

Table S2. *Cont.*

Protein Id	BrasWt	BrasFtf1	PFAM annotation	GO annotation	BrasWt -> BrasFtf1: Up/Down	BrasWt -> BrasFtf1: q value	ratio (m/w)	BrasWt -> BrasFtf1: >2-fold and significant difference?	log2 ratio
2285	111.728	256.582	PF05254:Protein of unknown function UPF0203;		Up	0.00613797	2.296487899	Yes	1.199429182
2291	3.86712	8.92201			Up	0.0157998	2.307145886	Yes	1.206109232
2308	83.0568	310.403			Up	1.43E-06	3.737237649	Yes	1.901972307
2382	17.4334	42.4479			Up	0.0319317	2.43486067	Yes	1.283839219
2409	16.0985	44.9355			Up	0.00233016	2.791284902	Yes	1.480929386
2427	69.8277	175.305			Up	0.000205088	2.510536649	Yes	1.327995786
2435	2.27902	7.56178	PF06422:CDR ABC transporter; PF00005:ABC transporter;	GO:0016021-integral to membrane; GO:0042626-ATPase activity, coupled to transmembrane movement of substances; GO:0017111-nucleoside-triphosphatase activity; GO:0006810-transport; GO:0000166-nucleotide binding; GO:0016887-ATPase activity; GO:0005524-ATP binding;	Up	4.24E-06	3.317996332	Yes	1.730312291
2440	20.4165	41.9262	PF00295:Glycoside hydrolase, family 28;	GO:0005975-carbohydrate metabolic process; GO:0004650-polygalacturonase activity;	Up	0.0102329	2.053544927	Yes	1.038116511
2513	39.5868	121.599			Up	1.01E-05	3.07170572	Yes	1.619040008
2584	22.9773	68.4537			Up	7.52E-06	2.979188155	Yes	1.574919242
2587	1.38914	4.99938	PF00657:Lipolytic enzyme, G-D-S-L;	GO:0016788-hydrolase activity, acting on ester bonds; GO:0006629-lipid metabolic process;	Up	0.0058555	3.598902918	Yes	1.847557186

Table S2. *Cont.*

Protein Id	BrasWt	BrasFtf1	PFAM annotation	GO annotation	BrasWt -> BrasFtf1: Up/Down	BrasWt -> BrasFtf1: q value	ratio (m/w)	BrasWt -> BrasFtf1: >2-fold and significant difference?	log2 ratio
2610	98.8331	241.601			Up	0.000325825	2.444535282	Yes	1.289560228
2624	4.39837	8.85526			Up	0.0432341	2.013304929	Yes	1.009565695
2632	6.38647	25.8098	PF06985:Heterokaryon incompatibility;		Up	1.41E-06	4.041324863	Yes	2.014828328
2637	10.7693	25.588			Up	0.0204888	2.376013297	Yes	1.24854291
2656	16.8177	48.8967			Up	0.000468882	2.907454646	Yes	1.539756687
2757	429.537	875.576	PF00226:Heat shock protein DnaJ, N-terminal;	GO:0006457-protein folding; GO:0031072-heat shock protein binding; GO:0051082-unfolded protein binding;	Up	0.0156706	2.038418111	Yes	1.027450001
2795	5.46192	12.4828	PF06966:Protein of unknown function DUF1295;	GO:0016021-integral to membrane; GO:0005737-cytoplasm; GO:0006629-lipid metabolic process; GO:0016627-oxidoreductase activity, acting on the CH-CH group of donors;	Up	0.00990394	2.285423441	Yes	1.192461491
2848	31.5373	84.3434	PF00096:Zn-finger, C2H2 type;	GO:0008270-zinc ion binding; GO:0003676-nucleic acid binding; GO:0005622-intracellular;	Up	0.000459476	2.674401423	Yes	1.419216028
2854	30.7704	65.0723			Up	0.00381933	2.114769389	Yes	1.080500349
2855	40.8262	105.275	PF02987:Late embryogenesis abundant protein;	GO:0003910-DNA ligase (ATP) activity;	Up	0.000635431	2.578613733	Yes	1.366595679
2856	49.5411	164.164			Up	7.69E-07	3.313693075	Yes	1.728439982

Table S2. *Cont.*

Protein Id	BrasWt	BrasFtf1	PFAM annotation	GO annotation	BrasWt -> BrasFtf1: Up/Down	BrasWt -> BrasFtf1: q value	ratio (m/w)	BrasWt -> BrasFtf1: >2-fold and significant difference?	log2 ratio
2868	532.35	1277.73	PF03647:Protein of unknown function UPF0136;	GO:0016020-membrane;	Up	0.00130894	2.400169062	Yes	1.263136029
2889	25.738	93.8372		GO:0016068-type I hypersensitivity;	Up	7.41E-07	3.645862149	Yes	1.866260014
2899	127.417	257.838			Up	0.00909738	2.023576132	Yes	1.016907128
2918	22.3696	56.3018		GO:0004316-3-oxoacyl-[acyl-carrier-protein] reductase activity; GO:0016491-oxidoreductase activity; GO:0008152-metabolic process;	Up	0.000550378	2.516888992	Yes	1.331641588
2923	52.5491	257.647			Up	1.47E-09	4.902976454	Yes	2.293657833
2933	54.9549	143.695			Up	9.83E-05	2.614780484	Yes	1.386689834
3002	6.74893	14.0977	PF00293:NUDIX hydrolase;	GO:0016787-hydrolase activity;	Up	0.0207125	2.088879274	Yes	1.062729115
3010	26.5831	54.0534		GO:0016831-carboxy-lyase activity;	Up	0.0186106	2.033374588	Yes	1.023876012
3012	117.21	289.951	PF00857:Isochorismatase hydrolase;	GO:0003824-catalytic activity; GO:0008152-metabolic process;	Up	0.000456034	2.473773569	Yes	1.306713453
3063	1.37684	3.32968	PF01055:Glycoside hydrolase, family 31;	GO:0004553-hydrolase activity, hydrolyzing O-glycosyl compounds; GO:0005975-carbohydrate metabolic process;	Up	0.0100917	2.418349264	Yes	1.274022617
3112	3.72156	7.76544	PF00096:Zn-finger, C2H2 type;	GO:0008270-zinc ion binding; GO:0003676-nucleic acid binding; GO:0005622-intracellular;	Up	0.0155429	2.086608841	Yes	1.061160176
3121	141.278	440.179			Up	0.000292949	3.11569388	Yes	1.639553494

Table S2. *Cont.*

Protein Id	BrasWt	BrasFtf1	PFAM annotation	GO annotation	BrasWt -> BrasFtf1: Up/Down	BrasWt -> BrasFtf1: q value	ratio (m/w)	BrasWt -> BrasFtf1: >2-fold and significant difference?	log2 ratio
3131	2.09979	578.182		GO:0006118-electron transport; GO:0004497-monooxygenase activity; GO:0020037-heme binding; GO:0005506-iron ion binding;	Up	0	275.3522971	Yes	8.105134834
3258	37.1827	78.1117			Up	0.00542915	2.100753845	Yes	1.070907125
3261	15.4154	35.085		GO:0003676-nucleic acid binding; GO:0004806-triacylglycerol lipase activity;	Up	0.00379578	2.275970782	Yes	1.186482037
3324	18.723	64.4381			Up	1.51E-06	3.441654649	Yes	1.783102338
3330	913.052	4837.37		GO:0051438-regulation of ubiquitin-protein ligase activity; GO:0051444-negative regulation of ubiquitin-protein ligase activity; GO:0051440-regulation of ubiquitin-protein ligase activity during meiotic cell cycle; GO:0051443-positive regulation of ubiquitin-protein ligase activity; GO:0004842-ubiquitin-protein ligase activity;	Up	0.00132698	5.298022457	Yes	2.405453959

Table S2. *Cont.*

Protein Id	BrasWt	BrasFtf1	PFAM annotation	GO annotation	BrasWt -> BrasFtf1: Up/Down	BrasWt -> BrasFtf1: q value	ratio (m/w)	BrasWt -> BrasFtf1: >2-fold and significant difference?	log2 ratio
3344	225.26	569.616	PF00704:Glycoside hydrolase, family 18;	GO:0004568-chitinase activity; GO:0003824-catalytic activity; GO:0008843-endochitinase activity; GO:0008152-metabolic process; GO:0006032-chitin catabolic process; GO:0004553-hydrolase activity, hydrolyzing O-glycosyl compounds; GO:0005975-carbohydrate metabolic process;	Up	0.00125121	2.528704608	Yes	1.338398518
3368	70.9836	143.98	PF01256:Protein of unknown function UPF0031;		Up	0.00826919	2.028355846	Yes	1.020310774
3382	48.0131	108.967			Up	0.00195113	2.269526442	Yes	1.182391297
3387	65.5927	150.39	PF00795:Nitrilase/ cyanide hydratase and apolipoprotein N-acyltransferase;	GO:0006807-nitrogen compound metabolic process; GO:0016810-hydrolase activity, acting on carbon-nitrogen (but not peptide) bonds;	Up	0.00135345	2.29278563	Yes	1.197101473
3427	65.4807	165.054			Up	0.000325825	2.520651123	Yes	1.333796452

Table S2. *Cont.*

Protein Id	BrasWt	BrasFtf1	PFAM annotation	GO annotation	BrasWt -> BrasFtf1: Up/Down	BrasWt -> BrasFtf1: q value	ratio (m/w)	BrasWt -> BrasFtf1: >2-fold and significant difference?	log2 ratio
3449	41.4035	104.805	PF00067:Cytochrome P450;	GO:0018588-tri-n-butyltin dioxygenase activity; GO:0018589-di-n-butyltin dioxygenase activity; GO:0018592-4-nitrocatechol 4-monooxygenase activity; GO:0018598-hydroxymethylsilanetriol oxidase activity; GO:0018597-ammonia monooxygenase activity; GO:0018591-methyl tertiary butyl ether 3-monooxygenase activity; GO:0043823-spheroidene monooxygenase activity; GO:0018585-fluorene oxygenase activity; GO:0005506-iron ion binding; GO:0018586-mono-butyltin dioxygenase activity; GO:0018600-alpha-pinene dehydrogenase activity; GO:0018593-4-chlorophenoxyacetate monooxygenase activity; GO:0018599-2-hydroxyisobutyrate 3-monooxygenase activity; GO:0018594-tert-butyl alcohol 2-monooxygenase activity; GO:0006118-electron transport; GO:0020037-heme binding; GO:0018596-dimethylsilanediol hydroxylase activity; GO:0018587-limonene 8-monooxygenase activity; GO:0004497-monooxygenase activity; GO:0018595-alpha-pinene monooxygenase activity; GO:0018590-methylsilanetriol hydroxylase activity;	Up	0.00015389	2.53130774	Yes	1.339882911

Table S2. *Cont.*

Protein Id	BrasWt	BrasFtf1	PFAM annotation	GO annotation	BrasWt -> BrasFtf1: Up/Down	BrasWt -> BrasFtf1: q value	ratio (m/w)	BrasWt -> BrasFtf1: >2-fold and significant difference?	log2 ratio
3477	21.9969	45.1673	PF00083:General substrate transporter;	GO:0016021-integral to membrane; GO:0016020-membrane; GO:0005215-transporter activity; GO:0008643-carbohydrate transport; GO:0006810-transport; GO:0005351-sugar:hydrogen symporter activity;	Up	0.0118531	2.053348426	Yes	1.037978455
3478	20.4671	44.9966		GO:0016021-integral to membrane; GO:0016020-membrane; GO:0005215-transporter activity; GO:0008643-carbohydrate transport; GO:0006810-transport; GO:0005351-sugar:hydrogen symporter activity;	Up	0.00441109	2.198484397	Yes	1.136509293
3509	17.4855	61.3291		GO:0019538-protein metabolic process; GO:0016706-oxidoreductase activity, acting on paired donors, with incorporation or reduction of molecular oxygen, 2-oxoglutarate as one donor, and incorporation of one atom each of oxygen into both donors;	Up	9.57E-07	3.507426153	Yes	1.810412729

Table S2. *Cont.*

Protein Id	BrasWt	BrasFtf1	PFAM annotation	GO annotation	BrasWt -> BrasFtf1: Up/Down	BrasWt -> BrasFtf1: q value	ratio (m/w)	BrasWt -> BrasFtf1: >2-fold and significant difference?	log2 ratio
3527	8.58077	20.8754	PF06985:Heterokaryon incompatibility; PF00069:Protein kinase;	GO:0004713-protein-tyrosine kinase activity; GO:0003676-nucleic acid binding; GO:0008270-zinc ion binding; GO:0004674-protein serine/threonine kinase activity; GO:0004672-protein kinase activity; GO:0006468-protein amino acid phosphorylation; GO:0005622-intracellular; GO:0005524-ATP binding;	Up	0.000500323	2.432811974	Yes	1.282624822
3533	16.9362	40.6633			Up	0.00568349	2.400969521	Yes	1.263617089
3576	3.78826	20.4945			Up	0.00110466	5.410003537	Yes	2.435629537
3596	1297.93	3260.75			Up	0.00307637	2.512269537	Yes	1.328991257
3656	112.98	292.351			Up	0.000154586	2.58763498	Yes	1.37163412
3676	78.7644	159.589		GO:0005515-protein binding;	Up	0.0123138	2.026156487	Yes	1.018745603
3693	42.0606	84.2616			Up	0.0169421	2.003338041	Yes	1.00240588
3789	480.23	1109.15	PF04419:Four F5 protein;		Up	0.0151799	2.309622473	Yes	1.20765705
3803	1119.21	2663.21			Up	0.00751579	2.3795445	Yes	1.250685435
3862	15.1336	34.1339			Up	0.00152141	2.255504308	Yes	1.173450042
3879	13.794	41.3174	PF00248:Aldo/ keto reductase;	GO:0005488-binding; GO:0050236-pyridoxine 4-dehydrogenase activity; GO:0016491-oxidoreductase activity;	Up	2.99E-05	2.995316804	Yes	1.5827086

Table S2. *Cont.*

Protein Id	BrasWt	BrasFtf1	PFAM annotation	GO annotation	BrasWt -> BrasFtf1: Up/Down	BrasWt -> BrasFtf1: q value	ratio (m/w)	BrasWt -> BrasFtf1: >2-fold and significant difference?	log2 ratio
3939	93.5968	283.652			Up	7.64E-06	3.030573695	Yes	1.599590925
3944	230.698	567.263			Up	0.000468386	2.458898647	Yes	1.29801227
3949	133.032	463.351			Up	3.40E-07	3.483004089	Yes	1.800332167
4202	7.53337	23.1969	PF00026:Peptidase A1, pepsin;	GO:0004194-pepsin A activity; GO:0004190-aspartic-type endopeptidase activity; GO:0006508-proteolysis; GO:0004196-saccharopepsin activity;	Up	2.62E-05	3.079219526	Yes	1.622564725
4228	9.47021	21.2183			Up	0.047174	2.240531097	Yes	1.16384075
4279	51.9903	104.704	PF04828:Glutathione-dependent formaldehyde-activating, GFA;	GO:0016846-carbon-sulfur lyase activity; GO:0008152-metabolic process;	Up	0.0308943	2.013914134	Yes	1.010002173
4305	10.5116	1244.02			Up	0	118.3473496	Yes	6.886883587
4306	6.47136	14.9838			Up	0.019303	2.315402018	Yes	1.211262707
4310	6.76028	17.2617			Up	0.00211226	2.553400155	Yes	1.352419647
4339	19.6718	39.8128	PF04791:LMBR1-like conserved region;	GO:0003676-nucleic acid binding;	Up	0.0082144	2.023851401	Yes	1.017103366
4350	51.8201	144.111			Up	0.000102765	2.780986528	Yes	1.475596756
4365	345.585	792.249			Up	0.0018792	2.292486653	Yes	1.196913334
4377	78.4714	165.915			Up	0.0100722	2.114337198	Yes	1.080205478
4393	66.6251	234.872			Up	1.24E-07	3.525278011	Yes	1.817737036

Table S2. *Cont.*

Protein Id	BrasWt	BrasFtf1	PFAM annotation	GO annotation	BrasWt -> BrasFtf1: Up/Down	BrasWt -> BrasFtf1: q value	ratio (m/w)	BrasWt -> BrasFtf1: >2-fold and significant difference?	log2 ratio
4420	22.9676	54.8463	PF01571:Glycine cleavage T protein (aminomethyl transferase); PF00753: Beta-lactamase-like;	GO:0006546-glycine catabolic process; GO:0004047-aminomethyltransferase activity; GO:0005737-cytoplasm; GO:0003824-catalytic activity; GO:0016787-hydrolase activity; GO:0047865-dimethylglycine dehydrogenase activity;	Up	0.000511332	2.387985684	Yes	1.255794188
4446	27.6008	134.88			Up	2.00E-11	4.886814875	Yes	2.288894453
4552	4.84354	12.5454			Up	0.0155265	2.590130359	Yes	1.373024709
4578	10.1356	25.6338	PF01979: Amidohydrolase;	GO:0016787-hydrolase activity;	Up	0.000357841	2.529085599	Yes	1.338615867
4658	7.72716	15.5885			Up	0.0289926	2.017364724	Yes	1.012471936
4663	5.58144	16.5758			Up	0.011768	2.969807075	Yes	1.570369213
4685	552.662	1256.44	PF02777:Manganese and iron superoxide dismutase; PF00081:Manganese and iron superoxide dismutase;	GO:0006801-superoxide metabolic process; GO:0046872-metal ion binding; GO:0004784-superoxide dismutase activity;	Up	0.00814168	2.273432948	Yes	1.184872454
4746	16.9687	43.5167	PF02798:Glutathione S-transferase, N-terminal;		Up	0.00056973	2.56452763	Yes	1.358693115

Table S2. *Cont.*

Protein Id	BrasWt	BrasFtf1	PFAM annotation	GO annotation	BrasWt -> BrasFtf1: Up/Down	BrasWt -> BrasFtf1: q value	ratio (m/w)	BrasWt -> BrasFtf1: >2-fold and significant difference?	log2 ratio
4768	3.16081	11.6532	PF06966:Protein of unknown function DUF1295;	GO:0016021-integral to membrane; GO:0005737-cytoplasm; GO:0006629-lipid metabolic process; GO:0016627-oxidoreductase activity, acting on the CH-CH group of donors;	Up	0.000199873	3.686776491	Yes	1.882359956
4771	10.6117	30.6232			Up	0.00069193	2.885795867	Yes	1.528969251
4862	11.3904	52.5479			Up	4.71E-07	4.613349838	Yes	2.205814699
4872	9.22856	22.3264	PF00795:Nitrilase/cyanide hydratase and apolipoprotein N-acyltransferase;	GO:0006807-nitrogen compound metabolic process; GO:0016810-hydrolase activity, acting on carbon-nitrogen (but not peptide) bonds;	Up	0.00348262	2.419272346	Yes	1.274573188
4886	18.3842	58.8184			Up	0.00133887	3.199399484	Yes	1.677801142
4900	1.53867	4.44146	PF00135:Carboxylesterase, type B;	GO:0004104-cholinesterase activity;	Up	0.00133206	2.886557871	Yes	1.52935015
4925	41.024	235.339	PF06422:CDR ABC transporter; PF00005:ABC transporter;	GO:0016021-integral to membrane; GO:0042626-ATPase activity, coupled to transmembrane movement of substances; GO:0017111-nucleoside-triphosphatase activity; GO:0006810-transport; GO:0000166-nucleotide binding; GO:0016887-ATPase activity; GO:0005524-ATP binding;	Up	9.30E-06	5.73661759	Yes	2.520200349
4964	14.8618	79.7366	PF00106:Short-chain dehydrogenase/reductase SDR;	GO:0016491-oxidoreductase activity; GO:0008152-metabolic process;	Up	5.17E-10	5.365204753	Yes	2.42363323

Table S2. *Cont.*

Protein Id	BrasWt	BrasFtf1	PFAM annotation	GO annotation	BrasWt -> BrasFtf1: Up/Down	BrasWt -> BrasFtf1: q value	ratio (m/w)	BrasWt -> BrasFtf1: >2-fold and significant difference?	log2 ratio
4965	14.8225	45.562	PF01425:Amidase;	GO:0004812-aminoacyl-tRNA ligase activity; GO:0005737-cytoplasm; GO:0006412-translation; GO:0000166-nucleotide binding; GO:0016884-carbon-nitrogen ligase activity, with glutamine as amido-N-donor; GO:0006418-tRNA aminoacylation for protein translation; GO:0005524-ATP binding;	Up	5.58E-06	3.073840445	Yes	1.620042281
4967	28.9073	126.845	PF01522:Polysaccharide deacetylase;	GO:0005975-carbohydrate metabolic process; GO:0016810-hydrolase activity, acting on carbon-nitrogen (but not peptide) bonds;	Up	5.28E-09	4.387991961	Yes	2.133560882
4968	2.32793	8.71879	PF01490:Amino acid/polyamine transporter II;		Up	2.75E-05	3.745297324	Yes	1.905080253
4995	15.2317	43.0393			Up	5.41E-05	2.825639948	Yes	1.498577645
4996	39.6422	96.5811	PF01255: Di-trans-poly-cis-decaprenylcistransferase;	GO:0008152-metabolic process; GO:0016740-transferase activity;	Up	0.000346484	2.436320386	Yes	1.284703866
5025	1.10267	5.07101	PF00179:Ubiquitin-conjugating enzymes;	GO:0051246-regulation of protein metabolic process; GO:0043687-post-translational protein modification; GO:0019787-small conjugating protein ligase activity;	Up	9.18E-07	4.598846436	Yes	2.201272024

Table S2. *Cont.*

Protein Id	BrasWt	BrasFtf1	PFAM annotation	GO annotation	BrasWt -> BrasFtf1: Up/Down	BrasWt -> BrasFtf1: q value	ratio (m/w)	BrasWt -> BrasFtf1: >2-fold and significant difference?	log2 ratio
5061	27.5248	659.812			Up	0	23.97154566	Yes	4.58325103
5123	15.9532	36.845	PF01753:Zn-finger, MYND type;	GO:0008270-zinc ion binding;	Up	0.021713	2.309567986	Yes	1.207623015
5128	8.56852	32.061	PF00141:Haem peroxidase, plant/fungal/bacterial;	GO:0006979-response to oxidative stress; GO:0004601-peroxidase activity; GO:0006118-electron transport; GO:0020037-heme binding;	Up	1.63E-07	3.74171969	Yes	1.903701484
5147	122.173	362.395			Up	5.00E-05	2.966244588	Yes	1.568637563
5148	83.9019	213.496			Up	0.000387218	2.544590766	Yes	1.347433653
5193	6.03826	79.0277	PF04479:RTA1 like protein;	GO:0006950-response to stress; GO:0016021-integral to membrane;	Up	0	13.08782663	Yes	3.710153637
5213	2.86078	6.91744		GO:0006118-electron transport; GO:0004497-monooxygenase activity; GO:0020037-heme binding; GO:0005506-iron ion binding;	Up	0.00786047	2.418025853	Yes	1.27382967
5238	9.21703	19.4928	PF00106:Short-chain dehydrogenase/reductase SDR;	GO:0004316-3-oxoacyl-[acyl-carrier-protein] reductase activity; GO:0016491-oxidoreductase activity; GO:0008152-metabolic process;	Up	0.0164893	2.114867804	Yes	1.080567487
5253	2.0134	7.36297			Up	0.00702563	3.656983212	Yes	1.870654004
5254	444.487	1969.04			Up	2.74E-07	4.429915836	Yes	2.147279289

Table S2. *Cont.*

Protein Id	BrasWt	BrasFtf1	PFAM annotation	GO annotation	BrasWt -> BrasFtf1: Up/Down	BrasWt -> BrasFtf1: q value	ratio (m/w)	BrasWt -> BrasFtf1: >2-fold and significant difference?	log2 ratio
5256	60.1834	127.265	PF07653:Variant SH3; PF00018:SH3; PF04366:Protein of unknown function DUF500;		Up	0.00376281	2.114619646	Yes	1.080398191
5317	19.627	43.2966			Up	0.00417698	2.205971366	Yes	1.141414065
5324	10.4732	28.1306	PF00644: Poly(ADP-ribose) polymerase, catalytic region; PF05406:WGR; PF02877: Poly(ADP-ribose) polymerase, regulatory region;	GO:0006471-protein amino acid ADP-ribosylation; GO:0005634-nucleus; GO:0003950-NAD+ ADP-ribosyltransferase activity;	Up	0.000201792	2.685960356	Yes	1.425438011
5364	18.2888	40.0788			Up	0.0038067	2.19143957	Yes	1.131878896
5406	160.246	479.226			Up	4.37E-06	2.990564507	Yes	1.580417837
5421	3.28771	7.13926			Up	0.0463716	2.171499311	Yes	1.118691495
5430	10.1364	22.2757			Up	0.011906	2.197594807	Yes	1.135925406
5433	111.213	413.431	PF01476:Peptidoglycan-binding LysM;	GO:0016998-cell wall catabolic process;	Up	1.92E-07	3.71747008	Yes	1.894321131
5453	385.658	795.697			Up	0.0089218	2.063219225	Yes	1.044897122
5463	25.4591	61.2575			Up	0.000651134	2.40611412	Yes	1.26670507
5466	0.981878	6.74928			Up	0.0064487	6.873847871	Yes	2.781117923

Table S2. *Cont.*

Protein Id	BrasWt	BrasFtf1	PFAM annotation	GO annotation	BrasWt -> BrasFtf1: Up/Down	BrasWt -> BrasFtf1: q value	ratio (m/w)	BrasWt -> BrasFtf1: >2-fold and significant difference?	log2 ratio
5493	103.931	225.068			Up	0.00533468	2.165552145	Yes	1.114734912
5494	134.879	300.316			Up	0.00148606	2.226558619	Yes	1.154815595
5507	45.7265	159.141	PF01565:FAD linked oxidase, N-terminal; PF00187:Chitin-binding, type 1;	GO:0050660-FAD binding; GO:0008061-chitin binding; GO:0016491-oxidoreductase activity;	Up	9.66E-07	3.480279488	Yes	1.799203168
5508	3.10034	12.1088	PF01966:Metal-dependent phosphohydrolase, HD subdomain;	GO:0003824-catalytic activity;	Up	4.31E-05	3.905636156	Yes	1.965557556
5513	16.6476	54.2651			Up	5.16E-05	3.259635022	Yes	1.704710437
5555	7.15729	27.1933			Up	1.81E-06	3.799384963	Yes	1.925765897
5612	389.371	1329.73			Up	4.28E-05	3.415072001	Yes	1.771915996
5623	8.79033	27.7423		GO:0016787-hydrolase activity; GO:0008152-metabolic process;	Up	0.000579784	3.156002107	Yes	1.658098168
5722	29.0226	74.1311			Up	0.000196934	2.554254271	Yes	1.35290215
5735	57.7983	199.87			Up	1.29E-07	3.458060185	Yes	1.789962978
5741	10.6164	30.6681		GO:0005634-nucleus;	Up	2.53E-05	2.888747598	Yes	1.530444155
5789	1.58447	15.3677	PF00249:Myb, DNA-binding;	GO:0003677-DNA binding;	Up	1.29E-10	9.698952962	Yes	3.277829011

Table S2. *Cont.*

Protein Id	BrasWt	BrasFtf1	PFAM annotation	GO annotation	BrasWt -> BrasFtf1: Up/Down	BrasWt -> BrasFtf1: q value	ratio (m/w)	BrasWt -> BrasFtf1: >2-fold and significant difference?	log2 ratio
5792	32.4813	147.495	PF00664:ABC transporter, transmembrane region; PF00005: ABC transporter;	GO:0016021-integral to membrane; GO:0042626-ATPase activity, coupled to transmembrane movement of substances; GO:0017111-nucleoside-triphosphatase activity; GO:0006810-transport; GO:0000166-nucleotide binding; GO:0016887-ATPase activity; GO:0005524-ATP binding;	Up	1.92E-07	4.540920468	Yes	2.182984769
5846	36.4396	109.687			Up	9.54E-06	3.010104392	Yes	1.589813521
5916	9.73818	19.5318			Up	0.0257954	2.005693056	Yes	1.004100838
5920	1522.22	4377.93			Up	0.00034477	2.876016607	Yes	1.524072007
5938	58.751	118.65	PF00026:Peptidase A1, pepsin;	GO:0004194-pepsin A activity; GO:0004190-aspartic-type endopeptidase activity; GO:0006508-proteolysis;	Up	0.00702563	2.019540093	Yes	1.014026787
5987	6.19655	16.0815			Up	0.00086989	2.595234445	Yes	1.375864873
6001	6.54617	22.4746		GO:0016021-integral to membrane; GO:0006810-transport; GO:0005215-transporter activity;	Up	2.61E-06	3.433244172	Yes	1.779572465
6057	1.80857	6.67734			Up	0.0114817	3.692055049	Yes	1.884424064
6060	25.5364	90.2928			Up	3.10E-07	3.53584687	Yes	1.822055796
6064	2.93144	6.90821		GO:0016021-integral to membrane; GO:0006810-transport; GO:0005215-transporter activity;	Up	0.00986584	2.356592664	Yes	1.236702411

Table S2. *Cont.*

Protein Id	BrasWt	BrasFtf1	PFAM annotation	GO annotation	BrasWt -> BrasFtf1: Up/Down	BrasWt -> BrasFtf1: q value	ratio (m/w)	BrasWt -> BrasFtf1: >2-fold and significant difference?	log2 ratio
6102	26.2385	57.9406			Up	0.00470941	2.208228367	Yes	1.142889378
6157	38.2146	122.962		GO:0004812-aminoacyl-tRNA ligase activity; GO:0006508-proteolysis; GO:0006464-protein modification process; GO:0005737-cytoplasm; GO:0006412-translation; GO:0000166-nucleotide binding; GO:0008237-metallopeptidase activity; GO:0008270-zinc ion binding; GO:0006418-tRNA aminoacylation for protein translation; GO:0005524-ATP binding;	Up	1.41E-06	3.217670733	Yes	1.686016701
6194	209.197	577.574			Up	6.55E-05	2.760909573	Yes	1.465143637
6235	36.3678	79.9173	PF00106:Short-chain dehydrogenase/ reductase SDR;	GO:0003858-3-hydroxybutyrate dehydrogenase activity; GO:0016491-oxidoreductase activity; GO:0008152-metabolic process;	Up	0.00603599	2.197474139	Yes	1.135846187
6295	3.37559	8.64106	PF00561:Alpha/beta hydrolase fold;	GO:0003824-catalytic activity; GO:0004301-epoxide hydrolase activity;	Up	0.00494249	2.559866571	Yes	1.356068614
6304	23.2129	67.8483			Up	0.000102765	2.922870473	Yes	1.547385897
6313	2000.15	4985.56	PF07249: Cerato-platanin;		Up	0.00902881	2.492593056	Yes	1.317647365
6321	3.09949	7.5855	PF00561:Alpha/beta hydrolase fold;	GO:0003824-catalytic activity;	Up	0.018908	2.44733811	Yes	1.29121343

Table S2. *Cont.*

Protein Id	BrasWt	BrasFtf1	PFAM annotation	GO annotation	BrasWt -> BrasFtf1: Up/Down	BrasWt -> BrasFtf1: q value	ratio (m/w)	BrasWt -> BrasFtf1: >2-fold and significant difference?	log2 ratio
6352	12.9686	26.8843	PF03417:Peptidase C45, acyl-coenzyme A:6-aminopenicillanic acid acyl-transferase;	GO:0042318-penicillin biosynthetic process;	Up	0.0163592	2.073030242	Yes	1.051741163
6353	46.2329	197.954	PF00153:Mitochondrial substrate carrier;	GO:0005488-binding; GO:0016020-membrane; GO:0006810-transport;	Up	1.07E-08	4.281669547	Yes	2.098173455
6359	32.6919	109.694	PF03746:LamB/YcsF;		Up	1.39E-06	3.355387726	Yes	1.746479484
6360	5.22933	14.9409	PF02786:Carbamoyl-phosphate synthase L chain, ATP-binding; PF00289:Carbamoyl-phosphate synthetase large chain, N-terminal; PF02626:Urea amidolyase-related; PF02785:Biotin carboxylase, C-terminal; PF02682:Protein of unknown function DUF213;	GO:0003824-catalytic activity; GO:0016874-ligase activity; GO:0008152-metabolic process; GO:0005524-ATP binding;	Up	3.22E-05	2.857134662	Yes	1.514569035

Table S2. *Cont.*

Protein Id	BrasWt	BrasFtf1	PFAM annotation	GO annotation	BrasWt -> BrasFtf1: Up/Down	BrasWt -> BrasFtf1: q value	ratio (m/w)	BrasWt -> BrasFtf1: >2-fold and significant difference?	log2 ratio
6396	1.5881	4.40974			Up	0.0297924	2.7767395	Yes	1.473391837
6428	2.06428	6.63252	PF00457:Glycoside hydrolase, family 11;	GO:0004553-hydrolase activity, hydrolyzing O-glycosyl compounds; GO:0005975-carbohydrate metabolic process;	Up	0.00944033	3.212994361	Yes	1.683918449
6439	10.1308	66.5241			Up	2.33E-12	6.566519919	Yes	2.715128983
6481	37.8241	93.4118	PF00578:Alkyl hydroperoxide reductase/Thiol specific antioxidant/ Mal allergen;	GO:0051920-peroxiredoxin activity; GO:0016491-oxidoreductase activity; GO:0016209-antioxidant activity;	Up	0.000468386	2.469637083	Yes	1.304299051
6495	38.6075	102.876	PF00441:Acyl-CoA dehydrogenase, C-terminal; PF00173:Cytochrome b5; PF02770: Acyl-CoA dehydrogenase, central region;	GO:0003995-acyl-CoA dehydrogenase activity; GO:0016937-short-branched-chain-acyl-CoA dehydrogenase activity; GO:0016627-oxidoreductase activity, acting on the CH-CH group of donors; GO:0046914-transition metal ion binding; GO:0019109-acyl-CoA reductase activity; GO:0043820-propionyl-CoA dehydrogenase activity; GO:0017099-very-long-chain-acyl-CoA dehydrogenase activity; GO:0043830-thiol-driven fumarate reductase activity; GO:0006118-electron transport; GO:0020037-heme binding;	Up	6.18E-05	2.664663602	Yes	1.413953413

Table S2. *Cont.*

Protein Id	BrasWt	BrasFtf1	PFAM annotation	GO annotation	BrasWt -> BrasFtf1: Up/Down	BrasWt -> BrasFtf1: q value	ratio (m/w)	BrasWt -> BrasFtf1: >2-fold and significant difference?	log2 ratio
6496	11.0261	26.8597			Up	0.0226509	2.43601092	Yes	1.2845206
6506	218.081	459.614	PF04588:Hypoxia induced protein conserved region;		Up	0.00782454	2.107538025	Yes	1.075558661
6545	20.3532	50.8851			Up	0.00154642	2.500103178	Yes	1.321987635
6581	2.42682	8.24317	PF00732:Glucose-methanol-choline oxidoreductase;	GO:0016614-oxidoreductase activity, acting on CH-OH group of donors; GO:0050660-FAD binding;	Up	0.00351084	3.396696088	Yes	1.764132142
6582	4.52387	13.6842		GO:0016491-oxidoreductase activity; GO:0005506-iron ion binding; GO:0009058-biosynthetic process;	Up	0.00141297	3.024887983	Yes	1.596881718
6621	2.80906	12.1733			Up	1.92E-05	4.3335849	Yes	2.115560969

Table S2. *Cont.*

Protein Id	BrasWt	BrasFtf1	PFAM annotation	GO annotation	BrasWt -> BrasFtf1: Up/Down	BrasWt -> BrasFtf1: q value	ratio (m/w)	BrasWt -> BrasFtf1: >2-fold and significant difference?	log2 ratio
6632	5.77829	16.8707	PF00106:Short-chain dehydrogenase/ reductase SDR;	GO:0033765-steroid dehydrogenase activity, acting on the CH-CH group of donors; GO:0018452-5-exo-hydroxycamphor dehydrogenase activity; GO:0016491-oxidoreductase activity; GO:0043713-(R)-2-hydroxyisocaproate dehydrogenase activity; GO:0004495-mevaldate reductase activity; GO:0004033-aldo-keto reductase activity; GO:0008152-metabolic process; GO:0018453-2-hydroxytetrahydrofuran dehydrogenase activity; GO:0033764-steroid dehydrogenase activity, acting on the CH-OH group of donors, NAD or NADP as acceptor; GO:0018451-epoxide dehydrogenase activity; GO:0016229-steroid dehydrogenase activity; GO:0000252-C-3 sterol dehydrogenase (C-4 sterol decarboxylase) activity; GO:0032442-phenylcoumaran benzylic ether reductase activity; GO:0000253-3-keto sterol reductase activity; GO:0048258-3-ketoglucose-reductase activity; GO:0008875-gluconate dehydrogenase activity; GO:0051990-(R)-2-hydroxyglutarate dehydrogenase activity; GO:0004448-isocitrate dehydrogenase activity;	Up	0.00031772	2.919670006	Yes	1.545805319

Table S2. *Cont.*

Protein Id	BrasWt	BrasFtf1	PFAM annotation	GO annotation	BrasWt -> BrasFtf1: Up/Down	BrasWt -> BrasFtf1: q value	ratio (m/w)	BrasWt -> BrasFtf1: >2-fold and significant difference?	log2 ratio
6682	4.38627	11.8211			Up	0.00227609	2.695023334	Yes	1.430297764
6707	179.072	371.805			Up	0.011768	2.07628775	Yes	1.054006399
6710	55.1987	161.42			Up	1.21E-05	2.924344233	Yes	1.548113145
6750	9.15899	23.4447			Up	0.00107457	2.559747308	Yes	1.356001398
6817	2.41238	5.80815	PF04253:Transferrin receptor-like dimerisation region; PF04389: Peptidase M28; PF02225: Protease-associated PA;	GO:0006508-proteolysis; GO:0043275-glutamate carboxypeptidase II activity; GO:0008233-peptidase activity;	Up	0.00542713	2.407643074	Yes	1.267621533
6824	18.0283	45.4681	PF00168:C2;		Up	0.000193647	2.522040348	Yes	1.334591356
6836	4.2264	22.7132			Up	0.00024545	5.37412455	Yes	2.426029758
6846	203.048	432.322	PF00227:20S proteasome, A and B subunits;	GO:0006511-ubiquitin-dependent protein catabolic process; GO:0004299-proteasome endopeptidase activity; GO:0004298-threonine endopeptidase activity; GO:0005839-proteasome core complex;	Up	0.00651788	2.129161578	Yes	1.090285437
6886	40.9965	124.98			Up	0.00580641	3.048552925	Yes	1.608124593
6912	27.7172	75.5453	PF01451:Low molecular weight phosphotyrosine protein phosphatase;	GO:0004725-protein tyrosine phosphatase activity; GO:0006470-protein amino acid dephosphorylation;	Up	0.000198564	2.725574733	Yes	1.446560479

Table S2. *Cont.*

Protein Id	BrasWt	BrasFtf1	PFAM annotation	GO annotation	BrasWt -> BrasFtf1: Up/Down	BrasWt -> BrasFtf1: q value	ratio (m/w)	BrasWt -> BrasFtf1: >2-fold and significant difference?	log2 ratio
6967	52.0435	120.75	PF00107:Zinc-containing alcohol dehydrogenase superfamily;	GO:0008106-alcohol dehydrogenase (NADP+) activity; GO:0008270-zinc ion binding; GO:0016491-oxidoreductase activity;	Up	0.000904502	2.320174469	Yes	1.214233295
6976	4.00742	13.8656	PF00023:Ankyrin;	GO:0003950-NAD+ ADP-ribosyltransferase activity;	Up	3.55E-06	3.459981734	Yes	1.790764422
7030	233.764	528.845			Up	0.00915006	2.262303006	Yes	1.177792173
7041	2.17756	9.28398			Up	3.63E-06	4.263478389	Yes	2.092030944
7069	95.6284	210.966			Up	0.00262591	2.206101953	Yes	1.141499465
7081	69.8287	331.375			Up	9.52E-09	4.74554159	Yes	2.246572746
7125	1.08832	2.83258	PF00135:Carboxylesterase, type B;	GO:0004091-carboxylesterase activity;	Up	0.00323023	2.602708762	Yes	1.380013886
7148	10.9646	34.7443			Up	8.20E-05	3.168770407	Yes	1.663923133
7150	5.34014	84.4447			Up	0	15.81319965	Yes	3.983057408
7219	107.431	241.947		GO:0005783-endoplasmic reticulum; GO:0006886-intracellular protein transport; GO:0005794-Golgi apparatus;	Up	0.00242258	2.252115311	Yes	1.171280697
7221	45.1999	107.994	PF00282:Pyridoxal-dependent decarboxylase;	GO:0004437-inositol or phosphatidylinositol phosphatase activity; GO:0004351-glutamate decarboxylase activity; GO:0030170-pyridoxal phosphate binding; GO:0019752-carboxylic acid metabolic process; GO:0006536-glutamate metabolic process; GO:0016831-carboxy-lyase activity;	Up	0.000484071	2.389253074	Yes	1.256559674

Table S2. *Cont.*

Protein Id	BrasWt	BrasFtf1	PFAM annotation	GO annotation	BrasWt -> BrasFtf1: Up/Down	BrasWt -> BrasFtf1: q value	ratio (m/w)	BrasWt -> BrasFtf1: >2-fold and significant difference?	log2 ratio
7239	962.428	2212.57	PF00187: Chitin-binding, type 1;	GO:0008061-chitin binding;	Up	0.0115946	2.298945999	Yes	1.200972578
7242	13.436	38.6074	PF00106:Short-chain dehydrogenase/ reductase SDR;	GO:0016491-oxidoreductase activity; GO:0008152-metabolic process;	Up	0.000134575	2.873429592	Yes	1.522773699
7245	292.068	811.545	PF00010:Basic helix-loop-helix dimerisation region bHLH;	GO:0045449-regulation of transcription; GO:0030528-transcription regulator activity; GO:0005634-nucleus;	Up	0.000268544	2.778616624	Yes	1.474366794
7286	33.498	81.9077		GO:0016491-oxidoreductase activity; GO:0008152-metabolic process;	Up	0.000305957	2.445151949	Yes	1.289924122
7323	22.1949	53.1798	PF00023:Ankyrin;		Up	0.00222595	2.396036927	Yes	1.260650143
7341	112.013	615.304			Up	3.72E-13	5.493148117	Yes	2.457633193
7346	927.047	3173.78			Up	1.76E-05	3.423537318	Yes	1.775487738
7373	5.0319	10.4692	PF06422:CDR ABC transporter; PF00005:ABC transporter;	GO:0016021-integral to membrane; GO:0042626-ATPase activity, coupled to transmembrane movement of substances; GO:0017111-nucleoside-triphosphatase activity; GO:0006810-transport; GO:0000166-nucleotide binding; GO:0016887-ATPase activity; GO:0005524-ATP binding;	Up	0.0121643	2.080565989	Yes	1.056976047
7374	2.24396	6.02179	PF00005:ABC transporter;	GO:0016887-ATPase activity; GO:0005524-ATP binding;	Up	0.00484686	2.683554965	Yes	1.424145438

Table S2. *Cont.*

Protein Id	BrasWt	BrasFtf1	PFAM annotation	GO annotation	BrasWt -> BrasFtf1: Up/Down	BrasWt -> BrasFtf1: q value	ratio (m/w)	BrasWt -> BrasFtf1: >2-fold and significant difference?	log2 ratio
7378	95.1705	191.949			Up	0.0381337	2.016895992	Yes	1.012136688
7379	104.288	220.268			Up	0.00626048	2.112112611	Yes	1.078686757
7451	2.19321	5.51617	PF00144: Beta-lactamase;	GO:0008800-beta-lactamase activity; GO:0030655-beta-lactam antibiotic catabolic process; GO:0046677-response to antibiotic;	Up	0.0130727	2.515112552	Yes	1.330622962
7468	119.934	243.365	PF05348:Proteasome maturation factor UMP1;		Up	0.0132623	2.029157703	Yes	1.020880994
7469	3.07157	8.93068			Up	0.00484543	2.907529374	Yes	1.539793767
7529	110.969	223.856			Up	0.0149372	2.017284106	Yes	1.012414282
7531	129.409	399.388			Up	3.99E-06	3.086245933	Yes	1.62585303
7569	4.66259	11.0382		GO:0005509-calcium ion binding;	Up	0.00519479	2.367396662	Yes	1.243301452
7644	28.1329	131.282		GO:0003700-transcription factor activity; GO:0046983-protein dimerization activity; GO:0043565-sequence-specific DNA binding; GO:0006355-regulation of transcription, DNA-dependent; GO:0005634-nucleus;	Up	8.05E-10	4.666493678	Yes	2.222338941
7651	22.6961	122.836	PF00394:Multicopper oxidase, type 1;	GO:0005507-copper ion binding; GO:0016491-oxidoreductase activity;	Up	1.76E-11	5.412207384	Yes	2.436217121
7652	34.3048	134.404			Up	3.84E-08	3.917935682	Yes	1.970093714
7667	3.20903	19.8582			Up	4.71E-07	6.188225102	Yes	2.629525677

Table S2. *Cont.*

Protein Id	BrasWt	BrasFtf1	PFAM annotation	GO annotation	BrasWt -> BrasFtf1: Up/Down	BrasWt -> BrasFtf1: q value	ratio (m/w)	BrasWt -> BrasFtf1: >2-fold and significant difference?	log2 ratio
7687	15.9871	64.8256			Up	3.11E-08	4.054869238	Yes	2.019655391
7733	0.962039	8.74742	PF01360:Flavoprotein monooxygenase;	GO:0016491-oxidoreductase activity; GO:0008152-metabolic process;	Up	1.55E-08	9.092583565	Yes	3.18469028
7741	119.61	541.34	PF00083:General substrate transporter;	GO:0016021-integral to membrane; GO:0006810-transport; GO:0005215-transporter activity;	Up	2.76E-07	4.525875763	Yes	2.178196983
7771	10.3978	29.1916	PF00795:Nitrilase/ cyanide hydratase and apolipoprotein N-acyltransferase;	GO:0006807-nitrogen compound metabolic process; GO:0016810-hydrolase activity, acting on carbon-nitrogen (but not peptide) bonds;	Up	0.000116945	2.807478505	Yes	1.489274977
7860	3.80473	8.59664			Up	0.0154571	2.25946125	Yes	1.175978815
7867	513.92	2073.01			Up	5.65E-07	4.033721202	Yes	2.012111373
7897	4.03878	9.58526			Up	0.00339454	2.3733058	Yes	1.246898004
7898	43.4932	169.553			Up	8.57E-09	3.898379517	Yes	1.962874547
7957	248.173	533.234	PF00011:Heat shock protein Hsp20;		Up	0.00779698	2.148638248	Yes	1.103422606
7983	33.2688	79.6317			Up	0.00101527	2.393584981	Yes	1.259173028
7994	12.6228	26.2898	PF03403:Platelet-activating factor acetylhydrolase, plasma/intracellular isoform II;	GO:0008247-2-acetyl-1-alkylglycerophosphocholine esterase complex; GO:0003847-1-alkyl-2-acetylglycerophosphocholine esterase activity; GO:0016042-lipid catabolic process;	Up	0.00796498	2.082723326	Yes	1.058471201

Table S2. *Cont.*

Protein Id	BrasWt	BrasFtf1	PFAM annotation	GO annotation	BrasWt -> BrasFtf1: Up/Down	BrasWt -> BrasFtf1: q value	ratio (m/w)	BrasWt -> BrasFtf1: >2-fold and significant difference?	log2 ratio
8007	172.597	539.437			Up	1.88E-05	3.125413536	Yes	1.644047091
8075	4.7722	9.75298	PF00083:General substrate transporter;	GO:0016021-integral to membrane; GO:0016020-membrane; GO:0005215-transporter activity; GO:0008643-carbohydrate transport; GO:0006810-transport; GO:0005351-sugar:hydrogen symporter activity; GO:0008733-L-arabinose isomerase activity;	Up	0.0221618	2.043707305	Yes	1.031188592
8077	242.101	672.82			Up	6.39E-05	2.779088067	Yes	1.474611553
8119	4.79078	10.5511	PF00501:AMP-dependent synthetase and ligase;	GO:0003824-catalytic activity; GO:0008152-metabolic process; GO:0016207-4-coumarate-CoA ligase activity;	Up	0.0083783	2.202376231	Yes	1.139060945
8124	294.852	626.302	PF00885:6,7-dimethyl-8-ribityllumazine synthase;	GO:0009349-riboflavin synthase complex; GO:0009231-riboflavin biosynthetic process; GO:0004746-riboflavin synthase activity;	Up	0.0164343	2.124123289	Yes	1.086867506
8182	110.934	284.46			Up	0.00148606	2.564227378	Yes	1.358524196
8187	2.81349	6.89185	PF00023:Ankyrin;		Up	0.0143564	2.449573306	Yes	1.292530466
8202	13.658	37.8052			Up	0.0265548	2.767989457	Yes	1.468838448

Table S2. *Cont.*

Protein Id	BrasWt	BrasFtf1	PFAM annotation	GO annotation	BrasWt -> BrasFtf1: Up/Down	BrasWt -> BrasFtf1: q value	ratio (m/w)	BrasWt -> BrasFtf1: >2-fold and significant difference?	log2 ratio
8203	22.3155	76.416	PF00083:General substrate transporter;	GO:0016021-integral to membrane; GO:0016020-membrane; GO:0006508-proteolysis; GO:0005215-transporter activity; GO:0008643-carbohydrate transport; GO:0006810-transport; GO:0008237-metallopeptidase activity; GO:0008270-zinc ion binding; GO:0005351-sugar:hydrogen symporter activity;	Up	3.30E-07	3.424346306	Yes	1.77582861
8205	11.9292	28.8267	PF01070: FMN-dependent alpha-hydroxy acid dehydrogenase;	GO:0008891-glycolate oxidase activity; GO:0003973-(S)-2-hydroxy-acid oxidase activity; GO:0009339-glycolate oxidase complex; GO:0006118-electron transport; GO:0016491-oxidoreductase activity;	Up	0.000906845	2.416482245	Yes	1.272908395
8235	1.23549	4.0479			Up	0.0363064	3.276351893	Yes	1.712090316
8284	22.4965	49.0433	PF02727:Copper amine oxidase;	GO:0005507-copper ion binding; GO:0009308-amine metabolic process; GO:0008131-amine oxidase activity; GO:0048038-quinone binding;	Up	0.00951009	2.18004134	Yes	1.124355493
8304	0.992637	6.98644			Up	0.011768	7.038262728	Yes	2.815219369
8306	44.23	96.365	PF00191:Annexin;	GO:0005544-calcium-dependent phospholipid binding; GO:0005509-calcium ion binding; GO:0047394-glycerophosphoinositol inositolphosphodiesterase activity;	Up	0.00304256	2.178724847	Yes	1.123484009

Table S2. *Cont.*

Protein Id	BrasWt	BrasFtf1	PFAM annotation	GO annotation	BrasWt -> BrasFtf1: Up/Down	BrasWt -> BrasFtf1: q value	ratio (m/w)	BrasWt -> BrasFtf1: >2-fold and significant difference?	log2 ratio
8313	58.9786	149.782			Up	0.000241335	2.539599109	Yes	1.344600777
8364	35.7377	143.179			Up	5.11E-08	4.006385414	Yes	2.002301215
8425	106.551	222.002	PF00107: Zinc-containing alcohol dehydrogenase superfamily;	GO:0008270-zinc ion binding; GO:0003960-NADPH:quinone reductase activity; GO:0016491-oxidoreductase activity;	Up	0.00716295	2.083528076	Yes	1.059028541
8486	75.8378	157.394			Up	0.00703547	2.075403031	Yes	1.053391527
8487	100.035	246.432			Up	0.000357981	2.46345779	Yes	1.300684752
8502	1.99504	4.20627	PF00501:AMP-dependent synthetase and ligase;	GO:0003824-catalytic activity; GO:0008152-metabolic process; GO:0016207-4-coumarate-CoA ligase activity;	Up	0.0447572	2.108363742	Yes	1.076123787
8515	54.5685	228.698			Up	4.15E-08	4.191025958	Yes	2.067303457
8517	14.7091	45.246	PF06766:Hydrophobin 2;	GO:0005576-extracellular region; GO:0007154-cell communication; GO:0005618-cell wall;	Up	0.00558966	3.076054959	Yes	1.62108128
8543	57.577	126.351			Up	0.00270298	2.194470014	Yes	1.133872557
8566	82.4965	353.103			Up	6.08E-08	4.280217949	Yes	2.097684261

Table S2. *Cont.*

Protein Id	BrasWt	BrasFtf1	PFAM annotation	GO annotation	BrasWt -> BrasFtf1: Up/Down	BrasWt -> BrasFtf1: q value	ratio (m/w)	BrasWt -> BrasFtf1: >2-fold and significant difference?	log2 ratio
8574	30.2424	68.7838			Up	0.00428723	2.274416052	Yes	1.185496186
8583	149.038	332.616	PF01145: Band 7 protein;	GO:0016020-membrane;	Up	0.00330813	2.231752976	Yes	1.15817735
8586	13.0367	29.6907			Up	0.00151686	2.277470526	Yes	1.187432383
8610	35.5326	79.3263			Up	0.00151686	2.232493541	Yes	1.158656001
8638	26.7944	67.1651	PF03853:YjeF-related protein, N-terminal;		Up	0.000665778	2.506684233	Yes	1.325780271
8644	17.7817	48.6572			Up	0.0128077	2.73636379	Yes	1.452260044
8655	24.9546	51.9934	PF01435:Peptidase M48, Ste24p;	GO:0008270-zinc ion binding; GO:0016020-membrane; GO:0004222-metalloendopeptidase activity; GO:0008237-metallopeptidase activity; GO:0006508-proteolysis;	Up	0.00737638	2.083519672	Yes	1.059022721
8728	6.41234	20.163			Up	0.000271775	3.144405942	Yes	1.652787481
8778	9.33913	28.2638			Up	5.28E-05	3.026384685	Yes	1.597595381
8793	9.30716	36.2555			Up	0.00421042	3.895441789	Yes	1.961786956
8833	14.8608	31.4497	PF00596:Class II aldolase/adducin, N-terminal;	GO:0046872-metal ion binding;	Up	0.0105813	2.116285799	Yes	1.081534473
8851	5.08129	21.6149			Up	1.63E-06	4.253821372	Yes	2.088759452

Table S2. *Cont.*

Protein Id	BrasWt	BrasFtf1	PFAM annotation	GO annotation	BrasWt -> BrasFtf1: Up/Down	BrasWt -> BrasFtf1: q value	ratio (m/w)	BrasWt -> BrasFtf1: >2-fold and significant difference?	log2 ratio
8855	19.8732	65.8492	PF01565:FAD linked oxidase, N-terminal;	GO:0018498-2,3-dihydroxy-2,3-dihydro-phenylpropionate dehydrogenase activity; GO:0016491-oxidoreductase activity; GO:0018500-trans-9R, 10R-dihydrodiolphenanthrene dehydrogenase activity; GO:0043786-cinnamate reductase activity; GO:0018499-cis-2,3-dihydrodiol DDT dehydrogenase activity; GO:0018503-trans-1,2-dihydrodiolphenanthrene dehydrogenase activity; GO:0018502-2,5-dichloro-2,5-cyclohexadiene-1,4-diol dehydrogenase activity; GO:0016631-enoyl-[acyl-carrier-protein] reductase activity; GO:0050660-FAD binding; GO:0018501-cis-chlorobenzene dihydrodiol dehydrogenase activity;	Up	4.71E-07	3.313467383	Yes	1.728341718

Table S2. *Cont.*

Protein Id	BrasWt	BrasFtf1	PFAM annotation	GO annotation	BrasWt -> BrasFtf1: Up/Down	BrasWt -> BrasFtf1: q value	ratio (m/w)	BrasWt -> BrasFtf1: >2-fold and significant difference?	log2 ratio
8863	74.9775	215.02			Up	2.08E-05	2.867793671	Yes	1.51994123
8942	33.0634	79.7284		GO:0047372-acylglycerol lipase activity; GO:0016787-hydrolase activity;	Up	0.000485095	2.41137935	Yes	1.269858629
8980	53.9136	423.714	PF01565:FAD linked oxidase, N-terminal;	GO:0019139-cytokinin dehydrogenase activity; GO:0050660-FAD binding; GO:0016491-oxidoreductase activity;	Up	0	7.859130164	Yes	2.974369646
9011	52.2082	120.731			Up	0.00716723	2.312491141	Yes	1.209447839
9025	6.65047	16.963			Up	0.00139453	2.550646796	Yes	1.350863134
9069	25.8727	58.4184	PF05032:Spo12;		Up	0.0143877	2.257916646	Yes	1.174992228
9096	3.35087	9.67599			Up	0.00363347	2.887605308	Yes	1.529873561
9101	1611	3683.38			Up	0.0017365	2.286393544	Yes	1.193073748
9261	65.3752	130.793			Up	0.00998452	2.000651623	Yes	1.00046997

Table S2. *Cont.*

Protein Id	BrasWt	BrasFtf1	PFAM annotation	GO annotation	BrasWt -> BrasFtf1: Up/Down	BrasWt -> BrasFtf1: q value	ratio (m/w)	BrasWt -> BrasFtf1: >2-fold and significant difference?	log2 ratio
9315	19.0079	70.4947	PF00069:Protein kinase;	GO:0043771-cytidine kinase activity; GO:0008443-phosphofructokinase activity; GO:0035004-phosphoinositide 3-kinase activity; GO:0042557-eukaryotic elongation factor-2 kinase activator activity; GO:0051731-polynucleotide kinase activity; GO:0043743-LPPG:FO 2-phospho-L-lactate transferase activity; GO:0043841-(S)-lactate 2-kinase activity; GO:0032942-inositol tetrakisphosphate 2-kinase activity; GO:0004672-protein kinase activity; GO:0008607-phosphorylase kinase regulator activity; GO:0051735-GTP-dependent polynucleotide kinase activity; GO:0005524-ATP binding; GO:0018720-phenol kinase activity; GO:0006468-protein amino acid phosphorylation; GO:0019914-cyclin-dependent protein kinase activating kinase regulator activity; GO:0016307-phosphatidylinositol phosphate kinase activity; GO:0051734-ATP-dependent polynucleotide kinase activity; GO:0008819-cobinamide kinase activity; GO:0043798-glycerate 2-kinase activity; GO:0004674-protein serine/threonine kinase activity; GO:0042556-eukaryotic elongation factor-2 kinase regulator activity; GO:0016538-cyclin-dependent protein kinase regulator activity;	Up	2.73E-08	3.708705328	Yes	1.890915644

Table S2. *Cont.*

Protein Id	BrasWt	BrasFtf1	PFAM annotation	GO annotation	BrasWt -> BrasFtf1: Up/Down	BrasWt -> BrasFtf1: q value	ratio (m/w)	BrasWt -> BrasFtf1: >2-fold and significant difference?	log2 ratio
9379	76.7457	229.363	PF01243: Pyridoxamine 5'-phosphate oxidase-related;	GO:0010181-FMN binding;	Up	2.01E-05	2.988610437	Yes	1.579474855
9384	18.4667	108.333			Up	6.80E-14	5.866397353	Yes	2.552474794
9393	47.4208	113.435			Up	0.00617255	2.392093765	Yes	1.258273941
9417	59.2758	146.246			Up	0.00148606	2.467212589	Yes	1.302882032
9437	140.245	358.017	PF00085: Thioredoxin-related;	GO:0004791-thioredoxin-disulfide reductase activity; GO:0045454-cell redox homeostasis;	Up	0.000418793	2.552796891	Yes	1.352078757
9517	86.7672	212.781			Up	0.00389573	2.452320693	Yes	1.294147654
9518	21.99	51.0836	PF00172:Fungal transcriptional regulatory protein, N-terminal;	GO:0003700-transcription factor activity; GO:0008270-zinc ion binding; GO:0006355-regulation of transcription, DNA-dependent; GO:0005634-nucleus;	Up	0.000947232	2.323037744	Yes	1.216012595
9527	922.625	4598.74			Up	1.43E-06	4.984408617	Yes	2.317422344
9528	5.4489	18.7363			Up	0.011017	3.43854723	Yes	1.781799162
9547	22.531	45.4273			Up	0.00905113	2.016213217	Yes	1.011648214
9548	20.9074	57.0519			Up	0.000163556	2.728789806	Yes	1.448261271
9569	7.23252	21.7739	PF01323:DSBA oxidoreductase;	GO:0015035-protein disulfide oxidoreductase activity; GO:0030288-outer membrane-bounded periplasmic space;	Up	4.59E-05	3.010555104	Yes	1.590029524
9572	6.58905	15.26			Up	0.0038067	2.315963606	Yes	1.211612583
9590	4.68001	11.3476			Up	0.0461716	2.424695674	Yes	1.277803684
9606	66.8381	286.213			Up	2.00E-08	4.282183365	Yes	2.098346574
9651	7.11166	18.4448	PF02586:Protein of unknown function DUF159;		Up	0.000828504	2.593599807	Yes	1.374955888

Table S2. *Cont.*

Protein Id	BrasWt	BrasFtf1	PFAM annotation	GO annotation	BrasWt -> BrasFtf1: Up/Down	BrasWt -> BrasFtf1: q value	ratio (m/w)	BrasWt -> BrasFtf1: >2-fold and significant difference?	log2 ratio
9654	4.26492	15.0537	PF06985:Heterokaryon incompatibility;		Up	3.87E-06	3.52965589	Yes	1.819527541
9700	1.79369	4.71388	PF00106:Short-chain dehydrogenase/ reductase SDR;	GO:0047038-D-arabinitol 2-dehydrogenase activity; GO:0016491-oxidoreductase activity; GO:0008152-metabolic process;	Up	0.0252732	2.628034945	Yes	1.393984459
9821	101.976	263.725			Up	0.000138749	2.586147721	Yes	1.370804684
9905	50.5775	126.14			Up	0.000357981	2.493994365	Yes	1.318458206
9951	403.852	914.916			Up	0.0017627	2.26547349	Yes	1.179812609
9986	332.131	984.072			Up	1.04E-05	2.962903192	Yes	1.567011489
10049	18.8171	80.3649	PF00664:ABC transporter, transmembrane region; PF00005: ABC transporter;	GO:0016021-integral to membrane; GO:0042626-ATPase activity, coupled to transmembrane movement of substances; GO:0017111-nucleoside-triphosphatase activity; GO:0006810-transport; GO:0000166-nucleotide binding; GO:0016887-ATPase activity; GO:0005524-ATP binding;	Up	7.24E-07	4.270844073	Yes	2.094521227
10062	5.64995	14.5374	PF04389:Peptidase M28; PF02225: Protease-associated PA;	GO:0006508-proteolysis; GO:0016284-alanine aminopeptidase activity; GO:0008233-peptidase activity;	Up	0.000482861	2.57301392	Yes	1.363459262
10117	4.66306	13.5722			Up	0.00698461	2.910578032	Yes	1.541305697
10142	1998.54	4347.92			Up	0.00366966	2.17554815	Yes	1.121378948
10149	72.9485	174.285			Up	0.00148606	2.389151251	Yes	1.25649819
10150	340.851	795.907			Up	0.0132603	2.335058427	Yes	1.223458649
10185	40.6913	103.438	PF01467: Cytidylyltransferase;	GO:0003824-catalytic activity; GO:0016779-nucleotidyltransferase activity; GO:0009058-biosynthetic process; GO:0016491-oxidoreductase activity;	Up	0.000149256	2.542017581	Yes	1.345974008

Table S2. *Cont.*

Protein Id	BrasWt	BrasFtf1	PFAM annotation	GO annotation	BrasWt -> BrasFtf1: Up/Down	BrasWt -> BrasFtf1: q value	ratio (m/w)	BrasWt -> BrasFtf1: >2-fold and significant difference?	log2 ratio
10218	1.6723	4.03972	PF06985: Heterokaryon incompatibility;		Up	0.00959741	2.415667045	Yes	1.27242162
10220	132.348	322.532			Up	0.000553149	2.436999426	Yes	1.285105911
10286	6.35497	17.0347			Up	0.00154642	2.68053193	Yes	1.42251932
10299	97.255	379.061			Up	1.46E-07	3.897599095	Yes	1.962585703
10370	42.1437	150.74	PF00708: Acylphosphatase;	GO:0003998-acylphosphatase activity;	Up	5.30E-05	3.57680982	Yes	1.838673412
10376	90.8928	247.003			Up	0.000811771	2.71751998	Yes	1.442290642
10409	230.958	785.123			Up	4.96E-07	3.399418942	Yes	1.76528817
10494	14.258	464.825			Up	0	32.60099593	Yes	5.026844133
10513	29.1683	79.97	PF01871:Protein of unknown function DUF51;		Up	8.53E-05	2.741675038	Yes	1.455057583
10529	19.3961	45.5842			Up	0.00390787	2.350173488	Yes	1.23276726
10557	179.679	421.716	PF02991:Light chain 3 (LC3);		Up	0.000792086	2.347052243	Yes	1.230849955
10592	74.3241	166.509			Up	0.00260391	2.240309671	Yes	1.163698165
10649	25.7559	433.785			Up	6.80E-14	16.84216044	Yes	4.074005308
10655	15.4266	42.7924			Up	0.000608285	2.773935929	Yes	1.471934465
10687	26.0407	75.7893			Up	0.000683004	2.910417155	Yes	1.541225952
45	2.37678	0.660062			Down	0.0127177	0.277712704	Yes	−1.848334921

Table S2. *Cont.*

Protein Id	BrasWt	BrasFtf1	PFAM annotation	GO annotation	BrasWt -> BrasFtf1: Up/Down	BrasWt -> BrasFtf1: q value	ratio (m/w)	BrasWt -> BrasFtf1: >2-fold and significant difference?	log2 ratio
53	83.7622	38.5821	PF01619:Proline dehydrogenase;	GO:0051700-fructosyl-amino acid oxidase activity; GO:0006562-proline catabolic process; GO:0006537-glutamate biosynthetic process; GO:0004657-proline dehydrogenase activity; GO:0004154-dihydropterin oxidase activity; GO:0019116-hydroxy-nicotine oxidase activity; GO:0051699-proline oxidase activity;	Down	0.00284031	0.460614693	Yes	−1.118367664
94	9.723	4.55026			Down	0.0262992	0.467989304	Yes	−1.095452539
106	15.4891	3.59738			Down	7.25E-05	0.232252358	Yes	−2.10623485
117	3.48999	1.32822	PF00172: Fungal transcriptional regulatory protein, N-terminal;	GO:0003700-transcription factor activity; GO:0008270-zinc ion binding; GO:0006355-regulation of transcription, DNA-dependent; GO:0005634-nucleus;	Down	0.00471843	0.380579887	Yes	−1.393728775
118	12.3805	4.08941	PF00773: Ribonuclease II;	GO:0004540-ribonuclease activity; GO:0003723-RNA binding;	Down	2.84E-05	0.330310569	Yes	−1.598104962
129	5.97962	2.84124		GO:0004274-dipeptidyl-peptidase IV activity;	Down	0.0176615	0.47515394	Yes	−1.073533104
152	3.33531	1.13035		GO:0003677-DNA binding;	Down	0.0071129	0.33890403	Yes	−1.561051303
159	6.2122	2.22047	PF05577:Peptidase S28;	GO:0008236-serine-type peptidase activity; GO:0006508-proteolysis; GO:0008234-cysteine-type peptidase activity;	Down	0.000620169	0.357436979	Yes	−1.484239198
166	8.21613	1.65932			Down	1.42E-09	0.20195883	Yes	−2.307866872

Table S2. *Cont.*

Protein Id	BrasWt	BrasFtf1	PFAM annotation	GO annotation	BrasWt -> BrasFtf1: Up/Down	BrasWt -> BrasFtf1: q value	ratio (m/w)	BrasWt -> BrasFtf1: >2-fold and significant difference?	log2 ratio
182	53.1051	25.2647	PF00649:Copper fist DNA-binding;	GO:0003700-transcription factor activity; GO:0003677-DNA binding; GO:0006355-regulation of transcription, DNA-dependent; GO:0005634-nucleus; GO:0005507-copper ion binding;	Down	0.00580641	0.475749034	Yes	−1.071727369
186	5.79254	0.98437			Down	0.0167203	0.16993754	Yes	−2.556923506
225	2.38925	0.234862			Down	0.000609196	0.098299466	Yes	−3.346672605
226	2.96122	0.237678			Down	0.0124689	0.08026354	Yes	−3.639111402
298	13.0517	5.43733	PF00134:Cyclin, N-terminal; PF02984:Cyclin, C-terminal;	GO:0005634-nucleus;	Down	0.0021136	0.41659937	Yes	−1.263267437
342	3.70395	1.63518		GO:0016021-integral to membrane; GO:0006810-transport; GO:0005215-transporter activity;	Down	0.033637	0.441469242	Yes	−1.179615168
377	52.6514	5.73863	PF03443:Glycoside hydrolase, family 61;		Down	0	0.108992923	Yes	−3.197693629
388	23.672	5.67021	PF03211: Pectate lyase;	GO:0005576-extracellular region; GO:0030570-pectate lyase activity;	Down	1.04E-05	0.239532359	Yes	−2.061707529
425	4.26014	1.39841		GO:0016021-integral to membrane; GO:0006810-transport; GO:0005215-transporter activity;	Down	0.0030225	0.328254471	Yes	−1.607113436
444	16.7908	6.97798			Down	0.00408024	0.415583534	Yes	−1.266789602
445	12.3855	4.501			Down	0.000715976	0.363408825	Yes	−1.460334641
462	42.6484	11.8863			Down	1.98E-07	0.278704477	Yes	−1.843191919
490	22.6974	11.0611		GO:0010181-FMN binding; GO:0016491-oxidoreductase activity;	Down	0.0117959	0.487328945	Yes	−1.03703218

Table S2. *Cont.*

Protein Id	BrasWt	BrasFtf1	PFAM annotation	GO annotation	BrasWt -> BrasFtf1: Up/Down	BrasWt -> BrasFtf1: q value	ratio (m/w)	BrasWt -> BrasFtf1: >2-fold and significant difference?	log2 ratio
526	37.5359	13.3411	PF00078: RNA-directed DNA polymerase (Reverse transcriptase);	GO:0006278-RNA-dependent DNA replication; GO:0003964-RNA-directed DNA polymerase activity; GO:0003723-RNA binding;	Down	0.00150437	0.355422409	Yes	−1.49239345
568	25.7511	12.2526			Down	0.0102604	0.4758088	Yes	−1.071546139
581	30.9105	13.9965	PF02629: CoA-binding; PF00549: ATP-citrate lyase/succinyl-CoA ligase;	GO:0042709-succinate-CoA ligase complex; GO:0003824-catalytic activity; GO:0004775-succinate-CoA ligase (ADP-forming) activity; GO:0009361-succinate-CoA ligase complex (ADP-forming); GO:0008152-metabolic process;	Down	0.00357751	0.452807298	Yes	−1.143030883
601	22.935	8.99909	PF03443:Glycoside hydrolase, family 61;		Down	0.000459476	0.392373665	Yes	−1.349699881
603	9.74483	4.24257	PF00271:Helicase, C-terminal;	GO:0003676-nucleic acid binding; GO:0004386-helicase activity; GO:0005524-ATP binding;	Down	0.00738329	0.43536624	Yes	−1.199698554
604	8.44589	2.87943	PF00270:DEAD/ DEAH box helicase, N-terminal;	GO:0008026-ATP-dependent helicase activity; GO:0003676-nucleic acid binding; GO:0005524-ATP binding;	Down	0.00136341	0.34092677	Yes	−1.552466207
612	3.34297	1.64132			Down	0.0349186	0.490976587	Yes	−1.026273867
630	35.7693	15.9704			Down	0.00287497	0.446483437	Yes	−1.163321438
636	28.7588	8.98683	PF00076: RNA-binding region RNP-1 (RNA recognition motif);	GO:0003676-nucleic acid binding;	Down	3.09E-06	0.312489742	Yes	−1.678119262
654	73.6263	27.5759	PF00988: Carbamoyl-phosphate synthase, small chain;	GO:0004086-carbamoyl-phosphate synthase activity; GO:0006807-nitrogen compound metabolic process;	Down	0.000849537	0.374538718	Yes	−1.41681323

Table S2. *Cont.*

Protein Id	BrasWt	BrasFtf1	PFAM annotation	GO annotation	BrasWt -> BrasFtf1: Up/Down	BrasWt -> BrasFtf1: q value	ratio (m/w)	BrasWt -> BrasFtf1: >2-fold and significant difference?	log2 ratio
655	66.7847	27.0111	PF00289: Carbamoyl-phosphate synthetase large chain, N-terminal; PF00117:Glutamine amidotransferase class-I;	GO:0004086-carbamoyl-phosphate synthase activity; GO:0006807-nitrogen compound metabolic process; GO:0003824-catalytic activity; GO:0008152-metabolic process; GO:0006541-glutamine metabolic process; GO:0009058-biosynthetic process; GO:0005524-ATP binding;	Down	0.00178003	0.404450421	Yes	−1.305965234
656	72.2983	27.801	PF02786: Carbamoyl-phosphate synthase L chain, ATP-binding; PF00289: Carbamoyl-phosphate synthetase large chain, N-terminal; PF02787: Carbamoyl-phosphate synthetase large chain, oligomerisation;	GO:0004086-carbamoyl-phosphate synthase activity; GO:0006807-nitrogen compound metabolic process; GO:0003824-catalytic activity; GO:0008152-metabolic process; GO:0005524-ATP binding;	Down	0.000654065	0.384531863	Yes	−1.378824947
684	143.761	52.7971		GO:0004254-acylaminoacyl-peptidase activity;	Down	3.49E-05	0.367256071	Yes	−1.445141756
706	4.04214	1.21472	PF01565:FAD linked oxidase, N-terminal;	GO:0050660-FAD binding; GO:0016491-oxidoreductase activity;	Down	0.000224248	0.300514084	Yes	−1.734495488

Table S2. *Cont.*

Protein Id	BrasWt	BrasFtf1	PFAM annotation	GO annotation	BrasWt -> BrasFtf1: Up/Down	BrasWt -> BrasFtf1: q value	ratio (m/w)	BrasWt -> BrasFtf1: >2-fold and significant difference?	log2 ratio
721	9.09194	3.99233	PF00004: AAA ATPase, central region;	GO:0042625-ATPase activity, coupled to transmembrane movement of ions; GO:0042626-ATPase activity, coupled to transmembrane movement of substances; GO:0004004-ATP-dependent RNA helicase activity; GO:0017111-nucleoside-triphosphatase activity; GO:0008026-ATP-dependent helicase activity; GO:0000166-nucleotide binding; GO:0015462-protein-transmembrane transporting ATPase activity; GO:0008094-DNA-dependent ATPase activity; GO:0004003-ATP-dependent DNA helicase activity; GO:0042624-ATPase activity, uncoupled; GO:0008186-RNA-dependent ATPase activity; GO:0017116-single-stranded DNA-dependent ATP-dependent DNA helicase activity; GO:0016887-ATPase activity; GO:0042623-ATPase activity, coupled; GO:0005524-ATP binding;	Down	0.00435103	0.439106505	Yes	−1.187357187
724	8.25111	3.23046	PF00930:Peptidase S9B, dipeptidylpeptidase IV N-terminal; PF00326:Peptidase S9, prolyl oligopeptidase active site region;	GO:0008236-serine-type peptidase activity; GO:0006508-proteolysis; GO:0008240-tripeptidyl-peptidase activity; GO:0016020-membrane; GO:0004274-dipeptidyl-peptidase IV activity; GO:0004287-prolyl oligopeptidase activity;	Down	0.000849537	0.391518232	Yes	−1.352848603

Table S2. *Cont.*

Protein Id	BrasWt	BrasFtf1	PFAM annotation	GO annotation	BrasWt -> BrasFtf1: Up/Down	BrasWt -> BrasFtf1: q value	ratio (m/w)	BrasWt -> BrasFtf1: >2-fold and significant difference?	log2 ratio
739	28.0932	11.6097		GO:0000166-nucleotide binding; GO:0017111-nucleoside-triphosphatase activity;	Down	0.00190545	0.413256589	Yes	−1.274890273
741	16.0888	5.9617			Down	0.00158828	0.370549699	Yes	−1.432261041
783	1059.12	510.398		GO:0033754-indoleamine 2,3-dioxygenase activity;	Down	0.0317308	0.481907621	Yes	−1.053171477
833	31.3268	11.1891	PF01728:Ribosomal RNA methyltransferase RrmJ/FtsJ;	GO:0016434-rRNA (cytosine) methyltransferase activity; GO:0016274-protein-arginine N-methyltransferase activity; GO:0019702-protein-arginine N5-methyltransferase activity; GO:0016427-tRNA (cytosine)-methyltransferase activity; GO:0030792-methylarsonite methyltransferase activity; GO:0008276-protein methyltransferase activity; GO:0018707-1-phenanthrol methyltransferase activity; GO:0016278-lysine N-methyltransferase activity; GO:0008650-rRNA (uridine-2'-O-)-methyltransferase activity; GO:0043803-hydroxyneurosporene-O-methyltransferase activity; GO:0009383-rRNA (cytosine-C5-967)-methyltransferase activity; GO:0043776-cobalt-precorrin-6B C5-methyltransferase activity;	Down	4.90E-05	0.357173411	Yes	−1.48530341

Table S2. *Cont.*

Protein Id	BrasWt	BrasFtf1	PFAM annotation	GO annotation	BrasWt -> BrasFtf1: Up/Down	BrasWt -> BrasFtf1: q value	ratio (m/w)	BrasWt -> BrasFtf1: >2-fold and significant difference?	log2 ratio
				GO:0016279-protein-lysine N-methyltransferase activity; GO:0016436-rRNA (uridine) methyltransferase activity; GO:0043827-tRNA (adenine-57, 58-N(1)-) methyltransferase activity; GO:0008649-rRNA methyltransferase activity; GO:0008170-N-methyltransferase activity; GO:0016433-rRNA (adenine) methyltransferase activity; GO:0003880-C-terminal protein carboxyl methyltransferase activity; GO:0016426-tRNA (adenine)-methyltransferase activity; GO:0016205-selenocysteine methyltransferase activity; GO:0016273-arginine N-methyltransferase activity; GO:0043834-trimethylamine methyltransferase activity; GO:0009008-DNA-methyltransferase activity; GO:0008171-O-methyltransferase activity; GO:0000179-rRNA (adenine-N6,N6-)-dimethyltransferase activity; GO:0008172-S-methyltransferase activity; GO:0008425-2-polyprenyl-6-methoxy-1,4-benzoquinone methyltransferase activity; GO:0043852-monomethylamine methyltransferase activity; GO:0008175-tRNA methyltransferase activity; GO:0008174-mRNA methyltransferase activity; GO:0043782-cobalt-precorrin-3 C17-methyltransferase activity;					

Table S2. *Cont.*

Protein Id	BrasWt	BrasFtf1	PFAM annotation	GO annotation	BrasWt -> BrasFtf1: Up/Down	BrasWt -> BrasFtf1: q value	ratio (m/w)	BrasWt -> BrasFtf1: >2-fold and significant difference?	log2 ratio
				GO:0016424-tRNA (guanosine) methyltransferase activity; GO:0018423-protein-leucine O-methyltransferase activity; GO:0008326-site-specific DNA-methyltransferase (cytosine-specific) activity; GO:0008169-C-methyltransferase activity; GO:0043780-cobalt-precorrin-5B C1-methyltransferase activity; GO:0016435-rRNA (guanine) methyltransferase activity; GO:0043790-dimethyladenosine transferase activity; GO:0016300-tRNA (uracil) methyltransferase activity; GO:0008173-RNA methyltransferase activity; GO:0016423-tRNA (guanine) methyltransferase activity; GO:0043833-methylamine-specific methylcobalamin:coenzyme M methyltransferase activity; GO:0043777-cobalt-precorrin-7 C15-methyltransferase activity; GO:0043851-methanol-specific methylcobalamin:coenzyme M methyltransferase activity; GO:0016431-tRNA (uridine) methyltransferase activity; GO:0043791-dimethylamine methyltransferase activity; GO:0043770-demethylmenaquinone methyltransferase activity;					

Table S2. *Cont.*

Protein Id	BrasWt	BrasFtf1	PFAM annotation	GO annotation	BrasWt -> BrasFtf1: Up/Down	BrasWt -> BrasFtf1: q value	ratio (m/w)	BrasWt -> BrasFtf1: >2-fold and significant difference?	log2 ratio
865	22.756	8.42221			Down	0.000763242	0.370109422	Yes	−1.433976233
891	15.8308	7.6961			Down	0.0463716	0.486147257	Yes	−1.040534713
915	14.564	5.51065			Down	0.000230011	0.37837476	Yes	−1.402112241
944	74.1245	32.7038	PF01536: S-adenosylmethionine decarboxylase;	GO:0004014-adenosylmethionine decarboxylase activity; GO:0008295-spermidine biosynthetic process; GO:0003824-catalytic activity; GO:0006597-spermine biosynthetic process;	Down	0.00147835	0.441200952	Yes	−1.18049219
949	2.02363	0.838805			Down	0.0125682	0.414505122	Yes	−1.270538166
958	28.2418	13.8073	PF00586:AIR synthase related protein;	GO:0003824-catalytic activity;	Down	0.0148586	0.488895892	Yes	−1.032400812
990	24.6878	8.84952			Down	0.000399267	0.358457214	Yes	−1.480127169
991	27.8426	11.4514			Down	0.00154636	0.411290612	Yes	−1.281769952
1047	52.3344	19.0039			Down	7.26E-05	0.363124446	Yes	−1.461464038
1063	1.84225	0.729247			Down	0.029691	0.395845841	Yes	−1.336989402

Table S2. *Cont.*

Protein Id	BrasWt	BrasFtf1	PFAM annotation	GO annotation	BrasWt -> BrasFtf1: Up/Down	BrasWt -> BrasFtf1: q value	ratio (m/w)	BrasWt -> BrasFtf1: >2-fold and significant difference?	log2 ratio
1076	27.245	13.4547	PF04851:Type III restriction enzyme, res subunit; PF00271:Helicase, C-terminal; PF00270:DEAD/DEAH box helicase, N-terminal;	GO:0008796-bis(5'-nucleosyl)-tetraphosphatase activity; GO:0008413-8-oxo-7,8-dihydroguanine triphosphatase activity; GO:0019177-dihydroneopterin triphosphate pyrophosphohydrolase activity; GO:0008758-UDP-2,3-diacylglucosamine hydrolase activity; GO:0004787-thiamin-pyrophosphatase activity; GO:0003676-nucleic acid binding; GO:0000810-diacylglycerol pyrophosphate phosphatase activity; GO:0008026-ATP-dependent helicase activity; GO:0016787-hydrolase activity; GO:0004386-helicase activity; GO:0043141-ATP-dependent 5'-3' DNA helicase activity; GO:0008828-dATP pyrophosphohydrolase activity; GO:0016462-pyrophosphatase activity; GO:0003677-DNA binding; GO:0019176-dihydroneopterin monophosphate phosphatase activity; GO:0005524-ATP binding;	Down	0.0104961	0.493841072	Yes	−1.017881267
1084	11.0614	3.53055			Down	0.000357981	0.3191775	Yes	−1.64756914
1085	14.7255	4.41621			Down	0.00018962	0.29990221	Yes	−1.737435939
1086	19.735	7.08476			Down	0.00026252	0.35899468	Yes	−1.477965632
1087	16.4966	1.58826	PF00295:Glycoside hydrolase, family 28;	GO:0005975-carbohydrate metabolic process; GO:0004650-polygalacturonase activity;	Down	0	0.096278021	Yes	−3.376649703

Table S2. *Cont.*

Protein Id	BrasWt	BrasFtf1	PFAM annotation	GO annotation	BrasWt -> BrasFtf1: Up/Down	BrasWt -> BrasFtf1: q value	ratio (m/w)	BrasWt -> BrasFtf1: >2-fold and significant difference?	log2 ratio
1108	0.925625	0.252722	PF01095: Pectinesterase;	GO:0005618-cell wall; GO:0030599-pectinesterase activity; GO:0042545-cell wall modification; GO:0004339-glucan 1,4-alpha-glucosidase activity;	Down	0.000469328	0.273028494	Yes	−1.872876571
1142	41.336	18.3148			Down	0.00417698	0.443071415	Yes	−1.174388842
1143	39.13	18.6475			Down	0.00605335	0.476552517	Yes	−1.069292883
1186	4.61174	0.978718			Down	9.83E-05	0.212223152	Yes	−2.236346041
1296	10.3311	4.48194	PF00248:Aldo/keto reductase;	GO:0043826-sulfur oxygenase reductase activity; GO:0043738-F420H2 dehydrogenase activity; GO:0043883-malolactic enzyme activity; GO:0051341-regulation of oxidoreductase activity; GO:0016491-oxidoreductase activity;	Down	0.010177	0.433829892	Yes	−1.204798632
1348	5.34877	2.25059			Down	0.00948218	0.420767765	Yes	−1.24890391
1362	120.885	53.205	PF00005:ABC transporter;	GO:0000166-nucleotide binding; GO:0017111-nucleoside-triphosphatase activity; GO:0016887-ATPase activity; GO:0005524-ATP binding;	Down	0.00164127	0.440129048	Yes	−1.184001503
1372	19.7112	7.96095	PF00400:G-protein beta WD-40 repeat;		Down	0.000569661	0.40387952	Yes	−1.308003102
1395	10.3753	4.99168	PF00246:Peptidase M14, carboxypeptidase A;	GO:0006508-proteolysis; GO:0008270-zinc ion binding; GO:0004182-carboxypeptidase A activity;	Down	0.0164674	0.481111871	Yes	−1.055555696
1397	59.7812	21.4285	PF01227:GTP cyclohydrolase I;	GO:0003934-GTP cyclohydrolase I activity; GO:0005737-cytoplasm; GO:0019438-aromatic compound biosynthetic process;	Down	5.32E-05	0.35844881	Yes	−1.480160992

Table S2. *Cont.*

Protein Id	BrasWt	BrasFtf1	PFAM annotation	GO annotation	BrasWt -> BrasFtf1: Up/Down	BrasWt -> BrasFtf1: q value	ratio (m/w)	BrasWt -> BrasFtf1: >2-fold and significant difference?	log2 ratio
1412	22.3942	8.623	PF00400:G-protein beta WD-40 repeat; PF04047: Periodic tryptophan protein-associated region;		Down	0.000185484	0.385055059	Yes	−1.376863344
1432	11.3547	5.55583			Down	0.0452514	0.489297824	Yes	−1.031215229
1445	21.6221	8.39203		GO:0003700-transcription factor activity; GO:0005622-intracellular; GO:0006355-regulation of transcription, DNA-dependent; GO:0043565-sequence-specific DNA binding;	Down	0.00533036	0.3881228	Yes	−1.365414908
1478	2.72459	0.720484			Down	0.0114736	0.264437585	Yes	−1.91900085
1479	3.961	1.06562			Down	0.00425362	0.269028023	Yes	−1.894171636
1487	3.71935	1.32231			Down	0.029531	0.355521798	Yes	−1.491990076
1488	87.5525	32.532	PF02219:Methylenetetrahydrofolate reductase;	GO:0004489-methylenetetrahydrofolate reductase (NADPH) activity; GO:0006555-methionine metabolic process;	Down	5.24E-05	0.371571343	Yes	−1.428288855
1507	56.1208	25.2701	PF06858:Nucleolar GTP-binding 1;	GO:0005525-GTP binding;	Down	0.00215106	0.450280466	Yes	−1.151104201
1545	199.772	99.0028			Down	0.00902654	0.49557896	Yes	−1.012813156
1562	14.5438	5.64588			Down	0.00327992	0.388198408	Yes	−1.365133895
1563	11.7696	4.47281			Down	0.00304256	0.380030757	Yes	−1.395811909

Table S2. *Cont.*

Protein Id	BrasWt	BrasFtf1	PFAM annotation	GO annotation	BrasWt -> BrasFtf1: Up/Down	BrasWt -> BrasFtf1: q value	ratio (m/w)	BrasWt -> BrasFtf1: >2-fold and significant difference?	log2 ratio
1565	147.009	62.1498	PF00109: Beta-ketoacyl synthase; PF02801: Beta-ketoacyl synthase; PF01648:4'-phosphopantetheinyl transferase;	GO:0009058-biosynthetic process; GO:0009059-macromolecule biosynthetic process; GO:0003824-catalytic activity; GO:0004315-3-oxoacyl-[acyl-carrier-protein] synthase activity; GO:0008897-phosphopantetheinyltransferase activity; GO:0000287-magnesium ion binding; GO:0006633-fatty acid biosynthetic process;	Down	0.0013016	0.422761872	Yes	−1.242082827
1566	88.4245	33.7212			Down	0.000443081	0.381355846	Yes	−1.390790279
1567	102.253	43.7215	PF00698:Acyl transferase region; PF01575:MaoC-like dehydratase;	GO:0005835-fatty acid synthase complex; GO:0016491-oxidoreductase activity; GO:0006633-fatty acid biosynthetic process; GO:0008152-metabolic process; GO:0016740-transferase activity; GO:0004312-fatty-acid synthase activity;	Down	0.00445441	0.427581587	Yes	−1.225728368
1569	20.461	9.75646	PF00067:Cytochrome P450;	GO:0050381-unspecific monooxygenase activity; GO:0005506-iron ion binding; GO:0006118-electron transport; GO:0020037-heme binding; GO:0004497-monooxygenase activity; GO:0016712-oxidoreductase activity, acting on paired donors, with incorporation or reduction of molecular oxygen, reduced flavin or flavoprotein as one donor, and incorporation of one atom of oxygen;	Down	0.0090603	0.476832022	Yes	−1.068446971

Table S2. *Cont.*

Protein Id	BrasWt	BrasFtf1	PFAM annotation	GO annotation	BrasWt -> BrasFtf1: Up/Down	BrasWt -> BrasFtf1: q value	ratio (m/w)	BrasWt -> BrasFtf1: >2-fold and significant difference?	log2 ratio
1701	11.3378	3.35379			Down	0.000293325	0.295806065	Yes	−1.757276464
1716	61.6196	25.5924			Down	0.00262415	0.415328889	Yes	−1.267673871
1747	22.7679	3.42208			Down	6.80E-14	0.150302839	Yes	−2.734055838
1758	7.01919	2.7575	PF00550:Phosphopantetheine-binding; PF00668:Condensation domain; PF00501:AMP-dependent synthetase and ligase;	GO:0031177-phosphopantetheine binding; GO:0048037-cofactor binding; GO:0003824-catalytic activity; GO:0008152-metabolic process; GO:0016874-ligase activity;	Down	0.00051205	0.392851597	Yes	−1.34794367
1765	14.1196	3.34068			Down	3.11E-08	0.236598771	Yes	−2.079485518
1812	726.868	127.558	PF03297:S25 ribosomal protein;		Down	3.68E-06	0.17548991	Yes	−2.51054001
1834	13.0587	5.46273	PF00860:Xanthine/uracil/vitamin C permease;	GO:0006810-transport; GO:0016020-membrane; GO:0005215-transporter activity; GO:0005515-protein binding;	Down	0.00202612	0.418321119	Yes	−1.25731726
1870	222.66	70.3307	PF00464:Glycine hydroxymethyltransferase;	GO:0006544-glycine metabolic process; GO:0004372-glycine hydroxymethyltransferase activity; GO:0006563-L-serine metabolic process;	Down	1.05E-06	0.315865894	Yes	−1.662615925
1873	82.8456	38.5089	PF00538:Histone H1/H5;	GO:0006334-nucleosome assembly; GO:0003677-DNA binding; GO:0000786-nucleosome; GO:0005634-nucleus;	Down	0.00445441	0.464827341	Yes	−1.105233163
1885	6.0688	2.6964			Down	0.0115191	0.444305299	Yes	−1.170376746

Table S2. *Cont.*

Protein Id	BrasWt	BrasFtf1	PFAM annotation	GO annotation	BrasWt -> BrasFtf1: Up/Down	BrasWt -> BrasFtf1: q value	ratio (m/w)	BrasWt -> BrasFtf1: >2-fold and significant difference?	log2 ratio
1897	201.636	86.3375	PF00271:Helicase, C-terminal;	GO:0003676-nucleic acid binding; GO:0004386-helicase activity; GO:0005524-ATP binding;	Down	0.00167615	0.428184947	Yes	−1.223694016
1898	177.212	72.8808			Down	0.000476473	0.411263346	Yes	−1.281865599
1916	4.1241	1.87464	PF00109: Beta-ketoacyl synthase; PF00107: Zinc-containing alcohol dehydrogenase superfamily; PF00698:Acyl transferase region; PF02801: Beta-ketoacyl synthase;	GO:0016491-oxidoreductase activity; GO:0031177-phosphopantetheine binding; GO:0003824-catalytic activity; GO:0008152-metabolic process; GO:0008270-zinc ion binding; GO:0009058-biosynthetic process; GO:0016740-transferase activity;	Down	0.00834118	0.454557358	Yes	−1.137465744
1928	16.7167	5.81717	PF00400:G-protein beta WD-40 repeat;		Down	7.03E-05	0.347985547	Yes	−1.522900706
1929	29.1667	10.759			Down	0.000339314	0.368879578	Yes	−1.438778173
1955	12.7049	3.86737	PF00400:G-protein beta WD-40 repeat; PF04192: Utp21 specific WD40-associated;	GO:0032040-small subunit processome; GO:0006364-rRNA processing;	Down	6.26E-06	0.304399877	Yes	−1.715960318
2026	15.364	7.5687		GO:0004221-ubiquitin thiolesterase activity;	Down	0.0143308	0.492625618	Yes	−1.02143644
2054	139.192	69.2598			Down	0.0148598	0.497584631	Yes	−1.006986169
2078	10.8902	4.78574			Down	0.0132218	0.439453821	Yes	−1.186216524

Table S2. *Cont.*

Protein Id	BrasWt	BrasFtf1	PFAM annotation	GO annotation	BrasWt -> BrasFtf1: Up/Down	BrasWt -> BrasFtf1: q value	ratio (m/w)	BrasWt -> BrasFtf1: >2-fold and significant difference?	log2 ratio
2081	12.3803	3.51247	PF03813: Nrap protein;		Down	4.30E-06	0.28371445	Yes	−1.817488466
2106	8.54127	3.6283			Down	0.0295918	0.424796312	Yes	−1.235156852
2119	10.1795	4.602	PF00271:Helicase, C-terminal; PF00270:DEAD/ DEAH box helicase, N-terminal; PF04408: Helicase-associated region;	GO:0042625-ATPase activity, coupled to transmembrane movement of ions; GO:0042626-ATPase activity, coupled to transmembrane movement of substances; GO:0003676-nucleic acid binding; GO:0004004-ATP-dependent RNA helicase activity; GO:0017116-single-stranded DNA-dependent ATP-dependent DNA helicase activity; GO:0008026-ATP-dependent helicase activity; GO:0015462-protein-transmembrane transporting ATPase activity; GO:0008094-DNA-dependent ATPase activity; GO:0004386-helicase activity; GO:0042624-ATPase activity, uncoupled; GO:0008186-RNA-dependent ATPase activity; GO:0005524-ATP binding; GO:0016887-ATPase activity; GO:0042623-ATPase activity, coupled; GO:0004003-ATP-dependent DNA helicase activity;	Down	0.00666261	0.452085073	Yes	−1.145333812
2126	0.957756	0.320497		GO:0004308-exo-alpha-sialidase activity;	Down	0.0495169	0.334633247	Yes	−1.579347306

Table S2. *Cont.*

Protein Id	BrasWt	BrasFtf1	PFAM annotation	GO annotation	BrasWt -> BrasFtf1: Up/Down	BrasWt -> BrasFtf1: q value	ratio (m/w)	BrasWt -> BrasFtf1: >2-fold and significant difference?	log2 ratio
2163	27.2075	10.5643	PF05148:Protein of unknown function DUF691;	GO:0003910-DNA ligase (ATP) activity;	Down	0.000270747	0.388286318	Yes	−1.364807222
2236	35.0946	13.5078		GO:0004308-exo-alpha-sialidase activity;	Down	0.000305992	0.384896822	Yes	−1.377456337
2256	21.3561	2.49	PF00734:Cellulose-binding region, fungal; PF03443:Glycoside hydrolase, family 61;	GO:0005576-extracellular region; GO:0004553-hydrolase activity, hydrolyzing O-glycosyl compounds; GO:0005975-carbohydrate metabolic process; GO:0030248-cellulose binding;	Down	1.82E-12	0.116594322	Yes	−3.100430562
2269	43.7146	17.4254			Down	0.000428358	0.398617396	Yes	−1.326923424
2276	4.89125	1.9997	PF00172:Fungal transcriptional regulatory protein, N-terminal; PF00096:Zn-finger, C2H2 type;	GO:0005488-binding; GO:0003676-nucleic acid binding; GO:0008270-zinc ion binding; GO:0003700-transcription factor activity; GO:0005622-intracellular; GO:0006355-regulation of transcription, DNA-dependent; GO:0005634-nucleus;	Down	0.00303284	0.408832098	Yes	−1.290419626
2334	29.5414	8.08793			Down	1.98E-07	0.273782895	Yes	−1.868895783
2347	24.1425	6.90859	PF01565:FAD linked oxidase, N-terminal;	GO:0050660-FAD binding; GO:0016491-oxidoreductase activity;	Down	7.60E-07	0.286158849	Yes	−1.805111877
2348	33.0254	13.8026			Down	0.00088359	0.41793892	Yes	−1.258635982
2414	39.3566	18.3225			Down	0.0263887	0.465550886	Yes	−1.102989228
2416	2.17794	0.80443			Down	0.0255276	0.369353609	Yes	−1.436925419
2424	1.61605	0.563241	PF02156:Glycoside hydrolase, family 26;	GO:0016985-mannan endo-1,4-beta-mannosidase activity; GO:0006080-substituted mannan metabolic process;	Down	0.0421647	0.348529439	Yes	−1.520647574
2510	8.52871	4.20726			Down	0.0395414	0.493305553	Yes	−1.019446568

Table S2. *Cont.*

Protein Id	BrasWt	BrasFtf1	PFAM annotation	GO annotation	BrasWt -> BrasFtf1: Up/Down	BrasWt -> BrasFtf1: q value	ratio (m/w)	BrasWt -> BrasFtf1: >2-fold and significant difference?	log2 ratio
2512	29.0406	10.6621	PF02347:Glycine cleavage system P-protein;	GO:0006544-glycine metabolic process; GO:0004374-glycine cleavage system; GO:0004375-glycine dehydrogenase (decarboxylating) activity; GO:0016887-ATPase activity; GO:0005524-ATP binding;	Down	5.58E-05	0.367144618	Yes	−1.445579642
2518	5.79374	2.67274			Down	0.0195326	0.461315144	Yes	−1.116175442
2519	74.7788	33.1491	PF03810:Importin-beta, N-terminal;	GO:0000059-protein import into nucleus, docking; GO:0006886-intracellular protein transport; GO:0008565-protein transporter activity; GO:0005737-cytoplasm; GO:0005643-nuclear pore; GO:0005634-nucleus;	Down	0.00220845	0.443295426	Yes	−1.173659618
2585	10.5497	4.78836	PF02005:N2,N2-dimethylguanosine tRNA methyltransferase;	GO:0004809-tRNA (guanine-N2-)-methyltransferase activity; GO:0003723-RNA binding; GO:0008033-tRNA processing;	Down	0.00637194	0.453885892	Yes	−1.139598447
2591	23.066	7.61615	PF04427:Brix;		Down	3.43E-05	0.330189456	Yes	−1.598634042
2593	24.7182	11.0185	PF03006:Hly-III related proteins;	GO:0016021-integral to membrane;	Down	0.00526969	0.445764659	Yes	−1.165645852
2603	15.2617	5.02896			Down	0.000160304	0.329515061	Yes	−1.601583689
2614	32.5452	15.3208		GO:0009573-chloroplast ribulose bisphosphate carboxylase complex; GO:0016984-ribulose-bisphosphate carboxylase activity; GO:0015977-carbon utilization by fixation of carbon dioxide;	Down	0.00917176	0.47075452	Yes	−1.086953148

Table S2. *Cont.*

Protein Id	BrasWt	BrasFtf1	PFAM annotation	GO annotation	BrasWt -> BrasFtf1: Up/Down	BrasWt -> BrasFtf1: q value	ratio (m/w)	BrasWt -> BrasFtf1: >2-fold and significant difference?	log2 ratio
2657	11.8842	3.56832			Down	0.000159851	0.300257485	Yes	−1.735727885
2662	199.176	93.9128		GO:0016491-oxidoreductase activity;	Down	0.0295845	0.471506607	Yes	−1.084650107
2663	100.723	36.9942		GO:0016787-hydrolase activity; GO:0008152-metabolic process;	Down	4.31E-05	0.367286518	Yes	−1.445022153
2674	29.9964	12.9112			Down	0.00163483	0.430424984	Yes	−1.216166272
2710	94.2391	38.7869	PF02854:Initiation factor eIF-4 gamma, middle;	GO:0016070-RNA metabolic process; GO:0005515-protein binding;	Down	0.000686188	0.411579695	Yes	−1.280756287
2711	66.8611	24.5787		GO:0004402-histone acetyltransferase activity;	Down	0.000449868	0.36760837	Yes	−1.443758479
2724	23.6306	2.38787	PF05199:GMC oxidoreductase; PF00732:Glucose-methanol-choline oxidoreductase;	GO:0008812-choline dehydrogenase activity; GO:0016614-oxidoreductase activity, acting on CH-OH group of donors; GO:0050660-FAD binding;	Down	0	0.10104991	Yes	−3.30686006
2725	2.92189	1.14081		GO:0005634-nucleus;	Down	0.00684574	0.390435643	Yes	−1.356843333
2751	7.28902	2.4488		GO:0004812-aminoacyl-tRNA ligase activity; GO:0005737-cytoplasm; GO:0006412-translation; GO:0000166-nucleotide binding; GO:0006418-tRNA aminoacylation for protein translation; GO:0005524-ATP binding;	Down	0.00255689	0.335957371	Yes	−1.573649909
2763	2.74451	0.885874			Down	0.00434419	0.322780387	Yes	−1.631375175
2767	17.479	8.51087			Down	0.0287946	0.486919732	Yes	−1.038244128
2782	4.73678	1.86054	PF00324:Amino acid permease-associated region;	GO:0006810-transport; GO:0016020-membrane; GO:0015171-amino acid transmembrane transporter activity; GO:0006865-amino acid transport;	Down	0.0060565	0.392785817	Yes	−1.34818526

Table S2. *Cont.*

Protein Id	BrasWt	BrasFtf1	PFAM annotation	GO annotation	BrasWt -> BrasFtf1: Up/Down	BrasWt -> BrasFtf1: q value	ratio (m/w)	BrasWt -> BrasFtf1: >2-fold and significant difference?	log2 ratio
2816	923.65	369.788	PF00012:Heat shock protein Hsp70;	GO:0005524-ATP binding;	Down	0.00342923	0.400355113	Yes	−1.320647864
2821	16.3016	7.67876			Down	0.0103715	0.471043333	Yes	−1.08606831
2897	192.495	70.5365		GO:0004338-glucan 1,3-beta-glucosidase activity;	Down	3.30E-05	0.366432894	Yes	−1.448379076
2904	8.62627	4.03976	PF00271:Helicase, C-terminal; PF00270:DEAD/ DEAH box helicase, N-terminal; PF04408: Helicase-associated region;	GO:0042625-ATPase activity, coupled to transmembrane movement of ions; GO:0042626-ATPase activity, coupled to transmembrane movement of substances; GO:0003676-nucleic acid binding; GO:0004004-ATP-dependent RNA helicase activity; GO:0017116-single-stranded DNA-dependent ATP-dependent DNA helicase activity; GO:0008026-ATP-dependent helicase activity; GO:0015462-protein-transmembrane transporting ATPase activity; GO:0008094-DNA-dependent ATPase activity; GO:0004386-helicase activity; GO:0042624-ATPase activity, uncoupled; GO:0008186-RNA-dependent ATPase activity; GO:0005524-ATP binding; GO:0016887-ATPase activity; GO:0042623-ATPase activity, coupled; GO:0004003-ATP-dependent DNA helicase activity;	Down	0.0106179	0.468309014	Yes	−1.094467287
2905	41.8749	16.5432	PF00400:G-protein beta WD-40 repeat;		Down	0.000305957	0.395062436	Yes	−1.339847419

Table S2. *Cont.*

Protein Id	BrasWt	BrasFtf1	PFAM annotation	GO annotation	BrasWt -> BrasFtf1: Up/Down	BrasWt -> BrasFtf1: q value	ratio (m/w)	BrasWt -> BrasFtf1: >2-fold and significant difference?	log2 ratio
2908	27.4663	10.6305	PF00400:G-protein beta WD-40 repeat; PF04003:Dip2/Utp12;	GO:0032040-small subunit processome; GO:0006364-rRNA processing;	Down	0.000187567	0.387037934	Yes	−1.369453122
2929	11.5301	5.55514	PF01426:Bromo adjacent region;	GO:0003677-DNA binding; GO:0004308-exo-alpha-sialidase activity;	Down	0.0242479	0.481794607	Yes	−1.05350985
2930	14.5507	7.26178	PF00628:Zn-finger-like, PHD finger;	GO:0008270-zinc ion binding; GO:0005515-protein binding;	Down	0.0354755	0.499067399	Yes	−1.002693431
2965	29.5547	13.3383	PF02889:Sec63; PF00271:Helicase, C-terminal; PF00270:DEAD/DEAH box helicase, N-terminal; PF04851:Type III restriction enzyme, res subunit;	GO:0008796-bis(5'-nucleosyl)-tetraphosphatase activity; GO:0008413-8-oxo-7,8-dihydroguanine triphosphatase activity; GO:0019177-dihydroneopterin triphosphate pyrophosphohydrolase activity; GO:0008758-UDP-2,3-diacylglucosamine hydrolase activity; GO:0004787-thiamin-pyrophosphatase activity; GO:0003676-nucleic acid binding; GO:0000810-diacylglycerol pyrophosphate phosphatase activity; GO:0008026-ATP-dependent helicase activity; GO:0016787-hydrolase activity; GO:0004386-helicase activity; GO:0043141-ATP-dependent 5'-3' DNA helicase activity; GO:0008828-dATP pyrophosphohydrolase activity; GO:0016462-pyrophosphatase activity; GO:0003677-DNA binding; GO:0019176-dihydroneopterin monophosphate phosphatase activity; GO:0005524-ATP binding;	Down	0.0035409	0.451308929	Yes	−1.147812773

Table S2. *Cont.*

Protein Id	BrasWt	BrasFtf1	PFAM annotation	GO annotation	BrasWt -> BrasFtf1: Up/Down	BrasWt -> BrasFtf1: q value	ratio (m/w)	BrasWt -> BrasFtf1: >2-fold and significant difference?	log2 ratio
2975	27.1636	9.1691	PF05127:Putative ATPase DUF699;		Down	1.67E-05	0.337550987	Yes	−1.566822656
2980	14.8916	6.68191	PF01137:RNA 3'-terminal phosphate cyclase;	GO:0003963-RNA-3'-phosphate cyclase activity;	Down	0.00946146	0.448703296	Yes	−1.156166314
2984	1.09563	0.216286	PF06985:Heterokaryon incompatibility;		Down	0.00338534	0.197407884	Yes	−2.340748486
3033	18.4761	7.1581			Down	0.000732658	0.387424835	Yes	−1.368011656
3046	94.4459	17.6063		GO:0016021-integral to membrane; GO:0016020-membrane; GO:0005215-transporter activity; GO:0008643-carbohydrate transport; GO:0006810-transport; GO:0005351-sugar:hydrogen symporter activity;	Down	1.34E-11	0.186416774	Yes	−2.423396413
3064	2.24326	0.612028		GO:0008413-8-oxo-7,8-dihydroguanine triphosphatase activity; GO:0008796-bis(5'-nucleosyl)-tetraphosphatase activity; GO:0004787-thiamin-pyrophosphatase activity; GO:0000810-diacylglycerol pyrophosphate phosphatase activity; GO:0008758-UDP-2,3-diacylglucosamine hydrolase activity; GO:0008828-dATP pyrophosphohydrolase activity; GO:0043141-ATP-dependent 5'-3' DNA helicase activity; GO:0019177-dihydroneopterin triphosphate pyrophosphohydrolase activity; GO:0016462-pyrophosphatase activity; GO:0019176-dihydroneopterin monophosphate phosphatase activity;	Down	0.00152547	0.272829721	Yes	−1.87392728

Table S2. *Cont.*

Protein Id	BrasWt	BrasFtf1	PFAM annotation	GO annotation	BrasWt -> BrasFtf1: Up/Down	BrasWt -> BrasFtf1: q value	ratio (m/w)	BrasWt -> BrasFtf1: >2-fold and significant difference?	log2 ratio
3077	7.97032	2.47885	PF00734: Cellulose-binding region, fungal; PF00331:Glycoside hydrolase, family 10;	GO:0005576-extracellular region; GO:0004553-hydrolase activity, hydrolyzing O-glycosyl compounds; GO:0005975-carbohydrate metabolic process; GO:0030248-cellulose binding;	Down	0.000201792	0.311010097	Yes	−1.684966674
3078	32.2597	15.7379	PF00171:Aldehyde dehydrogenase;	GO:0004029-aldehyde dehydrogenase (NAD) activity;	Down	0.00938551	0.487850166	Yes	−1.035489976
3079	19.2849	4.82905			Down	5.14E-05	0.250405758	Yes	−1.997660359
3085	10.5249	4.58964			Down	0.0141545	0.436074452	Yes	−1.197353624
3147	40.8761	16.4585	PF00076:RNA-binding region RNP-1 (RNA recognition motif);	GO:0003676-nucleic acid binding;	Down	0.000346484	0.402643599	Yes	−1.312424698
3157	22.815	8.64277			Down	0.000154586	0.378819636	Yes	−1.40041698
3162	6.95156	2.85139			Down	0.00240535	0.410179873	Yes	−1.285671391
3163	6.45426	1.50558	PF00445:Ribonuclease T2;	GO:0004521-endoribonuclease activity; GO:0003723-RNA binding; GO:0033897-ribonuclease T2 activity;	Down	8.31E-05	0.23326919	Yes	−2.099932327
3191	47.5516	19.7998			Down	0.00248643	0.416385569	Yes	−1.264008028
3192	48.5436	20.1459	PF01794:Ferric reductase-like transmembrane component;	GO:0016020-membrane; GO:0006118-electron transport; GO:0016491-oxidoreductase activity;	Down	0.000834409	0.415006304	Yes	−1.268794845

Table S2. *Cont.*

Protein Id	BrasWt	BrasFtf1	PFAM annotation	GO annotation	BrasWt -> BrasFtf1: Up/Down	BrasWt -> BrasFtf1: q value	ratio (m/w)	BrasWt -> BrasFtf1: >2-fold and significant difference?	log2 ratio
3207	14.6333	6.86472	PF00271:Helicase, C-terminal; PF00270:DEAD/DEAH box helicase, N-terminal;	GO:0008796-bis(5'-nucleosyl)-tetraphosphatase activity; GO:0008413-8-oxo-7,8-dihydroguanine triphosphatase activity; GO:0019177-dihydroneopterin triphosphate pyrophosphohydrolase activity; GO:0008758-UDP-2,3-diacylglucosamine hydrolase activity; GO:0004787-thiamin-pyrophosphatase activity; GO:0003676-nucleic acid binding; GO:0000810-diacylglycerol pyrophosphate phosphatase activity; GO:0008026-ATP-dependent helicase activity; GO:0004386-helicase activity; GO:0043141-ATP-dependent 5'-3' DNA helicase activity; GO:0008828-dATP pyrophosphohydrolase activity; GO:0016462-pyrophosphatase activity; GO:0019176-dihydroneopterin monophosphate phosphatase activity; GO:0005524-ATP binding;	Down	0.00673323	0.469116331	Yes	−1.091982371
3210	231.921	86.8577	PF00535:Glycosyl transferase, family 2;		Down	5.32E-05	0.374514166	Yes	−1.416907803
3257	691.602	94.696			Down	0	0.136922681	Yes	−2.868566649
3263	14.7293	5.17255	PF00646:Cyclin-like F-box;		Down	0.000202447	0.35117419	Yes	−1.509741278
3300	25.9228	11.6465	PF00096:Zn-finger, C2H2 type;	GO:0008270-zinc ion binding; GO:0003676-nucleic acid binding; GO:0005622-intracellular;	Down	0.00551436	0.449276313	Yes	−1.154325094

Table S2. *Cont.*

Protein Id	BrasWt	BrasFtf1	PFAM annotation	GO annotation	BrasWt -> BrasFtf1: Up/Down	BrasWt -> BrasFtf1: q value	ratio (m/w)	BrasWt -> BrasFtf1: >2-fold and significant difference?	log2 ratio
3306	84.3112	33.6934	PF02786: Carbamoyl-phosphate synthase L chain, ATP-binding; PF00289: Carbamoyl-phosphate synthetase large chain, N-terminal; PF02787: Carbamoyl-phosphate synthetase large chain, oligomerisation;	GO:0004086-carbamoyl-phosphate synthase activity; GO:0006807-nitrogen compound metabolic process; GO:0003824-catalytic activity; GO:0008152-metabolic process; GO:0004088-carbamoyl-phosphate synthase (glutamine-hydrolyzing) activity; GO:0005524-ATP binding;	Down	0.000305992	0.399631366	Yes	−1.323258275
3310	5.63091	2.66349			Down	0.0307237	0.473012355	Yes	−1.080050228
3322	44.3231	20.629	PF01000:RNA polymerase, insert; PF01193:RNA polymerase, dimerisation;	GO:0003899-DNA-directed RNA polymerase activity; GO:0003902-DNA-directed RNA polymerase III activity; GO:0003900-DNA-directed RNA polymerase I activity; GO:0006350-transcription; GO:0003901-DNA-directed RNA polymerase II activity; GO:0034062-RNA polymerase activity; GO:0046983-protein dimerization activity; GO:0003677-DNA binding; GO:0005634-nucleus;	Down	0.00492878	0.465423222	Yes	−1.103384901
3348	15.7889	4.67184	PF04147: Nop14-like protein;		Down	3.34E-06	0.295893951	Yes	−1.756847892

Table S2. *Cont.*

Protein Id	BrasWt	BrasFtf1	PFAM annotation	GO annotation	BrasWt -> BrasFtf1: Up/Down	BrasWt -> BrasFtf1: q value	ratio (m/w)	BrasWt -> BrasFtf1: >2-fold and significant difference?	log2 ratio
3361	6.11171	2.92037	PF00069:Protein kinase;	GO:0004713-protein-tyrosine kinase activity; GO:0005524-ATP binding; GO:0006468-protein amino acid phosphorylation; GO:0004674-protein serine/threonine kinase activity; GO:0004672-protein kinase activity;	Down	0.0210114	0.4778319	Yes	−1.065424924
3375	0.744668	0.0613345			Down	0.031581	0.082364893	Yes	−3.601826655
3390	26.6155	11.7507	PF00076: RNA-binding region RNP-1 (RNA recognition motif);	GO:0003676-nucleic acid binding;	Down	0.00266402	0.441498375	Yes	−1.179519967
3484	75.9193	13.3391			Down	3.12E-05	0.17570104	Yes	−2.50880536
3487	3.90274	0.700624			Down	0.000142401	0.179521054	Yes	−2.47777504
3493	45.5366	20.1167	PF00069:Protein kinase;	GO:0019199-transmembrane receptor protein kinase activity; GO:0004871-signal transducer activity; GO:0034211-GTP-dependent protein kinase activity; GO:0004704-NF-kappaB-inducing kinase activity; GO:0004698-calcium-dependent protein kinase C activity; GO:0004690-cyclic nucleotide-dependent protein kinase activity; GO:0004706-JUN kinase kinase kinase activity; GO:0004708-MAP kinase kinase activity; GO:0004713-protein-tyrosine kinase activity; GO:0004676-3-phosphoinositide-dependent protein kinase activity; GO:0004718-Janus kinase activity; GO:0008339-MP kinase activity; GO:0004672-protein kinase activity;	Down	0.00533036	0.441769917	Yes	−1.178632916

Table S2. *Cont.*

Protein Id	BrasWt	BrasFtf1	PFAM annotation	GO annotation	BrasWt -> BrasFtf1: Up/Down	BrasWt -> BrasFtf1: q value	ratio (m/w)	BrasWt -> BrasFtf1: >2-fold and significant difference?	log2 ratio
				GO:0004695-galactosyltransferase-associated kinase activity; GO:0016909-SAP kinase activity; GO:0004702-receptor signaling protein serine/threonine kinase activity; GO:0004677-DNA-dependent protein kinase activity; GO:0005524-ATP binding; GO:0004710-MAP/ERK kinase kinase activity; GO:0004705-JUN kinase activity; GO:0007165-signal transduction; GO:0019912-cyclin-dependent protein kinase activating kinase activity; GO:0004679-AMP-activated protein kinase activity; GO:0016908-MAP kinase 2 activity; GO:0004675-transmembrane receptor protein serine/threonine kinase activity; GO:0008338-MAP kinase 1 activity; GO:0004680-casein kinase activity; GO:0004711-ribosomal protein S6 kinase activity; GO:0004694-eukaryotic translation initiation factor 2alpha kinase activity; GO:0008349-MAP kinase kinase kinase kinase activity; GO:0004696-glycogen synthase kinase 3 activity; GO:0008545-JUN kinase kinase activity; GO:0004674-protein serine/threonine kinase activity; GO:0004681-casein kinase I activity; GO:0004700-atypical protein kinase C activity; GO:0006468-protein amino acid phosphorylation;					

Table S2. *Cont.*

Protein Id	BrasWt	BrasFtf1	PFAM annotation	GO annotation	BrasWt -> BrasFtf1: Up/Down	BrasWt -> BrasFtf1: q value	ratio (m/w)	BrasWt -> BrasFtf1: >2-fold and significant difference?	log2 ratio
3494	75.0011	33.4615	PF00069:Protein kinase;	GO:0005524-ATP binding; GO:0006468-protein amino acid phosphorylation; GO:0004674-protein serine/threonine kinase activity; GO:0004672-protein kinase activity;	Down	0.00923703	0.44614679	Yes	−1.164409636
3513	108.437	49.8118	PF00098:Zn-finger, CCHC type;	GO:0003676-nucleic acid binding; GO:0008270-zinc ion binding;	Down	0.00425362	0.459361657	Yes	−1.122297655
3526	19.728	9.25676			Down	0.0106195	0.469219384	Yes	−1.091665482
3539	75.1775	36.056			Down	0.00519479	0.479611586	Yes	−1.060061585
3635	3.1128	0.207671	PF01083:Cutinase;	GO:0016787-hydrolase activity; GO:0008152-metabolic process;	Down	0.00655351	0.066715176	Yes	−3.905841214
3640	24.1145	10.0968			Down	0.00316795	0.41870244	Yes	−1.256002768
3649	91.7939	43.3166			Down	0.00697378	0.471889744	Yes	−1.083478278
3695	2.6701	0.692873			Down	0.00771438	0.259493277	Yes	−1.946230931
3707	4.08275	1.66585	PF07519:Tannase and feruloyl esterase;		Down	0.0112714	0.408021554	Yes	−1.293282729
3734	9.23879	4.02503	PF04082:Fungal specific transcription factor;	GO:0008270-zinc ion binding; GO:0003677-DNA binding; GO:0006350-transcription; GO:0005634-nucleus;	Down	0.00437538	0.435666359	Yes	−1.198704379
3754	2.64748	0.624369	PF01565:FAD linked oxidase, N-terminal;	GO:0050660-FAD binding; GO:0016491-oxidoreductase activity;	Down	0.000923626	0.235835209	Yes	−2.084148971
3768	10.4949	1.73086	PF00544:Pectate lyase/Amb allergen;		Down	1.22E-08	0.164923915	Yes	−2.600127477

Table S2. *Cont.*

Protein Id	BrasWt	BrasFtf1	PFAM annotation	GO annotation	BrasWt -> BrasFtf1: Up/Down	BrasWt -> BrasFtf1: q value	ratio (m/w)	BrasWt -> BrasFtf1: >2-fold and significant difference?	log2 ratio
3792	46.0878	14.0301	PF00076:RNA-binding region RNP-1 (RNA recognition motif);	GO:0003906-DNA-(apurinic or apyrimidinic site) lyase activity; GO:0003676-nucleic acid binding;	Down	1.99E-06	0.304421127	Yes	−1.71585961
3793	4.59847	1.10333			Down	0.00651756	0.239934152	Yes	−2.059289571
3814	72.9746	34.8273	PF00330:Aconitate hydratase, N-terminal; PF00694:Aconitate hydratase, C-terminal;	GO:0006099-tricarboxylic acid cycle; GO:0003994-aconitate hydratase activity; GO:0008152-metabolic process; GO:0051539-4 iron, 4 sulfur cluster binding;	Down	0.00519479	0.477252359	Yes	−1.067175766
3860	129.956	47.4458	PF01063:Aminotransferase, class IV;	GO:0004084-branched-chain-amino-acid transaminase activity; GO:0003824-catalytic activity; GO:0008152-metabolic process;	Down	4.69E-05	0.365091262	Yes	−1.453670956
3889	18.9782	8.16438	PF03456:uDENN; PF03455:dDENN; PF02141:DENN;	GO:0007242-intracellular signaling cascade;	Down	0.00317429	0.430197806	Yes	−1.216927928
3898	4.51221	2.23799	PF04082:Fungal specific transcription factor;	GO:0008270-zinc ion binding; GO:0003677-DNA binding; GO:0006350-transcription; GO:0005634-nucleus;	Down	0.0428247	0.495985338	Yes	−1.011630623
3913	61.2634	20.0497			Down	1.15E-05	0.327270442	Yes	−1.611444786
3925	10.2702	3.19301			Down	1.04E-05	0.310900469	Yes	−1.6854753
3985	9.87121	3.30009	PF01565:FAD linked oxidase, N-terminal;	GO:0050660-FAD binding; GO:0016491-oxidoreductase activity;	Down	0.000160304	0.334314638	Yes	−1.580721569
3986	2.69192	0.29349			Down	0.000982372	0.109026271	Yes	−3.197252283
4007	2.98598	0.94571			Down	0.00445441	0.31671679	Yes	−1.658734745

Table S2. *Cont.*

Protein Id	BrasWt	BrasFtf1	PFAM annotation	GO annotation	BrasWt -> BrasFtf1: Up/Down	BrasWt -> BrasFtf1: q value	ratio (m/w)	BrasWt -> BrasFtf1: >2-fold and significant difference?	log2 ratio
4054	5.8393	0.7127	PF00734:Cellulose-binding region, fungal;	GO:0005576-extracellular region; GO:0004553-hydrolase activity, hydrolyzing O-glycosyl compounds; GO:0005975-carbohydrate metabolic process; GO:0030248-cellulose binding;	Down	8.97E-09	0.122052301	Yes	−3.034428603
4061	153.437	70.0673			Down	0.00323022	0.456651916	Yes	−1.130833209
4078	14.595	6.36199			Down	0.00723109	0.435902021	Yes	−1.197924202
4114	408.313	92.9466	PF00450:Peptidase S10, serine carboxypeptidase;	GO:0006508-proteolysis; GO:0004185-serine carboxypeptidase activity; GO:0004187-carboxypeptidase D activity;	Down	2.32E-10	0.227635662	Yes	−2.135201504
4115	195.103	63.7925			Down	8.42E-06	0.326968319	Yes	−1.612777238
4116	107.742	31.6804			Down	5.65E-07	0.294039465	Yes	−1.765918295
4120	53.0888	25.2066	PF00295:Glycoside hydrolase, family 28;	GO:0005975-carbohydrate metabolic process; GO:0004650-polygalacturonase activity;	Down	0.0117959	0.474800711	Yes	−1.074605999
4147	25.6189	10.5345	PF00076: RNA-binding region RNP-1 (RNA recognition motif);	GO:0003676-nucleic acid binding;	Down	0.00115771	0.411200325	Yes	−1.282086691
4161	57.3198	27.0492	PF00400:G-protein beta WD-40 repeat;		Down	0.00476705	0.471899762	Yes	−1.08344765

Table S2. *Cont.*

Protein Id	BrasWt	BrasFtf1	PFAM annotation	GO annotation	BrasWt -> BrasFtf1: Up/Down	BrasWt -> BrasFtf1: q value	ratio (m/w)	BrasWt -> BrasFtf1: >2-fold and significant difference?	log2 ratio
4171	4.69695	1.97045	PF00067: Cytochrome P450;	GO:0018588-tri-n-butyltin dioxygenase activity; GO:0018589-di-n-butyltin dioxygenase activity; GO:0018592-4-nitrocatechol 4-monooxygenase activity; GO:0018598-hydroxymethylsilanetriol oxidase activity; GO:0018597-ammonia monooxygenase activity; GO:0018591-methyl tertiary butyl ether 3-monooxygenase activity; GO:0043823-spheroidene monooxygenase activity; GO:0018585-fluorene oxygenase activity; GO:0005506-iron ion binding; GO:0018586-mono-butyltin dioxygenase activity; GO:0018600-alpha-pinene dehydrogenase activity; GO:0018593-4-chlorophenoxyacetate monooxygenase activity; GO:0018599-2-hydroxyisobutyrate 3-monooxygenase activity; GO:0018594-tert-butyl alcohol 2-monooxygenase activity; GO:0006118-electron transport; GO:0020037-heme binding; GO:0018596-dimethylsilanediol hydroxylase activity; GO:0018587-limonene 8-monooxygenase activity; GO:0004497-monooxygenase activity; GO:0018595-alpha-pinene monooxygenase activity; GO:0018590-methylsilanetriol hydroxylase activity;	Down	0.011906	0.419516921	Yes	−1.253199094

Table S2. *Cont.*

Protein Id	BrasWt	BrasFtf1	PFAM annotation	GO annotation	BrasWt -> BrasFtf1: Up/Down	BrasWt -> BrasFtf1: q value	ratio (m/w)	BrasWt -> BrasFtf1: >2-fold and significant difference?	log2 ratio
4176	1.77621	0.382979	PF00067:Cytochrome P450;	GO:0006118-electron transport; GO:0050381-unspecific monooxygenase activity; GO:0004497-monooxygenase activity; GO:0020037-heme binding; GO:0005506-iron ion binding;	Down	0.00502195	0.215615834	Yes	−2.213464969
4177	5.30782	1.07413			Down	0.000629387	0.20236745	Yes	−2.304950836
4235	3.40701	1.47604	PF00412:Zn-binding protein, LIM;	GO:0008270-zinc ion binding;	Down	0.0152215	0.433236181	Yes	−1.206774364
4301	2.55822	1.21811		GO:0004339-glucan 1,4-alpha-glucosidase activity;	Down	0.0324939	0.476155295	Yes	−1.070495917
4302	9.49555	1.52479	PF05199:GMC oxidoreductase; PF00732:Glucose-methanol-choline oxidoreductase;	GO:0008812-choline dehydrogenase activity; GO:0016614-oxidoreductase activity, acting on CH-OH group of donors; GO:0050660-FAD binding;	Down	1.55E-10	0.160579429	Yes	−2.638641004
4330	5.34643	2.09529			Down	0.00902654	0.391904505	Yes	−1.35142594
4345	1135.84	447.481		GO:0003900-DNA-directed RNA polymerase I activity; GO:0003902-DNA-directed RNA polymerase III activity; GO:0003899-DNA-directed RNA polymerase activity; GO:0034062-RNA polymerase activity; GO:0003901-DNA-directed RNA polymerase II activity;	Down	0.000136898	0.393964819	Yes	−1.343861292
4347	48.6653	23.6816			Down	0.0285675	0.486621885	Yes	−1.039126892

Table S2. *Cont.*

Protein Id	BrasWt	BrasFtf1	PFAM annotation	GO annotation	BrasWt -> BrasFtf1: Up/Down	BrasWt -> BrasFtf1: q value	ratio (m/w)	BrasWt -> BrasFtf1: >2-fold and significant difference?	log2 ratio
4348	21.2194	9.47195	PF00176: SNF2-related;	GO:0042626-ATPase activity, coupled to transmembrane movement of substances; GO:0016887-ATPase activity; GO:0004004-ATP-dependent RNA helicase activity; GO:0017116-single-stranded DNA-dependent ATP-dependent DNA helicase activity; GO:0008026-ATP-dependent helicase activity; GO:0015462-protein-transmembrane transporting ATPase activity; GO:0008094-DNA-dependent ATPase activity; GO:0004003-ATP-dependent DNA helicase activity; GO:0042624-ATPase activity, uncoupled; GO:0008186-RNA-dependent ATPase activity; GO:0042625-ATPase activity, coupled to transmembrane movement of ions; GO:0003677-DNA binding; GO:0042623-ATPase activity, coupled; GO:0005524-ATP binding;	Down	0.00379578	0.446381613	Yes	−1.163650493
4352	1321.81	654.975	PF00573:Ribosomal protein L4/L1e;	GO:0005840-ribosome; GO:0003735-structural constituent of ribosome; GO:0005622-intracellular; GO:0006412-translation;	Down	0.0317474	0.495513727	Yes	−1.013003069
4375	20.6495	8.60478		GO:0015078-hydrogen ion transmembrane transporter activity; GO:0016469-proton-transporting two-sector ATPase complex;	Down	0.00197752	0.416706458	Yes	−1.262896637

Table S2. *Cont.*

Protein Id	BrasWt	BrasFtf1	PFAM annotation	GO annotation	BrasWt -> BrasFtf1: Up/Down	BrasWt -> BrasFtf1: q value	ratio (m/w)	BrasWt -> BrasFtf1: >2-fold and significant difference?	log2 ratio
4381	9.3171	2.53282	PF00271:Helicase, C-terminal; PF00270:DEAD/ DEAH box helicase, N-terminal;	GO:0008796-bis(5'-nucleosyl)-tetraphosphatase activity; GO:0008413-8-oxo-7,8-dihydroguanine triphosphatase activity; GO:0019177-dihydroneopterin triphosphate pyrophosphohydrolase activity; GO:0008758-UDP-2,3-diacylglucosamine hydrolase activity; GO:0004787-thiamin-pyrophosphatase activity; GO:0003676-nucleic acid binding; GO:0000810-diacylglycerol pyrophosphate phosphatase activity; GO:0008026-ATP-dependent helicase activity; GO:0004386-helicase activity; GO:0043141-ATP-dependent 5'-3' DNA helicase activity; GO:0008828-dATP pyrophosphohydrolase activity; GO:0016462-pyrophosphatase activity; GO:0019176-dihydroneopterin monophosphate phosphatase activity; GO:0005524-ATP binding;	Down	9.16E-06	0.27184639	Yes	−1.879136425
4394	13.6734	6.01382			Down	0.00533036	0.439818918	Yes	−1.185018433
4405	7.64569	2.40128	PF00628:Zn-finger-like, PHD finger;	GO:0008270-zinc ion binding; GO:0005515-protein binding;	Down	1.43E-05	0.314069757	Yes	−1.670843068
4423	40.9489	1.31733	PF00067:Cytochrome P450;	GO:0006118-electron transport; GO:0004497-monooxygenase activity; GO:0020037-heme binding; GO:0005506-iron ion binding;	Down	0	0.032170095	Yes	−4.958135997
4424	11.4684	1.15728			Down	1.13E-10	0.100910328	Yes	−3.308854263

Table S2. *Cont.*

Protein Id	BrasWt	BrasFtf1	PFAM annotation	GO annotation	BrasWt -> BrasFtf1: Up/Down	BrasWt -> BrasFtf1: q value	ratio (m/w)	BrasWt -> BrasFtf1: >2-fold and significant difference?	log2 ratio
4425	25.816	8.37711	PF00575:RNA binding S1;	GO:0004654-polyribonucleotide nucleotidyltransferase activity; GO:0003723-RNA binding; GO:0006396-RNA processing; GO:0004553-hydrolase activity, hydrolyzing O-glycosyl compounds; GO:0005975-carbohydrate metabolic process; GO:0005622-intracellular;	Down	7.28E-06	0.32449295	Yes	−1.62374096
4431	24.0597	9.59974			Down	0.000329242	0.398996662	Yes	−1.325551416
4448	410.441	184.762			Down	0.00230288	0.450154833	Yes	−1.151506784
4484	48.2684	18.7984	PF05890:Eukaryotic rRNA processing;		Down	0.000235028	0.389455627	Yes	−1.360469131
4554	2.98351	0.72986			Down	5.67E-05	0.244631324	Yes	−2.031318951
4555	3.32881	0.640143	PF00096:Zn-finger, C2H2 type;	GO:0008270-zinc ion binding; GO:0003676-nucleic acid binding; GO:0005622-intracellular;	Down	0.00123904	0.192303856	Yes	−2.378540401
4556	0.976099	0.395657	PF00109:Beta-ketoacyl synthase; PF02801:Beta-ketoacyl synthase; PF00698:Acyl transferase region;	GO:0031177-phosphopantetheine binding; GO:0003824-catalytic activity; GO:0008152-metabolic process; GO:0009058-biosynthetic process; GO:0016740-transferase activity;	Down	0.00475167	0.405345155	Yes	−1.302777198
4590	8.38232	3.96633			Down	0.00729351	0.47317807	Yes	−1.079544883
4598	1907.24	485.439	PF00042:Globin; PF00175: Oxidoreductase FAD/ NAD(P)-binding;	GO:0016491-oxidoreductase activity; GO:0005506-iron ion binding; GO:0020037-heme binding; GO:0006118-electron transport;	Down	2.67E-06	0.254524339	Yes	−1.974124475
4606	4.0775	1.29572	PF00106:Short-chain dehydrogenase/reductase SDR;	GO:0016491-oxidoreductase activity; GO:0008152-metabolic process;	Down	0.02173	0.317773145	Yes	−1.653930886

Table S2. *Cont.*

Protein Id	BrasWt	BrasFtf1	PFAM annotation	GO annotation	BrasWt -> BrasFtf1: Up/Down	BrasWt -> BrasFtf1: q value	ratio (m/w)	BrasWt -> BrasFtf1: >2-fold and significant difference?	log2 ratio
4611	113.248	39.4257	PF02458:Transferase; PF00501: AMP-dependent synthetase and ligase; PF05141:Pyoverdine biosynthesis;	GO:0048037-cofactor binding; GO:0003824-catalytic activity; GO:0008152-metabolic process;	Down	2.82E-05	0.348135949	Yes	−1.522277297
4612	35.5649	10.2223			Down	1.15E-05	0.287426648	Yes	−1.79873427
4613	131.316	61.918			Down	0.00465408	0.471519084	Yes	−1.084611933
4690	10.522	4.05161			Down	0.0013282	0.385060825	Yes	−1.37684174
4697	29.1686	14.2669			Down	0.0247983	0.489118436	Yes	−1.031744252
4753	19.9054	8.69285	PF07535:Zn-finger, DBF type;	GO:0003676-nucleic acid binding; GO:0008270-zinc ion binding;	Down	0.00392384	0.436708129	Yes	−1.195258707
4766	50.2404	17.1398	PF05022:SRP40, C-terminal;		Down	2.01E-05	0.341155723	Yes	−1.551497676
4789	63.2392	14.0531			Down	1.55E-10	0.222221344	Yes	−2.169930705
4798	5.57082	2.70653	PF00550:Phosphopantetheine-binding; PF00668:Condensation domain; PF00501:AMP-dependent synthetase and ligase;	GO:0031177-phosphopantetheine binding; GO:0048037-cofactor binding; GO:0003824-catalytic activity; GO:0008152-metabolic process; GO:0016874-ligase activity;	Down	0.0132009	0.485840505	Yes	−1.041445322
4819	180.345	72.9401		GO:0003824-catalytic activity; GO:0008152-metabolic process;	Down	0.00123546	0.404447587	Yes	−1.305975343
4820	234.816	83.3207			Down	0.000339314	0.354833998	Yes	−1.49478385

Table S2. *Cont.*

Protein Id	BrasWt	BrasFtf1	PFAM annotation	GO annotation	BrasWt -> BrasFtf1: Up/Down	BrasWt -> BrasFtf1: q value	ratio (m/w)	BrasWt -> BrasFtf1: >2-fold and significant difference?	log2 ratio
4913	12.6509	5.39806	PF00271:Helicase, C-terminal; PF00270:DEAD/ DEAH box helicase, N-terminal;	GO:0008026-ATP-dependent helicase activity; GO:0003676-nucleic acid binding; GO:0004386-helicase activity; GO:0005524-ATP binding;	Down	0.00322785	0.426693753	Yes	−1.228727106
4932	74.8182	28.5018			Down	0.00898767	0.380947417	Yes	−1.392336224
4942	55.7309	16.9359	PF00658:Polyadenylate-binding protein/ HECT-associated; PF00076:RNA-binding region RNP-1 (RNA recognition motif);	GO:0003676-nucleic acid binding; GO:0003723-RNA binding; GO:0016071-mRNA metabolic process;	Down	2.72E-06	0.303887072	Yes	−1.718392796
4955	8.32898	3.39423			Down	0.014083	0.407520489	Yes	−1.2950555
5037	70.1028	31.8925	PF01189:Bacterial Fmu (Sun)/eukaryotic nucleolar NOL1/Nop2p;		Down	0.00263472	0.454939032	Yes	−1.136254876
5063	24.9516	8.87097			Down	0.000303345	0.3555271	Yes	−1.49196856
5073	64.4941	9.17481	PF00450:Peptidase S10, serine carboxypeptidase;	GO:0006508-proteolysis; GO:0004185-serine carboxypeptidase activity;	Down	0	0.142258129	Yes	−2.813417

Table S2. *Cont.*

Protein Id	BrasWt	BrasFtf1	PFAM annotation	GO annotation	BrasWt -> BrasFtf1: Up/Down	BrasWt -> BrasFtf1: q value	ratio (m/w)	BrasWt -> BrasFtf1: >2-fold and significant difference?	log2 ratio
5087	10.8135	3.98878	PF00271:Helicase, C-terminal; PF00270:DEAD/ DEAH box helicase, N-terminal;	GO:0008796-bis(5'-nucleosyl)-tetraphosphatase activity; GO:0008413-8-oxo-7,8-dihydroguanine triphosphatase activity; GO:0019177-dihydroneopterin triphosphate pyrophosphohydrolase activity; GO:0008758-UDP-2,3-diacylglucosamine hydrolase activity; GO:0004787-thiamin-pyrophosphatase activity; GO:0003676-nucleic acid binding; GO:0000810-diacylglycerol pyrophosphate phosphatase activity; GO:0008026-ATP-dependent helicase activity; GO:0004386-helicase activity; GO:0043141-ATP-dependent 5'-3' DNA helicase activity; GO:0008828-dATP pyrophosphohydrolase activity; GO:0016462-pyrophosphatase activity; GO:0019176-dihydroneopterin monophosphate phosphatase activity; GO:0005524-ATP binding;	Down	0.000208123	0.368870393	Yes	−1.438814096
5095	22.0393	8.99781			Down	0.000557951	0.408262059	Yes	−1.292432595
5106	3.29833	1.50189	PF00109:Beta-ketoacyl synthase; PF00107:Zinc-containing alcohol dehydrogenase superfamily; PF00698:Acyl transferase region; PF02801:Beta-ketoacyl synthase;	GO:0016491-oxidoreductase activity; GO:0003824-catalytic activity; GO:0008152-metabolic process; GO:0008270-zinc ion binding; GO:0009058-biosynthetic process; GO:0016740-transferase activity;	Down	0.00957283	0.455348616	Yes	−1.134956596

Table S2. *Cont.*

Protein Id	BrasWt	BrasFtf1	PFAM annotation	GO annotation	BrasWt -> BrasFtf1: Up/Down	BrasWt -> BrasFtf1: q value	ratio (m/w)	BrasWt -> BrasFtf1: >2-fold and significant difference?	log2 ratio
5135	128.83	50.3352	PF03810: Importin-beta, N-terminal;	GO:0000059-protein import into nucleus, docking; GO:0016303-1-phosphatidylinositol-3-kinase activity; GO:0006886-intracellular protein transport; GO:0008565-protein transporter activity; GO:0005737-cytoplasm; GO:0015073-phosphatidylinositol 3-kinase, class I, regulator activity; GO:0005643-nuclear pore; GO:0015072-phosphatidylinositol 3-kinase, class I, catalyst activity; GO:0005634-nucleus;	Down	0.000136297	0.390710238	Yes	−1.355829034
5151	19.0337	9.17963			Down	0.0109394	0.482283003	Yes	−1.052048128
5160	408.252	194.613	PF01794:Ferric reductase-like transmembrane component;	GO:0000293-ferric-chelate reductase activity;	Down	0.00901481	0.476698216	Yes	−1.068851869
5178	7.03271	2.97065	PF00174:Oxidoreductase, molybdopterin binding; PF03404: Mo-co oxidoreductase dimerisation domain;	GO:0008482-sulfite oxidase activity; GO:0030151-molybdenum ion binding; GO:0006118-electron transport; GO:0016491-oxidoreductase activity;	Down	0.00569847	0.422404734	Yes	−1.24330209
5186	377.195	129.571	PF02752:Arrestin, C-terminal;		Down	1.51E-05	0.343511977	Yes	−1.541567695

Table S2. *Cont.*

Protein Id	BrasWt	BrasFtf1	PFAM annotation	GO annotation	BrasWt -> BrasFtf1: Up/Down	BrasWt -> BrasFtf1: q value	ratio (m/w)	BrasWt -> BrasFtf1: >2-fold and significant difference?	log2 ratio
5208	11.4583	3.09661	PF02386:Cation transporter;	GO:0016021-integral to membrane; GO:0008324-cation transmembrane transporter activity; GO:0006508-proteolysis; GO:0006813-potassium ion transport; GO:0015079-potassium ion transmembrane transporter activity; GO:0008270-zinc ion binding; GO:0004182-carboxypeptidase A activity; GO:0006812-cation transport;	Down	1.29E-06	0.270250386	Yes	−1.887631415
5218	12.4751	5.76896	PF05199:GMC oxidoreductase; PF00732:Glucose-methanol-choline oxidoreductase;	GO:0008812-choline dehydrogenase activity; GO:0016614-oxidoreductase activity, acting on CH-OH group of donors; GO:0050660-FAD binding;	Down	0.00823412	0.462437976	Yes	−1.112668215
5221	53.2161	14.5313			Down	3.22E-05	0.2730621	Yes	−1.87269901
5223	14.4684	5.91168			Down	0.0139465	0.408592519	Yes	−1.291265306
5242	25.6017	7.23914	PF00109: Beta-ketoacyl synthase; PF00107: Zinc-containing alcohol dehydrogenase superfamily; PF00698:Acyl transferase region; PF02801: Beta-ketoacyl synthase;	GO:0016491-oxidoreductase activity; GO:0003824-catalytic activity; GO:0008152-metabolic process; GO:0008270-zinc ion binding; GO:0009058-biosynthetic process; GO:0016740-transferase activity;	Down	1.09E-07	0.282760129	Yes	−1.822349389

Table S2. *Cont.*

Protein Id	BrasWt	BrasFtf1	PFAM annotation	GO annotation	BrasWt -> BrasFtf1: Up/Down	BrasWt -> BrasFtf1: q value	ratio (m/w)	BrasWt -> BrasFtf1: >2-fold and significant difference?	log2 ratio
5250	16.3501	4.89558			Down	9.12E-06	0.299422022	Yes	−1.739747762
5277	44.9966	18.108	PF01207:Dihydrouridine synthase, DuS;	GO:0008033-tRNA processing; GO:0043826-sulfur oxygenase reductase activity; GO:0051341-regulation of oxidoreductase activity; GO:0043738-F420H2 dehydrogenase activity; GO:0050660-FAD binding; GO:0043883-malolactic enzyme activity; GO:0017150-tRNA dihydrouridine synthase activity;	Down	0.00034911	0.402430406	Yes	−1.313188782
5307	10.2525	2.76681			Down	1.32E-06	0.269866862	Yes	−1.889680263
5358	15.9257	4.29994	PF05285:SDA1;		Down	5.67E-07	0.270000063	Yes	−1.888968352
5400	4.54768	2.12374			Down	0.0214231	0.46699416	Yes	−1.098523588
5411	25.1123	11.8929	PF00271:Helicase, C-terminal; PF00270:DEAD/DEAH box helicase, N-terminal;	GO:0042625-ATPase activity, coupled to transmembrane movement of ions; GO:0042626-ATPase activity, coupled to transmembrane movement of substances; GO:0003676-nucleic acid binding; GO:0004004-ATP-dependent RNA helicase activity; GO:0017116-single-stranded DNA-dependent ATP-dependent DNA helicase activity; GO:0008026-ATP-dependent helicase activity; GO:0015462-protein-transmembrane transporting ATPase activity; GO:0008094-DNA-dependent ATPase activity; GO:0004386-helicase activity; GO:0042624-ATPase activity, uncoupled; GO:0008186-RNA-dependent ATPase activity; GO:0005524-ATP binding; GO:0016887-ATPase activity; GO:0042623-ATPase activity, coupled; GO:0004003-ATP-dependent DNA helicase activity;	Down	0.0053462	0.47358864	Yes	−1.07829362

Table S2. *Cont.*

Protein Id	BrasWt	BrasFtf1	PFAM annotation	GO annotation	BrasWt -> BrasFtf1: Up/Down	BrasWt -> BrasFtf1: q value	ratio (m/w)	BrasWt -> BrasFtf1: >2-fold and significant difference?	log2 ratio
5415	77.0606	25.3125	PF00083:General substrate transporter;	GO:0016021-integral to membrane; GO:0016020-membrane; GO:0005215-transporter activity; GO:0008643-carbohydrate transport; GO:0006810-transport; GO:0005351-sugar:hydrogen symporter activity; GO:0008733-L-arabinose isomerase activity;	Down	6.71E-06	0.328475252	Yes	−1.606143416
5417	36.4531	16.8966			Down	0.00605816	0.463516134	Yes	−1.109308537
5458	4.52444	2.00938	PF00172:Fungal transcriptional regulatory protein, N-terminal;	GO:0003700-transcription factor activity; GO:0008270-zinc ion binding; GO:0006355-regulation of transcription, DNA-dependent; GO:0005634-nucleus;	Down	0.0167691	0.444116841	Yes	−1.170988815
5514	108.73	19.7432	PF00544:Pectate lyase/Amb allergen;		Down	1.23E-11	0.181580061	Yes	−2.461322306
5538	11.4077	4.60786	PF06102:Protein of unknown function DUF947;		Down	0.00316741	0.403925419	Yes	−1.307839158
5539	25.1006	8.24175		GO:0016491-oxidoreductase activity;	Down	3.93E-05	0.328348725	Yes	−1.606699243
5548	11.7177	4.11023	PF00083:General substrate transporter;	GO:0016021-integral to membrane; GO:0006810-transport; GO:0005215-transporter activity;	Down	0.000262719	0.350771056	Yes	−1.511398388
5578	24.7445	11.0707	PF00753: Beta-lactamase-like;	GO:0004416-hydroxyacylglutathione hydrolase activity; GO:0016787-hydrolase activity ;	Down	0.00735466	0.447400432	Yes	−1.160361444

Table S2. *Cont.*

Protein Id	BrasWt	BrasFtf1	PFAM annotation	GO annotation	BrasWt -> BrasFtf1: Up/Down	BrasWt -> BrasFtf1: q value	ratio (m/w)	BrasWt -> BrasFtf1: >2-fold and significant difference?	log2 ratio
5595	27.122	10.3506			Down	0.000776638	0.381631148	Yes	−1.389749168
5631	7.19382	2.65789			Down	0.0218045	0.369468516	Yes	−1.436476663
5638	71.9916	35.2904			Down	0.00688876	0.490201635	Yes	−1.0285528
5717	992.233	465.655	PF03144:Elongation factor Tu, domain 2; PF00009:Protein synthesis factor, GTP-binding;	GO:0005525-GTP binding; GO:0008547-protein-synthesizing GTPase activity; GO:0003924-GTPase activity; GO:0006412-translation;	Down	0.0302346	0.469300054	Yes	−1.09141747
5724	61.3306	30.2795	PF05383: RNA-binding protein Lupus Lal; PF00076: RNA-binding region RNP-1 (RNA recognition motif);	GO:0042027-cyclophilin-type peptidyl-prolyl cis-trans isomerase activity; GO:0030051-FK506-sensitive peptidyl-prolyl cis-trans isomerase; GO:0003676-nucleic acid binding; GO:0004600-cyclophilin; GO:0003755-peptidyl-prolyl cis-trans isomerase activity;	Down	0.00923703	0.493709502	Yes	−1.018265682
5759	14.0612	4.97688		GO:0004339-glucan 1,4-alpha-glucosidase activity;	Down	0.000190872	0.353944187	Yes	−1.498406214
5764	3.04684	0.898624	PF00174:Oxidoreductase, molybdopterin binding; PF03404:Mo-co oxidoreductase dimerisation domain;	GO:0008482-sulfite oxidase activity; GO:0030151-molybdenum ion binding; GO:0006118-electron transport; GO:0016491-oxidoreductase activity;	Down	0.00154583	0.294936393	Yes	−1.761524243

Table S2. *Cont.*

Protein Id	BrasWt	BrasFtf1	PFAM annotation	GO annotation	BrasWt -> BrasFtf1: Up/Down	BrasWt -> BrasFtf1: q value	ratio (m/w)	BrasWt -> BrasFtf1: >2-fold and significant difference?	log2 ratio
5765	13.384	2.09219	PF00355:Rieske [2Fe-2S] region; PF03460:Nitrite/ sulfite reductase, hemoprotein beta-component, ferrodoxin-like; PF00175:Oxidoreductase FAD/ NAD(P)-binding; PF00070: FAD-dependent pyridine nucleotide-disulphide oxidoreductase; PF00970:Oxidoreductase FAD-binding region; PF04324:BFD-like [2Fe-2S]-binding region; PF00173:Cytochrome b5; PF01077:Nitrite and sulphite reductase 4Fe-4S region;	GO:0009703-nitrate reductase (NADH) activity; GO:0016491-oxidoreductase activity; GO:0046914-transition metal ion binding; GO:0051536-iron-sulfur cluster binding; GO:0006118-electron transport; GO:0020037-heme binding; GO:0050660-FAD binding;	Down	0	0.156320233	Yes	−2.677423571

Table S2. *Cont.*

Protein Id	BrasWt	BrasFtf1	PFAM annotation	GO annotation	BrasWt -> BrasFtf1: Up/Down	BrasWt -> BrasFtf1: q value	ratio (m/w)	BrasWt -> BrasFtf1: >2-fold and significant difference?	log2 ratio
5769	12.659	4.87618			Down	0.00213738	0.385194723	Yes	−1.376340155
5774	29.6677	10.2748			Down	3.82E-05	0.34632951	Yes	−1.529782774
5784	87.6391	38.6306			Down	0.00357751	0.440791838	Yes	−1.181830584
5813	23.8093	6.6003			Down	3.82E-05	0.277215206	Yes	−1.8509217
5831	30.3711	12.0912	PF02213:GYF;	GO:0004339-glucan 1,4-alpha-glucosidase activity;	Down	0.000465855	0.398115314	Yes	−1.328741729
5870	24.7046	8.61694	PF05178:Krr1;		Down	4.13E-05	0.348799009	Yes	−1.519532153
5872	5.4522	2.23457			Down	0.0032796	0.409847401	Yes	−1.286841245
5875	142.384	68.9993			Down	0.00752704	0.484600096	Yes	−1.045133406
5886	3.49638	1.12811	PF01055:Glycoside hydrolase, family 31;	GO:0004553-hydrolase activity, hydrolyzing O-glycosyl compounds; GO:0005975-carbohydrate metabolic process;	Down	0.000480012	0.322650856	Yes	−1.631954242
5887	24.4261	3.27118			Down	6.11E-09	0.133921502	Yes	−2.900540481
5919	19.7995	7.45895	PF04950:Protein of unknown function DUF663;	GO:0000166-nucleotide binding; GO:0017111-nucleoside-triphosphatase activity;	Down	0.000152439	0.37672416	Yes	−1.408419537
5922	8.29717	4.07057	PF02145:Rap/ ran-GAP;	GO:0005096-GTPase activator activity; GO:0051056-regulation of small GTPase mediated signal transduction; GO:0005622-intracellular;	Down	0.0156196	0.490597396	Yes	−1.027388517
5959	9.36131	3.6466			Down	0.0167691	0.389539498	Yes	−1.360158474
6026	6.75062	2.88658			Down	0.0139838	0.427602205	Yes	−1.225658801

Table S2. *Cont.*

Protein Id	BrasWt	BrasFtf1	PFAM annotation	GO annotation	BrasWt -> BrasFtf1: Up/Down	BrasWt -> BrasFtf1: q value	ratio (m/w)	BrasWt -> BrasFtf1: >2-fold and significant difference?	log2 ratio
6038	26.4117	12.8909	PF04565:RNA polymerase Rpb2, domain 3; PF04563:RNA polymerase beta subunit; PF04567:RNA polymerase Rpb2, domain 5; PF04560:RNA polymerase Rpb2, domain 7; PF00562:RNA polymerase Rpb2, domain 6; PF04566:RNA polymerase Rpb2, domain 4;	GO:0003899-DNA-directed RNA polymerase activity; GO:0003902-DNA-directed RNA polymerase III activity; GO:0003900-DNA-directed RNA polymerase I activity; GO:0006350-transcription; GO:0003901-DNA-directed RNA polymerase II activity; GO:0034062-RNA polymerase activity; GO:0003677-DNA binding;	Down	0.00980479	0.488075361	Yes	−1.034824173
6042	8.89714	4.09622	PF00083:General substrate transporter;	GO:0016021-integral to membrane; GO:0006810-transport; GO:0005215-transporter activity;	Down	0.0138704	0.460397386	Yes	−1.119048452

Table S2. *Cont.*

Protein Id	BrasWt	BrasFtf1	PFAM annotation	GO annotation	BrasWt -> BrasFtf1: Up/Down	BrasWt -> BrasFtf1: q value	ratio (m/w)	BrasWt -> BrasFtf1: >2-fold and significant difference?	log2 ratio
6052	39.3203	13.3001	PF01645: Ferredoxin-dependent glutamate synthase; PF04897:Glutamate synthase, amidotransferase region; PF04898:Glutamate synthase, central;	GO:0006807-nitrogen compound metabolic process; GO:0015930-glutamate synthase activity; GO:0016041-glutamate synthase (ferredoxin) activity; GO:0006537-glutamate biosynthetic process;	Down	5.17E-05	0.338250217	Yes	−1.563837236
6053	63.2929	22.8088	PF00070:FAD-dependent pyridine nucleotide-disulphide oxidoreductase; PF01493:Glutamate synthase, alpha subunit, C-terminal;	GO:0016491-oxidoreductase activity; GO:0006537-glutamate biosynthetic process; GO:0008152-metabolic process; GO:0006118-electron transport; GO:0016639-oxidoreductase activity, acting on the CH-NH2 group of donors, NAD or NADP as acceptor; GO:0050660-FAD binding;	Down	0.000102765	0.360369015	Yes	−1.472453125
6071	5.64237	2.67719			Down	0.0237654	0.474479696	Yes	−1.075581744
6100	5347.08	1963.59			Down	0.000829199	0.367226598	Yes	−1.445257537
6101	478.06	157.498	PF00005:ABC transporter;	GO:0000166-nucleotide binding; GO:0017111-nucleoside-triphosphatase activity; GO:0016887-ATPase activity; GO:0005524-ATP binding;	Down	0.000303345	0.32945237	Yes	−1.60185819

Table S2. *Cont.*

Protein Id	BrasWt	BrasFtf1	PFAM annotation	GO annotation	BrasWt -> BrasFtf1: Up/Down	BrasWt -> BrasFtf1: q value	ratio (m/w)	BrasWt -> BrasFtf1: >2-fold and significant difference?	log2 ratio
6103	34.3403	13.6053	PF00515:TPR repeat;		Down	0.00024137	0.396190482	Yes	−1.335733874
6105	70.4905	23.79	PF02524:RepA/ Rep+ protein KID;		Down	4.34E-05	0.337492286	Yes	−1.567073567
6126	18.69	1.99707	PF03443:Glycoside hydrolase, family 61;		Down	9.03E-13	0.106852327	Yes	−3.226309762
6145	3.20059	1.40789			Down	0.0433123	0.439884521	Yes	−1.184803258
6199	53.2127	16.2995	PF00734: Cellulose-binding region, fungal; PF00544:Pectate lyase/Amb allergen;	GO:0005576-extracellular region; GO:0004553-hydrolase activity, hydrolyzing O-glycosyl compounds; GO:0005975-carbohydrate metabolic process; GO:0030248-cellulose binding;	Down	5.52E-06	0.306308456	Yes	−1.706942898
6230	4.61322	1.82461	PF00249:Myb, DNA-binding;	GO:0043773-coenzyme F420-0 gamma-glutamyl ligase activity; GO:0018169-ribosomal S6-glutamic acid ligase activity; GO:0043774-coenzyme F420-2 alpha-glutamyl ligase activity; GO:0008766-UDP-N-acetylmuramoylalanyl-D-glutamyl-2,6-diaminopimelate-D-alanyl-D-alanine ligase activity; GO:0003677-DNA binding;	Down	0.00923703	0.395517664	Yes	−1.338185966
6256	110.715	15.2917			Down	0	0.13811769	Yes	−2.856029989
6257	15.4138	4.85078	PF00067:Cytochrome P450;	GO:0006118-electron transport; GO:0050381-unspecific monooxygenase activity; GO:0004497-monooxygenase activity; GO:0020037-heme binding; GO:0005506-iron ion binding;	Down	4.68E-05	0.314703707	Yes	−1.667933922

Table S2. *Cont.*

Protein Id	BrasWt	BrasFtf1	PFAM annotation	GO annotation	BrasWt -> BrasFtf1: Up/Down	BrasWt -> BrasFtf1: q value	ratio (m/w)	BrasWt -> BrasFtf1: >2-fold and significant difference?	log2 ratio
6258	5.30685	1.45549	PF05141:Pyoverdine biosynthesis;		Down	2.54E-05	0.274266278	Yes	−1.866350844
6259	97.0823	22.0076			Down	2.73E-08	0.226690138	Yes	−2.141206465
6275	6.20269	1.10501		GO:0000166-nucleotide binding; GO:0017111-nucleoside-triphosphatase activity;	Down	3.87E-10	0.178150125	Yes	−2.488834598
6330	27.7971	13.7079	PF00169: Pleckstrin-like;		Down	0.0198933	0.493141371	Yes	−1.019926806
6407	22.5743	8.48347			Down	0.000235124	0.375802129	Yes	−1.411954855
6502	13.0293	6.40532		GO:0008305-integrin complex; GO:0007155-cell adhesion; GO:0004339-glucan 1,4-alpha-glucosidase activity;	Down	0.0154571	0.491608912	Yes	−1.024417025
6504	20.7808	7.91595			Down	0.000277623	0.380926143	Yes	−1.39241679
6528	13.5894	4.6345	PF02535:Zinc transporter ZIP;	GO:0016021-integral to membrane; GO:0016020-membrane; GO:0006829-zinc ion transport; GO:0046873-metal ion transmembrane transporter activity; GO:0005385-zinc ion transmembrane transporter activity; GO:0030001-metal ion transport;	Down	0.000275163	0.341037868	Yes	−1.551996155
6580	9.32986	2.91942	PF00400:G-protein beta WD-40 repeat;		Down	6.95E-05	0.312911448	Yes	−1.676173655
6590	2.15245	0.810122	PF00248:Aldo/keto reductase;	GO:0016491-oxidoreductase activity;	Down	0.0237654	0.376372041	Yes	−1.409768634

Table S2. *Cont.*

Protein Id	BrasWt	BrasFtf1	PFAM annotation	GO annotation	BrasWt -> BrasFtf1: Up/Down	BrasWt -> BrasFtf1: q value	ratio (m/w)	BrasWt -> BrasFtf1: >2-fold and significant difference?	log2 ratio
6618	73.3276	19.2219	PF00324:Amino acid permease-associated region;	GO:0006810-transport; GO:0016020-membrane; GO:0015171-amino acid transmembrane transporter activity; GO:0006865-amino acid transport;	Down	2.18E-08	0.262137313	Yes	−1.931605374
6624	54.6134	24.5423			Down	0.014367	0.449382386	Yes	−1.153984517
6644	8.20041	2.60673			Down	0.00010387	0.317878009	Yes	−1.653454884
6658	1.69585	0.290162	PF00194:Carbonic anhydrase, eukaryotic;	GO:0006730-one-carbon compound metabolic process; GO:0008270-zinc ion binding; GO:0004089-carbonate dehydratase activity;	Down	0.0272654	0.171101218	Yes	−2.547078068
6663	15.064	6.7387		GO:0008270-zinc ion binding;	Down	0.00465428	0.447338024	Yes	−1.1605627
6671	35.5892	11.3581	PF00533:BRCT; PF06732:Pescadillo, N-terminal;	GO:0005730-nucleolus; GO:0005622-intracellular; GO:0008283-cell proliferation;	Down	5.52E-06	0.319144572	Yes	−1.647717984
6695	7.80379	2.12056			Down	0.0029817	0.271734632	Yes	−1.879729652
6697	11.8259	4.45724	PF00107:Zinc-containing alcohol dehydrogenase superfamily;	GO:0008270-zinc ion binding; GO:0003939-L-iditol 2-dehydrogenase activity; GO:0016491-oxidoreductase activity;	Down	0.00126508	0.376904929	Yes	−1.407727433
6702	4.20312	1.19929	PF05577:Peptidase S28;	GO:0008236-serine-type peptidase activity; GO:0006508-proteolysis; GO:0008234-cysteine-type peptidase activity;	Down	5.49E-05	0.285333276	Yes	−1.809280088
6745	26.7419	8.49937	PF00400:G-protein beta WD-40 repeat;		Down	3.09E-06	0.317829698	Yes	−1.653674159

Table S2. *Cont.*

Protein Id	BrasWt	BrasFtf1	PFAM annotation	GO annotation	BrasWt -> BrasFtf1: Up/Down	BrasWt -> BrasFtf1: q value	ratio (m/w)	BrasWt -> BrasFtf1: >2-fold and significant difference?	log2 ratio
6746	120.104	53.325	PF00134:Cyclin, N-terminal; PF02984:Cyclin, C-terminal;	GO:0004812-aminoacyl-tRNA ligase activity; GO:0005737-cytoplasm; GO:0006412-translation; GO:0000166-nucleotide binding; GO:0005634-nucleus; GO:0006418-tRNA aminoacylation for protein translation; GO:0005524-ATP binding;	Down	0.00255689	0.443990208	Yes	−1.171400234
6758	124.869	51.1648		GO:0004568-chitinase activity; GO:0008843-endochitinase activity;	Down	0.000939016	0.409747816	Yes	−1.287191836
6759	459.56	223.044			Down	0.0181979	0.485342502	Yes	−1.042924892
6767	8.70924	3.91581		GO:0016021-integral to membrane; GO:0006810-transport; GO:0005215-transporter activity;	Down	0.0197752	0.449615581	Yes	−1.153236064
6814	37.2494	11.7677	PF01965:ThiJ/PfpI;		Down	7.62E-05	0.315916498	Yes	−1.662384815
6874	22.3613	6.73206			Down	1.92E-05	0.301058525	Yes	−1.731884123
6877	27.6876	8.35445			Down	2.94E-06	0.301739768	Yes	−1.728623245
6878	10.6024	4.7545	PF00018:SH3;	GO:0003677-DNA binding;	Down	0.00580641	0.448436203	Yes	−1.15702534
6879	53.032	22.3013	PF04145:Ctr copper transporter;	GO:0016021-integral to membrane; GO:0006825-copper ion transport; GO:0005375-copper ion transmembrane transporter activity;	Down	0.00400852	0.420525343	Yes	−1.249735347
6880	26.7784	8.01653			Down	0.000156449	0.299365533	Yes	−1.740019964
6898	43.2116	18.0428			Down	0.00103458	0.417545289	Yes	−1.259995408
6918	25.0857	7.38975			Down	2.81E-05	0.294580179	Yes	−1.763267733
6928	10.3101	4.61406			Down	0.0463716	0.447528152	Yes	−1.159949656

Table S2. *Cont.*

Protein Id	BrasWt	BrasFtf1	PFAM annotation	GO annotation	BrasWt -> BrasFtf1: Up/Down	BrasWt -> BrasFtf1: q value	ratio (m/w)	BrasWt -> BrasFtf1: >2-fold and significant difference?	log2 ratio
6962	57.6315	20.5834	PF02882: Tetrahydrofolate dehydrogenase/ cyclohydrolase; PF01268:Formate-tetrahydrofolate ligase, FTHFS; PF00763: Tetrahydrofolate dehydrogenase/ cyclohydrolase;	GO:0003824-catalytic activity; GO:0009396-folic acid and derivative biosynthetic process; GO:0004329-formate-tetrahydrofolate ligase activity; GO:0005524-ATP binding;	Down	3.29E-05	0.357155375	Yes	−1.485376261
6971	432.494	214.954			Down	0.0314689	0.497010363	Yes	−1.008652161
6977	8.35518	3.82016	PF01302:CAP-Gly;		Down	0.00793477	0.457220551	Yes	−1.129037845
6995	25.7319	12.6999	PF00335:CD9/CD37/ CD63 antigen;	GO:0016021-integral to membrane;	Down	0.0251728	0.49354692	Yes	−1.01874085
7007	9.7683	3.37537	PF01565:FAD linked oxidase, N-terminal;	GO:0050660-FAD binding; GO:0005783-endoplasmic reticulum; GO:0016491-oxidoreductase activity;	Down	0.00026252	0.345543237	Yes	−1.533061853
7015	4.02795	0.649436			Down	4.25E-06	0.161232389	Yes	−2.63278651
7020	92.4362	39.6021	PF04145:Ctr copper transporter;	GO:0016021-integral to membrane; GO:0006825-copper ion transport; GO:0005375-copper ion transmembrane transporter activity;	Down	0.00227939	0.428426309	Yes	−1.222881018

Table S2. *Cont.*

Protein Id	BrasWt	BrasFtf1	PFAM annotation	GO annotation	BrasWt -> BrasFtf1: Up/Down	BrasWt -> BrasFtf1: q value	ratio (m/w)	BrasWt -> BrasFtf1: >2-fold and significant difference?	log2 ratio
7021	39.9294	11.3948	PF01794:Ferric reductase-like transmembrane component;		Down	3.42E-07	0.285373685	Yes	−1.809075791
7025	4.76185	2.20008		GO:0003824-catalytic activity; GO:0046872-metal ion binding;	Down	0.0106179	0.462022113	Yes	−1.113966191
7076	7.57639	3.24495		GO:0004339-glucan 1,4-alpha-glucosidase activity;	Down	0.00588	0.428297646	Yes	−1.223314348
7077	13.5413	3.60332	PF00264:Tyrosinase;	GO:0016491-oxidoreductase activity; GO:0008152-metabolic process;	Down	4.25E-06	0.266098528	Yes	−1.909967564
7099	6.01334	2.90975			Down	0.021144	0.483882501	Yes	−1.047271327
7102	1.25613	0.522142		GO:0003677-DNA binding;	Down	0.0416371	0.415675129	Yes	−1.266471664
7106	10.3806	5.00262	PF00083:General substrate transporter;	GO:0016021-integral to membrane; GO:0016020-membrane; GO:0005215-transporter activity; GO:0008643-carbohydrate transport; GO:0006810-transport; GO:0005351-sugar:hydrogen symporter activity;	Down	0.018908	0.48192012	Yes	−1.05313406
7131	52.7493	11.5628			Down	1.87E-09	0.219202909	Yes	−2.189661152

Table S2. *Cont.*

Protein Id	BrasWt	BrasFtf1	PFAM annotation	GO annotation	BrasWt -> BrasFtf1: Up/Down	BrasWt -> BrasFtf1: q value	ratio (m/w)	BrasWt -> BrasFtf1: >2-fold and significant difference?	log2 ratio
7137	54.0988	25.035	PF00271:Helicase, C-terminal; PF00270:DEAD/ DEAH box helicase, N-terminal;	GO:0008796-bis(5'-nucleosyl)-tetraphosphatase activity; GO:0008413-8-oxo-7,8-dihydroguanine triphosphatase activity; GO:0019177-dihydroneopterin triphosphate pyrophosphohydrolase activity; GO:0008758-UDP-2,3-diacylglucosamine hydrolase activity; GO:0004787-thiamin-pyrophosphatase activity; GO:0003676-nucleic acid binding; GO:0000810-diacylglycerol pyrophosphate phosphatase activity; GO:0008026-ATP-dependent helicase activity; GO:0004386-helicase activity; GO:0043141-ATP-dependent 5'-3' DNA helicase activity; GO:0008828-dATP pyrophosphohydrolase activity; GO:0016462-pyrophosphatase activity; GO:0019176-dihydroneopterin monophosphate phosphatase activity; GO:0005524-ATP binding;	Down	0.00334192	0.462764424	Yes	−1.111650138
7154	1455.2	701.821	PF00318:Ribosomal protein S2;	GO:0005840-ribosome; GO:0003735-structural constituent of ribosome; GO:0005622-intracellular; GO:0015935-small ribosomal subunit; GO:0006412-translation;	Down	0.0210584	0.482284909	Yes	−1.052042426

Table S2. *Cont.*

Protein Id	BrasWt	BrasFtf1	PFAM annotation	GO annotation	BrasWt -> BrasFtf1: Up/Down	BrasWt -> BrasFtf1: q value	ratio (m/w)	BrasWt -> BrasFtf1: >2-fold and significant difference?	log2 ratio
7196	73.2437	27.1786		GO:0004812-aminoacyl-tRNA ligase activity; GO:0006429-leucyl-tRNA aminoacylation; GO:0003700-transcription factor activity; GO:0006412-translation; GO:0000166-nucleotide binding; GO:0043565-sequence-specific DNA binding; GO:0006418-tRNA aminoacylation for protein translation; GO:0005737-cytoplasm; GO:0004823-leucine-tRNA ligase activity; GO:0006355-regulation of transcription, DNA-dependent; GO:0005634-nucleus; GO:0005524-ATP binding;	Down	8.78E-05	0.371070822	Yes	−1.430233529
7228	26.408	3.64167	PF00067:Cytochrome P450;	GO:0006118-electron transport; GO:0004497-monooxygenase activity; GO:0020037-heme binding; GO:0005506-iron ion binding;	Down	0	0.137900257	Yes	−2.858302944
7250	4.6715	1.37284			Down	3.10E-05	0.293875629	Yes	−1.766722374

Table S2. *Cont.*

Protein Id	BrasWt	BrasFtf1	PFAM annotation	GO annotation	BrasWt -> BrasFtf1: Up/Down	BrasWt -> BrasFtf1: q value	ratio (m/w)	BrasWt -> BrasFtf1: >2-fold and significant difference?	log2 ratio
7256	35.9321	16.9109	PF00632:HECT;	GO:0051438-regulation of ubiquitin-protein ligase activity; GO:0051440-regulation of ubiquitin-protein ligase activity during meiotic cell cycle; GO:0004842-ubiquitin-protein ligase activity; GO:0006464-protein modification process; GO:0006512-ubiquitin cycle; GO:0051444-negative regulation of ubiquitin-protein ligase activity; GO:0051443-positive regulation of ubiquitin-protein ligase activity; GO:0005622-intracellular;	Down	0.0156196	0.470634892	Yes	−1.087319812
7257	28.1641	11.9894			Down	0.0138395	0.425697963	Yes	−1.232097908
7258	19.8978	8.60714	PF06025:Protein of unknown function DUF913; PF06012:Protein of unknown function DUF908;	GO:0051438-regulation of ubiquitin-protein ligase activity; GO:0051440-regulation of ubiquitin-protein ligase activity during meiotic cell cycle; GO:0004842-ubiquitin-protein ligase activity; GO:0006512-ubiquitin cycle; GO:0005737-cytoplasm; GO:0051444-negative regulation of ubiquitin-protein ligase activity; GO:0051028-mRNA transport; GO:0051443-positive regulation of ubiquitin-protein ligase activity;	Down	0.00845837	0.43256742	Yes	−1.209003088

Table S2. *Cont.*

Protein Id	BrasWt	BrasFtf1	PFAM annotation	GO annotation	BrasWt -> BrasFtf1: Up/Down	BrasWt -> BrasFtf1: q value	ratio (m/w)	BrasWt -> BrasFtf1: >2-fold and significant difference?	log2 ratio
7285	8.21154	3.62199	PF00082:Peptidase S8 and S53, subtilisin, kexin, sedolisin;	GO:0004289-subtilase activity; GO:0006508-proteolysis; GO:0004285-proprotein convertase 1 activity;	Down	0.00321209	0.441085351	Yes	−1.180870249
7318	402.292	154.545	PF01522:Polysaccharide deacetylase; PF00187: Chitin-binding, type 1;	GO:0005975-carbohydrate metabolic process; GO:0016810-hydrolase activity, acting on carbon-nitrogen (but not peptide) bonds; GO:0008061-chitin binding;	Down	0.000399267	0.384161256	Yes	−1.380216069
7319	691.296	307.125			Down	0.00129846	0.444274233	Yes	−1.170477625
7328	18.6683	5.83427		GO:0042027-cyclophilin-type peptidyl-prolyl cis-trans isomerase activity; GO:0030051-FK506-sensitive peptidyl-prolyl cis-trans isomerase; GO:0004600-cyclophilin; GO:0003755-peptidyl-prolyl cis-trans isomerase activity;	Down	1.38E-05	0.312522833	Yes	−1.677966498
7349	196.174	72.7879			Down	0.0387355	0.371037446	Yes	−1.430363299
7350	28.2593	11.474			Down	0.000859465	0.406025627	Yes	−1.300357307
7432	37.0763	12.8465	PF00400:G-protein beta WD-40 repeat;		Down	2.01E-05	0.346488188	Yes	−1.529121925

Table S2. *Cont.*

Protein Id	BrasWt	BrasFtf1	PFAM annotation	GO annotation	BrasWt -> BrasFtf1: Up/Down	BrasWt -> BrasFtf1: q value	ratio (m/w)	BrasWt -> BrasFtf1: >2-fold and significant difference?	log2 ratio
7453	28.5716	12.6002	PF00106:Short-chain dehydrogenase/ reductase SDR; PF00550:Phosphopant etheine-binding; PF00501: AMP-dependent synthetase and ligase;	GO:0016491-oxidoreductase activity; GO:0031177-phosphopantetheine binding; GO:0003824-catalytic activity; GO:0008152-metabolic process; GO:0048037-cofactor binding; GO:0004043-L-aminoadipate-semialdehyde dehydrogenase activity; GO:0005949-aminoadipate-semialdehyde dehydrogenase complex;	Down	0.00172009	0.441004354	Yes	−1.181135196
7478	12.8535	5.35479	PF00018:SH3;		Down	0.00268697	0.416601704	Yes	−1.263259355
7496	694.331	130.442	PF05221:S-adenosyl-L-homocysteine hydrolase; PF00670:S-adenosyl-L-homocysteine hydrolase;	GO:0006730-one-carbon compound metabolic process; GO:0004013-adenosylhomocysteinase activity;	Down	2.64E-13	0.18786717	Yes	−2.412215119
7520	236.806	80.1496	PF00171:Aldehyde dehydrogenase;	GO:0006561-proline biosynthetic process; GO:0005759-mitochondrial matrix; GO:0003842-1-pyrroline-5-carboxylate dehydrogenase activity;	Down	6.10E-06	0.338461019	Yes	−1.56293841
7521	52.1581	21.5309	PF01926: GTP-binding protein, HSR1-related;	GO:0005525-GTP binding; GO:0005622-intracellular;	Down	0.000543717	0.412800696	Yes	−1.27648269

Table S2. *Cont.*

Protein Id	BrasWt	BrasFtf1	PFAM annotation	GO annotation	BrasWt -> BrasFtf1: Up/Down	BrasWt -> BrasFtf1: q value	ratio (m/w)	BrasWt -> BrasFtf1: >2-fold and significant difference?	log2 ratio
7524	12.3445	4.90207	PF00271:Helicase, C-terminal; PF00270:DEAD/ DEAH box helicase, N-terminal;	GO:0008796-bis(5'-nucleosyl)-tetraphosphatase activity; GO:0008413-8-oxo-7,8-dihydroguanine triphosphatase activity; GO:0019177-dihydroneopterin triphosphate pyrophosphohydrolase activity; GO:0008758-UDP-2,3-diacylglucosamine hydrolase activity; GO:0004787-thiamin-pyrophosphatase activity; GO:0003676-nucleic acid binding; GO:0000810-diacylglycerol pyrophosphate phosphatase activity; GO:0008026-ATP-dependent helicase activity; GO:0004386-helicase activity; GO:0043141-ATP-dependent 5'-3' DNA helicase activity; GO:0008828-dATP pyrophosphohydrolase activity; GO:0016462-pyrophosphatase activity; GO:0019176-dihydroneopterin monophosphate phosphatase activity; GO:0005524-ATP binding;	Down	0.0015656	0.397105594	Yes	−1.332405412

Table S2. *Cont.*

Protein Id	BrasWt	BrasFtf1	PFAM annotation	GO annotation	BrasWt -> BrasFtf1: Up/Down	BrasWt -> BrasFtf1: q value	ratio (m/w)	BrasWt -> BrasFtf1: >2-fold and significant difference?	log2 ratio
7527	25.986	10.0256	PF00271:Helicase, C-terminal; PF00270:DEAD/DEAH box helicase, N-terminal;	GO:0008796-bis(5'-nucleosyl)-tetraphosphatase activity; GO:0008413-8-oxo-7,8-dihydroguanine triphosphatase activity; GO:0019177-dihydroneopterin triphosphate pyrophosphohydrolase activity; GO:0008758-UDP-2,3-diacylglucosamine hydrolase activity; GO:0004787-thiamin-pyrophosphatase activity; GO:0003676-nucleic acid binding; GO:0000810-diacylglycerol pyrophosphate phosphatase activity; GO:0008026-ATP-dependent helicase activity; GO:0004386-helicase activity; GO:0043141-ATP-dependent 5'-3' DNA helicase activity; GO:0008828-dATP pyrophosphohydrolase activity; GO:0016462-pyrophosphatase activity; GO:0019176-dihydroneopterin monophosphate phosphatase activity; GO:0005524-ATP binding;	Down	0.00017833	0.385807743	Yes	−1.374045998
7533	1.57318	0.537366	PF00860:Xanthine/uracil/vitamin C permease;	GO:0006810-transport; GO:0016020-membrane; GO:0005215-transporter activity;	Down	0.0226348	0.341579476	Yes	−1.549706803
7601	35.5955	15.675			Down	0.00235034	0.440364653	Yes	−1.183229424
7637	17.5894	7.21362			Down	0.00240943	0.410111772	Yes	−1.285910939

Table S2. *Cont.*

Protein Id	BrasWt	BrasFtf1	PFAM annotation	GO annotation	BrasWt -> BrasFtf1: Up/Down	BrasWt -> BrasFtf1: q value	ratio (m/w)	BrasWt -> BrasFtf1: >2-fold and significant difference?	log2 ratio
7647	30.8536	12.7677	PF00172:Fungal transcriptional regulatory protein, N-terminal;	GO:0003700-transcription factor activity; GO:0008270-zinc ion binding; GO:0006355-regulation of transcription, DNA-dependent; GO:0005634-nucleus;	Down	0.000768177	0.413815568	Yes	−1.272940175
7659	30.4003	13.6278	PF05185:Skb1 methyltransferase;	GO:0005737-cytoplasm; GO:0008168-methyltransferase activity;	Down	0.00353437	0.448278471	Yes	−1.157532881
7663	11.0008	2.94388	PF00067:Cytochrome P450;	GO:0006118-electron transport; GO:0004497-monooxygenase activity; GO:0020037-heme binding; GO:0005506-iron ion binding;	Down	3.84E-06	0.267605992	Yes	−1.901817673
7695	11.5856	5.41824			Down	0.00915006	0.467670211	Yes	−1.096436557
7716	25.5598	10.9147			Down	0.00492924	0.427026033	Yes	−1.22760407
7719	42.9281	17.581	PF05383: RNA-binding protein Lupus La1;		Down	0.00151419	0.409545263	Yes	−1.287905188
7720	40.3977	12.6421		GO:0006265-DNA topological change; GO:0005694-chromosome; GO:0003918-DNA topoisomerase (ATP-hydrolyzing) activity; GO:0003677-DNA binding; GO:0005524-ATP binding;	Down	5.00E-05	0.312941083	Yes	−1.676037025
7753	15.3187	7.23968			Down	0.00835258	0.472604072	Yes	−1.081296035
7756	3.03242	1.11977			Down	0.0058535	0.369266131	Yes	−1.437267151
7758	3.63405	0.696704			Down	6.55E-05	0.191715579	Yes	−2.382960518
7789	351.405	113.775	PF01565:FAD linked oxidase, N-terminal;	GO:0050660-FAD binding; GO:0016491-oxidoreductase activity;	Down	6.20E-06	0.323771716	Yes	−1.626951133

Table S2. *Cont.*

Protein Id	BrasWt	BrasFtf1	PFAM annotation	GO annotation	BrasWt -> BrasFtf1: Up/Down	BrasWt -> BrasFtf1: q value	ratio (m/w)	BrasWt -> BrasFtf1: >2-fold and significant difference?	log2 ratio
7791	24.3428	10.2239		GO:0042027-cyclophilin-type peptidyl-prolyl cis-trans isomerase activity; GO:0030051-FK506-sensitive peptidyl-prolyl cis-trans isomerase; GO:0004600-cyclophilin; GO:0003755-peptidyl-prolyl cis-trans isomerase activity;	Down	0.00130168	0.419996878	Yes	−1.251549491
7800	17.3505	8.03986	PF04179:Initiator tRNA phosphoribosyl transferase;	GO:0016763-transferase activity, transferring pentosyl groups;	Down	0.011218	0.463379153	Yes	−1.109734954
7827	7.67188	2.16181			Down	2.08E-05	0.281783605	Yes	−1.827340423
7871	38.7321	13.1206	PF00172:Fungal transcriptional regulatory protein, N-terminal;	GO:0003700-transcription factor activity; GO:0008270-zinc ion binding; GO:0006355-regulation of transcription, DNA-dependent; GO:0005634-nucleus;	Down	1.44E-05	0.338752611	Yes	−1.561696029
7872	19.3194	5.91837	PF00536:Sterile alpha motif SAM; PF00169: Pleckstrin-like; PF07647:Sterile alpha motif homology 2;		Down	1.98E-06	0.306343365	Yes	−1.706778492
7877	10.2939	4.68984			Down	0.0170992	0.45559409	Yes	−1.134179063
7884	18.3278	8.42867			Down	0.00375026	0.459884438	Yes	−1.120656716

Table S2. *Cont.*

Protein Id	BrasWt	BrasFtf1	PFAM annotation	GO annotation	BrasWt -> BrasFtf1: Up/Down	BrasWt -> BrasFtf1: q value	ratio (m/w)	BrasWt -> BrasFtf1: >2-fold and significant difference?	log2 ratio
7899	10.9593	0	PF00172:Fungal transcriptional regulatory protein, N-terminal; PF04082: Fungal specific transcription factor;	GO:0003700-transcription factor activity; GO:0006350-transcription; GO:0008270-zinc ion binding; GO:0003677-DNA binding; GO:0006355-regulation of transcription, DNA-dependent; GO:0005634-nucleus;	Down	4.28E-40	0	Yes	#NUM!
7906	17.1983	6.6181			Down	0.00235034	0.384811289	Yes	−1.37777697
7920	27.7147	7.14133	PF00400:G-protein beta WD-40 repeat;	GO:0000250-lanosterol synthase activity;	Down	1.04E-07	0.257673004	Yes	−1.956386699
7953	25.4235	11.6232	PF01585:D111/ G-patch;	GO:0003676-nucleic acid binding; GO:0005622-intracellular;	Down	0.0053859	0.457183315	Yes	−1.129155343
7958	64.5972	12.0966	PF00067:Cytochrome P450;	GO:0018588-tri-n-butyltin dioxygenase activity; GO:0018589-di-n-butyltin dioxygenase activity; GO:0018592-4-nitrocatechol 4-monooxygenase activity; GO:0018598-hydroxymethylsilanetriol oxidase activity; GO:0018597-ammonia monooxygenase activity; GO:0018591-methyl tertiary butyl ether 3-monooxygenase activity; GO:0043823-spheroidene monooxygenase activity; GO:0018585-fluorene oxygenase activity; GO:0005506-iron ion binding; GO:0018586-mono-butyltin dioxygenase activity;	Down	9.36E-11	0.187261987	Yes	−2.416870027

Table S2. *Cont.*

Protein Id	BrasWt	BrasFtf1	PFAM annotation	GO annotation	BrasWt -> BrasFtf1: Up/Down	BrasWt -> BrasFtf1: q value	ratio (m/w)	BrasWt -> BrasFtf1: >2-fold and significant difference?	log2 ratio
				GO:0018600-alpha-pinene dehydrogenase activity; GO:0018593-4-chlorophenoxyacetate monooxygenase activity; GO:0018599-2-hydroxyisobutyrate 3-monooxygenase activity; GO:0018594-tert-butyl alcohol 2-monooxygenase activity; GO:0006118-electron transport; GO:0020037-heme binding; GO:0018596-dimethylsilanediol hydroxylase activity; GO:0018587-limonene 8-monooxygenase activity; GO:0004497-monooxygenase activity; GO:0018595-alpha-pinene monooxygenase activity; GO:0018590-methylsilanetriol hydroxylase activity;					
7959	11.3396	1.05797			Down	7.44E-11	0.093298705	Yes	−3.421999127
7961	1.74683	0.518213		GO:0016021-integral to membrane; GO:0005215-transporter activity; GO:0015904-tetracycline transport; GO:0046677-response to antibiotic; GO:0006810-transport; GO:0015520-tetracycline:hydrogen antiporter activity;	Down	0.00955407	0.296659091	Yes	−1.7531221
7980	19.421	8.20673	PF00860:Xanthine/ uracil/vitamin C permease;	GO:0006810-transport; GO:0016020-membrane; GO:0005215-transporter activity;	Down	0.00163549	0.422569899	Yes	−1.242738093

Table S2. *Cont.*

Protein Id	BrasWt	BrasFtf1	PFAM annotation	GO annotation	BrasWt -> BrasFtf1: Up/Down	BrasWt -> BrasFtf1: q value	ratio (m/w)	BrasWt -> BrasFtf1: >2-fold and significant difference?	log2 ratio
7986	9.26676	2.50029	PF01565:FAD linked oxidase, N-terminal;	GO:0050660-FAD binding; GO:0016491-oxidoreductase activity;	Down	4.25E-06	0.26981275	Yes	−1.88996957
7990	27.5658	12.5818	PF04427:Brix;		Down	0.00423448	0.456427893	Yes	−1.131541137
8015	114.928	56.6203			Down	0.0102842	0.492658882	Yes	−1.021339027
8025	30.272	4.65759			Down	0	0.153858021	Yes	−2.700328442
8048	125.856	12.804	PF00067:Cytochrome P450;	GO:0006118-electron transport; GO:0004497-monooxygenase activity; GO:0020037-heme binding; GO:0005506-iron ion binding;	Down	0	0.101735317	Yes	−3.297107509
8049	76.6924	9.82144	PF00067:Cytochrome P450;	GO:0006118-electron transport; GO:0004497-monooxygenase activity; GO:0020037-heme binding; GO:0005506-iron ion binding;	Down	0	0.128062755	Yes	−2.965077148
8065	36.8663	16.8401			Down	0.00666261	0.456788449	Yes	−1.130401926
8102	2.81629	0.836381	PF00750:Arginyl-tRNA synthetase, class Ic; PF05746:Arginyl tRNA synthetase anticodon binding;	GO:0005737-cytoplasm; GO:0006412-translation; GO:0000166-nucleotide binding; GO:0004814-arginine-tRNA ligase activity; GO:0006420-arginyl-tRNA aminoacylation; GO:0005524-ATP binding;	Down	8.16E-05	0.296979714	Yes	−1.751563705
8156	47.7299	18.0507			Down	0.00012925	0.378184325	Yes	−1.402838528
8191	136.143	54.3805			Down	0.000268154	0.399436622	Yes	−1.323961485
8215	5.12871	1.7062	PF01761:3-dehydroquinate synthase;	GO:0003856-3-dehydroquinate synthase activity; GO:0009073-aromatic amino acid family biosynthetic process;	Down	0.00215106	0.332676248	Yes	−1.587809229

Table S2. *Cont.*

Protein Id	BrasWt	BrasFtf1	PFAM annotation	GO annotation	BrasWt -> BrasFtf1: Up/Down	BrasWt -> BrasFtf1: q value	ratio (m/w)	BrasWt -> BrasFtf1: >2-fold and significant difference?	log2 ratio
8220	16.2578	4.83498	PF00400:G-protein beta WD-40 repeat;	GO:0003847-1-alkyl-2-acetylglycerophosphocholine esterase activity;	Down	2.32E-06	0.297394481	Yes	−1.749550219
8230	1.60605	0.563029	PF01179:Copper amine oxidase;	GO:0005507-copper ion binding; GO:0009308-amine metabolic process; GO:0008131-amine oxidase activity; GO:0048038-quinone binding;	Down	0.014558	0.350567541	Yes	−1.51223567
8245	10.2438	4.22359			Down	0.00423448	0.412306956	Yes	−1.278209292
8292	17.2431	8.43364		GO:0005509-calcium ion binding;	Down	0.0143877	0.489102308	Yes	−1.031791824
8299	44.0418	19.8938	PF02786:Carbamoyl-phosphate synthase L chain, ATP-binding; PF00289:Carbamoyl-phosphate synthetase large chain, N-terminal; PF02785:Biotin carboxylase, C-terminal;	GO:0003824-catalytic activity; GO:0016874-ligase activity; GO:0004485-methylcrotonoyl-CoA carboxylase activity; GO:0008152-metabolic process; GO:0005524-ATP binding;	Down	0.00852327	0.451702701	Yes	−1.146554556
8314	19.7621	6.89632			Down	7.11E-05	0.348966962	Yes	−1.518837637
8322	14.4521	5.83052			Down	0.00188533	0.403437563	Yes	−1.30958268
8334	11.9002	1.75678	PF03443:Glycoside hydrolase, family 61;		Down	1.41E-07	0.14762609	Yes	−2.75998038
8339	20.1905	2.07293			Down	0	0.102668582	Yes	−3.283933334

Table S2. *Cont.*

Protein Id	BrasWt	BrasFtf1	PFAM annotation	GO annotation	BrasWt -> BrasFtf1: Up/Down	BrasWt -> BrasFtf1: q value	ratio (m/w)	BrasWt -> BrasFtf1: >2-fold and significant difference?	log2 ratio
8344	45.313	21.0453	PF00450:Peptidase S10, serine carboxypeptidase;	GO:0006508-proteolysis; GO:0004185-serine carboxypeptidase activity;	Down	0.00629904	0.464442875	Yes	−1.106426934
8352	1.4046	0.2542	PF00664:ABC transporter, transmembrane region; PF00005:ABC transporter;	GO:0016021-integral to membrane; GO:0042626-ATPase activity, coupled to transmembrane movement of substances; GO:0017111-nucleoside-triphosphatase activity; GO:0006810-transport; GO:0000166-nucleotide binding; GO:0016887-ATPase activity; GO:0005524-ATP binding;	Down	2.63E-06	0.180976791	Yes	−2.466123405
8354	2.78291	0.980677	PF00550:Phosphopantetheine-binding; PF00668:Condensation domain; PF00501:AMP-dependent synthetase and ligase;	GO:0048037-cofactor binding; GO:0003824-catalytic activity; GO:0008152-metabolic process; GO:0016874-ligase activity;	Down	0.00154854	0.352392639	Yes	−1.504744305
8355	4.59816	1.28104			Down	0.0345297	0.278598396	Yes	−1.843741144

Table S2. *Cont.*

Protein Id	BrasWt	BrasFtf1	PFAM annotation	GO annotation	BrasWt -> BrasFtf1: Up/Down	BrasWt -> BrasFtf1: q value	ratio (m/w)	BrasWt -> BrasFtf1: >2-fold and significant difference?	log2 ratio
8356	9.47043	3.68725	PF00550:Phosphopantetheine-binding; PF00668:Condensation domain; PF00501:AMP-dependent synthetase and ligase;	GO:0031177-phosphopantetheine binding; GO:0003824-catalytic activity; GO:0005949-aminoadipate-semialdehyde dehydrogenase complex; GO:0003910-DNA ligase (ATP) activity; GO:0008152-metabolic process; GO:0006281-DNA repair; GO:0006310-DNA recombination; GO:0048037-cofactor binding; GO:0006260-DNA replication; GO:0016874-ligase activity; GO:0004043-L-aminoadipate-semialdehyde dehydrogenase activity; GO:0005524-ATP binding;	Down	0.000737001	0.389343462	Yes	−1.360884696
8384	39.025	19.0344	PF04158:Sof1-like protein; PF00400: G-protein beta WD-40 repeat;		Down	0.00875904	0.487748879	Yes	−1.035789538
8386	134.864	33.0014			Down	5.46E-09	0.244701329	Yes	−2.03090616
8387	46.6693	20.2088			Down	0.00242258	0.43302128	Yes	−1.207490171

Table S2. *Cont.*

Protein Id	BrasWt	BrasFtf1	PFAM annotation	GO annotation	BrasWt -> BrasFtf1: Up/Down	BrasWt -> BrasFtf1: q value	ratio (m/w)	BrasWt -> BrasFtf1: >2-fold and significant difference?	log2 ratio
8388	26.7746	9.93314	PF04997:RNA polymerase Rpb1, domain 1; PF04983:RNA polymerase Rpb1, domain 3; PF00623:RNA polymerase, alpha subunit; PF05000:RNA polymerase Rpb1, domain 4; PF04998:RNA polymerase Rpb1, domain 5;	GO:0003899-DNA-directed RNA polymerase activity; GO:0003902-DNA-directed RNA polymerase III activity; GO:0003900-DNA-directed RNA polymerase I activity; GO:0006350-transcription; GO:0003901-DNA-directed RNA polymerase II activity; GO:0034062-RNA polymerase activity; GO:0003677-DNA binding;	Down	0.000202093	0.370991163	Yes	−1.430543272
8409	51.0921	18.2102			Down	8.71E-05	0.356419094	Yes	−1.488353468
8429	14.822	6.85543	PF00023:Ankyrin;	GO:0030731-guanidinoacetate N-methyltransferase activity;	Down	0.0102905	0.462517204	Yes	−1.112421065
8439	116.495	43.3681	PF00083:General substrate transporter;	GO:0016021-integral to membrane; GO:0006810-transport; GO:0005215-transporter activity;	Down	9.83E-05	0.372274347	Yes	−1.425561892
8442	6.90568	1.82031			Down	1.88E-05	0.263596054	Yes	−1.92359932

Table S2. *Cont.*

Protein Id	BrasWt	BrasFtf1	PFAM annotation	GO annotation	BrasWt -> BrasFtf1: Up/Down	BrasWt -> BrasFtf1: q value	ratio (m/w)	BrasWt -> BrasFtf1: >2-fold and significant difference?	log2 ratio
8447	46.156	22.1981	PF00117:Glutamine amidotransferase class-I;	GO:0003922-GMP synthase (glutamine-hydrolyzing) activity; GO:0003824-catalytic activity; GO:0006177-GMP biosynthetic process; GO:0006541-glutamine metabolic process; GO:0009058-biosynthetic process; GO:0005524-ATP binding;	Down	0.00846797	0.48093639	Yes	−1.056082004
8451	35.9158	14.9029			Down	0.00152145	0.414939943	Yes	−1.269025555
8465	15.1404	3.14765			Down	0.018475	0.207897414	Yes	−2.266056285
8466	5.31053	0.957159	PF02535:Zinc transporter ZIP;	GO:0046873-metal ion transmembrane transporter activity; GO:0016020-membrane; GO:0030001-metal ion transport;	Down	8.42E-07	0.180237942	Yes	−2.472025346
8501	9.19723	1.38311		GO:0004339-glucan 1,4-alpha-glucosidase activity;	Down	3.72E-13	0.150383322	Yes	−2.733283519
8516	148.893	70.1182	PF00324:Amino acid permease-associated region;	GO:0016021-integral to membrane; GO:0006810-transport; GO:0016020-membrane; GO:0015171-amino acid transmembrane transporter activity; GO:0006865-amino acid transport;	Down	0.00352832	0.470930131	Yes	−1.086415063

Table S2. *Cont.*

Protein Id	BrasWt	BrasFtf1	PFAM annotation	GO annotation	BrasWt -> BrasFtf1: Up/Down	BrasWt -> BrasFtf1: q value	ratio (m/w)	BrasWt -> BrasFtf1: >2-fold and significant difference?	log2 ratio
8528	24.6045	11.3868	PF00271:Helicase, C-terminal; PF00270:DEAD/ DEAH box helicase, N-terminal;	GO:0008796-bis(5'-nucleosyl)-tetraphosphatase activity; GO:0008413-8-oxo-7,8-dihydroguanine triphosphatase activity; GO:0019177-dihydroneopterin triphosphate pyrophosphohydrolase activity; GO:0008758-UDP-2,3-diacylglucosamine hydrolase activity; GO:0004787-thiamin-pyrophosphatase activity; GO:0003676-nucleic acid binding; GO:0000810-diacylglycerol pyrophosphate phosphatase activity; GO:0008026-ATP-dependent helicase activity; GO:0004386-helicase activity; GO:0043141-ATP-dependent 5'-3' DNA helicase activity; GO:0008828-dATP pyrophosphohydrolase activity; GO:0016462-pyrophosphatase activity; GO:0019176-dihydroneopterin monophosphate phosphatase activity; GO:0005524-ATP binding;	Down	0.00437538	0.462793391	Yes	−1.111559832
8537	1.13339	0.450387	PF00023:Ankyrin;		Down	0.00846973	0.397380425	Yes	−1.331407288
8641	18.7478	0.440608	PF00043:Glutathione S-transferase, C-terminal; PF02798:Glutathione S-transferase, N-terminal;		Down	5.38E-11	0.023501851	Yes	−5.411081809

Table S2. *Cont.*

Protein Id	BrasWt	BrasFtf1	PFAM annotation	GO annotation	BrasWt -> BrasFtf1: Up/Down	BrasWt -> BrasFtf1: q value	ratio (m/w)	BrasWt -> BrasFtf1: >2-fold and significant difference?	log2 ratio
8671	13.1091	5.60987	PF06027:Eukaryotic protein of unknown function DUF914;		Down	0.00435103	0.427937082	Yes	−1.224529397
8677	13.2484	3.06181			Down	9.52E-09	0.231107907	Yes	−2.113361472
8682	74.1757	20.3061			Down	2.69E-05	0.273756769	Yes	−1.869033455
8683	395.303	183.169			Down	0.00317365	0.463363546	Yes	−1.109783545
8684	276.432	117.628			Down	0.00103778	0.425522371	Yes	−1.232693115
8714	23.9885	11.5952			Down	0.0253896	0.483364946	Yes	−1.048815244
8716	5.18921	2.51255		GO:0016021-integral to membrane; GO:0015103-inorganic anion transmembrane transporter activity; GO:0015698-inorganic anion transport; GO:0005215-transporter activity; GO:0006810-transport;	Down	0.0298631	0.484187381	Yes	−1.046362615
8741	5.84647	1.98037	PF00096:Zn-finger, C2H2 type;	GO:0008270-zinc ion binding; GO:0003676-nucleic acid binding; GO:0005622-intracellular;	Down	0.000663146	0.338729182	Yes	−1.561795813
8746	6.95071	1.70308			Down	0.000201792	0.245022451	Yes	−2.029014148
8757	2.44339	1.07281	PF01565:FAD linked oxidase, N-terminal;	GO:0050660-FAD binding; GO:0016491-oxidoreductase activity;	Down	0.0349186	0.439066215	Yes	−1.187489567
8803	43.9283	13.7027	PF00076: RNA-binding region RNP-1 (RNA recognition motif);	GO:0003676-nucleic acid binding;	Down	3.01E-06	0.31193331	Yes	−1.680690477

Table S2. *Cont.*

Protein Id	BrasWt	BrasFtf1	PFAM annotation	GO annotation	BrasWt -> BrasFtf1: Up/Down	BrasWt -> BrasFtf1: q value	ratio (m/w)	BrasWt -> BrasFtf1: >2-fold and significant difference?	log2 ratio
8812	3.22242	1.44029		GO:0005634-nucleus;	Down	0.0116518	0.446959118	Yes	−1.161785217
8964	637.702	196.772			Down	4.47E-06	0.308564188	Yes	−1.696357461
8969	19.8842	5.67985			Down	1.72E-06	0.285646393	Yes	−1.807697784
8998	14.896	6.19759			Down	0.0025448	0.416057331	Yes	−1.265145756
9003	36.3767	14.5507	PF00560: Leucine-rich repeat;	GO:0005515-protein binding;	Down	0.000482821	0.40000055	Yes	−1.321926112
9026	52.7695	16.6804	PF02265:S1/P1 nuclease;	GO:0006308-DNA catabolic process; GO:0003676-nucleic acid binding; GO:0004519-endonuclease activity;	Down	8.26E-06	0.316099262	Yes	−1.661550428
9077	40.2302	16.0981		GO:0004339-glucan 1,4-alpha-glucosidase activity;	Down	0.000248533	0.400149639	Yes	−1.321388488
9087	3.31719	1.41047	PF01179:Copper amine oxidase;	GO:0005507-copper ion binding; GO:0009308-amine metabolic process; GO:0008131-amine oxidase activity; GO:0048038-quinone binding;	Down	0.0161979	0.425200245	Yes	−1.233785667
9100	6.14268	2.55063			Down	0.00493829	0.415230811	Yes	−1.268014595
9105	41.5078	14.0953			Down	1.93E-05	0.339581958	Yes	−1.558168284
9128	6.52979	2.6932			Down	0.0307898	0.41244818	Yes	−1.277715225

Table S2. *Cont.*

Protein Id	BrasWt	BrasFtf1	PFAM annotation	GO annotation	BrasWt -> BrasFtf1: Up/Down	BrasWt -> BrasFtf1: q value	ratio (m/w)	BrasWt -> BrasFtf1: >2-fold and significant difference?	log2 ratio
9129	8.76625	3.98573		GO:0043773-coenzyme F420-0 gamma-glutamyl ligase activity; GO:0043774-coenzyme F420-2 alpha-glutamyl ligase activity; GO:0008766-UDP-N-acetylmuramoylalanyl-D-glutamyl-2,6-diaminopimelate-D-alanyl-D-alanine ligase activity; GO:0018169-ribosomal S6-glutamic acid ligase activity;	Down	0.0394842	0.454667617	Yes	−1.13711584
9130	12.4582	5.78906	PF00249:Myb, DNA-binding;	GO:0043773-coenzyme F420-0 gamma-glutamyl ligase activity; GO:0018169-ribosomal S6-glutamic acid ligase activity; GO:0043774-coenzyme F420-2 alpha-glutamyl ligase activity; GO:0008766-UDP-N-acetylmuramoylalanyl-D-glutamyl-2,6-diaminopimelate-D-alanyl-D-alanine ligase activity; GO:0003677-DNA binding;	Down	0.0160715	0.464678686	Yes	−1.105694624
9183	10.2482	0.969116			Down	3.97E-11	0.094564509	Yes	−3.402557364

Table S2. *Cont.*

Protein Id	BrasWt	BrasFtf1	PFAM annotation	GO annotation	BrasWt -> BrasFtf1: Up/Down	BrasWt -> BrasFtf1: q value	ratio (m/w)	BrasWt -> BrasFtf1: >2-fold and significant difference?	log2 ratio
9207	30.8262	14.0069	PF06870:A49-like RNA polymerase I associated factor;	GO:0003899-DNA-directed RNA polymerase activity; GO:0003902-DNA-directed RNA polymerase III activity; GO:0003900-DNA-directed RNA polymerase I activity; GO:0006350-transcription; GO:0003901-DNA-directed RNA polymerase II activity; GO:0034062-RNA polymerase activity; GO:0003677-DNA binding; GO:0005634-nucleus;	Down	0.00364066	0.45438296	Yes	−1.138019362
9213	14.3849	4.61256			Down	1.26E-05	0.320652907	Yes	−1.640915609
9273	2.58729	0.144069	PF03330:Rare lipoprotein A;	GO:0050832-defense response to fungus; GO:0042742-defense response to bacterium;	Down	0.000202093	0.05568336	Yes	−4.166609927
9290	40.6225	13.5983			Down	2.37E-05	0.334747984	Yes	−1.578852726
9291	43.0611	15.3991			Down	0.00115193	0.357610465	Yes	−1.483539139
9321	32.9689	16.1723	PF00856:Nuclear protein SET;	GO:0018024-histone-lysine N-methyltransferase activity; GO:0005634-nucleus;	Down	0.00998452	0.490531986	Yes	−1.027580881

Table S2. *Cont.*

Protein Id	BrasWt	BrasFtf1	PFAM annotation	GO annotation	BrasWt -> BrasFtf1: Up/Down	BrasWt -> BrasFtf1: q value	ratio (m/w)	BrasWt -> BrasFtf1: >2-fold and significant difference?	log2 ratio
9361	1525.1	735.348		GO:0003900-DNA-directed RNA polymerase I activity; GO:0003902-DNA-directed RNA polymerase III activity; GO:0003899-DNA-directed RNA polymerase activity; GO:0034062-RNA polymerase activity; GO:0003901-DNA-directed RNA polymerase II activity;	Down	0.0147148	0.482163793	Yes	−1.052404777
9411	2.86951	0.381405			Down	0.00088359	0.132916421	Yes	−2.91140874
9450	18.8773	6.62952	PF07653:Variant SH3; PF00018:SH3;	GO:0043773-coenzyme F420-0 gamma-glutamyl ligase activity; GO:0016020-membrane; GO:0018169-ribosomal S6-glutamic acid ligase activity; GO:0016491-oxidoreductase activity; GO:0003700-transcription factor activity; GO:0043774-coenzyme F420-2 alpha-glutamyl ligase activity; GO:0043565-sequence-specific DNA binding; GO:0006118-electron transport; GO:0046983-protein dimerization activity; GO:0008766-UDP-N-acetylmuramoylalanyl-D-glutamyl-2,6-diaminopimelate-D-alanyl-D-alanine ligase activity; GO:0006355-regulation of transcription, DNA-dependent; GO:0005634-nucleus;	Down	0.00022671	0.351190054	Yes	−1.509676109
9451	17.9575	7.05554			Down	0.00962212	0.39290213	Yes	−1.347758106

Table S2. *Cont.*

Protein Id	BrasWt	BrasFtf1	PFAM annotation	GO annotation	BrasWt -> BrasFtf1: Up/Down	BrasWt -> BrasFtf1: q value	ratio (m/w)	BrasWt -> BrasFtf1: >2-fold and significant difference?	log2 ratio
9486	30.8897	14.5795	PF04565:RNA polymerase Rpb2, domain 3; PF06883:RNA polymerase I, Rpa2 specific; PF04561:RNA polymerase Rpb2, domain 2; PF04563:RNA polymerase beta subunit; PF04567:RNA polymerase Rpb2, domain 5; PF04560:RNA polymerase Rpb2, domain 7; PF00562:RNA polymerase Rpb2, domain 6;	GO:0003899-DNA-directed RNA polymerase activity; GO:0003902-DNA-directed RNA polymerase III activity; GO:0003900-DNA-directed RNA polymerase I activity; GO:0006350-transcription; GO:0003901-DNA-directed RNA polymerase II activity; GO:0034062-RNA polymerase activity; GO:0003677-DNA binding; GO:0005634-nucleus;	Down	0.00535819	0.471985808	Yes	−1.083184616

Table S2. *Cont.*

Protein Id	BrasWt	BrasFtf1	PFAM annotation	GO annotation	BrasWt -> BrasFtf1: Up/Down	BrasWt -> BrasFtf1: q value	ratio (m/w)	BrasWt -> BrasFtf1: >2-fold and significant difference?	log2 ratio
9515	9.65228	2.25372	PF07539: Down-regulated in metastasis;		Down	6.69E-07	0.233490947	Yes	−2.098561479
9516	14.1229	3.34303			Down	1.29E-07	0.236709883	Yes	−2.078808156
9535	13.436	4.90129			Down	0.000668686	0.364787883	Yes	−1.454870285
9566	41.2346	16.5034	PF00271:Helicase, C-terminal; PF00270:DEAD/DEAH box helicase, N-terminal;	GO:0008796-bis(5'-nucleosyl)-tetraphosphatase activity; GO:0008413-8-oxo-7,8-dihydroguanine triphosphatase activity; GO:0019177-dihydroneopterin triphosphate pyrophosphohydrolase activity; GO:0008758-UDP-2,3-diacylglucosamine hydrolase activity; GO:0004787-thiamin-pyrophosphatase activity; GO:0003676-nucleic acid binding; GO:0000810-diacylglycerol pyrophosphate phosphatase activity; GO:0008026-ATP-dependent helicase activity; GO:0004386-helicase activity; GO:0043141-ATP-dependent 5'-3' DNA helicase activity; GO:0008828-dATP pyrophosphohydrolase activity; GO:0016462-pyrophosphatase activity; GO:0019176-dihydroneopterin monophosphate phosphatase activity; GO:0005524-ATP binding;	Down	0.000335188	0.400231844	Yes	−1.321092136

Table S2. *Cont.*

Protein Id	BrasWt	BrasFtfl	PFAM annotation	GO annotation	BrasWt -> BrasFtf1: Up/Down	BrasWt -> BrasFtf1: q value	ratio (m/w)	BrasWt -> BrasFtf1: >2-fold and significant difference?	log2 ratio
9571	26.1015	8.87041	PF06628: Catalase-related; PF00199:Catalase;	GO:0006979-response to oxidative stress; GO:0020037-heme binding; GO:0004096-catalase activity; GO:0005506-iron ion binding; GO:0006118-electron transport;	Down	1.88E-05	0.339842921	Yes	−1.557060024
9660	74.1273	31.1344			Down	0.00147861	0.4200126	Yes	−1.251495487
9677	6.16882	2.31164	PF05922:Proteinase inhibitor I9, subtilisin propeptide; PF00082:Peptidase S8 and S53, subtilisin, kexin, sedolisin;	GO:0006508-proteolysis; GO:0016808-proprotein convertase activity; GO:0008132-pancreatic elastase activity; GO:0008243-plasminogen activator activity; GO:0008991-serine-type signal peptidase activity; GO:0042802-identical protein binding; GO:0043086-negative regulation of catalytic activity; GO:0004289-subtilase activity;	Down	0.00351084	0.374729689	Yes	−1.416077811
9685	36.0138	16.9911	PF02548: Ketopantoate hydroxymethyltransferase;	GO:0015940-pantothenate biosynthetic process; GO:0003864-3-methyl-2-oxobutanoate hydroxymethyltransferase activity;	Down	0.00701346	0.471794146	Yes	−1.083770578
9762	5.48278	2.39757	PF01055:Glycoside hydrolase, family 31;	GO:0004558-alpha-glucosidase activity; GO:0004553-hydrolase activity, hydrolyzing O-glycosyl compounds; GO:0005975-carbohydrate metabolic process;	Down	0.00938551	0.437290936	Yes	−1.193334649
9763	34.86	17.2741	PF01171:PP-loop;		Down	0.011768	0.495527826	Yes	−1.012962023

Table S2. *Cont.*

Protein Id	BrasWt	BrasFtf1	PFAM annotation	GO annotation	BrasWt -> BrasFtf1: Up/Down	BrasWt -> BrasFtf1: q value	ratio (m/w)	BrasWt -> BrasFtf1: >2-fold and significant difference?	log2 ratio
9818	10.9173	5.32807	PF02736:Myosin, N-terminal, SH3-like; PF00063:Myosin head, motor region; PF00612:IQ calmodulin-binding region;	GO:0008570-myosin ATPase activity; GO:0016459-myosin complex; GO:0005524-ATP binding; GO:0003774-motor activity;	Down	0.0166858	0.488039167	Yes	−1.03493116
9831	809.996	184.234	PF02772: S-adenosylmethionine synthetase; PF02773: S-adenosylmethionine synthetase; PF00438: S-adenosylmethionine synthetase;	GO:0048269-methionine adenosyltransferase complex; GO:0006730-one-carbon compound metabolic process; GO:0004478-methionine adenosyltransferase activity; GO:0048270-methionine adenosyltransferase regulator activity; GO:0005524-ATP binding;	Down	3.61E-10	0.227450506	Yes	−2.136375451
9870	122.867	34.6173		GO:0004308-exo-alpha-sialidase activity;	Down	3.27E-07	0.281746116	Yes	−1.827532374
9876	11.1274	2.43941	PF04479:RTA1 like protein;	GO:0006950-response to stress; GO:0016021-integral to membrane;	Down	2.06E-06	0.219225515	Yes	−2.189512373
9913	244.727	66.8994			Down	7.18E-08	0.27336338	Yes	−1.871108101
9925	1.58414	0.16456	PF00291:Pyridoxal-5'-phosphate-dependent enzyme, beta subunit;	GO:0003824-catalytic activity; GO:0008152-metabolic process; GO:0030170-pyridoxal phosphate binding;	Down	0.000459476	0.103879708	Yes	−3.267014237
9927	83.6726	26.6107			Down	2.83E-06	0.318033622	Yes	−1.652748804
9933	17.6774	7.44883			Down	0.00414528	0.421375881	Yes	−1.246820356

Table S2. *Cont.*

Protein Id	BrasWt	BrasFtf1	PFAM annotation	GO annotation	BrasWt -> BrasFtf1: Up/Down	BrasWt -> BrasFtf1: q value	ratio (m/w)	BrasWt -> BrasFtf1: >2-fold and significant difference?	log2 ratio
9936	24.628	9.03773			Down	0.000169264	0.366969709	Yes	−1.446267111
9938	129.661	59.6979			Down	0.00353834	0.460415237	Yes	−1.118992517
9940	71.8729	10.7764	PF01565:FAD linked oxidase, N-terminal;	GO:0050660-FAD binding; GO:0016491-oxidoreductase activity;	Down	0	0.149936903	Yes	−2.737572591
9941	5.21651	1.1831			Down	4.14E-05	0.226799143	Yes	−2.140512903
9956	73.5828	31.8829	PF00005:ABC transporter;	GO:0000166-nucleotide binding; GO:0017111-nucleoside-triphosphatase activity; GO:0016887-ATPase activity; GO:0005524-ATP binding;	Down	0.00207914	0.433292835	Yes	−1.206585716
10003	30.9831	11.4225	PF01794:Ferric reductase-like transmembrane component;	GO:0000293-ferric-chelate reductase activity;	Down	6.39E-05	0.368668726	Yes	−1.439603057
10011	1.53149	0.519311		GO:0016021-integral to membrane; GO:0006810-transport; GO:0005215-transporter activity;	Down	0.0211886	0.339088731	Yes	−1.560265257
10035	24.6497	9.69561			Down	0.000370058	0.393335822	Yes	−1.346166515
10036	17.1746	6.44964	PF00400:G-protein beta WD-40 repeat;		Down	0.000333335	0.375533637	Yes	−1.412985958
10039	21.7409	6.61289			Down	0.000763242	0.304168181	Yes	−1.717058855
10125	212.828	54.2884	PF03928:Protein of unknown function DUF336;		Down	3.97E-07	0.255081098	Yes	−1.970972096

Table S2. *Cont.*

Protein Id	BrasWt	BrasFtf1	PFAM annotation	GO annotation	BrasWt -> BrasFtf1: Up/Down	BrasWt -> BrasFtf1: q value	ratio (m/w)	BrasWt -> BrasFtf1: >2-fold and significant difference?	log2 ratio
10190	4.18266	1.08609	PF02458:Transferase;		Down	0.000103671	0.259664902	Yes	−1.94527707
10199	22.3917	8.20601			Down	0.000441096	0.366475524	Yes	−1.448211246
10221	174.852	52.7274		GO:0019131-tripeptidyl-peptidase I activity;	Down	4.01E-07	0.301554457	Yes	−1.729509534
10226	335.421	56.8231	PF01717:Methionine synthase, vitamin-B12 independent;	GO:0009086-methionine biosynthetic process; GO:0003871-5-methyltetrahydropteroyltriglutamate-homocysteine S-methyltransferase activity;	Down	0	0.169408296	Yes	−2.56142357
10230	6.8258	3.22196	PF06165:Glycosyltransferase 36;		Down	0.0135608	0.472026722	Yes	−1.08305956
10251	17.5512	7.02186	PF04006:Mpp10 protein;		Down	0.000653986	0.400078627	Yes	−1.321644535
10288	41.3201	14.865	PF00487:Fatty acid desaturase;	GO:0050184-phosphatidylcholine desaturase activity; GO:0016020-membrane; GO:0006629-lipid metabolic process; GO:0016717-oxidoreductase activity, acting on paired donors, with oxidation of a pair of donors resulting in the reduction of molecular oxygen to two molecules of water; GO:0016491-oxidoreductase activity;	Down	6.76E-05	0.359752276	Yes	−1.474924282
10304	1.19637	0.295468	PF00023:Ankyrin;	GO:0003950-NAD+ ADP-ribosyltransferase activity;	Down	0.00686401	0.246970419	Yes	−2.017589843

Table S2. *Cont.*

Protein Id	BrasWt	BrasFtf1	PFAM annotation	GO annotation	BrasWt -> BrasFtf1: Up/Down	BrasWt -> BrasFtf1: q value	ratio (m/w)	BrasWt -> BrasFtf1: >2-fold and significant difference?	log2 ratio
10331	18.7264	5.29312	PF04615:Utp14 protein;	GO:0003918-DNA topoisomerase (ATP-hydrolyzing) activity; GO:0005694-chromosome; GO:0003910-DNA ligase (ATP) activity; GO:0006364-rRNA processing; GO:0006265-DNA topological change; GO:0003677-DNA binding; GO:0032040-small subunit processome; GO:0005524-ATP binding;	Down	8.70E-07	0.282655502	Yes	−1.822883313
10335	76.63	36.5383			Down	0.0100982	0.476814563	Yes	−1.068499794
10353	13.9413	5.93482		GO:0006979-response to oxidative stress; GO:0004601-peroxidase activity; GO:0006118-electron transport; GO:0020037-heme binding;	Down	0.00304256	0.425700616	Yes	−1.232088917
10355	24.0578	8.52503	PF00078: RNA-directed DNA polymerase (Reverse transcriptase);	GO:0006278-RNA-dependent DNA replication; GO:0003964-RNA-directed DNA polymerase activity; GO:0003723-RNA binding;	Down	0.000678851	0.354356176	Yes	−1.496727903
10358	16.4491	7.78329	PF00026:Peptidase A1, pepsin;	GO:0004194-pepsin A activity; GO:0004190-aspartic-type endopeptidase activity; GO:0006508-proteolysis;	Down	0.0208218	0.473174216	Yes	−1.079556633

Table S2. *Cont.*

Protein Id	BrasWt	BrasFtf1	PFAM annotation	GO annotation	BrasWt -> BrasFtf1: Up/Down	BrasWt -> BrasFtf1: q value	ratio (m/w)	BrasWt -> BrasFtf1: >2-fold and significant difference?	log2 ratio
10359	20.6941	10.0442	PF00271:Helicase, C-terminal; PF00270:DEAD/ DEAH box helicase, N-terminal;	GO:0008796-bis(5'-nucleosyl)-tetraphosphatase activity; GO:0008413-8-oxo-7,8-dihydroguanine triphosphatase activity; GO:0019177-dihydroneopterin triphosphate pyrophosphohydrolase activity; GO:0008758-UDP-2,3-diacylglucosamine hydrolase activity; GO:0004787-thiamin-pyrophosphatase activity; GO:0003676-nucleic acid binding; GO:0000810-diacylglycerol pyrophosphate phosphatase activity; GO:0008026-ATP-dependent helicase activity; GO:0004386-helicase activity; GO:0043141-ATP-dependent 5'-3' DNA helicase activity; GO:0008828-dATP pyrophosphohydrolase activity; GO:0016462-pyrophosphatase activity; GO:0019176-dihydroneopterin monophosphate phosphatase activity; GO:0005524-ATP binding;	Down	0.0102604	0.485365394	Yes	−1.042856845
10378	12.9459	6.039	PF00929:Exonuclease ;	GO:0004527-exonuclease activity; GO:0005622-intracellular;	Down	0.0103715	0.466479735	Yes	−1.100113687
10379	583.28	70.4902			Down	0	0.120851392	Yes	−3.048694003
10394	115.15	50.0322			Down	0.0149372	0.434495875	Yes	−1.202585615
10395	25.7915	7.8487			Down	0.000430899	0.304313437	Yes	−1.716370059

Table S2. *Cont.*

Protein Id	BrasWt	BrasFtf1	PFAM annotation	GO annotation	BrasWt -> BrasFtf1: Up/Down	BrasWt -> BrasFtf1: q value	ratio (m/w)	BrasWt -> BrasFtf1: >2-fold and significant difference?	log2 ratio
10404	32.1771	7.39248	PF03211:Pectate lyase;	GO:0005576-extracellular region; GO:0030570-pectate lyase activity;	Down	8.64E-07	0.229743513	Yes	−2.121903967
10411	72.4703	12.3519	PF00083:General substrate transporter;	GO:0016021-integral to membrane; GO:0016020-membrane; GO:0005215-transporter activity; GO:0008643-carbohydrate transport; GO:0006810-transport; GO:0005351-sugar:hydrogen symporter activity;	Down	1.23E-12	0.170440856	Yes	−2.552656889
10425	38.2385	15.3808			Down	0.00249269	0.402233351	Yes	−1.313895387
10429	41.7716	14.4502			Down	0.00088359	0.345933601	Yes	−1.531432944
10436	79.7371	26.8934			Down	0.00107457	0.337275873	Yes	−1.567998977
10458	542.748	257		GO:0003676-nucleic acid binding;	Down	0.0070394	0.473516254	Yes	−1.078514145
10460	45.7739	22.8324			Down	0.0262247	0.498808273	Yes	−1.003442702
10473	57.7929	21.4812			Down	0.0157834	0.371692717	Yes	−1.427817676
10475	52.3029	21.2215			Down	0.00148606	0.40574232	Yes	−1.301364307
10497	44.7427	17.2431			Down	0.00235034	0.385383537	Yes	−1.37563315
10505	11.4852	3.36424			Down	5.17E-05	0.292919583	Yes	−1.771423445
10514	10.3811	3.10163	PF00271:Helicase, C-terminal;	GO:0008026-ATP-dependent helicase activity; GO:0003676-nucleic acid binding; GO:0004386-helicase activity; GO:0005524-ATP binding;	Down	9.65E-05	0.298776623	Yes	−1.742860823

Table S2. *Cont.*

Protein Id	BrasWt	BrasFtf1	PFAM annotation	GO annotation	BrasWt -> BrasFtf1: Up/Down	BrasWt -> BrasFtf1: q value	ratio (m/w)	BrasWt -> BrasFtf1: >2-fold and significant difference?	log2 ratio
10578	44.2368	16.1966	PF00043:Glutathione S-transferase, C-terminal; PF02798:Glutathione S-transferase, N-terminal;	GO:0004364-glutathione transferase activity;	Down	0.00122963	0.366134078	Yes	−1.449556034
10601	26.8354	11.0303			Down	0.00364066	0.411035423	Yes	−1.282665363
10622	4.23214	1.4758			Down	0.0449103	0.348712472	Yes	−1.519890132
10647	4.59805	0	PF04110:Autophagy protein Apg12;	GO:0005737-cytoplasm; GO:0000045-autophagic vacuole formation;	Down	0.0179714	0	Yes	#NUM!
10651	13.2598	5.78303			Down	0.0272349	0.436132521	Yes	−1.197161524
10658	7.47474	1.3916	PF00096:Zn-finger, C2H2 type;	GO:0008270-zinc ion binding; GO:0003676-nucleic acid binding; GO:0005622-intracellular;	Down	5.58E-05	0.1861737	Yes	−2.425278814
10662	93.8337	36.2808			Down	0.00415148	0.386649999	Yes	−1.370899886

qRT-PCR results were calculated by delta delta Ct = [(Ct (gene i)-C t(Ef1-alpha) in Bdtf1 mutant -(Ct (gene i) - Ct (Ef1-alpha)) in wild type.

Table S3. Differentially expressed genes in both experimental conditions.

YORF	NAME	Log2Brassinin	Log2-44hpi
AB00715		1.67579972	1.116270645
AB03063	GO:0004553-hydrolase activity, hydrolyzing O-glycosyl compounds; GO:0005975-carbohydrate metabolic process;	1.274022617	1.378752976
AB04305		6.886883587	2.514336528
AB04350		1.475596756	1.163084323
AB05507	GO:0050660-FAD binding; GO:0008061-chitin binding; GO:0016491-oxidoreductase activity;	1.799203168	1.248669899
AB06313	PF07249:Cerato-platanin;	1.317647365	2.208640959
AB06321	GO:0003824-catalytic activity;	1.29121343	1.407199944
AB06545		1.321987635	1.118180568
AB06581	GO:0016614-oxidoreductase activity, acting on CH-OH group of donors; GO:0050660-FAD binding;	1.764132142	1.254261961
AB06582	GO:0016491-oxidoreductase activity; GO:0005506-iron ion binding; GO:0009058-biosynthetic process;	1.596881718	1.360414807
AB06621		2.115560969	1.081926308
AB06750		1.356001398	1.295178041
AB07373	GO:0016021-integral to membrane; GO:0042626-ATPase activity, coupled to transmembrane movement of substances; GO:0017111-nucleoside-triphosphatase activity; GO:0006810-transport; GO:0000166-nucleotide binding; GO:0016887-ATPase activity; GO:0005524-ATP binding;	1.056976047	1.426467702
AB07374	GO:0016887-ATPase activity; GO:0005524-ATP binding;	1.424145438	1.406582463
AB09101		1.193073748	1.045927474
AB01239		1.587084184	−1.083223604
AB02056		1.280036817	−1.252472448
AB03010	GO:0016831-carboxy-lyase activity;	1.023876012	−1.112104269
AB03012	GO:0003824-catalytic activity; GO:0008152-metabolic process;	1.306713453	−1.363519561
AB03789	PF04419:Four F5 protein;	1.20765705	−2.578593883
AB03939		1.599590925	−1.127070359
AB04228		1.16384075	−1.130433622
AB08566		2.097684261	−1.133610751
AB00053	GO:0051700-fructosyl-amino acid oxidase activity; GO:0006562-proline catabolic process; GO:0006537-glutamate biosynthetic process; GO:0004657-proline dehydrogenase activity; GO:0004154-dihydropterin oxidase activity; GO:0019116-hydroxy-nicotine oxidase activity; GO:0051699-proline oxidase activity;	−1.118367664	1.474363111
AB00094		−1.095452539	1.152944849
AB01487		−1.491990076	1.273764433
AB02424	GO:0016985-mannan endo-1,4-beta-mannosidase activity; GO:0006080-substituted mannan metabolic process;	−1.520647574	1.209984086

Table S3. *Cont.*

YORF	NAME	Log2Brassinin	Log2-44hpi
AB04330		−1.35142594	1.10712669
AB05539	GO:0016491-oxidoreductase activity;	−1.606699243	1.399863753
AB06658	GO:0006730-one-carbon compound metabolic process; GO:0008270-zinc ion binding; GO:0004089-carbonate dehydratase activity;	−2.547078068	1.058166552
AB06928		−1.159949656	1.146580914
AB09087	GO:0005507-copper ion binding; GO:0009308-amine metabolic process; GO:0008131-amine oxidase activity; GO:0048038-quinone binding;	−1.233785667	1.243864121
AB09677	GO:0006508-proteolysis; GO:0016808-proprotein convertase activity; GO:0008132-pancreatic elastase activity; GO:0008243-plasminogen activator activity; GO:0008991-serine-type signal peptidase activity; GO:0042802-identical protein binding; GO:0043086-negative regulation of catalytic activity; GO:0004289-subtilase activity;	−1.416077811	1.017398004
AB09762	GO:0004558-alpha-glucosidase activity; GO:0004553-hydrolase activity, hydrolyzing O-glycosyl compounds; GO:0005975-carbohydrate metabolic process;	−1.193334649	1.11184155
AB09925	GO:0003824-catalytic activity; GO:0008152-metabolic process; GO:0030170-pyridoxal phosphate binding;	−3.267014237	1.424642798
AB10039		−1.717058855	1.030826011
AB10379		−3.048694003	1.016679319
AB01296	GO:0043826-sulfur oxygenase reductase activity; GO:0043738-F420H2 dehydrogenase activity; GO:0043883-malolactic enzyme activity; GO:0051341-regulation of oxidoreductase activity; GO:0016491-oxidoreductase activity;	−1.204798632	−1.108489659
AB01545		−1.012813156	−1.024619333
AB02662	GO:0016491-oxidoreductase activity;	−1.084650107	−1.183557059
AB02663	GO:0016787-hydrolase activity; GO:0008152-metabolic process;	−1.445022153	−1.65925687
AB03046	GO:0016021-integral to membrane; GO:0016020-membrane; GO:0005215-transporter activity; GO:0008643-carbohydrate transport; GO:0006810-transport; GO:0005351-sugar:hydrogen symporter activity;	−2.423396413	−1.142734226
AB03484		−2.50880536	−1.563628317
AB03487		−2.47777504	−1.018454917
AB03793		−2.059289571	−1.319647901
AB04177		−2.304950836	−1.066806113
AB05764	GO:0008482-sulfite oxidase activity; GO:0030151-molybdenum ion binding; GO:0006118-electron transport; GO:0016491-oxidoreductase activity;	−1.761524243	−2.421054615

Table S3. *Cont.*

YORF	NAME	Log2Brassinin	Log2-44hpi
AB07899	GO:0003700-transcription factor activity; GO:0006350-transcription; GO:0008270-zinc ion binding; GO:0003677-DNA binding; GO:0006355-regulation of transcription, DNA-dependent; GO:0005634-nucleus;	−9	−9.113878724
AB08215	GO:0003856-3-dehydroquinate synthase activity; GO:0009073-aromatic amino acid family biosynthetic process;	−1.587809229	−2.740463459
AB08641	PF00043:Glutathione S-transferase, C-terminal; PF02798:Glutathione S-transferase, N-terminal;	−5.411081809	−3.606161107
AB09938		−1.118992517	−1.294929711
AB10411	GO:0016021-integral to membrane; GO:0016020-membrane; GO:0005215-transporter activity; GO:0008643-carbohydrate transport; GO:0006810-transport; GO:0005351-sugar:hydrogen symporter activity;	−2.552656889	−1.120649638

qRT-PCR results were calculated by delta delta Ct = [(Ct (gene i)-C t(Efl-alpha) in Bdtfl mutant -(Ct (gene i) - Ct (Efl-alpha)) in wild type.

Acknowledgments

We thank Fred Brooks for insightful discussions on the roles of the *Bdtf1* gene in brassinin digestion. This research was supported by USDA-TSTAR 2009-34135-20197 to YC, administered by the College of Tropical Agriculture and Human Resources, University of Hawaii at Manoa, Honolulu, HI and WCI2009-002 to BYK, and GRDC, NRF-2010-0079 to JSA administered by the Nation Research Foundation of Korea. This research was also support by Grants from the Chuncheongbuk-do and the KRIBB Research Initiative Program.

Author Contributions

YC, SBL, BYK, and JSA designed the study, analyzed data, and wrote the manuscript. YC, RO, and VI analyzed data.

Conflicts of Interest

The authors declare no conflict of interest.

References

1. Takasugi, M.; Katsui, N.; Shirata, A., Isolation of three novel sulphur-containing phytoalexins from the chinese cabbage *Brassica campestris* L. ssp. *pekinensis* (cruciferae). *J. Chem. Soc. Chem. Commun.* **1986**, 1077–1078.
2. Pedras, M.S.; Chumala, P.B.; Jin, W.; Islam, M.S.; Hauck, D.W., The phytopathogenic fungus *Alternaria brassicicola*: Phytotoxin production and phytoalexin elicitation. *Phytochemistry* **2009**, *70*, 394–402.

3. Govrin, E.M.; Levine, A., Infection of *Arabidopsis* with a necrotrophic pathogen, *Botrytis cinerea*, elicits various defense responses but does not induce systemic acquired resistance (SAR). *Plant Mol. Biol.* **2002**, *48*, 267–276.
4. Sellam, A.; Iacomi-Vasilescu, B.; Hudhomme, P.; Simoneau, P. *In vitro* antifungal activity of brassinin, camalexin and two isothiocyanates against the crucifer pathogens *Alternaria brassicicola* and *Alternaria brassicae*. *Plant Pathol.* **2007**, *56*, 296–301.
5. Sellam, A.; Dongo, A.; Guillemette, T.; Hudhomme, P.; Simoneau, P. Transcriptional responses to exposure to the brassicaceous defence metabolites camalexin and allyl-isothiocyanate in the necrotrophic fungus *Alternaria brassicicola*. *Mol. Plant. Pathol.* **2007**, *8*, 195–208.
6. Joubert, A.; Bataille-Simoneau, N.; Campion, C.; Guillemette, T.; Hudhomme, P.; Iacomi-Vasilescu, B.; Leroy, T.; Pochon, S.; Poupard, P.; Simoneau, P. Cell wall integrity and high osmolarity glycerol pathways are required for adaptation of *Alternaria brassicicola* to cell wall stress caused by brassicaceous indolic phytoalexins. *Cell Microbiol.* **2011**, *13*, 62–80.
7. Rogers, E.E.; Glazebrook, J.; Ausubel, F.M. Mode of action of the *Arabidopsis thaliana* phytoalexin camalexin and its role in Arabidopsis-pathogen interactions. *Mol. Plant. Microbe Interact.* **1996**, *9*, 748–757.
8. Pedras, M.S.; Ahiahonu, P.W.; Hossain, M. Detoxification of the cruciferous phytoalexin brassinin in *Sclerotinia sclerotiorum* requires an inducible glucosyltransferase. *Phytochemistry* **2004**, *65*, 2685–2694.
9. Pedras, M.S.; Minic, Z.; Jha, M. Brassinin oxidase, a fungal detoxifying enzyme to overcome a plant defense—Purification, characterization and inhibition. *FEBS J.* **2008**, *275*, 3691–3705.
10. Pedras, M.S.; Yaya, E.E.; Glawischnig, E. The phytoalexins from cultivated and wild crucifers: Chemistry and biology. *Nat. Prod. Rep.* **2011**, *28*, 1381–1405.
11. Srivastava, A.; Cho, I.K.; Cho, Y. The *Bdtf1* gene in *Alternaria brassicicola* is important in detoxifying brassinin and maintaining virulence on Brassica species. *Mol. Plant. Microbe Interact.* **2013**, *26*, 1429–1440.
12. Pedras, M.S.; Minic, Z.; Sarma-Mamillapalle, V.K. Substrate specificity and inhibition of brassinin hydrolases, detoxifying enzymes from the plant pathogens *Leptosphaeria maculans* and *Alternaria brassicicola*. *FEBS J.* **2009**, *276*, 7412–7428.
13. Cho, Y.; Davis, J.W.; Kim, K.H.; Wang, J.; Sun, Q.H.; Cramer, R.A.J.; Lawrence, C.B. A high throughput targeted gene disruption method for *Alternaria brassicicola* functional genomics using linear minimal element (LME) constructs. *Mol. Plant. Microbe Interact.* **2006**, *19*, 7–15.
14. Ohm, R.A.; Feau, N.; Henrissat, B.; Schoch, C.L.; Horwitz, B.A.; Barry, K.W.; Condon, B.J.; Copeland, A.C.; Dhillon, B.; Glaser, F.; *et al.* Diverse lifestyles and strategies of plant pathogenesis encoded in the genomes of eighteen Dothideomycetes fungi. *PLoS Pathog.* **2012**, *8*, e1003037.
15. Pedras, M.S.; Minic, Z. Differential protein expression in response to the phytoalexin brassinin allows the identification of molecular targets in the phytopathogenic fungus *Alternaria brassicicola*. *Mol. Plant Pathol* **2012**, *13*, 483–493.

16. Pedras, M.S.; Minic, Z.; Sarma-Mamillapalle, V.K. Synthetic inhibitors of the fungal detoxifying enzyme brassinin oxidase based on the phytoalexin camalexin scaffold. *J. Agric. Food Chem.* **2009**, *57*, 2429–2435.
17. Coleman, J.J.; Wasmann, C.C.; Usami, T.; White, G.J.; Temporini, E.D.; McCluskey, K.; VanEtten, H.D. Characterization of the gene encoding pisatin demethylase (*FoPDA1*) in *Fusarium oxysporum*. *Mol. Plant. Microbe Interact.* **2011**, *24*, 1482–1491.
18. Schafer, W.; Straney, D.; Ciuffetti, L.; VanEtten, H.D.; Yoder, O.C. One enzyme makes a fungal pathogen, but not a saprophyte, virulent on a new host plant. *Science* **1989**, *246*, 247–249.
19. Pedras, M.S.; Hossain, S.; Snitynsky, R.B. Detoxification of cruciferous phytoalexins in *Botrytis cinerea*: Spontaneous dimerization of a camalexin metabolite. *Phytochemistry* **2011**, *72*, 199–206.
20. Pedras, M.S.; Ahiahonu, P.W. Metabolism and detoxification of phytoalexins and analogs by phytopathogenic fungi. *Phytochemistry* **2005**, *66*, 391–411.
21. Cho, Y.; Kim, K.H.; la Rota, M.; Scott, D.; Santopietro, G.; Callihan, M.; Lawrence, C.B. Identification of virulence factors by high throughput targeted gene deletion of regulatory genes in *Alternaria brassicicola*. *Mol. Microbiol.* **2009**, *72*, 1316–1333.
22. Cho, Y.; Ohm, R.A.; Grigoriev, I.V.; Srivastava, A. Fungal-specific transcription factor *AbPf2* activates pathogenicity in *Alternaria brassicicola*. *Plant J.* **2013**, *75*, 498–514.
23. Grigoriev, I.V.; Nordberg, H.; Shabalov, I.; Aerts, A.; Cantor, M.; Goodstein, D.; Kuo, A.; Minovitsky, S.; Nikitin, R.; Ohm, R.A.; *et al.* The genome portal of the Department of Energy Joint Genome Institute. *Nucleic Acids Res.* **2012**, *40*, D26–32.
24. JGI fungal portal MycoCosm. Available online: http://jgi.doe.gov/Abrassicicola (accessed on 9 July 2014).
25. Trapnell, C.; Pachter, L.; Salzberg, S.L. TopHat: Discovering splice junctions with RNA-Seq. *Bioinformatics* **2009**, *25*, 1105–1111.
26. Langmead, B.; Trapnell, C.; Pop, M.; Salzberg, S.L. Ultrafast and memory-efficient alignment of short DNA sequences to the human genome. *Genome Biol.* **2009**, *10*, R25.
27. Trapnell, C.; Williams, B.A.; Pertea, G.; Mortazavi, A.; Kwan, G.; van Baren, M.J.; Salzberg, S.L.; Wold, B.J.; Pachter, L. Transcript assembly and quantification by RNA-Seq reveals unannotated transcripts and isoform switching during cell differentiation. *Nat. Biotechnol.* **2010**, *28*, 511–515.
28. Roberts, A.; Trapnell, C.; Donaghey, J.; Rinn, J.L.; Pachter, L. Improving RNA-Seq expression estimates by correcting for fragment bias. *Genome Biol.* **2011**, *12*, R22.
29. Cho, Y.; Srivastava, A.; Ohm, R.A.; Lawrence, C.B.; Wang, K.H.; Grigoriev, I.V.; Marahatta, S.P. Transcription factor *Amr1* induces melanin biosynthesis and suppresses virulence in *Alternaria brassicicola*. *PLoS Pathog.* **2012**, *8*, e1002974.
30. Srivastava, A.; Ohm, R.A.; Oxiles, L.; Brooks, F.; Lawrence, C.B.; Grigoriev, I.V.; Cho, Y. A zinc-finger-family transcription factor, *AbVf19*, is required for the induction of a gene subset important for virulence in *Alternaria brassicicola*. *Mol. Plant Microbe Interact.* **2012**, *25*, 443–452.

Sample availability: Contact YC, E-Mail: yangraec@kribb.re.kr; Tel.: +82-43-240-6255; Fax: +82-43-240-6259.

MALDI Mass Spectrometry Imaging for the Simultaneous Location of Resveratrol, Pterostilbene and Viniferins on Grapevine Leaves

Loïc Becker, Vincent Carré, Anne Poutaraud, Didier Merdinoglu and Patrick Chaimbault

Abstract: To investigate the *in-situ* response to a stress, grapevine leaves have been subjected to mass spectrometry imaging (MSI) experiments. The Matrix Assisted Laser Desorption/Ionisation (MALDI) approach using different matrices has been evaluated. Among all the tested matrices, the 2,5-dihydroxybenzoic acid (DHB) was found to be the most efficient matrix allowing a broader range of detected stilbene phytoalexins. Resveratrol, but also more toxic compounds against fungi such as pterostilbene and viniferins, were identified and mapped. Their spatial distributions on grapevine leaves irradiated by UV show their specific colocation around the veins. Moreover, MALDI MSI reveals that resveratrol (and piceids) and viniferins are not specifically located on the same area when leaves are infected by *Plasmopara viticola*. Results obtained by MALDI mass spectrometry imaging demonstrate that this technique would be essential to improve the level of knowledge concerning the role of the stilbene phytoalexins involved in a stress event.

Reprinted from *Molecules*. Cite as: Becker, L.; Carré, V.; Poutaraud, A.; Merdinoglu, D.; Chaimbault, P. MALDI Mass Spectrometry Imaging for the Simultaneous Location of Resveratrol, Pterostilbene and Viniferins on Grapevine Leaves. *Molecules* **2014**, *19*, 10587-10600.

1. Introduction

Mass spectrometry succeeds in providing a lot of qualitative and quantitative data on plant omics [1]. Generally, to get the information, the plant material needs to be sampled and extracted by solvents before chemical analysis. Using this approach, the sensitivity of detection is very high but the exact location of compounds of interest in tissues is lost. In just a few years, mass spectrometry imaging (MSI) made the dream of the location of molecular compounds at the micron scale to come true [2,3]. First developed on animal slices [4,5], this emerging technique finds a growing interest in plant proteomic [6] and metabolomic [7–9]. Indeed, MSI is able to provide a specific spatial distribution of proteins or metabolites in tissues. Depending on the ion source, the sample preparation and the mass spectrometer used, the spatial distribution of molecules on a plant sample is specifically dedicated to some compound families [10,11].

In this context, we developed a selective method to map several stilbene phytoalexins on grapevine leaves using a time-of-flight mass spectrometer (TOFMS) fitted with a laser ion source operated at 266 nm [12]. Phytoalexins are of great interest because they are biosynthetized in response to biotic or abiotic stress. They are also well known for inducing an antifungal activity [13–18] and, for example, we observed by MSI experiment a co-localization of resveratrol and pterostilbene (trimethoxystilbene) at the infection site on the leaf abaxial side (Cabernet Sauvignon) [12] after infection by downy mildew. MSI allowed the identification and mapping of resveratrol and pterostilbene but the biosynthesis of some other stilbenes is also expected to be induced in grapevine leaves and even berries in response to pathogen infection or UV irradiation [19–21]. Besides the glycosylation of stilbenes (e.g., piceids), oxidative oligomerization of resveratrol catalyzed by plant peroxidases may occur and several of them have been identified in stressed grapevine leaves [22–24]. For example, the δ-viniferin has been identified as the main dimer of *trans*-resveratrol synthesized in *Vitis vinifera* leaves infected by *Plasmopara viticola* [25]. These stilbene phytoalexins were found to be more or less toxic against fungi according to their chemical structure. In this field, the viniferins are known to be more active than resveratrol and thus are suspected to be highly involved against pathogen proliferation [13,26]. Resveratrol is synthesized in a large amount regardless of the cultivar, the susceptibility or the resistance to fungi. Even if resveratrol has itself an antimicrobial effect [27], its transformation to other stilbene phytoalexins could be decisive in the defense mechanisms of the grapevine. Consequently, it could be extremely informative to observe the spatial distribution of more stilbene phytoalexins than resveratrol or pterostilbene in stressed plant organs. In this paper, we also present the first MSI experiments allowing the mapping of viniferins on grapevine leaves.

In a previous paper, we reported the imaging of metabolites from *Vitis vinifera* leaves by laser desorption/ionization (LDI) time-of-flight mass spectrometry [12]. The relatively high laser power density (around 10^8 W·cm^{-1}) allowed a sufficient ion yield to highlight species of interest in the mass spectra at the grapevine leaf surface. In this configuration, we identified different molecules among which the resveratrol and the pterostilbene but also some additional compounds such as diacyl and triacyl glycerols. The gain of sensitivity to resveratrol and pterostilbene by the biphotonic ionization process at 266 nm wavelength was outstanding, but it may be counterbalanced by the signal loss of laser-sensitive substances such as viniferins. To overcome this limitation and access to the spatial distribution of other stilbene phytoalexins on plant material, we evaluate the potential of a MALDI approach which consists in the addition of a layer of an exogen compound called "matrix", deposited on the leaf surface before the LDI process at 337 nm wavelength.

2. Results and Discussion

For mass spectrometry imaging purpose, MALDI has already been used to map different metabolites on plant organs [28] such as lipids on leaf surface [29], flavonoids or glycoalkaloid in roots and root nodules [30,31], sugars within seed or stem sections [32,33]. Different matrices are described in the literature for MALDI MS analysis of vegetable material such as 1,5 diamino naphthalene (DAN) or 9-amino acridine (9-AA) in negative ion mode, and tri-hydroxy acetophenone (THAP), α-cyano-4-hydroxy cinamic acid (CHCA) or 2,5 dihydroxy-benzoic acid (DHB) in positive ion mode. Graphite was also used to avoid high matrix ion background in the *m/z* range of metabolites [34] but it may lead to overmuch carbon deposit on ion lenses during experiments. Consequently, we firstly proceeded to a matrix selection for MSI of phytoalexin by analyzing the standard stilbene compounds by MALDI-TOFMS with different matrices. To induce phytoalexin synthesis in high amount, grapevine leaves were irradiated with UV-C. They were firstly extracted by methanol to control phytoalexin content using HPLC-MS/MS and MALDI-TOFMS analyses. Finally, MSI experiments were conducted on *Vitis vinefera* leaves either irradiated by UV-C or infected by *Plasmopara viticola.*

2.1. Stilbene Analyses by MALDI-TOFMS

2.1.1. Matrix Selection for Stilbene Analyses by MALDI-TOFMS

To evaluate the most appropriate matrix for stilbene analysis, *trans*-resveratrol, pterostilbene and *trans*-δ-viniferin have been analyzed by MALDI-TOFMS in positive and negative ion mode using different matrices (Table 1). Laser desorption ionization (LDI) experiments which is our reference ionization method was carried out by depositing 1 μL of pure stilbene standard solution at 10^{-5} M on the target plate without matrix. For MALDI experiments, matrix solution at 10^{-1} M in acetonitrile/water (50/50 with or without 0.1% of TFA) was mixed with standard compound solution (matrix/analyte ratio of 1000) and deposited on the target plate (dried-droplet method). Results are given in Table 1 which represents, for each experimental condition, the signal-to-noise ratio (S/N) values of MS peaks corresponding to stilbene molecular ions.

Whether in negative or positive ion mode, resveratrol and pterostilbene are detected without any matrix. The highest sensitivity is reached for deprotonated molecular ions at *m/z* 227 and 255 respectively unlike what we observed at 266 nm [12]. As expected, *trans*-δ-viniferin is never detected in LDI conditions.

Table 1. Signal-to-noise ratio (S/N) values of Matrix Assisted Laser Desorption/Ionisation (MALDI)-time-of-flight mass spectrometer (TOFMS) peaks corresponding to stilbene molecular ions (* radical ion, otherwise, each peak corresponds to protonated ion (positive mode) or deprotonated ion (negative mode); N/A, not applicable because of the presence of an interfering peak).

Ion mode	Matrix	S/N ratio		
		trans-Resveratrol	Pterostilbene	*trans*-δ-Viniferin
Negative	without	307	166	/
	9-AA	68	114	11
	DAN	349	230	N/A
	CHCA	/	/	/
	THAP	/	/	/
	DHB	/	/	/
Positive	without	91	451	/
	DAN	/	/	/
	DAN+TFA	/	/	/
	DHB	25	55	139/145 *
	DHB+TFA	259	358	231/239 *
	CHCA	N/A	246	146 *
	CHCA+TFA	N/A	708	501 *
	THAP	/	/	/
	THAP+TFA	38	/	44
	9-AA	/	/	/

The matrix contribution to the detection of the viniferin is noteworthy. In positive ion detection, with CHCA or DHB in acid medium (TFA), the protonated molecular ion $[M+H]^+$ and radical molecular ion $M^{\bullet+}$ of the viniferin are detected with the highest signal-to-noise ratios at *m/z* 455 and 454. With a less pronounced effect, THAP allows the radical molecular ion $M^{\bullet+}$ to be detected and in negative detection mode, deprotonated molecular ion is detected at *m/z* 453 by using 9-AA. Due to the matrix ion interference at *m/z* 453, DAN cannot be used as a matrix for viniferin analysis.

CHCA matrix in positive mode generates an intense mass peak at *m/z* 228 corresponding to $[M+K]^+$ which interferes with the resveratrol molecular ion at the same *m/z* value. For its part, DHB matrix allows both resveratrol and pterostilbene to be detected as protonated molecular ions with a high S/N value. The addition of trifluoroacetic acid significantly improves the signal. Thus, the signal of resveratrol with DHB is only 15% lower from that obtained without matrix but it is around two times higher for pterostilbene compared to LDI. Therefore, DHB with TFA addition provides *m/z* peaks related to each tested stilbene with high sensitivity making it a matrix of choice for stilbene analysis by MALDI-TOF.

2.1.2. MALDI-TOFMS Analysis of Stressed Leaf Extract

To investigate resveratrol, pterostilbene and viniferins in a real sample, a grapevine leaf has been first irradiated by UV-C. Two days after this treatment, the production of stilbene phytoalexin is expected to be induced. Stilbenes were then extracted from leaves using methanol. The leaf was then extracted with methanol. The presence of stilbenes in the leaf extract was controlled by HPLC-ESI/MS and MS/MS in negative ionization mode (more sensitive detection than in the positive mode).

Several stilbenes are detected by LC-ESI/MS using a reversed-phase support (RPLC). The identification was confirmed by the retention time and the fragmentation pattern of standard compounds compared to the literature (MS/MS, Table 2) [25,35,36]. Ion extract chromatograms of deprotonated molecular ions are displayed in the Figure 1. Piceids are the most polar stilbenes due to their glucose moiety. Thus, they are the less retained stilbenes as expected in RPLC. The *trans*-isomer is eluted at 13.68 min and *cis*-isomer at 13.01 min. The peaks at 17.02 and 18.07 min are associated with the *cis*- and *trans*-ε-viniferin respectively. The following peaks at 21.60 and 22.65 min are in turn associated with *trans*- and *cis*- δ-viniferin respectively. This elution order of viniferin isomers have been described by Pezet using a C_{18} column [25]. Resveratrol is identified at 42.72 min. Finally, pterostilbene, the most apolar stilbene, is detected at 47.94 min.

Table 2. List of retention times, molecular formulae and MS/MS product ions of stilbenes detected by LC-ESI/MS (see also Figure 1). The relative intensities of fragments are indicated in parenthesis.

Compound	Retention Time (min)	$[M\text{-}H]^-$	Precursor Ion	MS/MS (CID)
cis-piceid	13.01	$C_{16}H_{15}O_3^-$	389	$228_{(100)}$
trans-piceid	13.68	$C_{16}H_{15}O_3^-$	389	$228_{(100)}$
cis-ε-viniferin	17.02	$C_{28}H_{21}O_6^-$	453	$435_{(20)}$; $411_{(10)}$; $369_{(12)}$; $359_{(42)}$; $347_{(100)}$; $333_{(44)}$
trans-ε-viniferin	18.07	$C_{28}H_{21}O_6^-$	453	$435_{(24)}$; $411_{(14)}$; $359_{(100)}$; $347_{(58)}$; $333_{(18)}$
trans-δ-viniferin	21.60	$C_{28}H_{21}O_6^-$	453	$435_{(22)}$; $411_{(15)}$; $359_{(100)}$; $347_{(40)}$; $333_{(8)}$
cis-δ-viniferin	22.65	$C_{28}H_{21}O_6^-$	453	$435_{(100)}$; $411_{(68)}$; $369_{(62)}$; $359_{(38)}$; $347_{(40)}$; $333_{(50)}$; $317_{(14)}$; $307_{(20)}$; $267_{(12)}$; $251_{(13)}$
trans-resveratrol	42.72	$C_{14}H_{11}O_3^-$	227	$185_{(100)}$; $183_{(38)}$; $159_{(32)}$; $157_{(29)}$; $143_{(11)}$
trans-pterostilbene	47.94	$C_{14}H_{11}O_3^-$	255	$240_{(100)}$; $239_{(5)}$

Figure 1. Ion extract chromatograms from HPLC-ESI/MS analysis of stressed leaf extract (**a**) piceid isomers at *m/z* 389 (**b**) viniferin isomers at *m/z* 453 (**c**) resveratrol at *m/z* 227 and (**d**) *m/z* pterostilbene at 255 – peak assignments were confirmed by MS/MS experiments of the 10 more abundant ions in each mass spectrum, the molecular structures of *trans*- isomers of each stilbene are displayed.

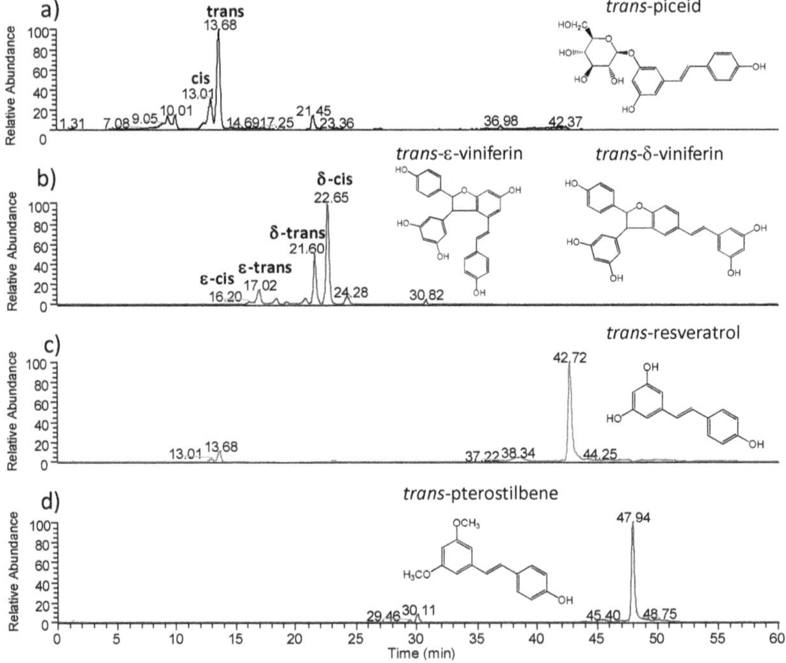

The extraction sample of grapevine leaf irradiated by UV was also investigated by MALDI-TOFMS with DHB as the matrix (and 0.1% of TFA). The average mass spectrum obtained from 50 mass spectra is displayed in the Figure 2. In the low mass region, the mass spectrum of a methanolic extract of leaf is similar to the one obtained with pure DHB matrix with mass peaks observed at *m/z* 137, 155, 177, 193 and 273. At *m/z* higher than 500, mass peaks correspond to diacylglycerol compounds as already reported in LDI-TOFMS analysis of *Vitis vinifera* leaves. With DHB, stilbenes are detected as protonated ions. Resveratrol and pterostilbene are also detected as protonated ions but the detection of viniferin as molecular ions is worth to note. They could be observed at *m/z* 454 and 455 (right window in Figure 2). Note that viniferin isomers cannot be distinguished from each other by MS and that they all contribute to the same MS signal. Other metabolites are jointly detected as for example flavonol aglycone at *m/z* 303 for quercetin and *m/z* 319 for myricetin (left window in Figure 2). In our experimental conditions, the loss of sugar moiety systemically occurs. Thus, piceids are solely detected without sugar moiety and contribute to the same *m/z* signal as resveratrol (in the next sections, *m/z* peak at 229 will be assigned to resveratrol/piceids).

Figure 2. MALDI-TOF mass spectrum of methanol extract from grapevine stressed leaf – the blue peak labels correspond to matrix ions (DHB); in inserts, a zoom of *m/z* peaks corresponding to flavonol aglycone ([M+H]$^+$) and viniferins (M$^{\bullet+}$ and [M+H]$^+$).

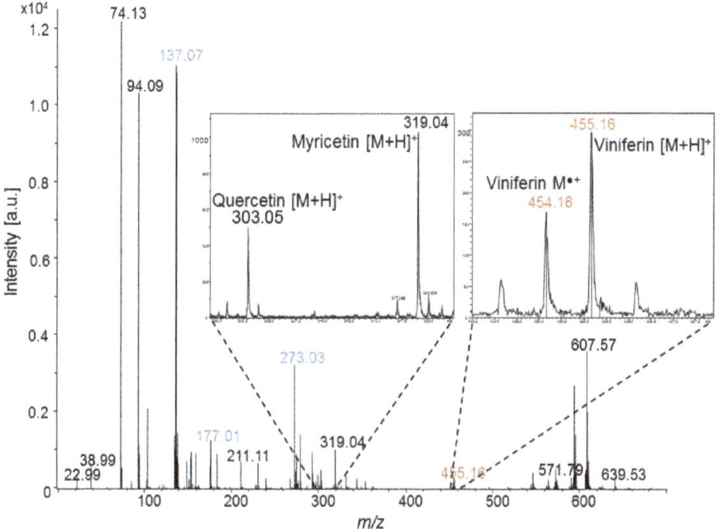

2.2. *In-Situ MALDI-MSI of Stilbenes on Stressed Leaves*

Using DHB matrix for the successful detection of viniferins by MALDI-MS experiment of leaf extract is thus a promising route for the MSI investigation of stilbene compounds. 4 µL of a DHB solution were deposited with a micropipette over a 10.6 mm² area on the abaxial side of a UV-C-stressed leaf. The drying time was short enough to prevent needle formation during matrix crystallization and to avoid the distribution of the studied metabolites in the native sample to be disturbed. Moreover, the size of the image is smaller than the surface of the deposited droplet containing the matrix. Thus, the imaged area is corresponding to the center of the dried droplet where the coffee ring effect does not exist. To support this affirmation, the repartition of the ion *m/z* 137

$[DHB-H_2O]^+$ (the most intense ion of the matrix) shows a rather homogenous deposition (Figure 3). MSI experiment was then conducted on this sample. A mass spectrum corresponding to one pixel is displayed on the Figure 3. DHB mass peaks contribute to a large part of the signal (peak labels in blue). However, all previously investigated stilbenes are detected as protonated molecular ions: resveratrol/piceids at *m/z* 229, pterostilbene at *m/z* 257 and viniferins at *m/z* 455 and also radical ion at *m/z* 454 (insert of the Figure 4). Compared to MALDI-TOFMS analysis of methanolic extract, the intensity of viniferins significantly increases meaning that the MALDI-MSI improves the sensitivity of their detection. This may be explained by the fact that there is no dilution effect in MSI (metabolites are detected where they are located) whereas in MALDI-TOFMS, methanol extracts rather provide an average response of metabolites contained in the sampled leaf disc.

Figure 3. MALDI-TOF mass spectrum of stressed grapevine leaf: the blue peak labels correspond to matrix ions (DHB). Stilbene phytoalexins are detected as protonated and radical ions as it is observed for the viniferin in insert.

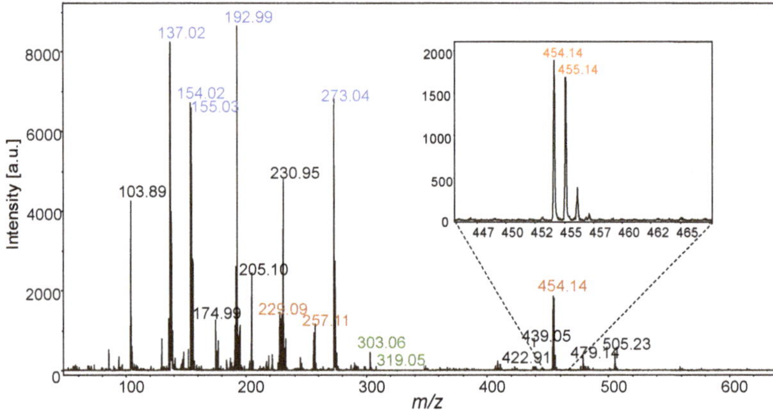

The spatial distributions of stilbene phytoalexins on the leaf are given in the Figure 4. The color scale represents the relative intensity of each ion. For each stilbene, the black color is used when nothing is detected in the corresponding pixel whereas the white color represents the maximum of intensity detected in the map.

Examining the ion extracted MS images of resveratrol/piceids at *m/z* 229, pterostilbene at *m/z* 257 and viniferins at *m/z* 454 and 455, the heterogeneity of their surface distributions is clearly highlighted. Moreover, they are almost exclusively localized on the same areas of the leaf. Their spatial location evidences a clear relationship in their synthesis under UV-stress conditions. Even if the image resolution is not sufficient to explore cells at the organelle scale, stilbenes seem to accumulate themselves in the network of small veins and more precisely in the dense parenchyma tissue. This is in good agreement with UV irradiated leaf analyzed by fluorescence for which global stilbene fluorescence has been mainly located in vein and lignified tissues [37,38]. Notably, none of these stilbenes was detected in the control samples (not irradiated) leading to black images (data not shown).

Figure 4. MALDI-MSI of UV stressed grapevine leaf (**a**) optical image; (**b**) ion extracted image related to one peak of the matrix [DHB-H_2O+H]$^+$ (*m/z* 137); (**c**) ion extracted image of resveratrol and piceids (*m/z* 229); (**d**) ion extracted image of pterostilbene (*m/z* 257); (**e**) ion extracted image of viniferin (*m/z* 454 and 455). The color scale indicates the absolute intensity of each pixel (arbitrary units).

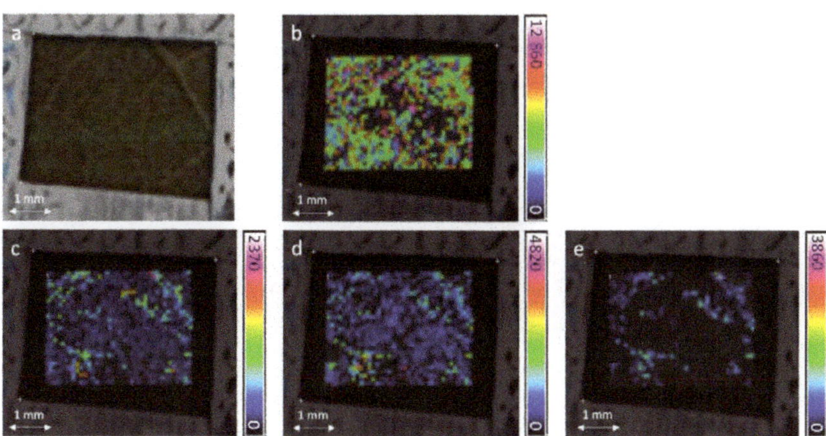

The MSI experiment was then conducted on Cabernet Sauvignon leaf infected by *Plasmopara viticola* (Figure 5). Before analysis, 4 µL of DHB matrix solution were deposited with a micropipette over a 9.1 mm² area on the grapevine leaf five days after infection. The MS image of the control leaves did not exhibit any stilbene (data not shown) because their concentration is under the detection limits. The Figure 4 displays ion extraction images of resveratrol/piceids and viniferins. As it was observed under a UV stress, their spatial distribution on the grapevine leaf is non-homogeneous. However, some differences appear between resveratrol/piceids and viniferin localizations (for this sample, the pterostilbene level was under the detection limit). While viniferins are mainly located around the veins, the distribution of resveratrol/piceids is more scattered on the leaf surface. Its distribution should correspond to the infection sites. Grapevine reacts to *P. viticola* infection by producing high amounts of stilbenes at the infection site [15] and MALDI-MSI brings new chemical details on their spatial distributions. Resveratrol and piceid are much less toxic against *P. viticola* than viniferins [14]. Viniferins are supposed to be involved depending on the cultivar resistance to the pathogen [22]. For the Cabernet Sauvignon, which has a low degree of resistance to *P. viticola*, the less toxic compounds for pathogen are localized at the leaf infection sites where the viniferins are not accumulated. It suggests that for a susceptible variety, the viniferins may be too far away from infection sites to play a real antifungal role. Their specific locations need to be understood according to the involved biosynthesize pathways and their isomeric composition [21]. Consequently, the investigation of viniferins locations from a range of susceptible to resistant grapevine species is now possible by MALDI-MSI to understand their role in the constitutive and inducible defenses of grapevine against fungi.

Figure 5. MALDI-MSI of grapevine leaf 5 days after infection by *Plasmopara viticola* (**a**) optical image; (**b**) ion extracted image of resveratrol/piceids at *m/z* 229; (**c**) ion extracted image of viniferins at *m/z* 454 and 455. The color scale indicates the absolute intensity of each pixel (arbitrary units). The dotted line highlights a small vein and the arrows point out pixels corresponding to a high intensity of viniferins.

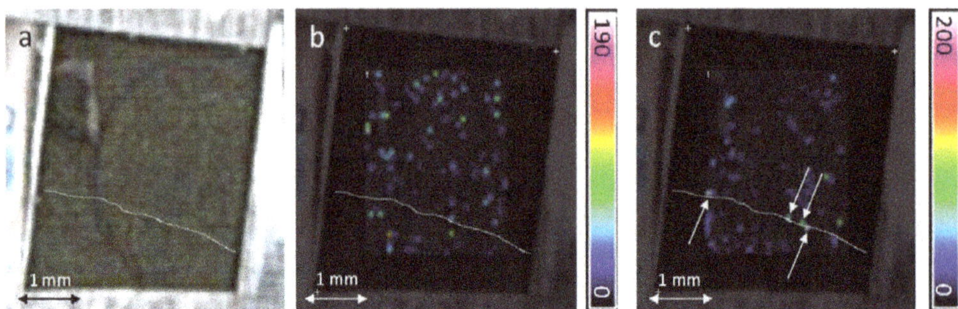

3. Experimental Section

3.1. Plant Material and Leaf Sample Preparation

The study was carried out on Cabernet Sauvignon, a grape variety of *Vitis vinifera* highly sensitive to *Plasmopara viticola.* Plants were grown in greenhouse. The sixth leaf, counted from the apex of 3.5 months old plants having 12–14 fully expanded leaves, was harvested and washed with demineralized water. To induce stilbene synthesis, grapevine leaves were irradiated by UV-C lamp at the 254 nm wavelength (Osram, 30 W, 90 $\mu W \cdot cm^{-2}$, Molsheim, France) for 180 s or were infected by spraying a 2×10^4 sporangia·mL^{-1} solution of *Plasmopara viticola* sporangia on the abaxial side. After inoculation or irradiation, leaves were transferred to wet paper with the abaxial surface up in trays closed by transparent plastic bags. Leaves were stored in a culture chamber for an initial period at a temperature of 23 °C in the dark for 24 h, then 18 h of light (about 200 $\mu mol \cdot m^{-2} \cdot s^{-1}$) and 6 h of darkness until analysis.

Two days after irradiation or five days after infection, foliar discs were cut out using a 2 cm-diameter hollow-punch. Leaf discs were placed in a high vacuum to stop phytoalexin synthesis and stored at 4 °C before imaging experiments. For MALDI experiments, matrix solution at 10^{-1} $mol \cdot L^{-1}$ in a solvent mix (acetonitrile/water—50/50) acidified by TFA was deposited on the leaf sample by using a P20 micropipette (Eppendorf).

3.2. Leaf Extraction

Solid-liquid extraction of the leaf was performed as follows: leaf disc placed in 0.5 mL of methanol and heated at 60 °C for 45 min under stirring. Then, the leaf disc was removed from the extract which was centrifuged at 12,000 rpm before MS analysis.

3.3. Standard and Solvents

Pure *trans*-resveratrol and *trans*-pterostilbene were purchased from Sigma-Aldrich. Viniferins were collected from semi-preparative LC of methanol extraction of stressed *Vitis* leaves. All matrices, 9-amino-acridine (9-AA), 1,5-diaminonaphtalene (DAN), 2,5-dihydroxybenzoic acid (DHB), α-cyano-4-hydroxycinnamic acid (CHCA) and trihydroxyacetophenone (THAP) were purchased from Sigma-Aldrich. Trifluoroacetic acid (TFA) was purchased from Sigma-Aldrich. HPLC-grade methanol, acetonitrile and water were purchased from VWR. ESI positive and negative calibration kit from Thermo was used to achieve mass calibration of ESI-LTQ system.

3.4. LC-MS and MS/MS

For LC-MS analysis, high performance liquid chromatography system (Dionex Ultimate 3000, Dionex, France) was connected to a dual-pressure linear ion trap mass spectrometer (LTQ Velos Pro, Thermo Fisher Scientific, San José, CA, USA). For stilbene separation, C18 reverse phase column was used (Symmetry Shield, 4.6 × 50 mm, 3.5 μm, Waters). 20 μL of sample were injected. The flow rate was kept to 500 $\mu L \cdot min^{-1}$ and a constant elution gradient was applied from 0 (5% acetonitrile/95% water) to 55 min (100% acetonitrile) during the LC run. HESI (*Heated Electrospray Ionization Source*) interface was plugged to the ion source of the LTQ mass spectrometer. MS system was running from 110 to 2000 *m/z* at MS scan rate of 9 Hz. To confirm chromatographic peak assignment, MS/MS by CID was systematically conducted on the most intense 10 mass peaks of each mass spectrum.

3.5. MALDI Mass Spectrometer

A Bruker Reflex IV MALDI–TOF mass spectrometer (Bruker Daltonics, Bremen, Germany) was used to perform *in situ* MALDI analysis and imaging experiments. The nitrogen laser generates a laser pulse at a wavelength of 337 nm with a pulse duration of 4 ns and a 9 Hz repetition rate (Science Inc., Boston, MA, USA). Positive mass spectra were acquired in the 0–1000 *m/z* range. The mass spectrometer was operated in the reflectron mode at a total acceleration voltage of 20 kV and a reflecting voltage of 23 kV. A delay time of 200 ns was used prior ion extraction. The used laser fluence was kept at ~0.5 J/cm^2.

3.6. Mass Spectrometry Imaging (MSI)

For MSI experiments, leaf discs were fixed on a metal MALDI target plate with aluminized tape. FlexImaging software (Bruker) allowed the tracking of the leaf sample on MALDI target plate, the image pixel features and the treatment of post-acquisition image to be achieved. The mass spectrum obtained for each pixel of the images corresponds to the averaged mass spectrum of 50 consecutive mass spectra on the same location. Approximately 3 h were required to achieve an image of 7.5 mm^2 area with a 75 μm spatial resolution (1100 pixels).

4. Conclusions

For the first time, MALDI was successfully conducted on stilbene phytoalexins and more particularly on the viniferins. From all tested matrices, the DHB allows the viniferins, resveratrol and pterostilbene to be detected with a higher sensitivity. The *in-situ* grapevine leaf response to a stress was then investigated using MALDI imaging mass spectrometry experiment. For this purpose, stress was generated by UV-C irradiation. The ion images of resveratrol/piceids, pterostilbene and viniferins exhibit heterogeneous distribution on leaf surface but also demonstrate their colocalization around the leaf veins. Moreover, the MALDI-MSI investigation of a Cabernet Sauvignon leaf infected by *Plasmopara viticola* allows different spatial distributions between resveratrol/piceids and viniferins to be highlighted. Only resveratrol/piceids, the less toxic compounds for fungi are detected on the infection sites of this susceptible cultivar. This result suggests that viniferin locations may influence the resistance level to a pathogen. The MALDI mass spectrometry imaging of stilbene phytoalexins provides a new level of understanding the plant response to a biotic or an abiotic stress.

Acknowledgments

The authors thank the "Conseil Interprofessionnel du Vin de Bordeaux" (CIVB, Bordeaux, France) for the financial support and the European Union for funding the LC-MS/MS equipment.

Author Contributions

V.C., A.P., D.M. and P.C. designed research; L.B., V.C., A.P. and P.C. performed research and analyzed data; L.B., V.C. and P.C. wrote the paper. All authors read and approved the final manuscript.

Conflicts of Interest

The authors declare no conflict of interest.

References

1. Glinski, M.; Weckwerth, W. The role of mass spectrometry in plant systems biology. *Mass Spectrom. Rev.* **2006**, *25*, 173–214.
2. Caprioli, R.M.; Farmer, T.B.; Gile, J. Molecular imaging of biological samples: Localization of peptides and proteins using MALDI-TOF MS. *Anal. Chem.* **1997**, *69*, 4751–4760.
3. McDonnell, L.A.; Heeren, R.M.A. Imaging mass spectrometry. *Mass Spectrom. Rev.* **2007**, *26*, 606–643.
4. Chaurand, P. Imaging mass spectrometry of thin tissue sections: A decade of collective efforts. *J. Proteomics* **2012**, *75*, 4883–4892.
5. Angel, P.M.; Caprioli, R.M. Matrix-assisted laser desorption ionization imaging mass spectrometry: In situ molecular mapping. *Biochemistry (Mosc.)* **2013**, *52*, 3818–3828.
6. Grassl, J.; Taylor, N.L.; Millar, A.H. Matrix-assisted laser desorption/ionisation mass spectrometry imaging and its development for plant protein imaging. *Plant Methods* **2011**, *7*, 21–21.

7. Esquenazi, E.; Yang, Y.-L.; Watrous, J.; Gerwick, W.H.; Dorrestein, P.C. Imaging mass spectrometry of natural products. *Nat. Prod. Rep.* **2009**, *26*, 1521–1534.
8. Lee, Y.J.; Perdian, D.C.; Song, Z.; Yeung, E.S.; Nikolau, B.J. Use of mass spectrometry for imaging metabolites in plants. *Plant J.* **2012**, *70*, 81–95.
9. Bjarnholt, N.; Li, B.; D'Alvise, J.; Janfelt, C. Mass spectrometry imaging of plant metabolites – principles and possibilities. *Nat. Prod. Rep.* **2014**, *31*, 818–837.
10. Bhardwaj, C.; Hanley, L. Ion sources for mass spectrometric identification and imaging of molecular species. *Nat. Prod. Rep.* **2014**, *31*, 756–767.
11. Matros, A.; Mock, H.-P. Mass spectrometry based imaging techniques for spatially resolved analysis of molecules. *Front. Plant Sic.* **2013**, *4*, 89, doi:10.3389/fpls.2013.00089.
12. Hamm, G.; Carré, V.; Poutaraud, A.; Maunit, B.; Frache, G.; Merdinoglu, D.; Muller, J.-F. Determination and imaging of metabolites from Vitis vinifera leaves by laser desorption/ionisation time-of-flight mass spectrometry. *Rapid Commun. Mass Spectrom.* **2010**, *24*, 335–342.
13. Jeandet, P.; Douillet-Breuil, A.-C.; Bessis, R.; Debord, S.; Sbaghi, M.; Adrian, M. Phytoalexins from the Vitaceae: Biosynthesis, phytoalexin gene expression in transgenic plants, antifungal activity, and metabolism. *J. Agric. Food Chem.* **2002**, *50*, 2731–2741.
14. Chong, J.; Poutaraud, A.; Hugueney, P. Metabolism and roles of stilbenes in plants. *Plant Sci.* **2009**, *177*, 143–155.
15. Alonso-Villaverde, V.; Voinesco, F.; Viret, O.; Spring, J.-L.; Gindro, K. The effectiveness of stilbenes in resistant Vitaceae: Ultrastructural and biochemical events during Plasmopara viticola infection process. *Plant Physiol. Biochem.* **2011**, *49*, 265–274.
16. Jeandet, P.; Delaunois, B.; Conreux, A.; Donnez, D.; Nuzzo, V.; Cordelier, S.; Clément, C.; Courot, E. Biosynthesis, metabolism, molecular engineering, and biological functions of stilbene phytoalexins in plants. *BioFactors* **2010**, *36*, 331–341.
17. Jeandet, P.; Delaunois, B.; Aziz, A.; Donnez, D.; Vasserot, Y.; Cordelier, S.; Courot, E. Metabolic engineering of yeast and plants for the production of the biologically active hydroxystilbene, resveratrol. *J. Biomed. Biotechnol.* **2012**, *2012*, doi:10.1155/2012/579089.
18. Jeandet, P.; Clément, C.; Courot, E.; Cordelier, S. Modulation of phytoalexin biosynthesis in engineered plants for disease resistance. *Int. J. Mol. Sci.* **2013**, *14*, 14136–14170.
19. Langcake, P.; Pryce, R.J. A new class of phytoalexins from grapevines. *Experientia* **1977**, *33*, 151–152.
20. Langcake, P.; Pryce, R.J. The production of resveratrol and the viniferins by grapevines in response to ultraviolet irradiation. *Phytochemistry* **1977**, *16*, 1193–1196.
21. Jeandet, P.; Breuil, A.C.; Adrian, M.; Weston, L.A.; Debord, S.; Meunier, P.; Maume, G.; Bessis, R. HPLC analysis of grapevine phytoalexins coupling photodiode array detection and fluorometry. *Anal Chem.* **1997**, *69*, 5172–5177.
22. Pezet, R.; Gindro, K.; Viret, O.; Spring, J.-L. Glycosylation and oxidative dimerization of resveratrol are respectively associated to sensitivity and resistance of grapevine cultivars to downy mildew. *Physiol. Mol. Plant Pathol.* **2004**, *65*, 297–303.

23. Schnee, S.; Viret, O.; Gindro, K. Role of stilbenes in the resistance of grapevine to powdery mildew. *Physiol. Mol. Plant Pathol.* **2008**, *72*, 128–133.
24. Mattivi, F.; Vrhovsek, U.; Malacarne, G.; Masuero, D.; Zulini, L.; Stefanini, M.; Moser, C.; Velasco, R.; Guella, G. Profiling of resveratrol oligomers, important stress metabolites, accumulating in the leaves of hybrid vitis vinifera (merzling × teroldego) genotypes infected with plasmopara viticola. *J. Agric. Food Chem.* **2011**, *59*, 5364–5375.
25. Pezet, R.; Perret, C.; Jean-Denis, J.B.; Tabacchi, R.; Gindro, K.; Viret, O. δ-Viniferin, a Resveratrol dehydrodimer: One of the major stilbenes synthesized by stressed grapevine leaves. *J. Agric. Food Chem.* **2003**, *51*, 5488–5492.
26. Pezet, R.; Gindro, K.; Viret, O.; Richter, H. Effects of resveratrol, viniferins and pterostilbene on Plasmopara viticola zoospore mobility and disease development. *Vitis* **2004**, *43*, 145–148.
27. Adrian, M.; Jeandet, P. Effects of resveratrol on the ultrastructure of Botrytis cinerea conidia and biological significance in plant/pathogen interactions. *Fitoterapia* **2012**, *83*, 1345–1350.
28. Kaspar, S.; Peukert, M.; Svatos, A.; Matros, A.; Mock, H.-P. MALDI-imaging mass spectrometry — An emerging technique in plant biology. *Proteomics* **2011**, *11*, 1840–1850.
29. Vrkoslav, V.; Muck, A.; Cvačka, J.; Svatoš, A. MALDI imaging of neutral cuticular lipids in insects and plants. *J. Am. Soc. Mass Spectrom.* **2010**, *21*, 220–231.
30. Ye, H.; Gemperline, E.; Venkateshwaran, M.; Chen, R.; Delaux, P.-M.; Howes-Podoll, M.; Ané, J.-M.; Li, L. MALDI mass spectrometry-assisted molecular imaging of metabolites during nitrogen fixation in the Medicago truncatula–Sinorhizobium meliloti symbiosis. *Plant J.* **2013**, *75*, 130–145.
31. Ha, M.; Kwak, J.H.; Kim, Y.; Zee, O.P. Direct analysis for the distribution of toxic glycoalkaloids in potato tuber tissue using matrix-assisted laser desorption/ionization mass spectrometric imaging. *Food Chem.* **2012**, *133*, 1155–1162.
32. Robinson, S.; Warburton, K.; Seymour, M.; Clench, M.; Thomas-Oates, J. Localization of water-soluble carbohydrates in wheat stems using imaging matrix-assisted laser desorption ionization mass spectrometry. *New Phytol.* **2007**, *173*, 438–444.
33. Burrell, M.M.; Earnshaw, C.J.; Clench, M.R. Imaging matrix assisted laser desorption ionization mass spectrometry: A technique to map plant metabolites within tissues at high spatial resolution. *J. Exp. Bot.* **2007**, *58*, 757–763.
34. Cha, S.; Zhang, H.; Ilarslan, H.I.; Wurtele, E.S.; Brachova, L.; Nikolau, B.J.; Yeung, E.S. Direct profiling and imaging of plant metabolites in intact tissues by using colloidal graphite-assisted laser desorption ionization mass spectrometry. *Plant J.* **2008**, *55*, 348–360.
35. Stella, L.; de Rosso, M.; Panighel, A.; Vedova, A.D.; Flamini, R.; Traldi, P. Collisionally induced fragmentation of $[M–H]^-$ species of resveratrol and piceatannol investigated by deuterium labelling and accurate mass measurements. *Rapid Commun. Mass Spectrom.* **2008**, *22*, 3867–3872.
36. Mazzotti, F.; di Donna, L.; Benabdelkamel, H.; Gabriele, B.; Napoli, A.; Sindona, G. The assay of pterostilbene in spiked matrices by liquid chromatography tandem mass spectrometry and isotope dilution method. *J. Mass Spectrom.* **2010**, *45*, 358–363.

37. Poutaraud, A.; Latouche, G.; Martins, S.; Meyer, S.; Merdinoglu, D.; Cerovic, Z.G. Fast and local assessment of stilbene content in grapevine leaf by *in vivo* fluorometry. *J. Agric. Food Chem.* **2007**, *55*, 4913–4920.
38. Jean-Denis, J.B.; Pezet, R.; Tabacchi, R. Rapid analysis of stilbenes and derivatives from downy mildew-infected grapevine leaves by liquid chromatography–atmospheric pressure photoionisation mass spectrometry. *J. Chromatogr. A* **2006**, *1112*, 263–268.

Sample Availability: Not available.

Study of Leaf Metabolome Modifications Induced by UV-C Radiations in Representative *Vitis*, *Cissus* and *Cannabis* Species by LC-MS Based Metabolomics and Antioxidant Assays

Guillaume Marti, Sylvain Schnee, Yannis Andrey, Claudia Simoes-Pires, Pierre-Alain Carrupt, Jean-Luc Wolfender and Katia Gindro

Abstract: UV-C radiation is known to induce metabolic modifications in plants, particularly to secondary metabolite biosynthesis. To assess these modifications from a global and untargeted perspective, the effects of the UV-C radiation of the leaves of three different model plant species, *Cissus antarctica* Vent. (Vitaceae), *Vitis vinifera* L. (Vitaceae) and *Cannabis sativa* L. (Cannabaceae), were evaluated by an LC-HRMS-based metabolomic approach. The approach enabled the detection of significant metabolite modifications in the three species studied. For all species, clear modifications of phenylpropanoid metabolism were detected that led to an increased level of stilbene derivatives. Interestingly, resveratrol and piceid levels were strongly induced by the UV-C treatment of *C. antarctica* leaves. In contrast, both flavonoids and stilbene polymers were upregulated in UV-C-treated *Vitis* leaves. In *Cannabis*, important changes in cinnamic acid amides and stilbene-related compounds were also detected. Overall, our results highlighted phytoalexin induction upon UV-C radiation. To evaluate whether UV-C stress radiation could enhance the biosynthesis of bioactive compounds, the antioxidant activity of extracts from control and UV-C-treated leaves was measured. The results showed increased antioxidant activity in UV-C-treated *V. vinifera* extracts.

Reprinted from *Molecules*. Cite as: Marti, G.; Schnee, S.; Andrey, Y.; Simoes-Pires, C.; Carrupt, P.-A.; Wolfender, J.-L.; Gindro, K. Study of Leaf Metabolome Modifications Induced by UV-C Radiations in Representative *Vitis*, *Cissus* and *Cannabis* Species by LC-MS Based Metabolomics and Antioxidant Assays. *Molecules* **2014**, *19*, 14004-14021.

1. Introduction

Naturally occurring compounds play an essential role in drug discovery. From 1981 to 2010, 64% of new approved therapeutic agents were inspired or directly derived from natural products [1,2]. The exploration of natural biodiversity has led to the identification of a remarkable variety of chemical entities that possess highly selective and specific biological activities and unique modes of action [3]. Bioprospecting of natural sources is still of great interest for the discovery of new scaffolds; only 1% of tropical species have been investigated for their biological activities [4]. Another aspect increasing the potential of bioresources is the ability of organisms to respond to biotic and abiotic stress by inducing biosynthetic pathways that create an array of secondary metabolites not otherwise detected in steady-state conditions [5]. Recent advances in genomics have highlighted gene clusters that remain silent in the absence of a specific trigger [6]. This hidden chemodiversity has been primarily explored in microorganisms through growth media alterations, and various stresses and genetic manipulations to unlock overlooked biosynthetic pathways and thus provide new metabolic diversity [7]. Some studies have also uncovered the ability of plants to enhance the biosynthesis of bioactive compounds upon stress or stimuli, thus improving their chemical defences, which in turn can be exploited to generate new stress-induced chemical entities of potential therapeutic value [8]. For example, the exposure of the roots of hydroponically grown plants to certain chemical agents induced the production of bioactive compounds, and the corresponding crude extracts were twice as likely to have *in vitro* activity against bacteria, fungi, or cancer in screening programs [9]. Abiotic stresses, such as ultraviolet (UV) radiation, are also known to stimulate plant defences and efficiently increase resistance to pathogens [10]. Phytochemical investigations of plant responses to UV stress have revealed the induction of phenolics such as stilbenoids in various *Vitis* sp., presumably antioxidants, that may protect cells against UV-induced oxidative damage [11,12]. Interestingly, other classes of secondary metabolites were also up-regulated, such as sesquiterpenes (rishitin) in tomato fruits [13], or phenylamides in rice leaves [14], and the benzolactam derivative wasalexins in the leaves of Salt cress (*Thellungiella halophila*) [15] or alkaloid derivatives like brachycerine in *Psychotria brachyceras* [16]. These few examples demonstrate the ability of UV radiation to induce phytoalexin biosynthesis, thus improving the chemical diversity of treated extracts, which in turn could improve their effects on given biological targets [17]. A recent study has shown the contrasting effects of UV-C irradiation of the leaves of several plant species on their antifungal activities. Interestingly, of the eighteen species tested, five species demonstrated a net increase of antifungal properties against a clinical *Fusarium solani* strain, whereas the activity of three extracts was decreased [18]. While most of the studies have focused on the biosynthesis of some characteristic phytoalexins, global metabolome perturbations caused by this intense abiotic stress have not yet been studied. Metabolomics provides a holistic overview of the global changes occurring after stress, chemical induction or genetic manipulation [19]. In particular, untargeted liquid chromatography-mass spectrometry (LC-MS)-based approaches are well suited to reveal the effects of biotic or abiotic stresses in plants at both primary and secondary metabolite levels. As an example, this approach has been used to assess the metabolic response of maize leaves after infestation by *Spodoptera frugiperda* larvae [20]. To evaluate metabolic changes upon UV treatment in leaves, three species known to produce a variety of phenolic secondary metabolites were chosen. *Vitis vinifera* leaves were first analysed, because stilbene biosynthesis upon abiotic or biotic stresses is well documented for this genus [21,22]. As with *V. vinifera*, the genus *Cissus* belongs to the Vitaceae family and is also known to produce various stilbenoids [23]. Several secondary metabolites have been isolated from *C. sativa*, including cannabinoids, flavonoids, alkaloids and stilbenoids, with characteristic structural backbones such as spirans, phenanthrenes and bibenzyls [24]. The goal of this study was to assess if leaves response to UV-C radiation could induce new chemical entities able to improve the antioxidant activity of the crude extract. Firstly, a differential LC-MS-based metabolomics approach was used to provide an overview of all the UV-induced modifications of the chemical compositions of the extracts studied. Several biomarkers, including characteristic phytoalexins of each species, were putatively identified based on molecular formula assignment from the high-resolution MS (HRMS) data recorded. Secondly, variations in the antioxidant activity were used to obtain an initial indication of the UV-C-induced chemical changes in a crude extract.

Radical-scavenging activity was assessed with both DPPH and ABTS assays, and antioxidant activity was assessed with the oxygen radical absorbance capacity (ORAC) assay.

2. Results and Discussion

2.1. LC-MS-Based Metabolomics Approach

Preliminary experiments of UV-C radiation on entire plants and detached leaves of *Vitis vinifera* have shown a similar induction of stilbenoid polymers [12,25]. However, UV-C radiation of entire plants induced a strong water stress within a few hours and displayed poor reproducibility between biological replicates. Thus, fresh leaves of *Vitis vinifera, Cissus antarctica* and *Cannabis sativa* were harvested from different plantlets and exposed to UV-C radiation for 10 min since an extended period of exposition to UV will trigger drastic and irremediable damages to tissue integrity. Seven biological replicates per case were profiled by reversed-phase ultra high performance liquid chromatography-time of flight mass spectrometry (UHPLC-TOFMS) using a generic MeCN-H_2O gradient after simple sample preparation on solid phase extraction (SPE) to remove pigments and very apolar compounds [26]. The LC-MS data, recorded in both positive (PI) and negative (NI) electrospray ESI ionisation modes, were processed using MZmine 2.10 to extract features characterised by their *m/z* ratio, retention time and area. Then, principal component analysis (PCA) was used as an exploratory step prior to multivariate supervised analysis (Figure 1B).

To evaluate rapidly when major modifications occur, a few leaves were analysed 24, 48 and 72 h after UV-C exposure. The quenching and extraction of leaves 48 hours after UV-C radiation revealed the most significant metabolomic variations. Differences between control and treated leaves were less pronounced when sampling was performed after 24 h. However, 72 h after radiation, degradation of the leaf surface was noticeable at many locations, and the variation between biological replicates was significant (data not shown). Thus, a single time point 48 h after UV-C treatment was chosen, because a reproducible and reliable metabolic response was observed without apparent material degradation (Figure 1A).

After PCA, a discriminant analysis approach (orthogonal projections to latent structures-discriminant analysis, OPLS-DA) was applied to obtain classification models and to highlight putative features involved in the stress response according to their variable important projection (VIP) values and position on an S-plot (Figure 1C) [27].

Figure 1. UHPLC-TOFMS metabolomics approach: (**A**) Leaves from independent plantlets were irradiated with UV-C and then extracted using IPA (isopropanol) after 48 h. Pictures of *Vitis vinifera* leaves are displayed. As shown, morphological changes can already be observed after 48 h (**B**) 2D ion maps of NI ESI UHPLC-TOFMS rapid metabolite profiling for each extract; (**C**) Principal component analysis (PCA) of *V. vinifera* LC-MS data in NI mode and S-plot display after OPLS-DA for the selection of the most important features for further annotation.

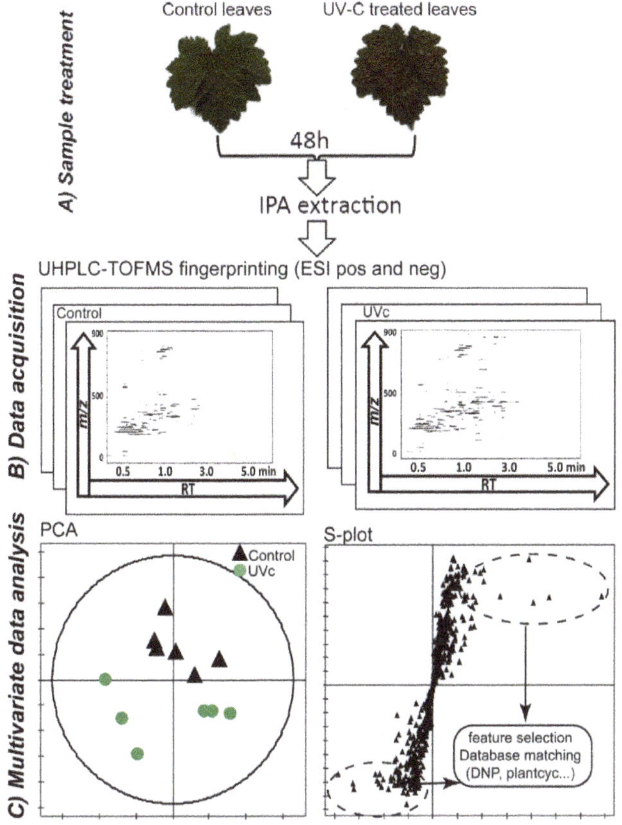

2.2. Data Treatment and Analysis

To evaluate LC-MS data, the PI and NI datasets were compared (RT: ±0.2 min, *m/z*: ±10 ppm) after removing features identified as adducts or ion complexes (Venn diagram, Figure 2A). The number of features detected varied significantly according to the species studied and the ionisation mode used. It was interesting that only 8% to 12% of detected features were common to PI and NI modes, thus emphasising the complementary information provided by both ionisation processes (Figure 2A). To incorporate all these data, the PI and NI datasets from each species were concatenated, and multivariate data analysis (low level data fusion) was performed [28,29]. Unit variance scaling was used to reduce the effect of differential sensitivity between the ionisation modes. Overall, PCA displayed two well-separated clusters corresponding to control and UV-C-treated leaves along the first principal component for the three species studied (Figure 2B). This result indicated that the leaves of all three species react strongly to UV-C treatment at the metabolome level.

Figure 2. (**A**) Venn diagram showing the distribution of detected features in PI (grey circle) and NI (white circle) ESI UHPLC-TOFMS modes. The intercept shows the number of features detected in both modes; (**B**) Principal component analysis of concatenated PI and NI datasets scaled in unit variance.

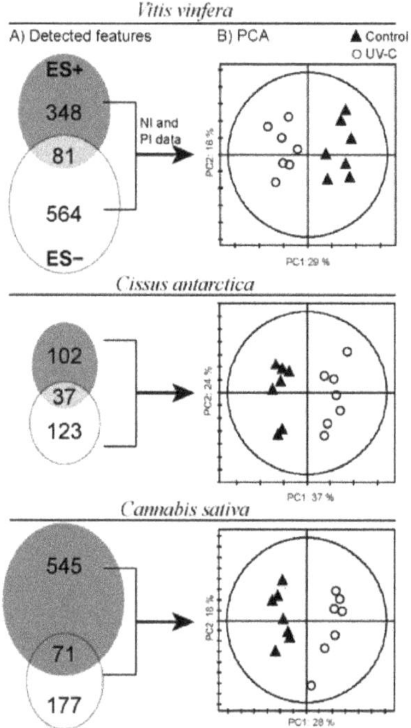

2.3. Global Estimation of the Leaf Metabolome Modifications upon UV-C Treatment

Following PCA, LC-MS datasets underwent OPLS-DA. A 7-fold procedure was used to cross-validate discriminant models of each species (*V. vinifera* model: R2Y = 0.98, Q2Y = 0.92; *C. Antarctica* model: R2Y = 0.99, Q2Y = 0.96; *C. sativa* model: R2Y = 0.99, Q2Y = 0.96). The VIP scores were used to rank the features according to their contribution to the discriminant model. Only VIP scores greater than one were retained for further analysis, because they are the most relevant for explaining the Y response (*i.e.*, UV-C leaf treatment response) [30]. Then, significant features were subjected to an unpaired T-test ($\alpha = 0.05$), and only fold changes greater than five between control and UV-C treated samples were retained for further analysis. Thus, a combination of multivariate and univariate analyses was used to estimate the number of up- and down-regulated features in each species.

In *V. vinifera* leaves, 100 features were significantly up-regulated upon UV-C radiation, and 55 were down-regulated (bar plot, Figure 3A). This high number corresponds to approximately 15% of the total number of features detected. In *C. antarctica* leaves, 60 features were up-regulated and three were significantly down-regulated, which represents 23% of total features detected in this species (bar plot, Figure 3B). In the case of *C. sativa*, 12% of the total features responded significantly, with 48 features elicited and 49 down-regulated (bar plot, Figure 3C).

Figure 3. Right: LC-MS chromatogram of control (BPI trace) (**bottom**) and UV-C-treated leaves (**top**). Superimposed are the main features plotted according to their mean log-fold change between control and UV-C treatment. The area of the circle is inversely proportional to the *p*-value of an unpaired T-test, $\alpha = 0.05$. **Left**: bar plot showing the number of features up- and down-regulated in UV-treated leaves (>5-fold intensity changes). Blank bar shows unidentified features; grey bar shows features with a database match.

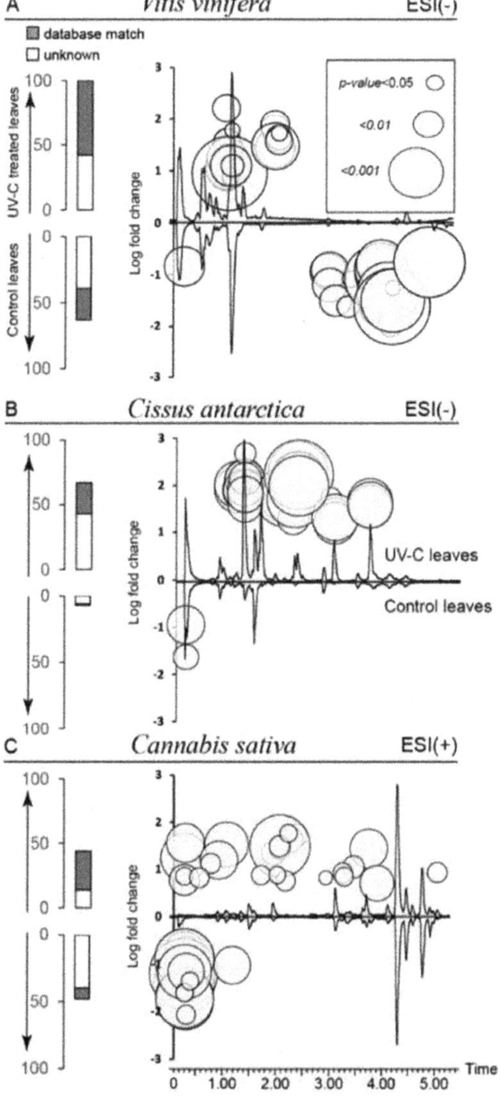

To provide an overview of the major changes observed, all significant features were superimposed on short LC-MS profiles of each species (Figure 3, UV-C-treated upper trace and control bottom trace). The features were positioned according to their log-fold changes (Y) and their retention times (X), with point sizes calculated from unpaired T-test *p*-values on the bubble chart. In *V. vinifera* (Figure 3A), the bubble chart clearly show that major up-regulated features are moderately polar and eluted between 1 and 2 min, whereas most of the down-regulated features had longer retention times (less polar metabolites). Interestingly, in the case of *C. antarctica* (Figure 3B), most of the induced features appeared in the same area as for *V. vinifera*, but the down-regulated features were detected close to the injection peak. The up-regulated features of *C. sativa* were distributed all along the chromatogram, whereas the down-regulated ones eluted close to the injection peak (Figure 3C). These charts display the response specificity of each species and give an idea of the polarity of significant features.

2.4. Biomarkers of UV-C Radiation

Significant features were annotated based on accurate HRMS exact mass spectral data and the use of heuristic filters [31] after deconvolution and adduct removal. Molecular formula determination and further cross-searching based on chemotaxonomy information were then performed. Putative identification was achieved using the molecular formula together with the botanical genus and family as queries in the Dictionary of Natural Products database [32]. Unidentified metabolites were then matched using Lipidmaps [33] and the Plant Metabolic Network [34], generic databases related to plant metabolism. These labels correspond to level 2 and 3 IDs according to the Metabolite Identification Task Group [35]. Altogether, the putative identification of approximately 50% of the significant features (>5-fold changes) in each species was achieved (bar plots, Figure 3).

In the three studied species, several down-regulated metabolites were detected close to the injection peak (possibly polar or charged metabolites) (bubble charts, Figure 3). Feature annotation highlighted that several simple organic acids, such as malic acid and fumaric acid, could be down-regulated in *V. vinifera* and *C. antarctica* leaves. In addition, a few phosphorylated metabolites, such as adenosine diphosphate and O-phospho-L-homoserine, which are involved in methionine biosynthesis, were down-regulated in *V. vinifera*. In *C. sativa*, α-iminosuccinate, which is involved in cofactor biosynthesis, and N-acetylglutamyl phosphate were also down-regulated (data not shown). A detailed identification of such polar metabolites would require the use of alternative profiling methods that focus on primary metabolites, such as hydrophilic interaction liquid chromatography (HILIC) [36] or gas chromatography-mass spectrometry (GC-MS) [37]. However, this pursuit was beyond the scope of this study, which is primarily dedicated to secondary metabolite induction.

Late-eluting metabolites (after 3 min) were putatively identified as glycerophospholipids (GPLs). Several GPLs were detected in reduced quantities in UV-C treated *V. vinifera* leaves compared to control. As major constituents of plant cell membranes, GPLs are particularly sensitive to denaturation upon UV radiation. This behaviour is mainly due to the oxidative damage by reactive oxygen species (ROS) of the methylene groups in unsaturated fatty acids, which leads to a chain reaction of peroxidation [10,38]. Interestingly, GPLs were induced in *C. antarctica* and *C. sativa* UV-C treated leaves (Table 1). Damages caused by UV-C do not involve specific cellular receptors but stimulate a metabolic response similar to that caused by wounding [39]. For instance, a systemic induction of phosphatidic acid and lysophospholipids was found in wounded tomato leaves [40]. The possible connection of phospholipids to jasmonates is illustrated by the fact that silencing phospholipase D in rice limits the induction of jasmonic acid levels [41]. Although the untargeted LC-MS profiling conducted here did not reveal any induction of jasmonates, our data reveal some evidence of cell membrane reconfiguration upon UV-C radiation.

Table 1. Putative identification of induced compounds in UV-C treated leaves.

Mode	HR-MS	RT (min.)	MF	Chemical Class	Database (hit) [a]	Putative ID [b]	Error (mDa)	Isotope Pattern Score (%)	Fold Change (UV/C)
					Vitis vinifera L.				
NI	405.1178	1.22	$C_{20}H_{22}O_9$	stilbene	Lipidmaps (5)	astringin	1.2	95	150
NI	453.1327	1.41	$C_{28}H_{22}O_6$	stilbene polymer	DNP (6)	ε-viniferin	1.6	97	120
NI	919.2451	1.74	$C_{56}H_{40}O_{13}$	stilbene polymer	DNP (1)	amurensin K	0.3	96	110
NI	471.1455	1.47	$C_{28}H_{24}O_7$	stilbene polymer	DNP (1)	amurensin A	0.6	95	90
NI	679.2027	1.88	$C_{42}H_{32}O_9$	stilbene polymer	DNP (6)	vitisin E	1.2	95	70
NI	597.1815	0.88	$C_{27}H_{34}O_{15}$	Flavonoid	Lipidmaps (2)	Catechin 3-O-rutinoside	3.3	95	60
NI	231.1013	1.18	$C_{28}H_{22}O_7$	stilbene polymer	DNP (1)	ampelopsin A	2	95	40
NI	227.0710	1.62	$C_{14}H_{12}O_3$	stilbene	DNP (1)	resveratrol *	0.2	96	30
					Cissus antarctica Vent.				
NI	227.0709	1.62	$C_{14}H_{12}O_3$	stilbene	DNP (1)	resveratrol *	0.7	98	110
NI	435.1295	1.17	$C_{21}H_{24}O_{10}$	dihydrochalcone flavonoids	DNP (1)	trilobatin	0.1	97	100
NI	453.1336	1.58	$C_{28}H_{22}O_6$	stilbene polymer	DNP (2)	pallidol	0.3	95	100
PI	637.4055	3.74	$C_{32}H_{61}O_{10}P$	glycerophospholipids	Lipidmaps (2)	PG(12:0/14:1(9Z))	2.0	95	90
NI	389.1229	1.31	$C_{20}H_{22}O_8$	stilbene	DNP (1)	piceid *	1.2	96	40
					Cannabis sativa L.				
PI	259.1348	3.20	$C_{16}H_{18}O_3$	stilbene	Lipidmaps (1)	3-O-methylbatatasin	1.9	95	100
NI	407.1881	0.72	$C_{25}H_{28}O_5$	Chalcone flavonoid	DNP (2)	3′-geranyl-2′,4,4′,6′-tetrahydroxychalcone	1.7	95	15
PI	625.2543	2.43	$C_{36}H_{36}N_2O_8$	cinnamic acid amide	DNP (4)	cannabisin D	0.1	96	10
PI	235.1697	2.77	$C_{15}H_{22}O_2$	aliphatic	DNP (1)	p-hydroxynonanophenone	0.5	97	8
PI	284.1289	1.70	$C_{17}H_{17}NO_3$	cinnamic acid amide	DNP (1)	N-*p*-trans-coumaroyltyramine *	0.8	96	7
PI	219.1343	0.49	$C_{14}H_{18}O_2$	spirans	DNP (1)	5,7-dihydroxy[indan-1-spirocyclohexane]	3.0	97	6
PI	454.2935	3.97	$C_{21}H_{43}NO_7P$	glycerophospholipids	Lipidmaps (2)	PE(16:0/0:0)	0.7	95	6
PI	496.3399	3.98	$C_{24}H_{50}NO_7P$	glycerophospholipids	Lipidmaps (5)	PC(16:0/0:0)	0.1	96	5

[a] For the DNP database, the molecular formula was crossed-filtered using the genus name or family name. The number of hits is indicated in brackets; [b] In the case of more than one hit, the annotation is indicative of a compound characteristic of the class; * Comparison with pure standard. The fold change indicates the ratio of intensity (up-regulation) of a given feature in the UV-C leaves compared to control.

Interestingly, most of the induced metabolites detected in Vitaceae species eluted between 1 and 2 min, denoting compounds with similar physicochemical properties (Figure 3A,B). Following

UV-C treatment, several stilbenoids were identified as major up-regulated compounds in *V. vinifera* leaves. The metabolites that were the most significantly induced were putatively annotated as astringin (3-OH-piceid) which has previously been purified from *Vitis* cell cultures [42]. Other strongly induced compounds are stilbene polymers, such as the dehydrodimer ε-viniferin, which was previously detected in UV-C-treated grapevine leaves [12]. Some features putatively identified as glycosylated flavonoids were also up-regulated upon UV-C treatment. As expected, the induction of resveratrol was detected in both *V. vinifera* and *C. antarctica* treated leaves, thus indicating a common genetic background. However, the diversity of stilbene polymers was less pronounced in *C. Antarctica* compared to *V. vinifera*, because only pallidol was detected [43]. Indeed, other phenylpropanoids were identified in this species, such as trilobatin, a dihydrochalcone flavonoid, along with the glycosylated stilbene, piceid. In contrast, *C. sativa* UV-C-treated leaves displayed a different pattern of induction compared to the Vitaceae species studied. Several other classes of induced compounds were putatively identified (GPLs, cinnamic acids, spirans), but the most significantly induced compound was interestingly also found to belong to stilbenes and was putatively identified as 3-hydroxy-5,4'-dimethoxybibenzyl (3-*O*-methylbatatasin). This compound could be biosynthesised from dihydroresveratrol as has already been reported to occur in *C. sativa* [44]. The coupling of dihydrostilbenes followed by reductive steps could also lead to spirans, such as the induced 5,7-dihydroxy[indan-1-spirocyclohexane] detected in UV-C treated leaves [45]. Cinnamic acid amides represented another class of induced compounds, such as cannabisin D, a lignamide resulting from the dimerisation of *N-trans*-feruloyltyramine [46], and *N-trans*-coumaroyltyramine. The latter has been identified as a wound biomarker and was also up-regulated after the UV-C radiation of *Capsicum annuum* leaves [47] and after herbivore attack in maize leaves [20].

Overall, our untargeted differential metabolomics approach revealed that approximately 15% of the detected features were affected by UV-C radiation. Some of these compounds remain unknown since they are not yet listed in databases and would require more extensive investigations to be identified. For instance, *de novo* induction of oxidised products compounds is expected since UV-C are known to increase ROS levels [10]. This is illustrated by the high level of astringin detected in *Vitis vinifera* which is the oxydised product of piceid. Several of the identified compounds are secondary metabolites with assessed biological activities. For instance, induced cinnamic acid amides in *C. sativa* leaves are known to play a role in cell wall reinforcement after tissue disruption [48]. Stilbenes induced in Vitaceae species contribute to a constitutive defence against microbial diseases [49] and powdery mildew infestation [50]. The antioxidant properties of stilbenes are also well documented [51], and in particular, several studies have shown stronger antiradical scavenging activities for cyclised stilbenes [52].

2.5. Antioxidant Activity of Extracts

To determine whether the metabolomic modifications observed could be linked to noticeable changes in bioactivity, the antioxidant properties of crude extracts obtained from control and UV-C-treated leaves were assessed. For this analysis, three radical scavenging assays were performed (ABTS, DPPH, and ORAC; see experimental section for details). No activity was found in *C. sativa* extracts, but a significant activity was measured for the *C. antarctica* extracts independently of UV treatment (Figure 4). Interestingly, the radical scavenging activity of the *V. vinifera* UV-C-treated extract was significantly higher compared to the controls in all assays. The EC_{50} values for the radical scavenging activity based on DPPH were determined through dose-response experiments with *V. vinifera* control extract, UV-C-treated *V. vinifera* extract, and resveratrol (Figure 5). The EC_{50} for *V. vinifera* control was 45.9 ± 2.5 µg/mL.

Figure 4. Antioxidant activity of plant extracts compared to resveratrol as positive control. UV indicates plants exposed to UV light, and control indicates non-exposed plants. * Significantly different from control ($p \leq 0.05$); ** Significantly different from control ($p \leq 0.001$).

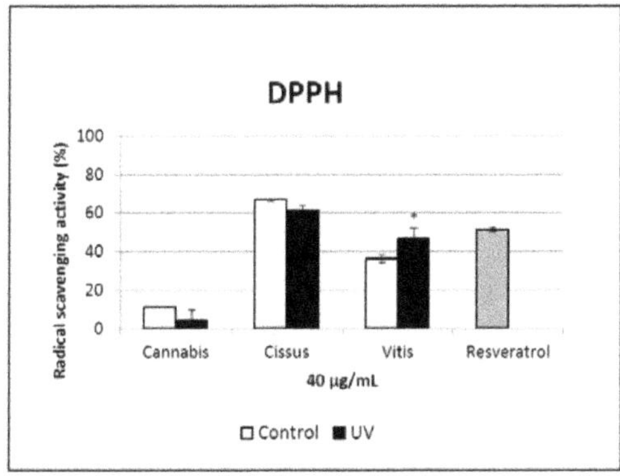

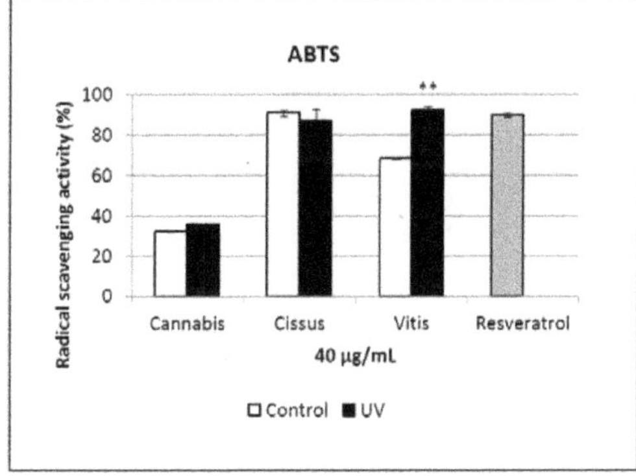

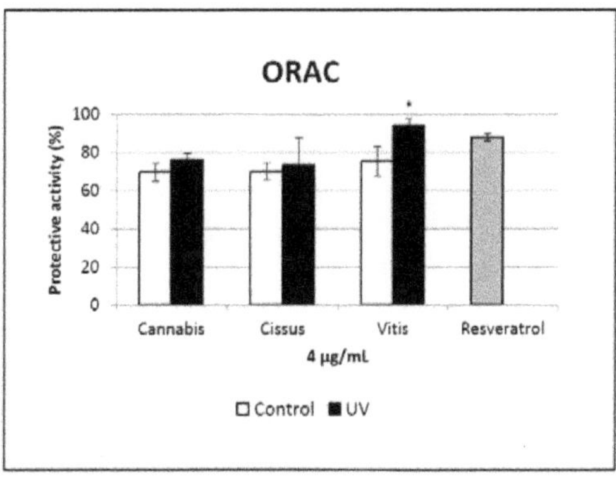

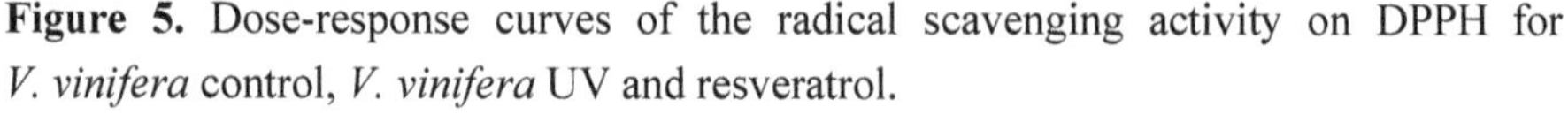

Figure 5. Dose-response curves of the radical scavenging activity on DPPH for *V. vinifera* control, *V. vinifera* UV and resveratrol.

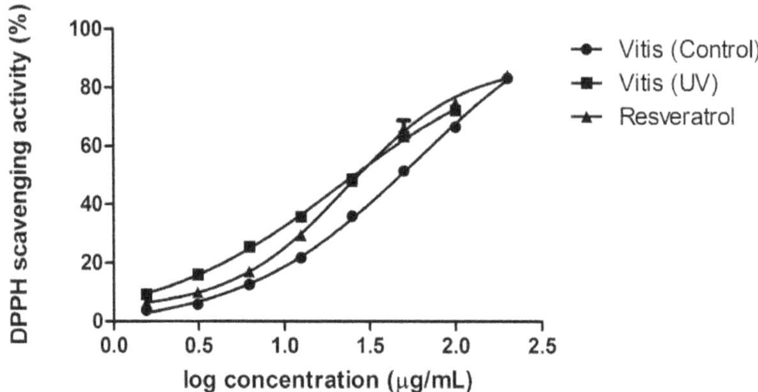

When *V. vinifera* was exposed to UV light, the scavenging potential of the plant extract was significantly higher ($p \leq 0.001$) with an EC_{50} of 25.12 ± 0.5 μg/mL. This value is similar to the EC_{50} obtained for resveratrol (28.8 ± 1.0 μg/mL). According to the metabolomic variation that was measured in this species, the enhancement of radical scavenging activity in the *V. vinifera* UV-C-treated extract could be explained by its higher level of resveratrol polymers [52]. Moreover, the induced astringin is also known to possess stronger antioxidant activity than resveratrol [53].

3. Experimental Section

3.1. Plant Growth and UV-C Treatment

Plantlets from the *Vitis vinifera* cultivar Chasselas and *Cissus antarctica* were cultivated in a greenhouse in accordance with the methods described by Pezet *et al.* [54]. Briefly, two-eyes woody cuttings of *Vitis vinifera* cultivar Chasselas and herbaceous cuttings of *Cissus antarctica* were cultivated in a mix of Perlite and potting compost. Liquid fertilizer (Vegesan mega, Hauert, Switzerland) was added weekly, and growing conditions were similar for the two species (20 °C, 70% of relative humidity and a daily photoperiod of 16 h using a sodium lamp at 120 watts/m^2). Seeds of *Cannabis sativa* (birdseeds produced in Switzerland and distributed by Coop) were sowed and cultivated in a greenhouse under a sodium lamp (400 watts/m^2). For all species studied, the stage "15 leaves fully developed" was required for further experimentation. All the leaves of each plant were detached and immediately transferred with the abaxial face up into large square Petri dishes (24 cm) containing wet blotting paper (180 g, papyrus, Thalwill, Lausanne, Switzerland). UV-C treatment was performed according to Jean-Denis *et al.* with slight modifications [25]: leaf were exposed during 10 min radiation at 253 nm at 21 °C in the dark. The lamp was placed at 13 cm from the leaves and delivered 0.18 Kj/min (TUV 30W, 92 μW·cm^{-2}, Philips, Seynod, France). After UV-C exposure, Petri dishes were sealed and incubated in a growth chamber under alternating light and dark conditions (16 h at 22 °C and 8 h at 18 °C, respectively) for 48 h. The controls consisted of leaves of each plant species that underwent the same treatment lacking the UV-C exposure.

3.2. Leaf Extraction

Each sample was ground to a powder using a mortar previously frozen in liquid nitrogen. The frozen powder was weighed (300 mg ± 2 mg), and 1.5 mL of isopropanol was immediately added for metabolite extraction. Samples were vortexed, sonicated in a bath at room temperature (5200 Bransonic, Danbury, CT, USA) for 20 min, vortexed again and centrifuged at 10,000 rpm for 2 min (Hettich mikrolitter D 7200, Buford, GA, USA). The supernatant was recovered, and the extraction procedure was repeated. Each isopropanol extract was dried under vacuum (Genevac HT-4X, Ipswich, UK) and suspended in a mixture of 85:15 (v/v) methanol:water for an SPE C_{18} enrichment procedure (100 mg C18 cartridge Sep-Pack®, Waters, Milford, MA, USA) to remove highly non-polar compounds. The filtered extracts were dried and dissolved to 1 mg/mL in 85:15 methanol/water for UHPLC-TOF-MS analysis. This protocol was adapted from Glauser and co-workers [26].

3.3. Short LC-MS Profiling

Metabolite analysis was performed on a UPLC-PDA-TOFMS instrument (LCT Premier, Waters) equipped with an electrospray ionisation (ESI) source. The LC-MS fingerprint of each extract was obtained using a short UPLC BEH C18 Acquity column (50 × 1.0 mm i.d., 1.7 μm, Waters). The mobile phase consisted of 0.1% formic acid (FA) in water (phase A) and 0.1% FA in acetonitrile (phase B). The linear gradient program was as follows: 98% A for 0.2 min to 100% B over 4.9 min, held at 100% B for a further 1.1 min, and then returned in 0.1 min to initial conditions (98% A) for 1.1 min of equilibration before the subsequent analysis. The flow rate was 0.3 mL/min. The column temperature was kept at 40 °C. Detection was performed by TOF-MS in W-mode in both electrospray (ESI) negative (NI) and positive ion (PI) modes in independent runs with the following settings: capillary voltage at 2.8 kV, cone voltage at 40 V, desolvation temperature at 250 °C, source temperature at 120 °C and desolvation gas flow at 600 L/h. The *m/z* range was 100–1000 Da with a scan time of 0.25 s. The MS was calibrated using sodium formate, and leucine enkephalin was used as an internal reference. The injection volume was 1 μL.

3.4. Data Processing and Data Analysis

The UHPLC-TOF-MS fingerprints were processed with MZmine 2.10 for mass signal extraction and alignment from 0 to 5 min with *m/z* values ranging from 100 to 1000 Da. The following parameters were employed: the chromatogram builder was set to a minimum time span of 0.06 min, a minimum height of 10 for NI mode and 100 for PI mode, and an *m/z* tolerance of 10 ppm. The local minimum search algorithm was applied for chromatogram deconvolution. Each peak list was de-isotoped and aligned using the RANSAC alignment method and then gap-filled. The resulting peak matrix from each sample containing areas of aligned peaks characterised by retention time and *m/z* ratio was exported into the ".csv" file-format prior to multivariate data analysis using SIMCA-P+ (version 12, Umetrics, Umeå, Sweden). Homemade Excel macros were used to compare data between PI and NI modes and between treated and control samples, and *p*-values were calculated by Student's T-test.

3.5. Standards

N-p-trans-coumaroyltyramine (NMR purity of 99%) was purified from *Zea mays* leaves according to a procedure described previously [55]. Resveratrol (GC purity ≥ 99%) and piceid (>95% HPLC) were purchased from Sigma-Aldrich (Sigma-Aldrich Chemie GmbH, Buchs, Switzerland).

3.6. Antioxidant Assays

DPPH radical scavenging assay: the capacity of samples to scavenge the stable radical 2,2-diphenyl-1-picrylhydrazyl (DPPH) was determined spectrophotometrically by measuring the loss of absorbance of DPPH at 515 nm [56]. Clear polystyrene flat-bottom 96-well microplates were filled with test sample solution (in ethanol containing up to 2% DMSO) or vehicle for the DPPH control. The reaction was initiated by the addition of 80 μM DPPH (in ethanol). The decrease in absorbance at 515 nm was monitored at room temperature after 10 min to determine the percentage of scavenged radical. Samples were tested in triplicate at 40 μg/mL, and dose-response experiments were performed with at least 6 concentrations.

ABTS radical scavenging assay: the capacity of samples to scavenge the monocation radical 2,2'-azinobis-(3-ethylbenzothiazoline-6-sulfonic acid) (ABTS) was determined spectrophotometrically by measuring the loss of absorbance of ABTS at 715 nm [57]. The same procedure as described for the DPPH assay was applied by using the ABTS solution instead of DPPH solution. The ABTS cation was produced by the reaction between 7 mM ABTS and 2.45 mM potassium persulfate in water. The reaction was initiated by the addition of 67 μM ABTS radical (in ethanol). Samples were tested in triplicate at the same concentrations used for the DPPH assay.

ORAC assay: the antioxidant activity of the tested samples was determined by their ability to preserve the fluorescence of fluorescein exposed to peroxyl radicals generated by 2,2'-azobis(2-methylpropionamidine) dihydrochloride (AAPH) [58]. Black polypropylene 96-well plates were filled with 60 nM fluorescein (in glycine buffer pH 8.3), together with test samples or vehicle (in 2% DMSO), and pre-incubated at 40 °C for 15 min. The oxidative reaction was obtained by adding 5 mM AAPH (in glycine buffer pH 8.3) to wells containing samples, positive control, and oxidised fluorescein control. Non-oxidised fluorescein controls were added with the same volume of assay buffer. The plate was incubated at 40 °C for 90 min with continuous shaking at 150 rpm and cooled to room temperature (5 min) prior to fluorescence reading at 485/528 nm. Samples were tested in triplicate at 4 μg/mL.

4. Conclusions

This LC-HRMS-based metabolomic study revealed important metabolic changes upon the UV-C treatment of the leaves of three different plant species. The metabolomic modifications were found to be species-specific, but in all plants studied, the UV-C stress significantly induced greater than five-fold changes for more than 10% of the features detected. The LC-MS-based approach used provided a holistic overview of the changes related to a given stress and generated useful data for a rapid estimation of the magnitude of UV-C-induced metabolomic modifications and a preliminary identification of related biomarkers.

In the case of *C. sativa*, no remarkable modification of the cannabinoid content was observed, but dehydrostilbenes and cinnamic acid amide derivatives were strongly induced. In contrast, Vitaceae species responded in the same manner with a strong induction of stilbenes derived from resveratrol. Furthermore, it has been demonstrated that in some cases (e.g., for *V. vinifera*) such metabolome modifications could enhance the antioxidant activity of the extracts.

This generic approach may represent an interesting method to screen leaves for new bioactivities or new metabolites that would not be detected without stress induction (e.g., stilbene polymers from Vitaceae species) and that may be related to cryptic biosynthetic pathways. Because common metabolites could be observed after UV-C stress treatment and other biotic or abiotic stresses (e.g., wound response, pathogen infection, *etc.*), such methods can also be used to indicate the overall metabolite induction potential of a given plant compared to more specific and relevant biological stresses.

Acknowledgments

We thank S. Bertrand for his help in coding a specific Excel macro. J.-L.W is thankful to the Swiss National Science Foundation (SNF) grant No. 200020_146200 for supporting his plant metabolomics studies. We gratefully acknowledge the first nine Grands Crus de Bordeaux for the salary of Sylvain Schnee: Château Ausone, Château Cheval blanc, Château Haut-Brion, Château Lafitte Rothschild, Château Latour, Château Margaux, Château Mouton Rothschild, Château Petrus and Château d'Yquem.

Author Contributions

K. Gindro proposed the study and designed the experiments with the help of S. Schnee. Y. Andrey performed the extraction of crude material under the supervision of G. Marti. G. Marti acquired and processed the LCMS profiles and performed multivariate data analysis under the supervision of J.-L. Wolfender. C. Simoes-Pires conducted the radical scavenging assays under the supervision of P.A. Carrupt. G. Marti analysed the results and wrote the manuscript.

Conflicts of Interest

The authors declare no conflict of interest.

References

1. Newman, D.J.; Cragg, G.M. Natural products as sources of new drugs over the 30 years from 1981 to 2010. *J. Nat. Prod.* **2012**, *75*, 311–335.
2. David, B.; Wolfender, J.-L.; Dias, D.A. The pharmaceutical industry and natural products: Historical status and new trends. *Phytochem. Rev.* **2014**, 1–17.
3. Cragg, G.M.; Newman, D.J. Natural products: A continuing source of novel drug leads. *Biochim. Biophys. Acta* **2013**, *1830*, 3670–3695.

4. Gurib-Fakim, A. Medicinal plants: Traditions of yesterday and drugs of tomorrow. *Mol. Aspects Med.* **2006**, *27*, 1–93.
5. Atkinson, N.J.; Urwin, P.E. The interaction of plant biotic and abiotic stresses: From genes to the field. *J. Exp. Bot.* **2012**, *63*, 3523–3543.
6. Winter, J.M.; Behnken, S.; Hertweck, C. Genomics-inspired discovery of natural products. *Curr. Opin. Chem. Biol.* **2011**, *15*, 22–31.
7. Bertrand, S.; Bohni, N.; Schnee, S.; Schumpp, O.; Gindro, K.; Wolfender, J.L. Metabolite induction via microorganism co-culture: A potential way to enhance chemical diversity for drug discovery. *Biotechnol. Adv.* **2014**, *32*, 1180–1204.
8. Wolfender, J.L.; Queiroz, E.F. New approaches for studying the chemical diversity of natural resources and the bioactivity of their constituents. *Chimia (Aarau)* **2012**, *66*, 324–329.
9. Poulev, A.; O'Neal, J.M.; Logendra, S.; Pouleva, R.B.; Timeva, V.; Garvey, A.S.; Gleba, D.; Jenkins, I.S.; Halpern, B.T.; Kneer, R.; *et al.* Elicitation, a new window into plant chemodiversity and phytochemical drug discovery. *J. Med. Chem.* **2003**, *46*, 2542–2547.
10. Kunz, B.A.; Cahill, D.M.; Mohr, P.G.; Osmond, M.J.; Vonarx, E.J. Plant responses to UV radiation and links to pathogen resistance. In *International Review of Cytology*; Jeon, K.W., Ed.; Academic Press: Knoxville, TN, USA, 2006; Volume 255, pp. 1–40.
11. Kostyuk, V.; Potapovich, A.; Suhan, T.; de Luca, C.; Pressi, G.; Dal Toso, R.; Korkina, L. Plant polyphenols against UV-C-induced cellular death. *Planta Med.* **2008**, *74*, 509–514.
12. Pezet, R.; Perret, C.; Jean-Denis, J.B.; Tabacchi, R.; Gindro, K.; Viret, O. Delta-viniferin, a resveratrol dehydrodimer: One of the major stilbenes synthesized by stressed grapevine leaves. *J. Agric. Food Chem.* **2003**, *51*, 5488–5492.
13. Charles, M.T.; Mercier, J.; Makhlouf, J.; Arul, J. Physiological basis of UV-C-induced resistance to botrytis cinerea in tomato fruit. *Postharvest Biol. Technol.* **2008**, *47*, 10–20.
14. Park, H.L.; Lee, S.W.; Jung, K.H.; Hahn, T.R.; Cho, M.H. Transcriptomic analysis of UV-treated rice leaves reveals UV-induced phytoalexin biosynthetic pathways and their regulatory networks in rice. *Phytochemistry* **2013**, *96*, 57–71.
15. Pedras, M.S.; Zheng, Q.A.; Schatte, G.; Adio, A.M. Photochemical dimerization of wasalexins in UV-irradiated thellungiellahalophila and *in vitro* generates unique cruciferous phytoalexins. *Phytochemistry* **2009**, *70*, 2010–2016.
16. Gregianini, T.S.; da Silveira, V.C.; Porto, D.D.; Kerber, V.A.; Henriques, A.T.; Fett-Neto, A.G. The alkaloid brachycerine is induced by ultraviolet radiation and is a singlet oxygen quencher. *Photochem. Photobiol.* **2003**, *78*, 470–474.
17. Jansen, M.A.K.; Hectors, K.; O'Brien, N.M.; Guisez, Y.; Potters, G. Plant stress and human health: Do human consumers benefit from uv-b acclimated crops? *Plant Sci.* **2008**, *175*, 449–458.
18. Schumpp, O.; Bruderhofer, N.; Monod, M.; Wolfender, J.L.; Gindro, K. Ultraviolet induction of antifungal activity in plants. *Mycoses* **2012**, *55*, 507–513.

19. Wolfender, J.L.; Marti, G.; Ferreira Queiroz, E. Advances in techniques for profiling crude extracts and for the rapid identificationof natural products: Dereplication, quality control and metabolomics. *Curr. Org. Chem.* **2010**, *14*, 1808–1832.
20. Marti, G.; Erb, M.; Boccard, J.; Glauser, G.; Doyen, G.R.; Villard, N.; Robert, C.A.; Turlings, T.C.; Rudaz, S.; Wolfender, J.L. Metabolomics reveals herbivore-induced metabolites of resistance and susceptibility in maize leaves and roots. *Plant Cell Environ.* **2013**, *36*, 621–639.
21. Gindro, K.; Alonso-Villaverde, V.; Viret, O.; Spring, J.-L.; Marti, G.; Wolfender, J.-L.; Pezet, R. Stilbenes: Biomarkers of grapevine resistance to disease of high relevance for agronomy, oenology and human health. In *Plant Defence: Biological Control*; Merillon, J.M., Ramawat, K.G., Eds.; Springer: Dordrecht, The Netherlands, 2012; Volume 12, pp. 25–54.
22. Jeandet, P.; Clement, C.; Courot, E.; Cordelier, S. Modulation of phytoalexin biosynthesis in engineered plants for disease resistance. *Int. J. Mol. Sci.* **2013**, *14*, 14136–14170.
23. Riviere, C.; Pawlus, A.D.; Merillon, J.M. Natural stilbenoids: Distribution in the plant kingdom and chemotaxonomic interest in vitaceae. *Nat. Prod. Rep.* **2012**, *29*, 1317–1333.
24. Flores-Sanchez, I.J.; Verpoorte, R. Secondary metabolism in cannabis. *Phytochem. Rev.* **2008**, *7*, 615–639.
25. Jean-Denis, J.B.; Pezet, R.; Tabacchi, R. Rapid analysis of stilbenes and derivatives from downy mildew-infected grapevine leaves by liquid chromatography-atmospheric pressure photoionisation mass spectrometry. *J. Chromatogr. A* **2006**, *1112*, 263–268.
26. Glauser, G.; Guillarme, D.; Grata, E.; Boccard, J.; Thiocone, A.; Carrupt, P.A.; Veuthey, J.L.; Rudaz, S.; Wolfender, J.L. Optimized liquid chromatography-mass spectrometry approach for the isolation of minor stress biomarkers in plant extracts and their identification by capillary nuclear magnetic resonance. *J. Chromatogr. A* **2008**, *1180*, 90–98.
27. Bylesjö, M.; Rantalainen, M.; Cloarec, O.; Nicholson, J.K.; Holmes, E.; Trygg, J. Opls discriminant analysis: Combining the strengths of pls-da and simca classification. *J. Chemom.* **2006**, *20*, 341–351.
28. Forshed, J.; Idborg, H.; Jacobsson, S.P. Evaluation of different techniques for data fusion of LC/MS and ^{1}H-NMR. *Chemom. Intell. Lab. Syst.* **2007**, *85*, 102–109.
29. Hall, D.L.; Llinas, J. An introduction to multisensor data fusion. *Proc. IEEE* **1997**, *85*, 6–23.
30. Wold, S.; Ruhe, A.; Wold, H.; Dunn, I.W.J. The collinearity problem in linear regression. The partial least squares (pls) approach to generalized inverses. *SIAM J. Sci. Stat. Comput.* **1984**, *5*, 735–743.
31. Kind, T.; Fiehn, O. Seven golden rules for heuristic filtering of molecular formulas obtained by accurate mass spectrometry. *BMC Bioinform.* **2007**, *8*, 105.
32. *Dictionary of Natural Products on DVD*, Version 20:2; CRC Press, Taylor & Francis: Boca Raton, FL, USA, 2014.
33. Sud, M.; Fahy, E.; Cotter, D.; Brown, A.; Dennis, E.A.; Glass, C.K.; Merrill, A.H., Jr.; Murphy, R.C.; Raetz, C.R.; Russell, D.W.; *et al.* Lmsd: Lipid maps structure database. *Nucleic Acids Res.* **2007**, *35*, D527–D532.

34. Caspi, R.; Altman, T.; Dreher, K.; Fulcher, C.A.; Subhraveti, P.; Keseler, I.M.; Kothari, A.; Krummenacker, M.; Latendresse, M.; Mueller, L.A.; *et al.* The metacyc database of metabolic pathways and enzymes and the biocyc collection of pathway/genome databases. *Nucleic Acids Res.* **2012**, *40*, D742–D753.
35. Creek, D.J.; Dunn, W.B.; Fiehn, O.; Griffin, J.L.; Hall, R.D.; Lei, Z.; Mistrik, R.; Neumann, S.; Schymanski, E.L.; Sumner, L.W.; *et al.* Metabolite identification: Are you sure? And how do your peers gauge your confidence? *Metabolomics* **2014**, *10*, 350–353.
36. Gika, H.G.; Theodoridis, G.A.; Plumb, R.S.; Wilson, I.D. Current practice of liquid chromatography-mass spectrometry in metabolomics and metabonomics. *J. Pharm. Biomed. Anal.* **2014**, *87*, 12–25.
37. t'Kindt, R.; Morreel, K.; Deforce, D.; Boerjan, W.; van Bocxlaer, J. Joint gc-ms and lc-ms platforms for comprehensive plant metabolomics: Repeatability and sample pre-treatment. *J. Chromatogr. B* **2009**, *877*, 3572–3580.
38. Hollosy, F. Effects of ultraviolet radiation on plant cells. *Micron* **2002**, *33*, 179–197.
39. Frohnmeyer, H.; Staiger, D. Ultraviolet-b radiation-mediated responses in plants. Balancing damage and protection. *Plant Physiol.* **2003**, *133*, 1420–1428.
40. Lee, S.; Suh, S.; Kim, S.; Crain, R.C.; Kwak, J.M.; Nam, H.-G.; Lee, Y. Systemic elevation of phosphatidic acid and lysophospholipid levels in wounded plants. *Plant J.* **1997**, *12*, 547–556.
41. Qi, J.; Zhou, G.; Yang, L.; Erb, M.; Lu, Y.; Sun, X.; Cheng, J.; Lou, Y. The chloroplast-localized phospholipases d alpha4 and alpha5 regulate herbivore-induced direct and indirect defenses in rice. *Plant Physiol.* **2011**, *157*, 1987–1999.
42. Fauconneau, B.; Waffo-Teguo, P.; Huguet, F.; Barrier, L.; Decendit, A.; Merillon, J.-M. Comparative study of radical scavenger and antioxidant properties of phenolic compounds from *Vitis vinifera* cell cultures using *in vitro* tests. *Life Sci.* **1997**, *61*, 2103–2110.
43. Khan, M.A.; Nabi, S.G.; Prakash, S.; Zaman, A. Pallidol, a resveratrol dimer from *Cissus pallida*. *Phytochemistry* **1986**, *25*, 1945–1948.
44. Crombie, L.; Crombie, W.M.L. Natural products of thailand high δ1-thc-strain cannabis. The bibenzyl-spiran-dihydrophenanthrene group: Relations with cannabinoids and canniflavones. *J. Chem. Soc. Perkin Trans. 1* **1982**, 1455–1466.
45. Crombie, L. Natural products of cannabis and khat. *Pure Appl. Chem.* **1986**, *58*, 693–700.
46. Sakakibara, I.; Ikeya, Y.; Hayashi, K.; Mitsuhashi, H. Three phenyldihydronaphthalene lignanamides from fruits of cannabis sativa. *Phytochemistry* **1992**, *31*, 3219–3223.
47. Back, K.; Jang, S.M.; Lee, B.C.; Schmidt, A.; Strack, D.; Kim, K.M. Cloning and characterization of a hydroxycinnamoyl-CoA:Tyramine *N*-(hydroxycinnamoyl)transferase induced in response to UV-C and wounding from *Capsicum annuum*. *Plant Cell Physiol.* **2001**, *42*, 475–481.
48. Clarke, D. The accumulation of cinnamic acid amides in the cell walls of potato tissue as an early response to fungal attack. *Act. Def. Mech. Plants* **1982**, *37*, 321–322.
49. Chong, J.; Poutaraud, A.; Hugueney, P. Metabolism and roles of stilbenes in plants. *Plant Sci.* **2009**, *177*, 143–155.

50. Schnee, S.; Viret, O.; Gindro, K. Role of stilbenes in the resistance of grapevine to powdery mildew. *Physiol. Mol. Plant Pathol.* **2008**, *72*, 128–133.
51. Waffo Teguo, P.; Fauconneau, B.; Deffieux, G.; Huguet, F.; Vercauteren, J.; Merillon, J.M. Isolation, identification, and antioxidant activity of three stilbene glucosides newly extracted from vitis vinifera cell cultures. *J. Nat. Prod.* **1998**, *61*, 655–657.
52. Privat, C.; Telo, J.P.; Bernardes-Genisson, V.; Vieira, A.; Souchard, J.-P.; Nepveu, F. Antioxidant properties oftrans-ε-viniferin as compared to stilbene derivatives in aqueous and nonaqueous media. *J. Agric. Food Chem.* **2002**, *50*, 1213–1217.
53. Merillon, J.M.; Fauconneau, B.; Teguo, P.W.; Barrier, L.; Vercauteren, J.; Huguet, F. Antioxidant activity of the stilbene astringin, newly extracted from vitis vinifera cell cultures. *Clin. Chem.* **1997**, *43*, 1092–1093.
54. Pezet, R.; Gindro, K.; Viret, O.; Spring, J.L. Glycosylation and oxidative dimerization of resveratrol are respectively associated to sensitivity and resistance of grapevine cultivars to downy mildew. *Physiol. Mol. Plant Pathol.* **2004**, *65*, 297–303.
55. Marti, G.; Erb, M.; Rudaz, S.; Turlings, T.; Wolfender, J.-L. Search for low-molecular-weight biomarkers in plant tissues and seeds using metabolomics: Tools, strategies, and applications. In *Seed Development: Omics Technologies toward Improvement of Seed Quality and Crop Yield*; Agrawal, G.K., Rakwal, R., Eds.; Springer: Dordrecht, The Netherlands, 2012; pp. 305–341.
56. Ancerewicz, J.; Migliavacca, E.; Carrupt, P.-A.; Testa, B.; Brée, F.; Zini, R.; Tillement, J.-P.; Labidalle, S.; Guyot, D.; Chauvet-Monges, A.-M.; *et al.* Structure–property relationships of trimetazidine derivatives and model compounds as potential antioxidants. *Free Radic. Biol. Med.* **1998**, *25*, 113–120.
57. Re, R.; Pellegrini, N.; Proteggente, A.; Pannala, A.; Yang, M.; Rice-Evans, C. Antioxidant activity applying an improved abts radical cation decolorization assay. *Free Radic. Biol. Med.* **1999**, *26*, 1231–1237.
58. Huang, D.; Ou, B.; Hampsch-Woodill, M.; Flanagan, J.A.; Prior, R.L. High-throughput assay of oxygen radical absorbance capacity (orac) using a multichannel liquid handling system coupled with a microplate fluorescence reader in 96-well format. *J. Agric. Food Chem.* **2002**, *50*, 4437–4444.

Sample Availability: Samples of the compounds are available from the authors.

Maslinic Acid, a Natural Phytoalexin-Type Triterpene from Olives — A Promising Nutraceutical?

Glòria Lozano-Mena, Marta Sánchez-González, M. Emília Juan and Joana M. Planas

Abstract: Maslinic acid is a pentacyclic triterpene found in a variety of natural sources, ranging from herbal remedies used in traditional Asian medicine to edible vegetables and fruits present in the Mediterranean diet. In recent years, several studies have proved that maslinic acid exerts a wide range of biological activities, *i.e.* antitumor, antidiabetic, antioxidant, cardioprotective, neuroprotective, antiparasitic and growth-stimulating. Experimental models used for the assessment of maslinic acid effects include established cell lines, which have been often used to elucidate the underlying mechanisms of action, and also animal models of different disorders, which have confirmed the effects of the triterpene *in vivo*. Overall, and supported by the lack of adverse effects in mice, the results provide evidence of the potential of maslinic acid as a nutraceutical, not only for health promotion, but also as a therapeutic adjuvant in the treatment of several disorders.

Reprinted from *Molecules*. Cite as: Lozano-Mena, G.; Sánchez-González, M.; Juan, M.E.; Planas, J.M. Maslinic Acid, a Natural Phytoalexin-Type Triterpene from Olives — A Promising Nutraceutical? *Molecules* **2014**, *19*, 11538-11559.

1. Introduction

Maslinic acid, also known as crategolic acid or (2α,3β)-2,3-dihydroxyolean-12-en-28-oic acid (Figure 1), is a pentacyclic triterpene widely distributed in the plant kingdom. In the last decades, and in response to an increasing interest to identify new natural molecules with beneficial effects on health, maslinic acid has been isolated not only from various plants used in traditional herbal medicine, but also from edible vegetables and fruits. In parallel, the biological activities of maslinic acid have been assessed in different experimental models, from tumor cell lines to animal models of several diseases, supported by the lack of adverse effects *in vivo* after the oral administration of the triterpene [1]. In summary, maslinic acid is arising as a novel natural and safe molecule with different biological targets, which might derive to considering it as a nutraceutical in the future.

Figure 1. Chemical structure of maslinic acid.

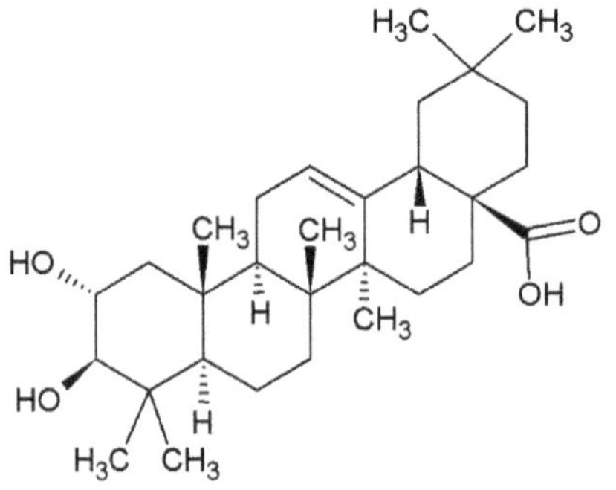

Historically, maslinic acid was named "crategolic acid", since it was first isolated from *Crataegus oxyacantha* L. [2] Tschesche *et al.* [3] described it as a triterpenoid carboxylic acid with molecular formula $C_{30}H_{48}O_4$, mainly found in the leaves of the abovementioned species, where it accounted for 25%–30% of the amount of triterpenoids in this tissue [4]. In the early 1960s, a series of studies by other authors reported the identification of a new triterpenic acid from *Olea europaea* L., although with some controversy. Caglioti *et al.* [5] isolated from olive husks a triterpenic acid with molecular formula and structure identical to those of crategolic acid, and named it maslinic acid. However, a few years later the study was questioned, since the results could not be reproduced, and maslinic acid was considered a product derived from the aging of the fruit [6]. In parallel to the work by Caglioti *et al.* [5], Vioque and Morris [7] found two triterpenic acids in the acetonic extract of the olive pomace, one of which was identified as oleanolic acid and the other was defined as a dihydroxytriterpenic acid, which could be maslinic acid. More than three decades later, Bianchi *et al.* [8] shed light about the composition of the olive fruit, quantifying maslinic acid together with oleanolic acid as the major lipidic compounds in the cuticle of the drupe.

1.1. Biosynthesis and Role as a Phytoalexin

Triterpenoids, such as maslinic acid, are a group of secondary metabolites derived from the cyclation of squalene, oxidosqualene or bis-oxidosqualene [9]. These precursors (C_{30}) are substrate of several types of triterpene synthases, which catalyze their cyclation through intermediate cations to a wide variety of triterpenes. Depending on the number of rings, the latter are classified as mono-, bi-, tri-, tetra- or pentacyclic triterpene alcohols [9]. Lupeol, α- and β-amyrin are examples of pentacyclic triterpene alcohols, which not only constitute secondary metabolites themselves, but also might undergo oxidation reactions to yield other derivatives, such as betulinic, ursolic and maslinic acids.

Not long after the identification in *Crataegus oxyacantha* L., Tschesche *et al.* [10] recognized maslinic acid as a derivative of the β-amyrin series, but it was Stiti *et al.* [11] who more recently postulated the biosynthetic pathway that leads to the formation of maslinic acid in the fruits of *Olea europaea* L., one of the main natural sources of this triterpene. The authors suggest that in the developing olive both the sterols (primary metabolites) and the non-steroidal triterpenoids (secondary metabolites) share oxidosqualene as a common precursor. The enzyme β-amyrin synthase catalyzes its cyclation into β-amyrin, and further oxidation steps give rise to the triterpenic dialcohol erythrodiol followed by the hydroxy pentacyclic triterpenic acids oleanolic and maslinic [11].

Regarding the function, plant secondary metabolites are not essential for the growth, development and reproduction of individuals, but might contribute to their survival or give them evolutionary advantages. Phytoalexins are a particular case of secondary metabolites, involved in the protection of the plant against pathogens, and maslinic acid can be considered as such, since different studies have proved its protective activity under adverse conditions. Kombargi *et al.* [12] observed that dipping *Olea europaea* L. fruits in solutions of maslinic acid prevented the oviposition of eggs from females of the olive fruit fly (*Bactrocera oleae*), which is the major insect pest of olives in the Mediterranean countries. Furthermore, the isolated triterpene is toxic after ingestion by rice weevil adults (*Sitophilus oryzae*) [13], a widespread and destructive pest of stored cereals.

1.2. Natural Sources

Maslinic acid was first detected in *Crataegus oxyacantha* L., but the growing interest in this triterpene because of its wide range of health-enhancing activities has led to its identification in other natural sources, being present in more than 30 plants worldwide. On one hand, the triterpene has been found in plants used in traditional Asian medicine for the treatment of diverse affections. To mention only a few examples, the leaves of loquat (*Eriobotrya japonica*) [14], which have been used as antitussive and anti-inflammatory for chronic bronchitis, and also as diuretic, digestive and antipyretic [15]; the flowers of *Campsis grandiflora*, employed for female disorders like uterine hemorrhage [16]; the whole plant of *Geum japonicum* [17], used as diuretic [18]; and *Agastache rugosa* [19], for the treatment of anorexia, vomiting and other intestinal disorders [20]. On the other hand, maslinic acid has recently been quantified in edible vegetables, such as table olives [21], spinach and eggplant [22], aromatic herbs like mustard and basil [22,23], legumes such as chickpeas and lentils [24], and to a lesser extent in some fruits like mandarin and pomegranate [25] (Table 1). Therefore, plant-based diets might provide a constant supply of maslinic acid, which could be considered, among many other factors, partly responsible of the health-enhancing properties of these dietary habits.

Table 1. Maslinic acid content in edible sources.

	Maslinic Acid (mg/kg Dry Weight)	References
Table olives		
Kalamata (plain black)	1318	[21]
Hojiblanca (plain green)	905	[21]
Gordal (plain green)	414	[21]
Manzanilla (plain green)	384	[21]
Cacereña (plain black)	295	[21]
Fresh vegetables		
Spinach	1260	[22]
Eggplant	840	[22]
Aromatic herbs		
Brown mustard	330	[23]
Leaf mustard	1740	[22]
Basil	350, 320	[22,23]
Cooked legumes		
Small lentils	26.3	[24]
Large lentils	39.5	[24]
Chickpeas	61.9	[24]
Fresh fruits		
Mandarin	1.18	[25]
Pomegranate	10.8	[25]

2. Biological Effects

2.1. Maslinic Acid and Cancer

The antitumor activity of maslinic acid has become remarkable in recent years, as evidenced by the higher number of studies that address this issue, compared to those about other biological effects. The vast majority of published references correspond to *in vitro* experiments that show the anti-proliferative and/or pro-apoptotic effect of maslinic acid, together with plausible mechanisms of action that involve different signaling pathways. Colon cancer cell lines have been extensively used with this aim, but there is no shortage of studies that prove the above-mentioned effects in a wide range of cell lines from other origins. Moreover, this antitumor effect has also been assessed in several animal models, with positive results that reinforce its potential as anticarcinogenic agent.

2.1.1. Maslinic Acid Exerts an Anti-Proliferative Activity through Arresting Cell Cycle and Activates Both the Intrinsic and the Extrinsic Apoptotic Pathways *in Vitro*

The study conducted by Juan *et al.* [26] demonstrated for the first time the potent anti-proliferative activity of maslinic in the human colorectal adenocarcinoma cell line HT-29. The triterpene did not show non-specific cytotoxicity up to 250 μM, but exerted a dose-dependent anti-proliferative activity with IC_{50} of 101.2 μM at 72 h of exposure [27]. Similar results were found by Reyes *et al.* [28] in both the colon cancer cell line HT-29 and Caco-2, in which incubation with the triterpene for 72 h resulted in inhibition of cell growth with IC_{50} of 61 μM and 85 μM, respectively. Further experiments by the same authors revealed that maslinic acid exerted its anti-proliferative activity by arresting cell cycle, since the cell population in the G0/G1 phases was significantly increased, while that in the S phase was reduced [28]. Remarkably, in both studies the effect of the compound on cell proliferation coincided with apoptotic cell death.

Apoptosis, also called programmed cell death, refers to a cascade of biochemical events that lead to the disintegration of the cell into fragments, which are further removed by phagocytic cells without eliciting an inflammatory response. This process might occur through death receptors, the so-called extrinsic pathway, or by means of an intrinsic pathway, in which mitochondria play a role. Both routes converge at the level of caspase-3, which is one of the effector caspases [29]. Interestingly, maslinic acid has been found to affect both pathways at different levels.

In the study of Juan *et al.* [27], the activation of caspase-3 was more than 60-fold at 24 h of exposure to 250 μM of the triterpene, compared to vehicle-treated cells. In order to know whether the activation of caspase-3 resulted from the extrinsic or the intrinsic pathway, the production of superoxide anions was evaluated, since it is one of the possible inductors of the latter [30]. Indeed, higher levels of O_2^- were found in cells incubated with maslinic acid (150 μM) for 4 h, compared to controls. The apoptotic process was further confirmed by the occurrence of plasma membrane disintegration and nuclear fragmentation [27]. Similarly, Reyes *et al.* [28] also reported that the apoptotic process observed in both HT-29 and Caco-2 cell lines occurred through activation of caspase-3, as evidenced by the observation of morphological changes, such as cell shrinkage or chromatin condensation.

Attention was then drawn to the molecular events underlying the induction of the mitochondrial apoptotic pathway. This organelle is a reservoir of several pro-apoptotic proteins that upon the proper stimulus are released to the cytosol, where the interaction with other elements finally triggers caspase-3 activation. An important set of regulators of this pathway is the Bcl-2 family, which includes both anti- and pro-apoptotic members [29].

Experiments performed by Reyes-Zurita *et al.* [31] with HT-29 cells showed that maslinic acid concomitantly activated the expression of Bax (pro-apoptotic protein) and inhibited the expression of Bcl-2 (anti-apoptotic protein), resulting in mitochondrial disruption and cytochrome-c release to the cytosol. It is known that once in the cytosol cytochrome-c binds to Apaf-1, which triggers the sequential activation of caspase-9 and caspase-3 [32]. Although in this study the formation of the complex was not directly assayed, a strong time- and dose-dependent cleavage of both caspases was observed [31].

More recently, the same authors postulated that the effect of maslinic acid on Bcl-2 family proteins could be mediated by the kinase JNK, since its expression was found increased in HT-29 cells after a 12 h treatment with the triterpene [33]. Actually, some of the effects of JNK had been previously described. Tsuruta *et al.* [34] found that JNK promotes Bax translocation to mitochondria through phosphorylation (inactivation) of a cytoplasmic anchor of Bax. Another consequence of JNK activation is the cleavage of Bid (pro-apoptotic protein), which results in translocation to mitochondria and Smac/DIABLO release to the cytosol [35]. This protein induces apoptosis through neutralizing inhibitors of apoptosis (IAPs) [36]. Apart from JNK activation, maslinic acid also enhanced the expression of p53, which is a well-known tumor-suppressor transcription factor that regulates the expression of genes involved in apoptosis, such as those coding for the above-mentioned Bcl-2 and Bax proteins [37].

In contrast with the intrinsic pathway, the extrinsic route is initiated by the binding of a ligand with a receptor of the tumor necrosis factor receptor (TNFR) superfamily. This results in the assembly of several elements, which constitute the so-called complex I [38]. Next, two possible ways trigger the regulation of apoptosis with opposite outcomes. On one hand, complex I can activate the kinase IKK, responsible of the phosphorylation of IKBα and its subsequent degradation. IKBα normally recruits NF-κB in the cytosol, but after its degradation the transcription factor is released and translocates to the nucleus [39], where it up-regulates anti-apoptotic genes [40]. On the other hand, some elements of the complex I can be exchanged, including the recruitment of procaspase-8, and this leads to the formation of a secondary complex (complex II) [38]. Activation of procaspase-8 results in the cleavage of downstream effector caspases, such as caspase-3, thus propagating the apoptotic signal [41].

The role of maslinic acid in the death-receptor pathway was first demonstrated by Li *et al.* [42] using the pancreatic cancer cells Panc-28. The compound exerted a synergistic effect together with TNF-α on both inhibition of cell proliferation (maslinic acid at 10 μM) and induction of cell death (25 μM), being the latter more than 55% higher, compared to control. The determination of activated caspase-3 in the cells confirmed the occurrence of apoptosis. Further experiments showed that maslinic acid affected the NF-κB pathway by inhibiting IKBα phosphorylation, thus preventing both NF-κB translocation to nucleus and its DNA binding activity.

The inhibitory effect of maslinic acid on NF-κB DNA-binding activity was also proved in the Raji B lymphoma cell line [43]. In this study, the impaired function of NF-κB was used to explain the dose-dependent reduction of COX-2 expression. COX-2 is well-known for its role in the inflammatory process and has been found overexpressed in a wide range of premalignant and malignant tissues [44].

The NF-κB transcriptional activity can be modulated through phosphorylation by various members of the mitogen-activated protein kinase family (MAPK), including JNK and p38 [45]. Wu *et al.* [46] described for the first time that maslinic acid also interacts with the p38 cascade so that ultimately triggers a pro-apoptotic effect. The experiments were performed in two cell lines of human salivary gland adenoid cystic carcinoma, ACC-2 and ACC-M, corresponding to low and high metastasis, respectively. The anti-proliferative activity after 24 h of incubation (IC_{50} of 43.6 and 45.8 μM, respectively) was attributed to an apoptotic process, as evidenced by the observation of both apoptotic bodies and microstructural changes, such as chromosomal DNA condensation and loss of microvilli. Cells exposed to the triterpene showed activated caspase-3, and this occurred as a consequence of p38 MAPK phosphorylation, which in turn was the result of an increase in the concentration of intracellular Ca^{2+}. The mechanism by which maslinic acid provokes intracellular Ca^{2+} overload remains to be investigated. On the contrary, the implication of p38 MAPK in maslinic acid-induced apoptosis is consistent with the results obtained in two cell lines of human urinary bladder carcinoma (T24 and 253J) [47]. Incubation with the triterpene dose- and time-dependently increased p38 phosphorylation, and this was correlated with reduced cell survival (IC_{50} of 33.0 and 71.8 μM in each cell line, respectively).

The latest assessment of the anti-proliferative activity of maslinic acid *in vitro* has been performed in the soft tissue sarcoma cell lines SW982 (human synovial sarcoma) and SK-UT-1 (leiomyosarcoma). IC_{50} values were of 45.3 and 59.1 μM, after incubating the cells with the triterpene for 24 h [48]. However, the most remarkable contribution of this study is the fact that maslinic acid is proposed as an adjuvant of the established anticancer drug doxorubicin, which constitutes a novel therapeutic approach for the treatment of cancer diseases. Concretely, cells treated simultaneously with both compounds showed higher sensitivity to doxorubicin as a consequence of an increased intracellular accumulation of the drug. Since doxorubicin is a well-known substrate of the efflux proteins P-gp and MRP1, a plausible mechanism behind the intracellular accumulation of the drug when co-incubated with maslinic acid could be that the triterpene inhibited these transporters. A kinetic study revealed that the parameters V_{max} and K_m (obtained by the Michaelis-Menten equation) of P-gp were not affected by maslinic acid, while those of MRP1 were dose-dependently lowered, thus indicating that maslinic behaved as a non-competitive inhibitor of MRP1 [48]. Table 2 summarizes the IC_{50} values of the anti-proliferative activity of maslinic acid found in different cell lines.

Table 2. *In vitro* anti-proliferative effect of maslinic acid.

Origin	Cell Line	IC_{50} (μM)	References
Human colorectal adenocarcinoma	HT-29	101.2	[27]
	HT-29	61	[28]
	Caco-2	85	[28]
	Caco-2	15.4	[49]
Human hepatocellular carcinoma	HepG2	69.1	[49]
Human breast adenocarcinoma	MCF-7	136.0	[49]
Human salivary gland adenoid cystic carcinoma	ACC-2 (low metastasis)	43.7	[46]
	ACC-M (high metastasis)	45.8	[46]
Human transitional cell urinary bladder carcinoma	T24	33.0	[47]
	253J	71.8	[47]
	TCCSUP	28.0	[47]
Human transitional cell urinary bladder papilloma	RT4	42.7	[47]
Human synovial sarcoma	SW 982	45.3	[48]
Human uterus leiomyosarcoma	SK-UT-1	59.1	[48]

2.1.2. Maslinic Acid Targets Other Cancer-Related Signaling Pathways

Besides the abnormal cell proliferation occurring in tumor growth, angiogenesis emerges in response to the hypoxic environment within the tumor and constitutes another therapeutic target for cancer diseases. The hypoxia inducible factor-1α (HIF-1α) is one of the pivotal regulators of angiogenesis in response to oxygen deficiency and has been found overexpressed in many human cancers [50]. This factor induces the expression of pro-angiogenic molecules, such as the vascular endothelial growth factor (VEGF) and its receptors [51], among others. The new blood vessels might be used by some cells detached from the primary tumor to reach systemic circulation, thus they would be distributed throughout the organism and are likely to ultimately colonize distant tissues. This process requires the action of proteins that degrade the extracellular matrix, such as matrix metalloproteinases (MMP) and urokinase-type plasminogen activator (uPA), which are secreted as inactive forms by either tumor or stroma cells [52,53]. Concretely, the expression of MMP-2, MMP-9 and uPA may be induced by the above-mentioned HIF-1α [51].

Park *et al.* [54] conducted an exhaustive study about the effect of maslinic acid on the metastatic capacity of the human prostate cancer cell DU145. Treatment with the triterpene resulted in a decrease of both basal and EGF-induced migration of cells in a dose-dependent manner (10–25 μM). This effect was correlated with both MMP and uPA systems; firstly, the triterpene reduced both the secretion of pro-MMP-2 and pro-MMP-9, and also MMP-9 mRNA levels. Secondly, a diminished secretion of pro- and active-uPA was observed, together with decreased uPA activity and mRNA levels, and reduced uPA receptor (uPAR) protein levels. Since MMP and uPA systems are regulated by HIF-1α, it was further assessed whether the effects of maslinic acid observed on the proteases took place through the alteration of HIF-1α levels. It was demonstrated that under hypoxic conditions the triterpene not only counteracted the increased expression of HIF-1α but also inhibited its translocation to the nucleus and decreased its half-life from 11.81 min to 4.96 min [54].

Similar results were obtained in three human liver cancer cell lines (Hep3B, Huh7 and HA227) [22]. In this study, however, the effects of maslinic acid were attributed to the antioxidant effect of the triterpene, since reduced levels of reactive oxygen species (ROS) and nitric oxide (NO) were observed in cells treated with maslinic acid. It had been previously reported that these molecules are natural enhancers of the expression of both HIF-1α and VEGF in cancer cells [55].

2.1.3. The Antitumor Activity of Maslinic Acid also Occurs *in Vivo*

Only a few studies up to now have assessed the antitumor activity of maslinic acid in animal models of cancer disorders, compared to the extensive number of references about its *in vitro* effects and their mechanisms. However, the positive outcomes achieved in all them are encouraging and stimulate further research in this field.

The first *in vivo* approach to the antitumor activity of maslinic acid was performed with athymic nu/nu mice in which xenograft pancreatic cells were implanted [42]. The subcutaneous administration of 10 and 50 mg/kg of the triterpene significantly decreased in a dose-dependent manner both the volume and the weight of the tumors, which in turn showed an increased number of apoptotic cells (from 8% in the control group to 21% and 38% in 10 mg/kg and 50 mg/kg groups, respectively) and a reduced expression of two NF-κB-regulated anti-apoptotic genes, Survivin and Bcl-xl.

More recently, Sánchez-Tena *et al.* [56] assessed the effect of a maslinic acid-enriched diet (100 mg/kg) in $Apc^{Min/+}$ mice, a common animal model of spontaneous intestinal polyposis. Results showed that, after a 6-week treatment period, maslinic acid inhibited the formation of polyps in the small intestine by 45%. Microarray analyses of gene expression profiles suggested that the compound inhibited cell-survival signaling and inflammation pathways.

Finally, bladder cancer has also been targeted by maslinic acid, after implanting T24 and 253J cells in nude mice. Both the size and the weight of the tumors were dose-dependently and significantly reduced in the animals treated with intraperitoneal injections of 20 mg/kg of the triterpene every other day over 35 days [47].

In summary, there is strong evidence that maslinic acid targets a variety of signaling pathways that finally trigger an anticarcinogenic effect, both *in vitro* and *in vivo*. Consequently, maslinic acid is emerging as a potential agent for the treatment of cancer disorders, either alone or in combination with other drugs.

2.2. Maslinic Acid and Diabetes

The role of maslinic acid in glucose metabolism has also been extensively studied. Wen *et al.* [57] provided the first evidence of the inhibitory effect of the triterpene on glycogen phosphorylases (GP), which catalyze the first step of glycogen breakdown. In a first *in vitro* assay using GPa (activated form of the enzyme) isolated from rat liver, maslinic acid inhibited the enzyme with an IC_{50} of 99 μM, being 6-fold more potent than caffeine, an established GP inhibitor. Based on this finding, the hypoglycemic activity of the triterpene was evaluated *in vivo*, using a mouse model of diabetes induced by adrenalin, which is known to indirectly stimulate glycogenolysis and thus increase glucose blood concentration. After the oral administration of maslinic acid (100 mg/kg) for 7 days, fasted plasma glucose appeared to be up to 46% lower, compared to animals that had received only the vehicle. Further work of the same authors went into detail about the mechanism of inhibition of maslinic acid on GP. The crystal structure of the complex GPb (inactivated form of the enzyme)-maslinic acid was determined, which revealed that the triterpene binds at the allosteric activator site, where the physiological activator AMP binds [58].

The *in vivo* antidiabetic effect of maslinic acid has been also proved in KK-A^y mice [59], an animal model for obesity and Type II non-insulin-dependent diabetes. Single oral administrations of the triterpene at doses of 10 and 30 mg/kg significantly diminished plasma glucose at 2 and 4 h after administration, and at the highest dose the effect was sustained up to 7 h. Similar results were obtained when maslinic acid was given daily for 2 weeks at the same doses, being the reduction in both cases of approximately 30%, with respect to control animals. Furthermore, after the repeated oral administration of 10 and 30 mg/kg of the triterpene, a dose-dependent reduction of plasma insulin levels was observed, as well as a decrease of blood glucose concentrations in the insulin tolerance test, *i.e.*, after the subcutaneous injection of insulin. The latter effect might be attributed to the normalization of plasma adiponectin levels, which was observed in groups treated with both 10 and 30 mg/kg doses [59].

Another animal model commonly used in the study of diabetes is the streptozotocin (STZ)-induced hyperglycemic rats. Khathi *et al.* [60] assessed the effect of maslinic acid (80 mg/kg, p.o.) on postprandial blood glucose in this model, and observed that the co-administration of the triterpene with either sucrose or starch significantly reduced the levels of glucose in plasma up until 120 min, in a similar way to that of acarbose, the positive control. Further research was carried out in order to dilucidate the mechanism by which maslinic acid exerted the hypoglycemic effect. On one hand, treatment with the triterpene reversed the higher expression of SGLT1 and GLUT2 found in diabetic animals compared to controls. These transporters are implicated in the intestinal absorption of glucose, thus their downregulation, which was similar to that produced by the standard drugs insulin and metformin, contributed to diminishing plasma glucose. Similarly, the expression of α-glucosidase and α-amylase, which are carbohydrates hydrolyzing enzymes, was attenuated in the small intestine of STZ-induced diabetic rats [60].

The lowering effect of maslinic acid on blood glucose of STZ-induced diabetic rats was consistent with that observed in a previous study [61], in which the triterpene was administered orally at a dose of 50 mg/kg for 28 days and the reduction of plasma glucose reached 66% at the end of the period. These results were obtained as part of a study about the beneficial effect of maslinic acid on cerebral ischemic injury, which will be discussed later.

Although the antidiabetic effect of maslinic acid has been extensively proved, little is known about the underlying mechanism of action. Liu *et al.* [62] confirmed the inhibitory activity of the triterpene on GPa (IC_{50} of 6.9 μM) using cell cultures of the hepatic cell line HepG2. More remarkably, the authors also hypothesized that maslinic acid targets the insulin signaling pathway [63], and found that incubation with the compound resulted in increased insulin receptor β (IRβ) phosphorylation [62]. Downstream events of IRβ activation include Akt phosphorylation, which in turn phosphorylates and inactivates glycogen synthase kinase 3β (GSK3β). GSK3β is a central enzyme in the regulation of glucose metabolism, since one of its targets is glycogen synthase. The lack of GSK3β activity allows glycogen synthase to be functional, thus resulting in glycogen build-up. Both Akt phosphorylation and GSK3β were increased in HepG2 cells in response to maslinic acid treatment, and the higher amount of glycogen content correlated with these findings. Interestingly, when maslinic acid was given orally to mice fed a high-fat diet, blood glucose concentration was markedly diminished at both doses (50 and 100 mg/kg). Moreover, the highest dose improved hyperinsulinemia and adiposity, and also increased hepatic glycogen [62].

All together, the results suggest that maslinic acid is a natural antidiabetic compound, which could be helpful to maintain the levels of blood glucose within the physiological range and thus contribute to the pharmacological treatment of the disease.

2.3. Maslinic Acid as Antioxidant and Anti-Inflammatory

The antioxidant effect of maslinic acid was first evaluated by Montilla *et al.* [64] in a model of oxidative status induced by CCl_4, which induces lipid peroxidation. Pre-treatment of the rats once daily for 3 days with the triterpene at doses of 50 and 100 mg/kg reduced by approximately 18% plasma levels of endogenous lipid peroxides, at both doses, and by 6.5% and 19%, respectively, the susceptibility of plasma to lipid peroxidation [64]. Similarly, the triterpene isolated from the flowers of *Punica granatum* prevented the $CuSO_4$-induced oxidation of rabbit plasma LDL, monitored by the formation of dienes, by 33.8% [65]. More recently, Allouche *et al.* [66] conducted an exhaustive study about the antioxidant properties of several pentacyclic triterpenic diols and acids on LDL particles isolated from human plasma. Maslinic acid not only retarded the initiation and decreased the rate of $CuSO_4$-induced LDL oxidation, but also showed peroxyl radical scavenging activity and a slight metal (copper) chelating effect.

Further research has been done in macrophages, which play a role in the defensive system of the organism in response to activation by a pathogen [67]. Cells were isolated from murine peritoneum and activated with lipopolysaccharide (LPS), a compound that gives rise to a potent inflammatory response mediated by the production of cytokines, such as TNF-α, and also by reactive nitrogen and oxygen species, among others. In this study, the effect of the triterpene was tested on the synthesis of NO, superoxide and hydrogen peroxide. Although maslinic acid did not exert any direct inhibitory effects on the formation of the first two species, the compound did reduce the generation of hydrogen peroxide (IC_{50} of 46.3 μM), in a way that was similar to that of catalase. In addition, the release of the pro-inflammatory cytokines IL-6 and TNF-α was significantly reduced after treatment with maslinic acid at concentrations of 50 and 100 μM [67].

The anti-inflammatory activity of maslinic acid has been also proved in primary cortical astrocytes [68], which could be translated to a neuroprotective effect if further confirmed *in vivo.* Cells were cultured with the triterpene (0.1, 1, 10 μM) for 24 h before being exposed to LPS. The focus here was the TNF-α signaling pathway, which is in part mediated by NF-κB. As previously described, this transcription factor is found in the cytosol, retained by IκBα. Under stimulation, IκBα is phosphorylated and then the p65 subunit of the transcription factor is released, which allows its migration to the nucleus [39]. Maslinic acid not only suppressed the expression of TNF-α, but also hampered p65 translocation to the nucleus, which was correlated with a lower phosphorylation of IκBα.

Additionally, the triterpene did inhibit the LPS-induced formation of NO, as well as mRNA and protein levels of iNOS and COX-2 [68].

Although several studies support the antioxidant activity of maslinic acid in terms of preventing LDL oxidation, the underlying mechanism remains to be clarified. In contrast, fewer assessments have been performed on the anti-inflammatory potential of the triterpene, but it seems to be driven by alterations in the TNF-α signaling pathway resulting in altered gene expression of enzymes involved in the inflammatory process.

2.4. Maslinic Acid and Cardioprotection

To date, the antitumor, antidiabetic and antioxidant effects of maslinic acid have focused the greatest attention, but other promising activities have been attributed to the triterpene, which contribute to raise the interest for this potential nutraceutical.

The protective effect of maslinic acid against cardiovascular diseases has been studied using different approaches, which include the assessment of the triterpene in controlling risk factors such as hypertension or hyperlipidemia.

On one hand, experiments with aortic rings isolated from spontaneously hypertensive rats showed that maslinic acid exerted a concentration-dependent relaxation (IC_{50} of 14.1 μM), after precontraction with phenylephrine [69]. The effect was endothelium-dependent, since the removal of the endothelium attenuated the relaxation. In order to elucidate the underlying mechanism, intact (with endothelium) aortic rings were pre-incubated with NG-nitro-L-arginine methyl ester (L-NAME), a NO synthase inhibitor. This resulted in a diminished relaxation in intact aortic rings, indicating that NO was involved in maslinic acid-induced vasodilation.

On the other hand, in rats fed a high-cholesterol diet for 30 days, the oral administration of maslinic acid (100 mg/kg) for the last two weeks resulted in a hypolipidemic effect, as evidenced by a reduction of more than 70% in serum triglycerides, total cholesterol and LDL-cholesterol [70]. The triterpene also restored the levels of the hepatic marker enzymes lactate dehydrogenase (LDH), alkaline phosphatase (ALP), aspartate aminotransferase (AST) and alanine aminotransferase (ALT). Similarly, both the glycogen content and the morphological alterations observed in hepatocytes were reversed in maslinic acid-treated animals, compared to controls.

The cardioprotective effect of maslinic acid has also been tested in isoproterenol-induced myocardial infarction in Wistar rats [71]. Animals that had been pre-treated with maslinic acid (15 mg/kg) for 7 days showed an improved serum lipid profile with significantly decreased levels of total cholesterol, triglycerides, LDL-cholesterol, VLDL-cholesterol and increased HDL-cholesterol. The activity of the cardiac marker enzymes creatine kinase (CK), ALT, AST and γ-glutamyl transferase (GGT) significantly decreased. Furthermore, the oxidative status of the animals was evaluated by measuring malondialdehyde (MDA), an indicator of lipid peroxidation, and paraoxonase (PON), an atheroprotective enzyme found in HDL particles [72]. MDA levels were significantly reduced, while the activity of PON increased remarkably in rats that had received maslinic acid, compared to non-treated animals [71].

In summary, maslinic acid, as a bioactive compound present in a wide variety of natural edible sources, may contribute to the beneficial effects ascribed to the Mediterranean diet on the prevention of cardiovascular diseases [73].

2.5. Maslinic Acid and Neuroprotection

A series of exhaustive studies have demonstrated that maslinic acid may confer neuroprotection in some pathological situations. In a first experiment with primary cultures of rat cortical neurons, cells were incubated with different concentrations of the triterpene (0.1, 1, 10 μM) and subjected to 1 h of oxygen-glucose deprivation followed by reoxygenation (24 h). Maslinic acid dose-dependently attenuated neuronal damage, which was evaluated through observation of morphological changes, release of lactate dehydrogenase (LDH) and neuronal viability [74], and this effect resulted from reduced activity of both caspase-9 and caspase-3. Upstream of caspases, high levels of NO might trigger apoptotic cell death [75]. This gaseous molecule is synthetized in great amounts by the inducible nitric oxide synthase (iNOS) in response to hypoxia [76], thus inhibition of this enzyme could be the mechanism underlying the protective effect of maslinic acid in oxygen-deprived cortical neurons. Qian *et al.* [74] observed that when challenged neurons were exposed to the triterpene (10 μM), the amount of NO in the culture medium was rescued to levels close to those found in normoxic conditions, which was correlated with reduced iNOS protein and mRNA levels.

In another study from the same authors, the neuroprotective effect of maslinic acid was assessed in front of glutamate-induced toxicity. Glutamate is the main excitatory neurotransmitter in the central nervous system, but excessive stimulation is associated with neuronal damage [77]. The removal of glutamate from the synaptic cleft takes place through the high-affinity transporters GLAST and GLT-1 located in astrocytes [78], thus ensuring the end of stimulation. In primary cultures of cortical neurons exposed to glutamate, maslinic acid did not exert any direct beneficial effects, since LDH release was comparable to that of vehicle-treated cells at all tested concentrations of maslinic acid (0.1, 1, 10 μM). However, a protective effect was indeed observed when neurons were cultured with conditioned medium obtained from astrocytes that had been incubated with maslinic acid (24 h) [79]. Further experiments evidenced that the triterpene dose-dependently increased the clearance of extracellular glutamate in cultures of astrocytes, and this was attributed to enhanced expression of both GLAST and GLT-1 after exposure to maslinic acid (10 μM). In a last assessment with co-cultures of astrocytes and neurons, maslinic acid significantly reversed the effects of glutamate in terms of LDH release, extracellular glutamate levels and neuron survival and morphology [79].

At this point it is convenient to recall the anti-inflammatory activity of maslinic acid in primary astrocytes, which has been described in a previous section [68]. All together, the results obtained from *in vitro* studies with primary cultures of neurons and astrocytes strongly support the hypothesis that maslinic acid exerts beneficial effects in the central nervous system, thus *in vivo* studies are the next step towards considering maslinic acid a neuroprotective agent.

Guan *et al.* [61] tested whether maslinic acid prevented brain damage after a transient ischemic episode in animals. Since hyperglycemia is a risk factor for stroke [80], streptozotocin-induced diabetic rats were given the triterpene orally at doses of 5 and 50 mg/kg for 14 days. Then, a transient middle cerebral artery occlusion was performed and the consequences of the infarction were evaluated. At both low and high doses, the triterpene decreased the infarct size in a range between 63.7% and 75.4%, depending on the dose and the time of reperfusion after the intervention (24 or 72 h). Moreover, maslinic acid treatment compensated the neurological deficits induced by the infarction, as showed by higher neurological scores recorded from animals that had received the triterpene [61].

To conclude, the recent interest for maslinic acid as a neuroprotective agent is supported not only by exhaustive *in vitro* studies on its mechanism of action but also by an *in vivo* assessment in infarcted diabetic rats. If proved in other species and pathological situations, the triterpene may be considered an adjuvant to lower the risk of occurrence of certain cerebral incidents.

2.6. Maslinic Acid as Antiparasitic

Historically, one of the first remarkable reports that focused the attention on the biological activities of maslinic acid was published by Xu *et al.* [17] and described the anti-HIV properties of several triterpenic acids isolated from the methanolic extract of *Geum japonicum*. Although the study did not provide mechanistic details of the inhibitory effect on HIV-1 protease, it is clearly stated that maslinic acid was the most potent compound [17]. More recently, the antibacterial activity of this triterpene was tested against different bacteria after its isolation from the methanolic extract of the leaves of *Symplocos lancifolia*. The lowest minimal inhibitory concentrations (MIC) of maslinic acid were found for *Enterococcus faecalis* (33.8 μM) and *Staphylococcus aureus* (135.4 μM) [81]. Although neither the antiviral nor the antibacterial activities of maslinic acid have been further studied exhaustively, the protective effect of the triterpene against parasitic infections has arisen much interest in recent years.

De Pablos *et al.* [82] observed that maslinic acid blocked the entrance of *Toxoplasma gondii* into Vero cells in a dose-dependent manner, with IC_{50} of 8 μM at 48 h of treatment. The underlying mechanism seemed to be the inhibitory activity of the triterpene against proteases secreted by the parasite, which are essential for the proteolytic processing of other proteins that participate in the invasion of host cells. Concretely, the gliding motility was suppressed by up to 100% by maslinic acid (50 μM). Moreover, the triterpene induced morphological alterations in the endomembrane systems of the parasite, such as a greater amount of apparently empty spaces that authors attribute to a possible collapse of the Golgi apparatus. Disruptions in external and nuclear membranes were also observed and attributed to a general blockage of protein turnover, which would hinder the functionality of those proteins necessary for the structural maintenance of the membranes. The same group evidenced the anti-parasitic effect of maslinic acid in *Gallus domesticus* chicks infected with *Eimeria tenella* [83]. The animals were fed a maslinic-acid supplemented diet (90 ppm) for 21 days, and this treatment resulted in a reduced release of oocysts in the faeces by 80.1%, being more effective than the positive control with sodium salinomycin (60 ppm). Histological evaluation of the caeca revealed that the characteristic lesions of this coccidiosis were less evident in the animals that had received maslinic acid. Furthermore, the body weight gain was significantly higher in treated animals compared not only to the positive control but also to the uninfected group, indicating that besides the anticoccidial activity, the triterpene enhanced weight gain [83].

Maslinic acid has also been found effective against different species of the genus *Plasmodium*, responsible of causing malaria. *In vitro* experiments using erythrocytes infected with *Plasmodium falciparum* demonstrated that maslinic acid (0.1–200 μM) inhibited the growth of the parasite in a dose-dependent manner [84]. At a concentration of 30 μM (close to the IC_{50}), the triterpene reduced parasitaemia to 4% (compared to 8% in untreated red blood cells) and slowed down the cell cycle, since only the infective (schizonts) and immature (new rings) forms, but not the mature forms (trophozoites), were observed in the erythrocytes. However, the removal of maslinic acid from the medium permitted the infection to resume, meaning that the triterpene acts as a parasitostatic agent [84]. This effect was further confirmed *in vivo* with ICR mice infected with the lethal strain of *Plasmodium yoelii* [85]. The intraperitoneal injection of 40 mg/kg for 4 days increased the survival rate of the animals to 80%, compared to 20% found in animals without any experimental intervention, and this was associated with an arrest of the maturation of the parasite in the erythrocytes. In addition, the animals that survived the primary infection were rechallenged with an identical second infection 40 days later. Parasitaemia was monitored for the following 30 days but no parasites were detected, indicating that mice were completely protected against the parasite [85]. Further research on the mechanism of action underlying the antimalarial activity of maslinic acid showed that the compound hampers the maturation of the parasite inside the erythrocytes by inhibiting different proteins [86].

To sum up, several lines of evidence point to maslinic acid as antiparasitic and/or parasitostatic agent. Further research is needed in order to confirm its efficacy in target species, which would allow the use of maslinic acid either alone or in combination with other therapeutic strategies for the treatment of parasitoses.

2.7. Maslinic Acid and Growth

The growth-stimulating activity of maslinic acid has been studied in rainbow trouts (*Oncorhynchus mykiss*) [87,88], in order to determine whether it can be used as a feed additive in pond aquaculture to increase production rates. In both reports, the animals were fed a maslinic acid-enriched diet (1, 5, 25 and 250 mg/kg diet) twice daily for 225 days. At the end of the period, trouts that had received the highest dose of the triterpene reached a body weight that was almost 30% higher compared to the group fed the standard diet. While the first study focused on the consequences of maslinic acid consumption on the liver, the second assessed the effects on white muscle. Both of them found similar results in all the variables analyzed. The weight of the liver and the white muscle from animals that ingested the highest amount of the triterpene was 52.1% and 39.8% higher, respectively, compared to the corresponding control groups. Protein, DNA and RNA levels were evaluated in order to get some insight into the nature of the increased weight. Total DNA, which is indicative of hyperplasia, was remarkably higher in liver and white muscle, as well as RNA content. These findings were correlated with a stimulation of the protein-synthesis efficiency in both cases. Observation of the hepatic structure under the light and electron microscopes revealed a larger degree of cell packaging in the parenchyma of livers from animals that were fed the diet containing 250 mg/kg of maslinic acid, together with a major proportion of rough endoplasmic reticulum, greater number of mitochondria and considerable quantities of peripheral glycogen granules [87].

The latest contribution in this field aimed at identifying the differences in liver protein profile between fish fed a maslinic acid-supplemented diet and fish fed a standard diet [89]. The experimental design was similar to that followed in the above-mentioned studies, except for the animal species, which was the gilthead sea bream (*Sparus aurata*). The diet contained 100 mg/kg of the triterpene and was supplied over 210 days. The proteomic analysis of the liver revealed that the expression of 19 proteins was altered, being either up- or down-regulated. These included proteins involved in a wide range of metabolic pathways, such as glucose, sterol and amino acid metabolism, protein synthesis and folding, oxidative stress, detoxification and xenobiotic metabolism, immune system and cell proliferation [89]. Beyond the effects of the triterpene on the liver protein profile, this study provides evidence of the validity of the method to characterize the differential expression of liver proteins after a nutritional intervention.

In conclusion, maslinic acid appears to be a promising compound to stimulate growth by means of affecting protein synthesis. It remains to be investigated whether this effect also occurs in other species, being those subjected to intensive animal farming of particular interest. If proved, maslinic acid may be considered a natural growth promoter and thus constitute another alternative to the use of hormones or antibiotics to increase production rates.

2.8. Other Biological Activities

To date, the previously described health-enhancing properties of maslinic acid have focused the major attention, as evidenced by the fact that each of them has been addressed by several studies. However, maslinic acid has also been attributed a variety of other biological effects, which include the inhibition of elastase [90] and tyrosinase [91] *in vitro*, the suppression of osteoclastogenesis in cell cultures and the prevention of ovariectomy-induced bone loss in mice [92], antinociceptive and antiallodynic effects in different pain models in mice [93], and the ability to alter the structural properties of biological membranes [94].

3. Conclusion and Future Prospects

Maslinic acid is a natural pentacyclic triterpene present in a variety of plant species, many of them being common ingredients of plant-based dietary patterns, such as the Mediterranean diet. In recent years, a number of studies assessing its biological effects have raised interest in this compound. These include not only health-enhancing properties, such as cardioprotective or neuroprotective, but also a therapeutic potential that may help in the treatment of several disorders, such as cancer, diabetes or parasitoses. However, the amount of maslinic acid in natural edible sources is low, and data about its pharmacokinetics, which we are currently assessing in our laboratory, show that the triterpene has a poor oral bioavailability. From this it would appear that dietary maslinic acid is not sufficient to reach effective concentrations in target organs, thus the compound should be supplied in pure form, *i.e.* as a nutraceutical. Nevertheless, maslinic acid is in the spotlight of research on this field. Further studies will surely provide new mechanisms of action to explain the effects already described or even widen the spectrum of biological activities of this pentacyclic triterpene.

Acknowledgments

This work was supported by grants AGL2009-12866 and AGL2013-41188-R from Ministerio de Ciencia e Innovación, Spain, and grants 2009SGR471 and 2014SGR1221 from Generalitat de Catalunya, Spain. G.L.M. was a recipient of a fellowship Ajuts de Personal Investigador en Formació de la Universitat de Barcelona (APIF-UB) and M.S.G. of a fellowship from project 2009SGR471.

Author Contributions

JMP and GLM designed the content of the review; JMP, MEJ, GLM and MSG compiled the data; JMP and GLM wrote the manuscript; JMP, MEJ, GLM and MSG revised the manuscript. All authors read and approved the final manuscript.

Conflicts of Interest

The authors declare no conflict of interest.

References

1. Sánchez-González, M.; Lozano-Mena, G.; Juan, M.E.; García-Granados, A.; Planas, J.M. Assessment of the safety of maslinic acid, a bioactive compound from *Olea europaea* L. *Mol. Nutr. Food Res.* **2013**, *57*, 339–346.
2. Bächler, L. *Chemische Untersuchungen über die Früchte von* Crataegus oxyacantha *L. (Monographie der Mehlbeeren)*; Universität Basel: Basel, Switzerland, 1927.
3. Tschesche, R.; Fugmann, R. Crataegolsäure, ein neues triterpenoid aus *Crataegus oxyacantha.* Ein beitrag zur konstitution der α-amyrine. *Chem. Ber.* **1951**, *84*, 810–826.
4. Tschesche, R.; Heesch, A.; Fugmann, R. Über triterpenoide, III. Mitteil.: Zur kenntnis der crataegolsäure. *Chem. Ber.* **1953**, *86*, 626–629.
5. Caglioti, L.; Cainelli, G.; Minutilli, F. Constitution of maslinic acid. *Chim. Ind.* **1961**, *43*, 278.
6. Caputo, R.; Mangoni, L.; Monaco, P.; Previtera, L. Triterpenes in husks of *Olea europaea. Phytochemistry* **1974**, *13*, 1551–1552.
7. Vioque, A.; Morris, L. Minor components of olive oils. I. Triterpenoid acids in an acetone-extracted orujo oil. *J. Am. Oil Chem. Soc.* **1961**, *38*, 458–488.
8. Bianchi, G.; Pozzi, N.; Vlahov, G. Pentacyclic triterpene acids in olives. *Phytochemistry* **1994**, *37*, 205–207.
9. Xu, R.; Fazio, G.C.; Matsuda, S.P.T. On the origins of triterpenoid skeletal diversity. *Phytochemistry* **2004**, *65*, 261–291.
10. Tschesche, R.; Poppel, G.; Über Triterpene, V. Zur Kenntnis der crataegolsäure und über zwei neue triterpencarbonsäuren aus *Crataegus oxyacantha* L. *Chem. Ber.* **1959**, *92*, 320–328.
11. Stiti, N.; Triki, S.; Hartmann, M.A. Formation of triterpenoids throughout *Olea europaea* fruit ontogeny. *Lipids* **2007**, *42,* 55–67.
12. Kombargi, W.S.; Michelakis, S.E.; Petrakis, C.A. Effect of olive surface waxes on oviposition by *Bactrocera oleae* (Diptera: Tephritidae). *J. Econ. Entomol.* **1998**, *91*, 993–998.

13. Pungitore, C.R.; García, M.; Gianello, J.C.; Sosa, M.E.; Tonn, C.E. Insecticidal and antifeedant effects of *Junellia aspera* (Verbenaceae) triterpenes and derivatives on *Sitophilus oryzae* (Coleoptera: Curculionidae). *J. Stored Prod. Res.* **2005**, *41*, 433–443.
14. Lu, H.; Xi, C.; Chen, J.; Li, W. Determination of triterpenoid acids in leaves of *Eriobotrya japonica* collected at in different seasons. *Zhongguo Zhong Yao Za Zhi* **2009**, *34*, 2353–2355.
15. Banno, N.; Akihisa, T.; Tokuda, H.; Yasukawa, K.; Taguchi, Y.; Akazawa, H.; Ukiya, M.; Kimura, Y.; Suzuki, T.; Nishino, H. Anti-inflammatory and antitumor-promoting effects of the triterpene acids from the leaves of *Eriobotrya japonica*. *Biol. Pharm. Bull.* **2005**, *28*, 1995–1999.
16. Kim, D.H.; Han, K.M.; Chung, I.S.; Kim, D.K.; Kim, S.H.; Kwon, B.M.; Jeong, T.S.; Park, M.H., Ahn, E.M.; Baek, N.I. Triterpenoids from the flower of *Campsis grandiflora* K. Schum. as human acyl-CoA: Cholesterol acyltransferase inhibitors. *Arch. Pharm. Res.* **2005**, *28*, 550–556.
17. Xu, H.X.; Zeng, F.Q.; Wan, M.; Sim, K.Y. Anti-HIV triterpene acids from *Geum japonicum*. *J. Nat. Prod.* **1996**, *59*, 643–645.
18. Yoshida, T.; Okuda, T.; Memon, M.U.; Shingu, T. Tannins of rosaceous medicinal plants. Part 2. Gemins A, B, and C, new dimeric ellagitannins from *Geum japonicum*. *J. Chem. Soc. Perkin Trans.* **1985**, *1*, 315–321.
19. Zou, Z.M.; Cong, P.Z. Studies on the chemical constituents from roots of *Agastache rugosa*. *Yao Xue Xue Bao* **1991**, *26*, 906–910.
20. Shin, S.; Kang, C.A. Antifungal activity of the essential oil of *Agastache rugosa* Kuntze and its synergism with ketoconazole. *Lett. Appl. Microbiol.* **2003**, *36*, 111–115.
21. Romero, C.; García, A.; Medina, E.; Ruiz-Méndez, M.V.; de Castro, A.; Brenes, M. Triterpenic acids in table olives. *Food Chem.* **2010**, *118*, 670–674.
22. Lin, C.C.; Huang, C.Y.; Mong, M.C.; Chan, C.Y.; Yin, M.C. Antiangiogenic potential of three triterpenic acids in human liver cancer cells. *J. Agric. Food Chem.* **2011**, *59*, 755–762.
23. Yin, M.C.; Lin, M.C.; Mong, M.C.; Lin, C.Y. Bioavailability, distribution, and antioxidative effects of selected triterpenes in mice. *J. Agric. Food Chem.* **2012**, *60*, 7697–7701.
24. Kalogeropoulos, N.; Chiou, A.; Ioannou, M.; Karathanos, V.T.; Hassapidou, M.; Andrikopoulos, N.K. Nutritional evaluation and bioactive microconstituents (phytosterols, tocopherols, polyphenols, triterpenic acids) in cooked dry legumes usually consumed in the Mediterranean countries. *Food Chem.* **2010**, *121*, 682–690.
25. Li, G.L.; You, J.M.; Song, C.H.; Xia, L.; Zheng, J.; Suo, Y.R. Development of a new HPLC method with precolumn fluorescent derivatization for rapid, selective and sensitive detection of triterpenic acids in fruits. *J. Agric. Food Chem.* **2011**, *59*, 2972–2979.
26. Juan, M.E.; Wenzel, U.; Ruiz-Gutiérrez, V.; Planas, J.M.; Daniel, H. Maslinic acid, a natural compound from olives, induces apoptosis in HT-29 human colon cancer cell line. In Proceedings of the Experimental Biology 2005 Meeting, 35th International Congress of Physiological Sciences, San Diego, CA, USA, 31 March–6 April 2005.
27. Juan, M.E.; Planas, J.M.; Ruiz-Gutiérrez, V.; Daniel, H.; Wenzel, U. Anti-proliferative and apoptosis-inducing effects of maslinic and oleanolic acids, two pentacyclic triterpenes from olives, on HT-29 colon cancer cells. *Br. J. Nutr.* **2008**, *100*, 36–43.

28. Reyes, F.J.; Centelles, J.J.; Lupiáñez, J.A.; Cascante, M. (2α,3β)-2,3-dihydroxyolean-12-en-28-oic acid, a new natural triterpene from *Olea europea*, induces caspase dependent apoptosis selectively in colon adenocarcinoma cells. *FEBS Lett.* **2006**, *580*, 6302–6310.
29. Hengartner, M.O. The biochemistry of apoptosis. *Nature* **2000**, *407*, 770–776.
30. Ott, M.; Gogvadze, V.; Orrenius, S.; Zhivotovsky, B. Mitochondria, oxidative stress and cell death. *Apoptosis* **2007**, *12*, 913–922.
31. Reyes-Zurita, F.J.; Rufino-Palomares, E.E.; Lupiáñez, J.A.; Cascante, M. Maslinic acid, a natural triterpene from *Olea europaea* L., induces apoptosis in HT29 human colon-cancer cells via the mitochondrial apoptotic pathway. *Cancer Lett.* **2009**, *273*, 44–54.
32. Li, P.; Nijhawan, D.; Budihardjo, I.; Srinivasula, S.M.; Ahmad, M.; Alnemri, E.S.; Wang, X. Cytochrome c and dATP-dependent formation of Apaf-1/caspase-9 complex initiates an apoptotic protease cascade. *Cell* **1997**, *91*, 479–489.
33. Reyes-Zurita, F.J.; Pachón-Peña, G.; Lizárraga, D.; Rufino-Palomares, E.E.; Cascante, M.; Lupiáñez, J.A. The natural triterpene maslinic acid induces apoptosis in HT29 colon cancer cells by a JNK-p53-dependent mechanism. *BMC Cancer* **2011**, 154:1–154:13.
34. Tsuruta, F.; Sunayama, J.; Mori, Y.; Hattori, S.; Shimizu, S.; Tsujimoto, Y.; Yoshioka, K.; Masuyama, N.; Gotoh, Y. JNK promotes Bax translocation to mitochondria through phosphorylation of 14-3-3 proteins. *EMBO J.* **2004**, *23*, 1889–1899.
35. Deng, Y.; Ren, X.; Yang, L.; Lin, Y.; Wu, X. A JNK-dependent pathway is required for TNFα-induced apoptosis. *Cell* **2003**, *115*, 61–70.
36. Wilkinson, J.C.; Wilkinson, A.S.; Scott, F.L.; Csomos, R.A.; Salvesen, G.S.; Duckett, C.S. Neutralization of Smac/Diablo by inhibitors of apoptosis (IAPs). A caspase-independent mechanism for apoptotic inhibition. *J. Biol. Chem.* **2004**, *279*, 51082–51090.
37. Miyashita, T.; Krajewski, S.; Krajewska, M.; Wang, H.G.; Lin, H.K.; Liebermann, D.A.; Hoffman, B.; Reed, J.C. Tumor suppressor p53 is a regulator of bcl-2 and bax gene expression *in vitro* and *in vivo*. *Oncogene* **1994**, *9*, 1799–1805.
38. Micheau, O.; Tschopp, J. Induction of TNF receptor I-mediated apoptosis via two sequential signaling complexes. *Cell* **2003**, *114*, 181–190.
39. Napetschnig, J.; Wu, H. Molecular basis of NF-κB signaling. *Annu. Rev. Biophys.* **2013**, *42*, 443–468.
40. Karin M. Nuclear factor-κB in cancer development and progression. *Nature* **2006**, *441*, 431–436.
41. Stennicke, H.R.; Jürgensmeier, J.M.; Shin, H.; Deveraux, Q.; Wolf, B.B.; Yang, X.; Zhou, Q.; Ellerby, H.M.; Ellerby, L.M.; Bredesen, D.; *et al.* Pro-caspase-3 is a major physiologic target of caspase-8. *J. Biol. Chem.* **1998**, *273*, 27084–27090.
42. Li, C.; Yang, Z.; Zhai, C.; Qiu, W.; Li, D.; Yi, Z.; Wang, L.; Tang, J.; Qian, M.; Luo, J.; *et al.* Maslinic acid potentiates the anti-tumor activity of tumor necrosis factor α by inhibiting NF-κB signaling pathway. *Mol. Cancer* **2010**, 73:1–73:13.
43. Hsum, Y.W.; Yew, W.T.; Hong, P.L.; Soo, K.K.; Hoon, L.S.; Chieng, Y.C.; Mooi, L.Y. Cancer chemopreventive activity of maslinic acid: Suppression of COX-2 expression and inhibition of NF-κB and AP-1 activation in Raji cells. *Planta Med.* **2011**, *77*, 152–157.

44. Dannenberg, A.J.; Subbaramaiah, K. Targeting cyclooxygenase-2 in human neoplasia: Rationale and promise. *Cancer Cell* **2003**, *4*, 431–436.
45. Schulze-Osthoff, K.; Ferrari, D.; Riehemann, K.; Wesselborg, S. Regulation of NF-kappa B activation by MAP kinase cascades. *Immunobiology* **1997**, *198*, 35–49.
46. Wu, D.M.; Zhao, D.; Li, D.Z.; Xu, D.Y.; Chu, W.F.; Wang, X.F. Maslinic acid induces apoptosis in salivary gland adenoid cystic carcinoma cells by Ca^{2+}-evoked p38 signaling pathway. *Naunyn Schmiedebergs Arch. Pharmacol.* **2011**, *383*, 321–330.
47. Zhang, S.; Ding, D.; Zhang, X.; Shan, L.; Liu, Z. Maslinic acid induced apoptosis in bladder cancer cells through activating p38 MAPK signaling pathway. *Mol. Cell Biochem.* **2014**, *392*, 281–287.
48. Villar, V.H.; Vögler, O.; Barceló, F.; Gómez-Florit, M.; Martínez-Serra, J.; Obrador-Hevia, A.; Martín-Broto, J.; Ruiz-Gutiérrez, V.; Alemany, R. Oleanolic and maslinic acid sensitize soft tissue sarcoma cells to doxorubicin by inhibiting the multidrug resistance protein MRP-1, but not P-glycoprotein. *J. Nutr. Biochem.* **2014**, *25*, 429–438.
49. He, X.; Liu, R.H. Triterpenoids isolated from apple peels have potent anti-proliferative activity and may be partially responsible for apple's anticancer activity. *J. Agric. Food Chem.* **2007**, *55*, 4366–4370.
50. Zhong, H.; de Marzo, A.M.; Laughner, E.; Lim, M.; Hilton, D.A.; Zagzag, D.; Buechler, P.; Isaacs, W.B.; Semenza, G.L.; Simons, J.W. Overexpression of hypoxia-inducible factor 1α in common human cancers and their metastases. *Cancer Res.* **1999**, *59*, 5830–5835.
51. Rankin, E.B.; Giaccia, A.J. The role of hypoxia-inducible factors in tumorigenesis. *Cell Death Differ.* **2008**, *15*, 678–685.
52. Bourboulia, D.; Stetler-Stevenson, W.G. Matrix metalloproteinases (MMPs) and tissue inhibitors of metalloproteinases (TIMPs): Positive and negative regulators in tumor cell adhesion. *Semin. Cancer Biol.* **2010**, *20*, 161–168.
53. Mekkawy, A.H.; Morris, D.L.; Pourgholami, M.H. Urokinase plasminogen activator system as a potential target for cancer therapy. *Future Oncol.* **2009**, *5*, 1487–1499.
54. Park, S.Y.; Nho, C.W.; Kwon, D.Y.; Kang, Y.H.; Lee, K.W.; Park, J.H. Maslinic acid inhibits the metastatic capacity of DU145 human prostate cancer cells: Possible mediation via hypoxia-inducible factor-1α signalling. *Br. J. Nutr.* **2013**, *109*, 210–222.
55. Pialoux, V.; Mounier, R.; Brown, A.D.; Steinback, C.D.; Rawling, J.M.; Poulin, M.J. Relationship between oxidative stress and HIF-1α mRNA during sustained hypoxia in humans. *Free Radic. Biol. Med.* **2009**, *46*, 321–326.
56. Sánchez-Tena, S.; Reyes-Zurita, F.J.; Díaz-Moralli, S.; Vinardell, M.P.; Reed, M.; García-García, F.; Dopazo, J.; Lupiáñez, J.A.; Günther, U.; Cascante, M. Maslinic acid-enriched diet decreases intestinal tumorigenesis in $Apc^{Min/+}$ mice through transcriptomic and metabolomic reprogramming. *PLoS One.* **2013**, *8*, e59392.
57. Wen, X.; Sun, H.; Liu, J.; Wu, G.; Zhang, L.; Wu, X.; Ni, P. Pentacyclic triterpenes. Part 1: The first examples of naturally occurring pentacyclic triterpenes as a new class of inhibitors of glycogen phosphorylases. *Bioorg. Med. Chem. Lett.* **2005**, *15*, 4944–4948.

58. Wen, X.; Sun, H.; Liu, J.; Cheng, K.; Zhang, P.; Zhang, L.; Hao, J.; Zhang, L.; Ni, P.; Zographos, S.E.; *et al.* Naturally occurring pentacyclic triterpenes as inhibitors of glycogen phosphorylase: Synthesis, structure-activity relationships, and X-ray crystallographic studies. *J. Med. Chem.* **2008**, *51*, 3540–3554.
59. Liu, J.; Sun, H.; Duan, W.; Mu, D.; Zhang, L. Maslinic acid reduces blood glucose in KK-A^y mice. *Biol. Pharm. Bull.* **2007**, *30*, 2075–2078.
60. Khathi, A.; Serumula, M.R.; Myburg, R.B.; van Heerden, F.R.; Musabayane, C.T. Effects of *Syzygium aromaticum*-derived triterpenes on postprandial blood glucose in streptozotocin-induced diabetic rats following carbohydrate challenge. *PLoS One* **2013**, *8*, e81632.
61. Guan, T.; Qian, Y.; Tang, X.; Huang, M.; Huang, L.; Li, Y.; Sun, H. Maslinic acid, a natural inhibitor of glycogen phosphorylase, reduces cerebral ischemic injury in hyperglycemic rats by GLT-1 up-regulation. *J. Neurosci. Res.* **2011**, *89*, 1829–1839.
62. Liu, J.; Wang, X.; Chen, Y.P.; Mao, L.F.; Shang, J.; Sun, H.B.; Zhang, L.Y. Maslinic acid modulates glycogen metabolism by enhancing the insulin signaling pathway and inhibiting glycogen phosphorylase. *Chin. J. Nat. Med.* **2014**, *12*, 259–265.
63. Saltiel, A.R.; Kahn, C.R. Insulin signalling and the regulation of glucose and lipid metabolism. *Nature* **2001**, *414*, 799–806.
64. Montilla, M.P.; Agil, A.; Navarro, M.C.; Jiménez, M.I.; García-Granados, A.; Parra, A.; Cabo, M.M. Antioxidant activity of maslinic acid, a triterpene derivative obtained from *Olea europaea*. *Planta Med.* **2003**, *69*, 472–474.
65. Wang, R.; Wang, W.; Wang, L.; Liu, R.; Ding, Y.; Du, L. Constituents of the flowers of *Punica granatum*. *Fitoterapia* **2006**, *77*, 534–537.
66. Allouche, Y.; Beltrán, G.; Gaforio, J.J.; Uceda, M.; Mesa, M.D. Antioxidant and antiatherogenic activities of pentacyclic triterpenic diols and acids. *Food Chem. Toxicol.* **2010**, *48*, 2885–2890.
67. Márquez-Martín, A.; de la Puerta, R.; Fernández-Arche, A.; Ruiz-Gutiérrez, V.; Yaqoob P. Modulation of cytokine secretion by pentacyclic triterpenes from olive pomace oil in human mononuclear cells. *Cytokine* **2006**, *36*, 211–217.
68. Huang, L.; Guan, T.; Qian, Y.; Huang, M.; Tang, X.; Li, Y.; Sun, H. Anti-inflammatory effects of maslinic acid, a natural triterpene, in cultured cortical astrocytes via suppression of nuclear factor-kappa B. *Eur. J. Pharmacol.* **2011**, *672*, 169–174.
69. Rodríguez-Rodríguez, R.; Perona, J.S.; Herrera, M.D.; Ruiz-Gutiérrez, V. Triterpenic compounds from "orujo" olive oil elicit vasorelaxation in aorta from spontaneously hypertensive rats. *J. Agric. Food Chem.* **2006**, *54*, 2096–2102.
70. Liu, J.; Sun, H.; Wang, X.; Mu, D.; Liao, H.; Zhang, L. Effects of oleanolic acid and maslinic acid on hyperlipidemia. *Drug Dev. Res.* **2007**, *68*, 261–266.
71. Hussain Shaik, A.; Rasool, S.N.; Abdul Kareem, M.; Krushna, G.S.; Akhtar, P.M.; Devi, K.L. Maslinic acid protects against isoproterenol-induced cardiotoxicity in albino Wistar rats. *J. Med. Food* **2012**, *15*, 741–746.
72. Mackness, M.; Mackness, B. Targeting paraoxonase-1 in atherosclerosis. *Expert Opin. Ther. Targets* **2013**, *17*, 829–837.

73. Ros, E.; Martínez-González, M.A.; Estruch, R.; Salas-Salvadó, J.; Fitó, M.; Martínez, J.A.; Corella, D. Mediterranean diet and cardiovascular health: Teachings of the PREDIMED Study. In Proceedings of the IUNS 20th Congress of Nutrition, Granada, Spain, 15–20 September 2013; pp. 330S–336S.
74. Qian, Y.; Guan, T.; Tang, X.; Huang, L.; Huang, M.; Li, Y.; Sun, H. Maslinic acid, a natural triterpenoid compound from *Olea europaea*, protects cortical neurons against oxygen-glucose deprivation-induced injury. *Eur. J. Pharmacol.* **2011**, *670*, 148–153.
75. Moncada, S.; Erusalimsky, J.D. Does nitric oxide modulate mitochondrial energy generation and apoptosis? *Nat. Rev. Mol. Cell Biol.* **2002**, *3*, 214–220.
76. Moro, M.A.; de Alba, J.; Leza, J.C.; Lorenzo, P.; Fernández, A.P.; Bentura, M.L.; Boscá, L.; Rodrigo, J.; Lizasoain, I. Neuronal expression of inducible nitric oxide synthase after oxygen and glucose deprivation in rat forebrain slices. *Eur. J. Neurosci.* **1998**, *10*, 445–456.
77. Lau, A.; Tymianski, M. Glutamate receptors, neurotoxicity and neurodegeneration. *Pflugers Arch.* **2010**, *460*, 525–542.
78. Kanai, Y.; Hediger, M.A. The glutamate/neutral amino acid transporter family SLC1: Molecular, physiological and pharmacological aspects. *Pflügers Archiv* **2004**, *447*, 469–479.
79. Qian, Y.; Guan, T.; Tang, X.; Huang, L.; Huang, M.; Li, Y.; Sun, H.; Yu, R.; Zhang, F. Astrocytic glutamate transporter-dependent neuroprotection against glutamate toxicity: An *in vitro* study of maslinic acid. *Eur. J. Pharmacol.* **2011**, *651*, 59–65.
80. Kagansky, N.; Levy, S.; Knobler, H. The role of hyperglycemia in acute stroke. *Arch. Neurol.* **2001**, *58*, 1209–1212.
81. Acebey-Castellón, I.L.; Voutquenne-Nazabadioko, L.; Mai, D.T.H.; Roseau, N.; Bouthagane, N.; Muhammad, D.; le Debar, M.E.; Gangloff, S.C.; Litaudon, M.; Sevenet, T.; *et al.* Triterpenoid saponins from *Symplocos lancifolia*. *J. Nat. Prod.* **2011**, *74*, 163–168.
82. De Pablos, L.M.; González, G.; Rodrigues, R.; García-Granados, A.; Parra, A.; Osuna, A. Action of a pentacyclic triterpenoid, maslinic acid, against *Toxoplasma gondii*. *J. Nat. Prod.* **2010**, *73*, 831–834.
83. De Pablos, L.M.; dos Santos, M.F.; Montero, E.; García-Granados, A.; Parra, A.; Osuna, A. Anticoccidial activity of maslinic acid against infection with *Eimeria tenella* in chickens. *Parasitol. Res.* **2010**, *107*, 601–604.
84. Moneriz, C.; Marín-García, P.; García-Granados, A.; Bautista, J.M.; Diez, A.; Puyet, A. Parasitostatic effect of maslinic acid. I. Growth arrest of *Plasmodium falciparum* intraerythrocytic stages. *Malar. J.* **2011**, 82:1–82:10.
85. Moneriz, C.; Marín-García, P.; Bautista, J.M.; Diez, A.; Puyet, A. Parasitostatic effect of maslinic acid. II. Survival increase and immune protection in lethal *Plasmodium yoelii*-infected mice. *Malar. J.* **2011**, 103:1–103:9.
86. Moneriz, C.; Mestres, J.; Bautista, J.M.; Diez, A.; Puyet, A. Multi-targeted activity of maslinic acid as an antimalarial natural compound. *FEBS J.* **2011**, *278*, 2951–2961.
87. Fernández-Navarro, M.; Peragón, J.; Esteban, F.J.; de la Higuera, M.; Lupiáñez, J.A. Maslinic acid as a feed additive to stimulate growth and hepatic protein-turnover rates in rainbow trout (*Onchorhynchus mykiss*). *Comp. Biochem. Physiol. C* **2006**, *144*, 130–140.

88. Fernández-Navarro, M.; Peragón, J.; Amores, V.; de la Higuera, M.; Lupiáñez, J.A. Maslinic acid added to the diet increases growth and protein-turnover rates in the white muscle of rainbow trout (*Oncorhynchus mykiss*). *Comp. Biochem. Physiol. C* **2008**, *147*, 158–167.
89. Rufino-Palomares, E.; Reyes-Zurita, F.J.; Fuentes-Almagro, C.A.; de la Higuera, M.; Lupiáñez, J.A.; Peragón, J. Proteomics in the liver of gilthead sea bream (*Sparus aurata*) to elucidate the cellular response induced by the intake of maslinic acid. *Proteomics* **2011**, *11*, 3312–3325.
90. Sultana, N.; Lee, N.H. Antielastase and free radical scavenging activities of compounds from the stems of *Cornus kousa*. *Phytother Res.* **2007**, *21*, 1171–1176.
91. Ullah, F.; Hussain, H.; Hussain, J.; Bukhari, I.A.; Khan, M.T.; Choudhary, M.I.; Gilani, A.H.; Ahmad, V.U. Tyrosinase inhibitory pentacyclic triterpenes and analgesic and spasmolytic activities of methanol extracts of *Rhododendron collettianum*. *Phytother Res.* **2007**, *21*, 1076–1081.
92. Li, C.; Yang, Z.; Li, Z.; Ma, Y.; Zhang, L.; Zheng, C.; Qiu W.; Wu, X.; Wang, X.; Li, H.; *et al.* Maslinic acid suppresses osteoclastogenesis and prevents ovariectomy-induced bone loss by regulating RANKL-mediated NF-κB and MAPK signaling pathways. *J. Bone Miner. Res.* **2011**, *26*, 644–656.
93. Nieto, F.R.; Cobos, E.J.; Entrena, J.M.; Parra, A.; García-Granados, A.; Baeyens, J.M. Antiallodynic and analgesic effects of maslinic acid, a pentacyclic triterpenoid from *Olea europaea*. *J. Nat. Prod.* **2013**, *76*, 737–740.
94. Prades, J.; Vögler, O.; Alemany, R.; Gómez-Florit, M.; Funari, S.S.; Ruiz-Gutiérrez, V.; Barceló, F. Plant pentacyclic triterpenic acids as modulators of lipid membrane physical properties. *Biochim. Biophys. Acta* **2011**, *1808*, 752–760.

ROS-Dependent Antiproliferative Effect of Brassinin Derivative Homobrassinin in Human Colorectal Cancer Caco2 Cells

Martin Kello, David Drutovic, Martina Chripkova, Martina Pilatova, Mariana Budovska, Lucia Kulikova, Peter Urdzik and Jan Mojzis

Abstract: This study was designed to examine the *in vitro* antiproliferative effect of brassinin and its derivatives on human cancer cell lines. Among seven tested compounds, homobrassinin (**K1**; N-[2-(indol-3-yl)ethyl]-S-methyldithiocarbamate) exhibited the most potent activity with $IC_{50} = 8.0$ μM in human colorectal Caco2 cells and was selected for further studies. The flow cytometric analysis revealed a **K1**-induced increase in the G_2/M phase associated with dysregulation of α-tubulin, α_1-tubulin and β_5-tubulin expression. These findings suggest that the inhibitory effect of **K1** can be mediated via inhibition of microtubule formation. Furthermore, simultaneously with G_2/M arrest, **K1** also increased population of cells with sub-G_1 DNA content which is considered to be a marker of apoptotic cell death. Apoptosis was also confirmed by annexin V/PI double staining, DNA fragmentation assay and chromatin condensation assay. The apoptosis was associated with the loss of mitochondrial membrane potential (MMP), caspase-3 activation as well as intracellular reactive oxygen species (ROS) production. Moreover, the antioxidant Trolox blocked ROS production, changes in MMP and decreased **K1** cytotoxicity, which confirmed the important role of ROS in cell apoptosis. Taken together, our data demonstrate that **K1** induces ROS-dependent apoptosis in Caco2 cells and provide the rationale for further *in vivo* anticancer investigation.

Reprinted from *Molecules*. Cite as: Kello, M.; Drutovic, D.; Chripkova, M.; Pilatova, M.; Budovska, M.; Kulikova, L.; Urdzik, P.; Mojzis, J. ROS-Dependent Antiproliferative Effect of Brassinin Derivative Homobrassinin in Human Colorectal Cancer Caco2 Cells. *Molecules* **2014**, *19*, 10877-10897.

1. Introduction

A plethora of epidemiological and animal studies show that consumption of cruciferous vegetables may lower the risk for variety of cancers [1–4]. It is suggested that the cancer-protective effects of cruciferous vegetables can be associated with the presence of glucosinolates, which are cleaved to biologically active compounds, such as indoles and isothiocyanates [5].

Another group of cruciferous-derived phytochemicals, the indole phytoalexins, have attracted scientists' interest because of their ability to modulate processes involved in oncogenic transformation, such as alterations of cell cycle control, apoptosis evasion and inhibition of different signalling pathways [6–8].

Generally, phytoalexins are low molecular weight secondary metabolites biosynthesized *de novo* by plants in response to stress caused by biotic or abiotic factors [9,10]. Although phytoalexins are part of general defense mechanisms used to ward off plant invaders, their chemical diversity suggest substantially broader biological activities. In addition to their antimicrobial activity, some phytoalexins also possess antiinflammatory [11], antioxidant [12], antiproliferative [13,14], as well as anticancer [15,16] properties.

Indole phytoalexins are structurally unique, sulfur-containing natural products isolated from plants of the family Cruciferae (syn. Brassicaceae). Besides their antimicrobial properties, several indole phytoalexins also exhibit antiproliferative/anticancer activity [17–21].

Brassinin ([3-(S-methyldithiocarbamoyl) aminomethyl indole]), first isolated from Chinese cabbage [22], is an indole phytoalexin with demonstrated antiproliferative/anticancer activity. Mehta and co-workers [23] documented dose-dependent inhibition of 7,12-dimethylbenz[a]anthracene (DMBA)-induced preneoplastic lesion formation by brassinin and cyclobrassinin in a mouse mammary gland organ culture model. Later, Csomós *et al.* [24] showed antiproliferative effects of brassinin, isobrassinin and isobrassinin derivatives in different cancer cell types. Recently, Izutani *et al.* [6] described the ability of brassinin to inhibit cell growth in human colon cancer cells by arresting the cell cycle at the G_1 phase via increased expression of p21 and p27. In the last decade we have also documented the antiproliferative effects of brassinin or its derivatives in different cancer cells [25–30].

Although the precise mechanism(s) of the antiproliferative activity of brassinin and its derivatives still remain unknown, inhibition of indoleamine 2,3-dioxygenase and inhibition of PI3K/Akt/mTOR signalling pathways may interfere with cancer cell survival and proliferation [7,31]. However, so far there is no published information about the antiproliferative molecular mechanisms of homobrassinin on cancer cells.

It is well known that oxidative stress may play role in the cytotoxicity of different natural compounds [32,33]. Recently, it was documented that the antiproliferative effect of some indole phytoalexins may be associated with ROS production [34,35] or glutathione depletion [19,30], which may lead to imbalance between antioxidant and prooxidant factors. This prompted us to explore the role of ROS in the antiproliferative effects of brassinin and its derivatives. Our results demonstrate that homobrassinin (**K1**) is the most active in inhibiting the growth of Caco2 cells among the compounds studied. Effect of **K1** is associated with ROS production leading to mitochondrial dysfunction, caspase 3 activation and apoptosis induction. The role of ROS in **K1**-induced cell death was analysed by intracellular ROS generation and ROS scavenger experiments. These findings generate a rationale for *in vivo* efficacy studies with this compound in preclinical cancer models.

2. Results and Discussion

2.1. Effect of Brassinin and Its Derivatives on Cell Proliferation

The antiproliferative effect of indole phytoalexins was evaluated on eight human cancer cell lines using the MTT assay. Survival of cancer cells exposed to the studied indole phytoalexins is shown in Table 1. Our data showed that brassinin (**1**, Figure 1) possesses relatively weak antiproliferative effect with $IC_{50} > 100$ μM in all cancer cell lines used. Similar results were obtained also with compounds **K10** and **47**. On the other hand, homobrassinin (**K1**, Figure 1) displayed the highest antiproliferative activity with IC_{50} from 8.0 to 35.0 μM with the greatest antiproliferative activity in Caco2 cells. Other indole phytoalexins (**K49**, **K124** and **K170**) were less potent.

Table 1. The IC_{50} (μM) of tested compounds in different cell lines after 72 h incubation. Results are presented as a mean ± SD of three independent experimental determinations performed in triplicate. The tested compounds: Brassinin (**1**), Homobrassinin (**K1**), *N*-{[1-(*tert*-Butoxycarbonyl)indol-3-yl]methyl}-*N*'-methyl-N'-phenylthiourea (**K10**), *N*-{[(1-*tert*-Butoxycarbonyl)indol-3-yl]metyl}-*N*'-(4-methoxyphenyl)thiourea (**K124**), 1-(β-D-glucopyranosyl)brassinin (**47**), 1-[(1*R*,2*S*,5*R*)-Menthoxycarbonyl]brassinin (**K49**); 1-[(1*R*,2*S*,5*R*)-8-Phenylmenthoxycarbonyl]brassinin (**K170**).

Compound	Cancer Cell Lines							
	Jurkat	Caco2	HepG2	HCT-116	A549	HeLa	MCF-7	MDA-MB-231
1	>100	>100	>100	>100	>100	>100	>100	>100
K1	28.2 ± 1.2	8.0 ± 0.6	21.3 ± 2.3	27.3 ± 1.8	33.4 ± 0.8	26.1 ± 2.4	35.0 ± 2.1	22.8 ± 1.4
K10	>100	>100	>100	>100	>100	>100	>100	>100
K124	30.0 ± 0.5	>100	>100	30.7 ± 2.9	>100	34.0 ± 0.7	92.0 ± 2.7	>100
K49	39.6 ± 1.8	89.4 ± 1.3	37.4 ± 3.7	76.0 ± 2.5	22.8 ± 0.5	45.6 ± 1.4	62.0 ± 3.4	>100
K170	44.3 ± 3.2	>100	>100	53.7 ± 2.1	96.5 ± 4.1	>100	>100	>100
47	>100	>100	>100	>100	>100	>100	>100	>100

Because only **K1** displayed interesting antiproliferative activity, it was selected for further mechanistic studies in Caco2 cells. The antiproliferative activity of compound **K1** was also compared with its activity against non-cancer cells, human umbilical vein endothelial cells (HUVEC). Our results shown that compound **K1** was more active against cancer cells than non-cancer cells (IC_{50} = 8.0 *vs.* IC_{50} = 46.2 μM).

Figure 1. Chemical structure of brassinin (**1**) and homobrassinin (**K1**).

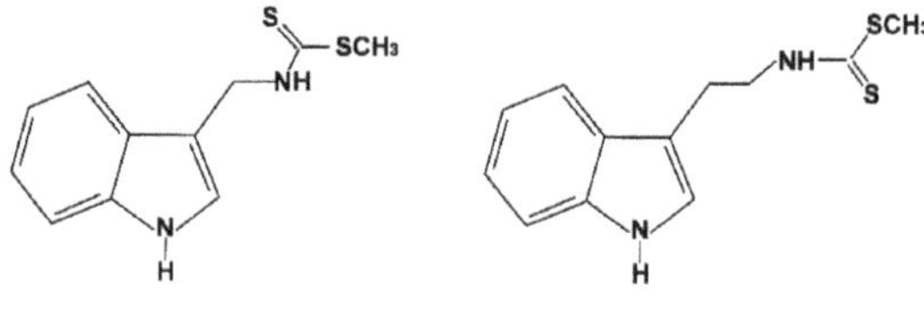

brassinin (1) **homobrassinin (K1)**

To confirm the potential antiproliferative effect of **K1**, the BrdU proliferation assay was used. The magnitude of the absorbance for the developed colour is proportional to the quantity of BrdU incorporated into cells, which is a direct indication of cell proliferation. As shown in Figure 2A, **K1** at concentrations of 100, 50 and 10 μM significantly decreased BrdU incorporation compared with the control (approximately 94%, 83% and 56% respectively) ($p < 0.001$; $p < 0.05$). Antiproliferative effect was not observed at the concentration of 1 μM.

Figure 2. Antiproliferative effect of compound **K1** on Caco2 cells. (**A**) Effect of 72 h **K1** treatment on proliferation of Caco2 cells, as measured by the BrdU ELISA proliferation assay, *** $p < 0.001$, * $p < 0.05$. Data were obtained from three independent replicate experiments with at least three wells per treatment group in each individual replicate. (**B**) Real-time monitoring of Caco2 cell proliferation after incubation with **K1** using the xCELLigence system. Appropriate concentrations of **K1** were added 24 h after seeding. Representative picture from three independent experiments is shown.

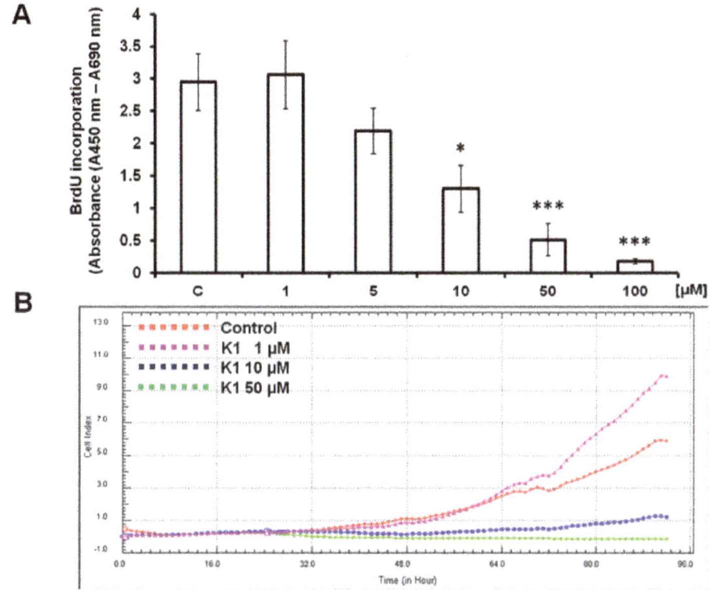

The antiproliferative effect of compound **K1** on Caco2 cells was also evaluated by xCELLigence system. As shown in Figure 2B the cell index of Caco2 cells decreased significantly after treatment with **K1** in a concentration-dependent manner. However, the cells treated with 1 µM of **K1** exhibited a similar proliferation rate as control for most of the duration of the experiment, which confirm results obtained by BrdU proliferation assay.

*2.2. **K1** Blocks Cell Cycle at G_2/M Transition*

To evaluate the effect of **K1** treatment on Caco2 cell cycle progression, we performed flow cytometric analysis on cells treated with 10 µM **K1** for 24, 48 and 72 h as described in the Experimental Section. Results obtained are summarized in Table 2. The significant enrichment in G_2/M cell populations was observed after 24 h (70.48%) of **K1** treatment, as compared to untreated cells (36.64%) ($p < 0.001$). As a consequence, a significant reduction of the G_0/G_1 phase cell population confirms the cell cycle arrest in G_2/M as an effect of colorectal cells exposure to **K1**. After 48 and 72 h incubation G_2/M arrest still persisted with addition of increased sub-G_1 population of cells. These results strongly suggest that **K1**-induced cell cycle arrest may represent one of the mechanisms by which this indole phytoalexin inhibits colorectal cancer cells growth.

Table 2. The distribution of cell cycle in Caco2 cells treated with **K1**. Cells were treated with **K1** derivate (c = 10 μM) for 24, 48 and 72 h. The distribution of cell cycle was assessed by flow cytometry. Each value is the mean ± SD of three independent experiments. The significant differences between control and **K1**-treated cells were signed as $p < 0.05$ (*), $p < 0.01$ (**), $p < 0.001$ (***).

Treatment	Time (h)	Sub-G_1	G_0/G_1	S	G_2/M
Control		0.29 ± 0.15	40.81 ± 3.29	18.25 ± 3.90	36.64 ± 2.29
K1	24	1.49 ± 0.37	21.34 ± 3.13 **	6.69 ± 1.56 *	70.48 ± 2.28 ***
K1	48	3.36 ± 1.67 *	30.89 ± 2.81 *	9.58 ± 2.43	56.20 ± 2.58 **
K1	72	5.58 ± 1.16 *	30.17 ± 1.65 *	11.47 ± 1.65	52.79 ± 2.77 **

*2.3. **K1** Induces Apoptotic Cell Death*

The increase of cell having sub-G_1 DNA content is considered to be a marker of apoptotic cell death. To confirm whether **K1** could induce apoptosis on colorectal cell lines, we performed Annexin V/PI staining, which detects an early stage of apoptosis and combined staining with PI, which detects a late stage of apoptosis or necrosis. The results showed that only a small percentage of untreated Caco2 cells (3.17%) bound with annexin V. In contrast, the percentage of annexin V binding Caco2 cells significantly increased in a time-dependent manner after 24, 48 and 72 h of treatment with 10 μM **K1** (18.54% to 37.07%, $p < 0.01$; $p < 0.001$). Simultaneously, percentage of annexin V/PI positive cells increase from 2.05% (untreated cells) to 20.85% (**K1** treated cells after 72 h of incubation). These experimental results demonstrate that **K1** induced apoptosis of Caco2 cells (Table 3; Figure 3).

Chromatin condensation is one of the most important markers for apoptotic cells. The nuclear morphological changes of Caco2 cells were analysed using DAPI staining. In control groups, cells appeared to be round and manifested homogeneous nuclei. In cells treated with **K1** at concentration 10 μM, cells displayed condensed nuclei. A significant increase of cells with typical apoptotic morphology was observed (Figure 4A), reaching 6.83% after 24 h of treatment, as evaluated by determination of apoptotic index (Figure 4B).

Table 3. Induction of apoptosis after **K1** treatment measured by Annexin V/PI staining Caco2 cells were treated with **K1** for 24, 48 and 72 h, stained with fluoresceinated Annexin V and PI, and analysed by flow cytometry. The percentage of events in the nonapoptotic (lower left, An^-/PI^-), early apoptotic (lower right, An^+/PI^-), and late apoptotic/necrotic (upper left + right, An^-/PI^+ + An^+/PI^+) quadrants (Figure 3) is indicated. The significant differences between control and **K1**-treated cells were signed as $p < 0.05$ (*), $p < 0.01$ (**), $p < 0.001$ (***).

Treatment	Time (h)	An^-/PI^-	An^+/PI^-	An^+/PI^+
Control		94.07 ± 1.32	3.17 ± 0.60	2.05 ± 0.50
K1	24	67.81 ± 1.18 **	18.54 ± 1.63 **	13.75 ± 2.43 *
K1	48	49.92 ± 3.20 ***	38.55 ± 2.47 ***	11.81 ± 2.24 *
K1	72	42.36 ± 3.65 ***	37.07 ± 2.66 ***	20.85 ± 3.69 **

Figure 3. Induction of apoptosis after **K1** treatment measured by Annexin V/PI staining. Caco2 cells were treated with **K1** for 24, 48 and 72 h, stained with fluoresceinated Annexin V and PI, and analysed by flow cytometry. One representative experiment out of three is shown. The percentage of events in the nonapoptotic (lower left, An^{-}/PI^{-}), early apoptotic (lower right, An^{+}/PI^{-}), and late apoptotic/necrotic (upper left + right, $An^{-}/PI^{+} + An^{+}/PI^{+}$) quadrants is indicated in Table 3.

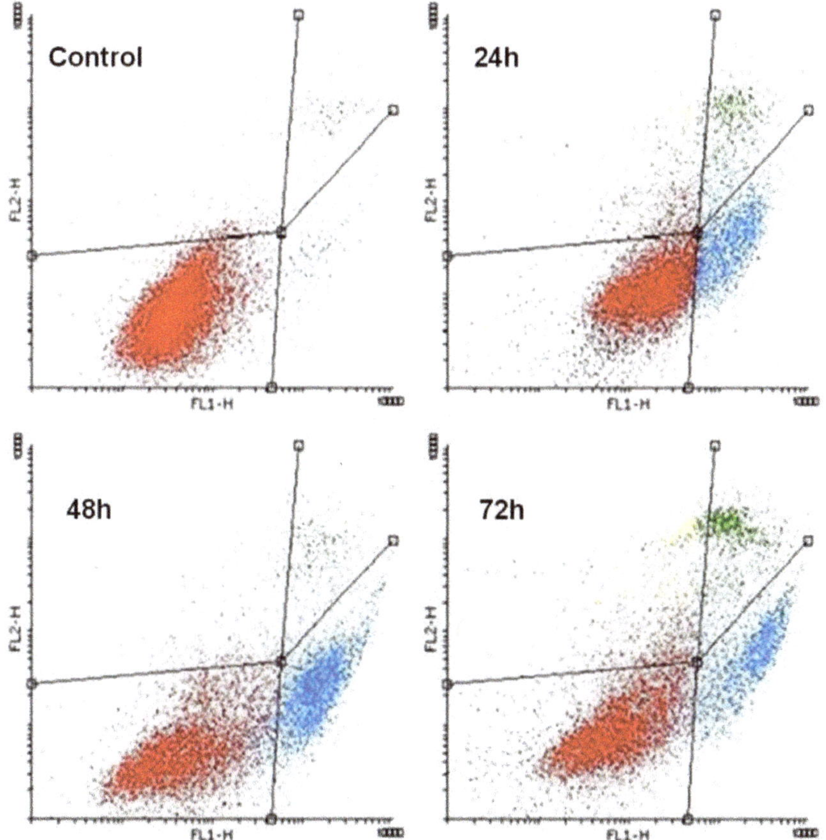

Figure 4. Apoptotic effect of **K1** on Caco2 cells. (**A**) Morphological changes of Caco2 cells treated with 10 μM of **K1** for 24 h. Cells were cultured in chamber slides and stained with DAPI to show typical apoptotic morphology. Magnification 400×. (**B**) Apoptotic index of Caco2 cells treated with 10 μM of **K1** for 24 h. The results (mean ± SD) of three independent experiments are shown as the apoptotic index evaluated as a percentage of cells with fragmented nuclei from a total number of minimum 300 cells, * $p < 0.05$. (**C**) DNA fragmentation of Caco2 cells after incubation with 10 μM of **K1** for 24, 48 and 72 h, C-control (untreated cells), PC-positive control (cells treated with 10 μM of cisplatin). A representative picture from three independent experiments is shown.

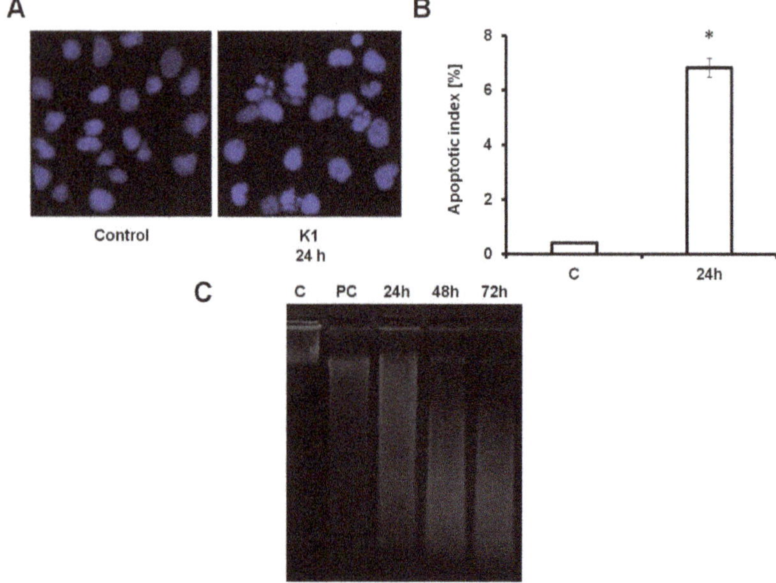

Analysis of DNA fragmentation by agarose gel electrophoresis is one of the most widely used biochemical markers for cell death. Fragmentation of DNA was observed in Caco2 cells after 24 h incubation with 10 μM of compound **K1** compared to control (untreated cells). This effect also persisted after 48 and 72 h of incubation (Figure 4C).

2.4. *Effect of **K1** on ROS Formation*

It has been reported that increased liberation of ROS can activate a cascade of events leading to apoptosis [36]. Therefore, we decided to detect whether K1 was able to trigger ROS generation by measuring the rhodamine 123 fluorescence intensity. The results showed that treatment of the cells with **K1** significantly increased ROS generation in a time-dependent manner. After 12 h incubation with **K1** the significant ROS accumulation was observed. Culminated accumulation was marked after 48 h treatment (Figure 5B). On the other hand, no increase in ROS production was observed after 1, 3 and 6 h of incubation (data not shown). In order to show that generation of ROS is a key step in the **K1**-induced apoptotic pathway, Caco2 cells were pretreated with Trolox, a water-soluble analogue of the free radical scavenger α-tocopherol. Trolox significantly decreased ROS accumulation in time dependent manner. These effects were associated also with recovered cell viability (Figure 6).

Figure 5. Effect of **K1** and Trolox on caspase-3, ROS accumulation and MMP. (**A**) Effect of **K1** and Trolox treatment on caspase-3 activation. Caspase-3 activation was measured 24, 48 and 72 h after treatment by quantifying the fluorescence intensity. (**B**) Measurement of ROS production in Caco2 cells treated with **K1**, Trolox or mutual combinations. Cytosolic ROS were measured 12, 24, 48 and 72 h after treatment by quantifying the fluorescence intensity of activate DHR-123. The results (mean ± SD) of three independent experiments are shown as multiples of the control group fluorescence. (**C**) MMP changes in Caco2 cells treated with **K1**, Trolox or mutual combinations were analysed 12, 24, 48 and 72 h after treatment.

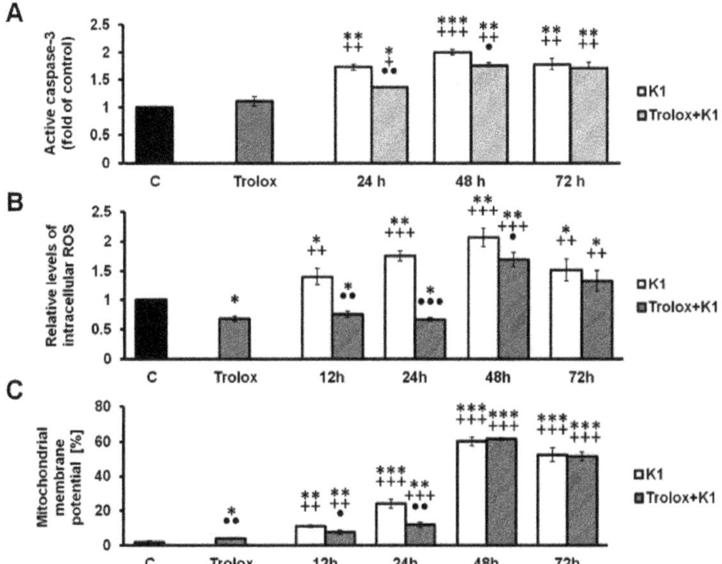

Results are expressed as mean values ± SD of three independent experiments. * $p < 0.05$, ** $p < 0.01$, *** $p < 0.001$ *vs.* untreated control; $^{+}p < 0.05$, $^{++}p < 0.01$, $^{+++}p < 0.001$ *vs.* Trolox; $^{\bullet}p < 0.05$, $^{\bullet\bullet}p < 0.01$, $^{\bullet\bullet\bullet}p < 0.001$ *vs.* **K1**.

Collectively, these findings suggest that an increase in ROS generation may take part in the **K1**-induced apoptosis in Caco2 cells.

*2.5. **K1**-Induced Mitochondrial Dysfunction*

A decrease in MMP is one of the earliest events in apoptosis. Mitochondrial membrane integrity was evaluated using the cationic dye TMRE, a highly specific probe for detecting changes in mitochondrial $\Delta\Psi_m$. In our experiments, changes in MMP were detected 12, 24, 48 and 72 h after **K1** treatment. Effect of **K1** treatment on Caco2 cells led to intensive and considerable destruction of mitochondrial function demonstrated with significant changes in the percentage of cells with dissipated MMP in time-dependent manner. On the other hand, co-treatment with Trolox significantly diminished effect of **K1** on MMP changes and prevented mitochondrial dysfunction (Figure 5C).

Figure 6. Relative survival of Caco2 cells treated with **K1**, Trolox or mutual combinations as evaluated by MTT assay. Cells were incubated with tested compounds for 24, 48 and 72 h.

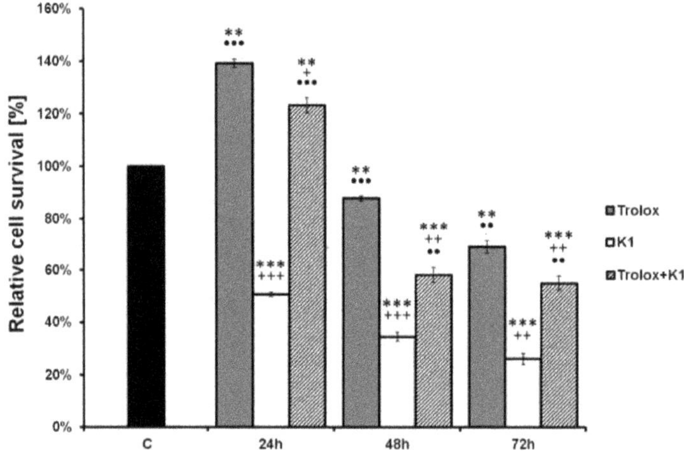

Data is presented as a mean ± SD of three independent experimental determinations performed in triplicate. ** $p < 0.01$, *** $p < 0.001$ *vs.* untreated control; $^{+}p < 0.05$, $^{++}p < 0.01$, $^{+++}p < 0.001$ *vs.* Trolox; $^{\bullet\bullet}p < 0.01$, $^{\bullet\bullet\bullet}p < 0.001$ *vs.* **K1**.

*2.6. Activation of Caspase-3 by **K1***

Analysis of caspase-3 activation (Figure 5A) clearly demonstrated caspase-dependent form of cell death in Caco2 cells line. The more than 2-fold higher increase of caspase-3 activation after 48 h treatment confirmed this suggestion. On the other hand, co-treatment with Trolox significantly decreased the effect of **K1** on caspase-3 activation. Significant caspase-3 activation correlated well with apoptosis occurrence frequency, ROS and loss of MMP leading to cell death after **K1** treatment. These results clearly indicate that caspase-3 activation plays an important role in Caco2 cell apoptosis induced by **K1**.

*2.7. Effect of **K1** on Gene Expression of Tubulins*

The cell cycle analysis showed that **K1** induces the G_2/M arrest. Based on this result, the alteration in selected gene expression was analysed after 24, 48 and 72 h incubation by real-time PCR. We found significant downregulation of β-tubulin expression and significant upregulation of α-tubulin expression after 24 h treatment with compound **K1** at concentration 10 μM. Similar results were also obtained after 72 h of incubation. After 48 h incubation with the tested compound an opposite but non-significant effect was noted (Table 4).

Table 4. Fold changes of specific genes after 24, 48 and 72 h treatment with **K1** (c = 10 μM). β-actin gene was used as a housekeeping gene to normalize each sample.

Genes	Normalized Ratio		
	24 h	48 h	72 h
α-tubulin	5.59	1.68	3.35
α_1-tubulin	1.89	0.65	1.47
β_5-tubulin	0.29	0.99	0.48

These results suggest that **K1** may affect microtubule and microtubule assembly. The results of real-time PCR analysis for selected genes are in direct correlation with the data obtained in cell cycle analysis.

*2.8. Effect of **K1** on Cytoskeletal Tubulins Assembly*

To determine whether the tubulins are involved in **K1**-mediated apoptosis of Caco2 cells, western blotting was used to detect cellular changes of α, α1c and β tubulins from whole cell lysates. All α, α1c and β tubulin levels showed a time-dependent decrease after treatment with **K1** (Figure 7).

Figure 7. Western blot detection of α and β tubulins in Caco2 cells treated with **K1**, Trolox or mutual combinations. Cells were incubated with tested compounds for 24, 48 and 72 h. Representative picture from three independent experiments is shown. β-Actin served as a loading control.

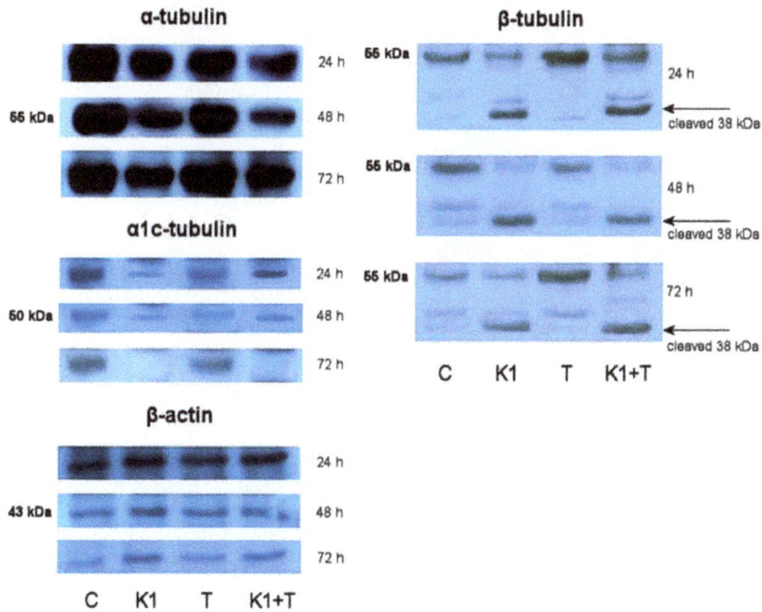

The β tubulin analysis also confirmed degradation and cleaving of presented protein to lower mass form. The results from both 48 and 72 h treatments indicated that tubulin levels decreased rapidly, and the effects proceeded for a long time. The correlation between changes of tubulins level and cell cycle distribution suggests that tubulin degradation may be involved in cell cycle arrest and apoptosis after **K1** treatment. To explain the possible involvement of oxidative stress in the tubulin degradation, disaggregation or deregulation we used Trolox as an antioxidant to prevent an oxidative burst after **K1** treatment. Data showed that Trolox alone has minimal or no effect on tubulin assembly. In combination, Trolox with **K1** treatment is not able to diminish the effect of **K1** on tubulin levels. Only short time treatment (24 h) with **K1** and Trolox at the β tubulins level pointed to a minor Trolox effect compared with **K1** treatment.

2.9. Discussion

Cancer, characterized by unregulated proliferation of cells, is the second major cause of death after cardiovascular disease [37]. Despite enormous progress in the understanding of carcinogenesis, the discovery of anticancer drugs remains a highly challenging endeavor. Natural products provide one of the most important sources of promising leads for the development of novel chemotherapeutics.

In the present study, the antiproliferative effects of brassinin and some of its synthetic derivatives on cancer cells were investigated. Our results revealed that the natural or synthesized compounds exhibited different levels of antiproliferative activity in cancer cells. Compound **K1**, in particular, showed much higher levels of bioactivity than naturally occurring brassinin, with IC_{50} = 8.0 μM in human colorectal adenocarcinoma Caco2 cells. Interestingly, **K1** at lower concentration stimulated cancer cell proliferation. Similar biphasic, dose-dependent effect of natural compounds or their chemical analogues have been documented only in a limited number of studies. Polyphenols such as apigenin, quercetin, genistein or red wine polyphenols at low concentrations stimulated cell growth. On the contrary, at the higher concentrations decrease in cell proliferation was observed [38–41].

To our knowledge, this is the first report of an investigation into the effect of **K1** on proliferation in cancer cells. We demonstrated that **K1**-induced antiproliferative effect in human colorectal Caco2 cells was related to cell cycle arrest and increase of cells with sub-G_1 DNA content and induction of apoptosis, as confirmed by annexin V/PI staining, measurement of chromatin condensation, DNA fragmentation and caspase 3 activation. Our data also indicated that **K1** induced apoptosis of Caco2 cells through generation of ROS and mitochondrial dysfunction, suggesting that ROS act as upstream signalling molecules for initiation of cell death.

Flow cytometric analysis showed that the **K1**-induced decrease in Caco2 cell viability is associated with cell cycle arrest in the G_2/M phase. This result prompted us to analyse expression of selected genes involved in microtubules formation. Microtubules are built up of a heterodimer of α- and β-tubulin. These constantly growing and shortening dynamic structures are involved in many important cellular processes particularly associated with cell division [42]. Their importance in mitotic spindle formation and chromosome movement during cell division which is uncontrolled in cancer cells makes tubulin an important target for cancer therapy. The effect of the most successful antineoplastic drugs called as tubulin inhibitors, such as the taxanes and the vinca alkaloids, interfere directly with the tubulin system [43]. The tested compound **K1** increased expression of α-tubulin and decreased expression of β-tubulin. Dysregulated expression of tubulins can lead to insufficiency of α/β hetero-dimer production, material necessary for mitotic spindle formation, resulting in the G_2/M cell cycle arrest, eventually resulting in apoptotic cell death [44]. Furthermore, **K1** treatment led to degradation or cleavage of corresponding proteins in time dependent manner (Figure 7). Presented destruction of cytoskeletal tubulin assembly also correlates with the observed apoptosis progression.

In addition to G_2/M cell cycle arrest, the increase of cells with sub-G_1 DNA content was observed. Compared with the control, cells with sub-G_1 DNA content were increased 5.1-, 11.5- and 19.2-fold after 24, 48 and 72 h of treatment, respectively. The appearance of the sub-G_1 fraction and conventional morphological signs of apoptosis provided support for the idea that **K1** induced apoptosis in Caco2 cells (Figure 4). Pro-apoptotic effect of **K1** was further confirmed by annexin V/PI staining. The percentage of apoptotic cells increased in a time-dependent manner (Figure 3). These findings also supported caspase 3 activation observed after **K1** treatment that could be one of crucial step to execution of apoptotic features.

Mitochondria are crucial for energy production, intermediary metabolism, and calcium homeostasis [45]. On the other hand, mitochondrial dysfunction has been recognized as one of the key events occurring at the initial stages of apoptosis [46]. Several experimental studies reported a fall in the MMP during apoptosis induced by various compounds with anticancer properties [47–49]. Therefore, MMP changes after **K1** treatment in Caco2 cells were measured.

In our study, the decrease of MMP was detected 12 h after **K1** treatment, suggesting **K1** induced cell apoptosis in Caco2 cells might be mitochondria dependent (Figure 5). Mitochondria are also the major sites for ROS production. It is well known that excessive generation of ROS may result in cell death [50,51]. Our results showed **K1** treatment significantly stimulated ROS generation in Caco2 cells (Figure 5). To confirm role of ROS in **K1**-induced apoptosis and cell death, Caco2 cells were pretreated with the antioxidant Trolox. Compared to **K1** treatment only, Trolox pretreatment caused a reduction in ROS levels and prevented loss of MMP as well as significantly rescued **K1**-induced Caco2 cytotoxicity (Figure 6). We have also demonstrated that peroxyl radical scavenging does not affect **K1**-mediated tubulin degradation, but we cannot disprove the involvement of other free radical forms in the shortened time of treatment shown in Figure 7.

3. Experimental Section

3.1. Test Compounds

Brassinin (**1**), homobrassinin (**K1**), N-{[1-(*tert*-butoxycarbonyl)indol-3-yl]methyl}-N'-methyl-N'-phenylthiourea (**K10**), N-{[(1-*tert*-butoxycarbonyl)indol-3-yl]methyl}-N'-(4-methoxyphenyl)thiourea (**K124**), 1-(β-D-glucopyranosyl)brassinin (**47**), 1-[(1R,2S,5R)-menthoxycarbonyl]brassinin (**K49**); 1-[(1R,2S,5R)-8-phenylmenthoxycarbonyl]brassinin (**K170**). The synthesis of tested compounds was described in the previous studies: **1**, **K1** [52]; **K10**, **K124** [28]; **47** [53]; **K49**, **K170** [54].

3.2. Cell Culture

The human cancer cell lines HCT116 (human colorectal carcinoma), HepG2 (human hepatocellular carcinoma), HeLa (human cervical adenocarcinoma), Jurkat (human leukemic T cell lymphoma) were cultured in RPMI 1640 medium (PAA Laboratories, Pasching, Austria) and Caco2 (human colorectal adenocarcinoma), A549 (human alveolar adenocarcinoma), MCF-7 (human Caucasian breast adenocarcinoma) and MDA-MB-231 (human mammary gland adenocarcinoma) were maintained in growth medium consisting of high glucose Dulbecco's Modified Eagle Medium (Invitrogen, Carlsbad, CA, USA). Both media were supplemented with a 10% fetal bovine serum (FBS), penicillin (100 IU/mL) and streptomycin (100 μg/mL) (all from Invitrogen). The cells (obtained from the American Tissue Culture Collection, ATCC, Rockville, MD, USA) were maintained under standard tissue culture conditions of 37 °C, 95% air/5% CO_2. Cell viability, estimated by trypan blue exclusion, was greater than 95% before each experiment. Human umbilical vein endothelial cells (HUVECs) were isolated and cultured as previously described by Ivanova *et al.* [55].

3.3. Growth Inhibition Assay

The antiproliferative effects of compounds were determined using colorimetric microculture assay with the MTT end-point [56]. Briefly, 3×10^3 cells were plated per well in 96-well polystyrene microplates (Sarstedt AG & Co, Nümbrecht, Germany) in the culture medium containing tested chemicals at final concentrations of 10^{-4}–10^{-6} mol/L or Trolox (6-hydroxy-2,5,7,8-tetramethylchroman-2-carboxylic acid; Fluka, Buchs, Schwitzerland) at final concentrations 300 μM, or at mutual combinations. After 72 h of incubation (for Trolox and mutual combinations also after 24 and 48 h), 10 μL of MTT (5 mg/mL) (Sigma-Aldrich Corporation, St. Louis, MO, USA) were added in each well. After an additional 4 h, during which insoluble formazan was produced, 100 μL of 10% sodium dodecyl sulfate were added in each well and another 12 h were allowed for the dissolution of formazan. The absorbance was measured at 540 nm using the automated uQuant ™ Universal Microplate Spectrophotometer (Biotek, Winooski, VT, USA). The blank-corrected absorbance of the control wells was taken as 100% and the results were expressed as a percentage of the control. All experiments were performed in triplicate. Due to spontaneous apoptosis, HUVEC cells were incubated only 48 h.

3.4. 5-Bromo-2'-Deoxyuridine (BrdU) Cell Proliferation Assay

Cell proliferation activity was directly monitored by quantification of BrdU incorporated into the genomic DNA during cell growth. DNA synthesis was assessed using colorimetric cell proliferation ELISA assay (Roche Diagnostics GmbH, Mannheim, Germany) following the vendor's protocol. Briefly, 2×10^3 cells/well in 80 μL medium were plated in a 96-well polystyrene microplates (Sarstedt AG & Co, Nümbrecht, Germany). Twenty-four hours after cell seeding different concentrations (10^{-4}–10^{-6} mol/L) of the compound were added. After 48 h of treatment, cells were incubated with BrdU labeling solution (10 μM final concentration) for another 24 h at 37 °C followed by fixation and incubation with anti-BrdU peroxidase conjugate for an additional 1.5 h at room temperature. Finally, after substrate reaction, the stop solution was added (25 μL 1 M H_2SO_4) and colour intensity was measured with multi-well microplate ELISA reader at 450 nm (reference wavelength: 690 nm).

3.5. xCELLigence Cell Analysis System

The xCELLigence system is a unique, impedance-based system for cell-based assays, allowing for label-free and real-time monitoring of cellular processes such as cell growth, proliferation, cytotoxicity, adhesion, morphological dynamics and modulation of barrier function. It measures impedance changes in a meshwork of interdigitated gold microelectrodes located at the well bottom (E-plate) or at the bottom side of a microporous membrane (CIM16-plate). These changes are caused by the gradual increase of electrode surface occupation by (proliferated/ migrated/invaded) cells during the course of time and thus can provide an index of cell viability, migration and invasion. This method of quantification is directly proportional to cellular morphology, spreading, ruffling and adhesion quality as well as cell number [57,58].

The xCELLigence RTCA system was initialized, as per manufacturer's instructions, prior to commencement of the experiment by filling all 16 wells of the E-plate (ACEA Biosciences, San Diego, CA, USA) with the growth medium (100 μL) and equilibrated at room temperature for 30 min. The plate was placed into the single plate (SP) station cradle (housed in a humidified incubator at 37 °C with a 5% CO_2 atmosphere) to establish a background reading. Then, Caco2 cells were seeded in E-plates at a density of 2×10^3 cells per well. After 24 h, **K1** was added at final concentrations of 1–50 μM and cells were allowed to grow for additional 72 h under label-free conditions. The electrical impedance was measured by the RTCA-integrated software of the xCELLigence system (ACEA Biosciences) as a dimensionless parameter termed CI.

3.6. Experimental Design for Flow Cytometry Analysis

Caco2 cells (3×10^5) were seeded in Petri dishes and cultivated 24 h in a complete medium with 10% FBS. Cells were treated with **K1** (c = 10 μM) for 12, 24, 48 and 72 h prior to analysis. The apoptosis, caspase 3 activation and cell cycle parameters were analysed 24, 48 and 72 h after treatment. Changes in MMP and ROS production were also analysed after 12 h of incubation. To evaluate ROS-dependent/independent mechanisms of **K1** treatment we used Trolox, a water-soluble analogue of vitamin E, as antioxidant. Trolox (300 μM) was used 1 h before **K1** treatment and after as a co-treatment with **K1** for 12, 24, 48 and 72 h before ROS and MMP analysis.

3.7. Analysis of Cell Cycle

For flow cytometric analysis (FCM) of the cell cycle, floating and adherent cells were harvested together 24, 48 and 72 h after treatment, washed in cold PBS, fixed in cold 70% ethanol and kept at −20 °C overnight. Prior to analysis, cells were washed twice in PBS, resuspended in staining solution (final concentration 0.1% Triton X-100, 0.5 mg/mL ribonuclease A and 0.025 mg/mL propidium iodide-PI), incubated in the dark at room temperature for 30 min and analysed using a FACSCalibur flow cytometer (Becton Dickinson, San Jose, CA, USA).

3.8. Annexin V-FITC Labelling

The plasma membrane changes characteristic of apoptosis were analysed by double staining with Annexin V-FITC and PI according to the manufacturer's instructions. Adherent and floating cells (1×10^5) were harvested together 24, 48 and 72 h after treatment and stained with Annexin V-FITC (BD Biosciences Pharmingen, San Diego, CA, USA) in binding buffer for 15 min, washed, stained with PI for 5 min and thereafter analysed using a BD FACSCalibur flow cytometer. Three populations of cells were observed: viable cells: Annexin V-FITC negative and PI negative; apoptotic cells: Annexin V-FITC positive and PI negative; late apoptotic/necrotic cells: Annexin V-FITC positive and PI positive or and Annexin V-FITC negative and PI positive.

3.9. Measurement of ROS

The intracellular production of ROS was detected with FCM analysis using dihydrorhodamine-123 (DHR-123, Fluka), which reacts with intracellular hydrogen peroxide. The cells treated with an appropriate agent were harvested, washed twice in PBS, and resuspended in PBS. DHR-123 was added at a final concentration of 0.2 μM. The samples were then incubated for 15 min in dark and after incubation samples were placed on ice. Fluorescence was detected with 530/30 (FL-1) optical filter. Forward and side scatters were used to gate the viable populations of cells.

3.10. Detection of MMP

The changes in MMP were analysed with FCM using tetramethylrhodamine ethyl ester per chlorate (TMRE, Molecular Probes, Eugene, OR, USA). The cells were washed with PBS, resuspended in 0.1 μM of TMRE in PBS, and incubated for 30 min at room temperature in the dark. The cells were then washed twice with PBS, resuspended in 500 μM of the total volume, and analysed (1×10^4 cell per sample). Fluorescence was detected with 585/42 (FL-2) optical filter.

3.11. Detection of Active Caspase 3

The changes in caspase 3 activation were analysed with FCM using BD Pharmingen Active Caspase-3 PE MAb Apoptosis kit (BD Bioscience, San Diego, CA, USA). The cells were prepared according to manufactory condition and stained with PE conjugated antibody and incubated for 30 min at room temperature in the dark. The cells were then washed twice with PBS, resuspended in 500 μM of the total volume, and analysed (1×10^4 cell per sample). Fluorescence was detected with 585/42 (FL-2) optical filter.

3.12. DNA Fragmentation Assay

The culture medium was removed from untreated (1×10^6) and treated Caco2 cells (**K1** 10 μM for 24, 48 and 72 h and 10 μM of cisplatin as positive control) and centrifuged at 1300 rpm for 5 min to collect them. Cells were washed twice with PBS calcium and magnesium free. Then cells were lysed in a lysis buffer containing 10 mmol/L EDTA, 0.5% Triton X-100. Proteinase K (1 mg/mL) was added and cells were incubated at 37 °C for 1 h followed by 10 min incubation at 70 °C. RNase

(200 μg/mL) was added and cells were incubated for another 1 h at 37 °C. Samples were transferred to 2% agarose gel and run with 40 V for 3 h. DNA fragments were visualized by a UV illuminator.

3.13. DAPI Staining

Twenty four hours after treatment (**K1** 10 μM), Caco2 cells grown on cover slips were fixed with 2% paraformaldehyde for 20 min at 4 °C. After incubation, the cells were washed briefly with PBS and incubated at room temperature with SlowFade® Gold antifade reagent with 4',6-diaminidino-2-phenyl-indole, dihydrochloride (DAPI) (Invitrogen) for nuclear visualization. The slides were analysed using fluorescence microscope Leica DMI6000 B (Leica Microsystems, Inc., Bannockburn, IL, USA) and evaluated as percentages of cells with a fragmented nucleus from a minimum of 300 cells.

3.14. RNA Isolation and cDNA Synthesis

Total RNA was isolated from Caco2 cells using the TRI Reagent (Molecular Research Center, Inc., Cincinnati, OH, USA) according to the manufacturer's instruction. Total RNA quality was verified on an agarose gel. Total RNA (0.5 μg) was reverse transcribed into cDNA by the RevertAid™ H Minus First Strand cDNA synthesis kit (Fermentas GmbH, St. Leon-Rot, Germany) according to the manufacturer's instruction, and used for quantitative real time PCR.

3.15. Quantitative Real-Time PCR

Quantitative real time PCR analysis was performed in Light Cycler (Roche, Mannheim, Germany) using iQTM SYBR Green Supermix (Bio-Rad Laboratories, Hercules, CA, USA) to verify the alterations of α-tubulin, α_1-tubulin, β_5-tubulin gene expression. The PCR program was initiated by 5 min at 95 °C before 40 thermal cycles, each of 30 s at 95 °C and 45 s at 55 °C. Data were analysed according to the comparative Ct method and were normalized by β-actin expression in each sample. Melting curves for each PCR reaction were generated to ensure the purity of the amplification product.

3.16. Western Blot Analysis

Caco2 cells were treated with compound **K1** (10 μM), Trolox (300 μM) and mutual combinations for 24, 48 and 72 h. Protein extracts were obtained using a lysis buffer containing 100 mM Tris (pH 7.4), 1% SDS and 10% glycerol in the presence of PIC (protease inhibitor cocktail), for 30 min on ice. After the insoluble materials were removed by centrifugation at 12,000 *g* for 10 min at 4 °C, total protein concentrations were quantified using the Pierce® BCA Protein Assay Kit (Thermo Fisher Scientific, Waltham, MA, USA). Twenty micrograms of total cellular proteins were separated on 10% SDS polyacrylamide gels and electrotransferred onto nitrocellulose membranes (Pall Corporation, Port Washington, NY, USA). Membranes were blocked in 5% skim milk in Tris-buffered saline (TBS) containing 0.1% Tween-20 for 1 h at room temperature to minimize non-specific binding and incubated with the primary antibodies overnight at 4 °C. Immunoblotting was carried out with α Tubulin (E-19), α1c Tubulin (MH-87), β Tubulin (H-235) and β-Actin (C4)

Antibody (all from Santa Cruz Biotechnology, Inc. Dallas, Texas, USA). After incubation with primary antibodies, membranes were washed 1 × 5 min with TBS-Tween followed by an incubation of 1 h at room temperature with the corresponding horseradish peroxidase-conjugated secondary antibodies (anti-rabbit IgG-HRP, anti-mouse IgG-HRP, all from Sigma-Aldrich). After washing 4 × 10 min with TBS-Tween expression was detected by chemiluminescence emission using ECL (Thermo Fisher Scientific) and then the blots were exposed to x-ray films.

3.17. Statistical Analysis

Results are expressed as mean ± SD. Statistical analyses of the data were performed using standard procedures, with one-way ANOVA followed by the Bonferroni multiple comparisons test. Differences were considered significant when p values were smaller than 0.05.

4. Conclusions

In summary, the results revealed that **K1** exhibited the highest levels of antiproliferative activity against Caco2 colon cancer cells and was selected for follow-up study. The cell cycle results demonstrated that **K1** induced time-dependent G_2/M arrest simultaneously with an increase in cells with sub-G_1 DNA content. Furthermore, the apoptosis data showed that **K1** induced cellular apoptosis as confirmed by annexin V/PI double staining, DNA fragmentation and chromatin condensation. Moreover, increased ROS generation and loss of MMP and caspase 3 activation supported these findings. Application of Trolox, a water-soluble analogue of vitamin E, diminished ROS production and loss of MMP and prevented cytotoxicity of **K1** in Caco2 cells. These results insinuate that **K1** may inhibit the growth of colon cancer cells by inducing apoptosis through ROS-mitochondrial pathway. Although many details of the effects of **K1** remain to be elucidated, the *in vitro* findings of the present study provide the basis for future *in vitro* and *in vivo* studies.

Acknowledgments

This study was supported (50%) by the project Medicínsky univerzitný park v Košiciach (MediPark, Košice) ITMS: 26220220185 (95%) supported by Operational Programme Research and Development (OP VaV-2012/2.2/08-RO) (Contract No. OPVaV/12/2013.) We would like also to thank the Slovak Grant Agency for Science (Grant No. 1/0322/14) for financial support of this work.

Author Contributions

Conceived and designed the experiments: JM, MK. Performed the experiments: MK, DD, MCh, MP. Synthesis of indole phytoalexins: MB. Analysed the data: JM, MK. Contributed reagents/materials/analysis tools: MK, DD, MCh, MP, LK, PU. Wrote the manuscript: JM, MK. All authors read and approved the final manuscript.

Conflicts of Interest

The authors declare no conflict of interest.

References

1. Han, B.; Li, X.; Yu, T. Cruciferous vegetables consumption and the risk of ovarian cancer: A meta-analysis of observational studies. *Diagn. Pathol.* **2014**, *9*, 1–7.
2. Tse, G.; Eslick, G.D. Cruciferous vegetables and risk of colorectal neoplasms: A systematic review and meta-analysis. *Nutr. Cancer* **2014**, *66*, 128–139.
3. Liu, B.; Mao, Q.; Wang, X.; Zhou, F.; Luo, J.; Wang, C.; Lin, Y.; Zheng, X.; Xie, L. Cruciferous vegetables consumption and risk of renal cell carcinoma: A meta-analysis. *Nutr. Cancer* **2013**, *65*, 668–676.
4. Aras, U.; Gandhi, Y.A.; Masso-Welch, P.A.; Morris, M.E. Chemopreventive and anti-angiogenic effects of dietary phenethyl isothiocyanate in an N-methyl nitrosourea-induced breast cancer animal model. *Biopharm. Drug Dispos.* **2013**, *34*, 98–106.
5. Abdull Razis, A.F.; Noor, N.M. Cruciferous vegetables: Dietary phytochemicals for cancer prevention. *Asian Pac. J. Cancer Prev.* **2013**, *14*, 1565–1570.
6. Izutani, Y.; Yogosawa, S.; Sowa, Y.; Sakai, T. Brassinin induces G1 phase arrest through increase of p21 and p27 by inhibition of the phosphatidylinositol 3-kinase signalling pathway in human colon cancer cells. *Int. J. Oncol* **2012**, *40*, 816–824.
7. Kim, S.M.; Park, J.H.; Kim, K.D.; Nam, D.; Shim, B.S.; Kim, S.H.; Ahn, K.S.; Choi, S.H. Brassinin induces apoptosis in PC-3 human prostate cancer cells through the suppression of PI3K/Akt/mTOR/S6K1 signalling cascades. *Phytother. Res.* **2014**, *28*, 423–431.
8. Smith, B.; Randle, D.; Mezencev, R.; Thomas, L.; Hinton, C.; Odero-Marah, V. Camalexin-induced apoptosis in prostate cancer cells involves alterations of expression and activity of lysosomal protease cathepsind. *Molecules* **2014**, *19*, 3988–4005.
9. Pedras, M.S.; Yaya, E.E.; Glawischnig, E. The phytoalexins from cultivated and wild crucifers: Chemistry and biology. *Nat. Prod. Rep.* **2011**, *28*, 1381–1405.
10. Jeandet, P.; Clement, C.; Courot, E.; Cordelier, S. Modulation of phytoalexin biosynthesis in engineered plants for disease resistance. *Int. J. Mol. Sci.* **2013**, *14*, 14136–14170.
11. Borriello, A.; Bencivenga, D.; Caldarelli, I.; Tramontano, A.; Borgia, A.; Zappia, V.; Della Ragione, F. Resveratrol: From basic studies to bedside. *Cancer Treat. Res.* **2014**, *159*, 167–184.
12. Kim, H.J.; Lim, J.S.; Kim, W.K.; Kim, J.S. Soyabean glyceollins: Biological effects and relevance to human health. *Proc. Nutr. Soc.* **2012**, *71*, 166–174.
13. Moody, C.J.; Roffey, J.R.; Stephens, M.A.; Stratford, I.J. Synthesis and cytotoxic activity of indolyl thiazoles. *Anticancer Drugs* **1997**, *8*, 489–499.
14. Yang, Q.; Wang, B.; Zang, W.; Wang, X.; Liu, Z.; Li, W.; Jia, J. Resveratrol inhibits the growth of gastric cancer by inducing G_1 phase arrest and senescence in a Sirt1-dependent manner. *PLoS One* **2013**, *8*, e70627.
15. Romagnolo, D.F.; Davis, C.D.; Milner, J.A. Phytoalexins in cancer prevention. *Front. Biosci.* **2012**, *17*, 2035–2058.
16. Yin, H.T.; Tian, Q.Z.; Guan, L.; Zhou, Y.; Huang, X.E.; Zhang, H. *In vitro* and *in vivo* evaluation of the antitumor efficiency of resveratrol against lung cancer. *Asian Pac. J. Cancer Prev.* **2013**, *14*, 1703–1706.

17. Mezencev, R.; Mojzis, J.; Pilatova, M.; Kutschy, P. Antiproliferative and cancer chemopreventive activity of phytoalexins: Focus on indole phytoalexins from crucifers. *Neoplasma* **2003**, *50*, 239–245.
18. Curillova, Z.; Kutschy, P.; Solcaniova, E.; Pilatova, M.; Mojzis, J.; Kovacik, V. Synthesis and antiproliferative activity of 1-methoxy-, 1-(α-D-ribofuranosyl)- and 1-(β-D-ribofuranosyl)brassenin B. *ARKIVOC* **2008**, *8*, 85–104.
19. Mezencev, R.; Kutschy, P.; Salayova, A.; Curillova, Z.; Mojzis, J.; Pilatova, M.; McDonald, J. Anticancer properties of 2-piperidyl analogues of the natural indole phytoalexin 1-methoxyspirobrassinol. *Chemotherapy* **2008**, *54*, 372–378.
20. Kutschy, P.; Salayova, A.; Curillova, Z.; Kozar, T.; Mezencev, R.; Mojzis, J.; Pilatova, M.; Balentova, E.; Pazdera, P.; Sabol, M.; *et al.* 2-(Substituted phenyl)amino analogs of 1-methoxyspirobrassinol methyl ether: synthesis and anticancer activity. *Bioorg. Med. Chem.* **2009**, *17*, 3698–3712.
21. Pilatova, M.; Ivanova, L.; Kutschy, P.; Varinska, L.; Saxunova, L.; Repovska, M.; Sarissky, M.; Seliga, R.; Mirossay, L.; Mojzis, J. *In vitro* toxicity of camalexin derivatives in human cancer and non-cancer cells. *Toxicol. In Vitro* **2013**, *27*, 939–944.
22. Takasugi, M.; Katsui, N.; Shirata, A. Isolation of 3 novel sulfur-containing phytoalexins from the Chinese-cabbage brassica-campestris L ssp pekinensis (cruciferae). *J. Chem Soc. Chem Commun.* **1986**, *14*, 1077–1078.
23. Mehta, R.G.; Liu, J.; Constantinou, A.; Thomas, C.F.; Hawthorne, M.; You, M.; Gerhuser, C.; Pezzuto, J.M.; Moon, R.C.; Moriarty, R.M. Cancer chemopreventive activity of brassinin, a phytoalexin from cabbage. *Carcinogenesis* **1995**, *16*, 399–404.
24. Csomos, P.; Zupko, I.; Rethy, B.; Fodor, L.; Falkay, G.; Bernath, G. Isobrassinin and its analogues: Novel types of antiproliferative agents. *Bioorg. Med. Chem. Lett.* **2006**, *16*, 6273–6276.
25. Pilatova, M.; Sarissky, M.; Kutschy, P.; Mirossay, A.; Mezencev, R.; Curillova, Z.; Suchy, M.; Monde, K.; Mirossay, L.; Mojzis, J. Cruciferous phytoalexins: Antiproliferative effects in T-Jurkat leukemic cells. *Leuk. Res.* **2005**, *29*, 415–421.
26. Monde, K.; Taniguchi, T.; Miura, N.; Kutschy, P.; Curillova, Z.; Pilatova, M.; Mojzis, J. Chiral cruciferous phytoalexins: Preparation, absolute configuration, and biological activity. *Bioorg. Med. Chem.* **2005**, *13*, 5206–5212.
27. Kutschy, P.; Sykora, A.; Curillova, Z.; Repovska, M.; Pilatova, M.; Mojzis, J.; Mezencev, R.; Pazdera, P.; Hromjakova, T. Glyoxyl analogs of indole phytoalexins: Synthesis and anticancer activity. *Collect. Czech. Chem. C* **2010**, *75*, 887–903.
28. Budovska, M.; Pilatova, M.; Varinska, L.; Mojzis, J.; Mezencev, R. The synthesis and anticancer activity of analogs of the indole phytoalexins brassinin, 1-methoxyspirobrassinol methyl ether and cyclobrassinin. *Bioorg. Med. Chem.* **2013**, *21*, 6623–6633.
29. Ocenas, P.; Tomasova, L.; Kutschy, P.; Pazdera, P.; Mojzis, J.; Pilatova, M. Spirocyclisation of phytoalexin 1-methoxybrassinin in the presence of Grignard reagents. *Chem. Pap.* **2013**, *67*, 631–642.

30. Chripkova, M.; Drutovic, D.; Pilatova, M.; Mikes, J.; Budovska, M.; Vaskova, J.; Brogginy, M.; Mirossay, L.; Mojzis, J. Brassinin and its derivatives as potential anticancer agents. *Toxicol. In Vitro* **2014**, *28*, 907–915.
31. Banerjee, T.; Duhadaway, J.B.; Gaspari, P.; Sutanto-Ward, E.; Munn, D.H.; Mellor, A.L.; Malachowski, W.P.; Prendergast, G.C.; Muller, A.J. A key *in vivo* antitumor mechanism of action of natural product-based brassinins is inhibition of indoleamine 2,3-dioxygenase. *Oncogene* **2008**, *27*, 2851–2857.
32. Patel, P.B.; Thakkar, V.R. L-Carvone induces p53, caspase 3 mediated apoptosis and inhibits the migration of breast cancer cell lines. *Nutr. Cancer* **2014**, *66*, 453–462.
33. Zhao, B.; Li, X. Altholactone induces reactive oxygen species-mediated apoptosis in bladder cancer T24 cells through mitochondrial dysfunction, MAPK-p38 activation and Akt suppression. *Oncol. Rep.* **2014**, *31*, 2769–2775.
34. Mezencev, R.; Updegrove, T.; Kutschy, P.; Repovska, M.; McDonald, J.F. Camalexin induces apoptosis in T-leukemia Jurkat cells by increased concentration of reactive oxygen species and activation of caspase-8 and caspase-9. *J. Nat. Med.* **2011**, *65* , 488–499.
35. Smith, B.A.; Neal, C.L.; Chetram, M.; Vo, B.; Mezencev, R.; Hinton, C.; Odero-Marah, V.A. The phytoalexin camalexin mediates cytotoxicity towards aggressive prostate cancer cells via reactive oxygen species. *J. Nat. Med.* **2013**, *67*, 607–618.
36. Circu, M.L.; Aw, T.Y. Reactive oxygen species, cellular redox systems, and apoptosis. *Free Radic. Biol. Med.* **2010**, *48*, 749–762.
37. Jemal, A.; Bray, F.; Center, M.M.; Ferlay, J.; Ward, E.; Forman, D. Global cancer statistics. *CA Cancer J. Clin.* **2011**, *61*, 69–90.
38. Long, X.; Fan, M.; Bigsby, R.M.; Nephew, K.P. Apigenin inhibits antiestrogen-resistant breast cancer cell growth through estrogen receptor-alpha-dependent and estrogen receptor-alpha-independent mechanisms. *Mol. Cancer Ther.* **2008**, *7*, 2096–2108.
39. ElAttar, T.M.; Virji, A.S. Modulating effect of resveratrol and quercetin on oral cancer cell growth and proliferation. *Anticancer Drugs* **1999**, *10*, 187–193.
40. Baron-Menguy, C.; Bocquet, A.; Guihot, A.L.; Chappard, D.; Amiot, M.J.; Andriantsitohaina, R.; Loufrani, L.; Henrion, D. Effects of red wine polyphenols on postischemic neovascularization model in rats: Low doses are proangiogenic, high doses anti-angiogenic. *FASEB J.* **2007**, *21*, 3511–3521.
41. Choi, E.J.; Kim, G.H. Antiproliferative activity of daidzein and genistein may be related to ERalpha/c-erbB-2 expression in human breast cancer cells. *Mol. Med. Rep.* **2013**, *7*, 781–784.
42. Downing, K.H.; Nogales, E. Tubulin and microtubule structure. *Curr. Opin. Cell. Biol.* **1998**, *10*, 16–22.
43. Perez, E.A. Microtubule inhibitors: Differentiating tubulin-inhibiting agents based on mechanisms of action, clinical activity, and resistance. *Mol. Cancer Ther.* **2009**, *8*, 2086–2095.
44. Jordan, M.A.; Wilson, L. Microtubules as a target for anticancer drugs. *Nat. Rev. Cancer* **2004**, *4*, 253–265.
45. Kang, J.; Pervaiz, S. Mitochondria: Redox metabolism and dysfunction. *Biochem. Res. Int.* **2012**, *2012*, 896751.

46. Green, D.R., Apoptotic pathways: Ten minutes to dead. *Cell* **2005**, *121*, 671–674.
47. Erejuwa, O.O.; Sulaiman, S.A.; Wahab, M.S. Effects of honey and its mechanisms of action on the development and progression of cancer. *Molecules* **2014**, *19*, 2497–2522.
48. Villena, J.; Madrid, A.; Montenegro, I.; Werner, E.; Cuellar, M.; Espinoza, L. Diterpenylhydroquinones from natural ent-labdanes induce apoptosis through decreased mitochondrial membrane potential. *Molecules* **2013**, *18*, 5348–5359.
49. Yang, T.; Li, M.H.; Liu, J.; Huang, N.; Li, N.; Liu, S.N.; Liu, Y.; Zhang, T.; Zou, Q.; Li, H. Benzimidazole derivative, BMT-1, induces apoptosis in multiple myeloma cells *via* a mitochondrial-mediated pathway involving H+/K+-ATPase inhibition. *Oncol. Rep.* **2014**, doi:10.3892/or.2014.3122.
50. Thangam, R.; Senthilkumar, D.; Suresh, V.; Sathuvan, M.; Sivasubramanian, S.; Pazhanichamy, K.; Gorlagunta, P.K.; Kannan, S.; Gunasekaran, P.; Rengasamy, R.; *et al.* Induction of ROS-dependent mitochondria-mediated intrinsic apoptosis in MDA-MB-231 cells by glycoprotein from codium decorticatum. *J. Agric. Food Chem.* **2014**, *62*, 3410–3421.
51. Duan, D.; Zhang, B.; Yao, J.; Liu, Y.; Fang, J. Shikonin targets cytosolic thioredoxin reductase to induce ROS-mediated apoptosis in human promyelocytic leukemia HL-60 cells. *Free Radic. Biol. Med.* **2014**, *70*, 182–193.
52. Gaspari, P.; Banerjee, T.; Malachowski, W.P.; Muller, A.J.; Prendergast, G.C.; DuHadaway, J.; Bennett, S.; Donovan, A.M. Structure-activity study of brassinin derivatives as indoleamine 2,3-dioxygenase inhibitors. *J. Med. Chem.* **2006**, *49*, 684–692.
53. Kutschy, P.; Sabol, M.; Maruskova, R.; Curillova, Z.; Dzurilla, M.; Geci, I.; Alfoldi, J.; Kovacik, V. A linear synthesis of 1-(beta-D-glucopyranosyl)brassinin, -brassenin A, -brassenin B and 9-(beta-D-glucopyranosyl)-cyclobrassinin. *Collect. Czech. Chem. Commun.* **2004**, *69*, 850–866.
54. Budovska, M.; Kutschy, P.; Kozar, T.; Gondova, T.; Petrovaj, J. Synthesis of spiroindoline phytoalexin (S)-(−)-spirobrassinin and its unnatural (R)-(+)-enantiomer. *Tetrahedron* **2013**, *69*, 1092–1104.
55. Ivanova, L.; Varinska, L.; Pilatova, M.; Gal, P.; Solar, P.; Perjesi, P.; Smetana, K., Jr.; Ostro, A.; Mojzis, J. Cyclic chalcone analogue KRP6 as a potent modulator of cell proliferation: An *in vitro* study in HUVECs. *Mol. Biol. Rep.* **2013**, *40*, 4571–4580.
56. Mosmann, T. Rapid colorimetric assay for cellular growth and survival: Application to proliferation and cytotoxicity assays. *J. Immunol. Meth.* **1983**, *65*, 55–63.
57. Limame, R.; Wouters, A.; Pauwels, B.; Fransen, E.; Peeters, M.; Lardon, F.; de Wever, O.; Pauwels, P. Comparative analysis of dynamic cell viability, migration and invasion assessments by novel real-time technology and classic endpoint assays. *PLoS One* **2012**, *7*, e46536.
58. Ke, N.; Wang, X.; Xu, X.; Abassi, Y.A. The xCELLigence system for real-time and label-free monitoring of cell viability. *Methods Mol. Biol.* **2011**, *740*, 33–43.

Sample Availability: Sample of the compound K1 is available from the authors.

Camalexin-Induced Apoptosis in Prostate Cancer Cells Involves Alterations of Expression and Activity of Lysosomal Protease Cathepsin D

Basil Smith, Diandra Randle, Roman Mezencev, LeeShawn Thomas, Cimona Hinton and Valerie Odero-Marah

Abstract: Camalexin, the phytoalexin produced in the model plant *Arabidopsis thaliana*, possesses antiproliferative and cancer chemopreventive effects. We have demonstrated that the cytostatic/cytotoxic effects of camalexin on several prostate cancer (PCa) cells are due to oxidative stress. Lysosomes are vulnerable organelles to Reactive Oxygen Species (ROS)-induced injuries, with the potential to initiate and or facilitate apoptosis subsequent to release of proteases such as cathepsin D (CD) into the cytosol. We therefore hypothesized that camalexin reduces cell viability in PCa cells via alterations in expression and activity of CD. Cell viability was evaluated by MTS cell proliferation assay in LNCaP and ARCaP Epithelial (E) cells, and their respective aggressive sublines C4-2 and ARCaP Mesenchymal (M) cells, whereby the more aggressive PCa cells (C4-2 and ARCaPM) displayed greater sensitivity to camalexin treatments than the lesser aggressive cells (LNCaP and ARCaPE). Immunocytochemical analysis revealed CD relocalization from the lysosome to the cytosol subsequent to camalexin treatments, which was associated with increased protein expression of mature CD; p53, a transcriptional activator of CD; BAX, a downstream effector of CD, and cleaved PARP, a hallmark for apoptosis. Therefore, camalexin reduces cell viability via CD and may present as a novel therapeutic agent for treatment of metastatic prostate cancer cells.

Reprinted from *Molecules*. Cite as: Smith, B.; Randle, D.; Mezencev, R.; Thomas, L.; Hinton, C.; Odero-Marah, V. Camalexin-Induced Apoptosis in Prostate Cancer Cells Involves Alterations of Expression and Activity of Lysosomal Protease Cathepsin D. *Molecules* **2014**, *19*, 3988-4005.

1. Introduction

Among cancer-related deaths in men, prostate cancer is the second-leading cause in the United States despite recently observed decrease in mortality [1]. Once cancers metastasize to other organs, a majority of patients die from their tumors as opposed to other causes [2], and this is due in part to the tumor ultimately becoming androgen-independent within a median of 18 to 24 months after castration [3]. Current treatments have proved inadequate in controlling prostate cancer, and search for novel therapeutic agents for the management of this disease has become a priority for researchers. The discoveries of the chemopreventive and chemotherapeutic properties of phytochemicals have generated considerable interest among cancer researchers. Camalexin (Figure 1) is an indole phytoalexin produced in various cruciferous plants upon exposure to environmental stress and plant pathogens [4–7].

Figure 1. The structural formula of camalexin.

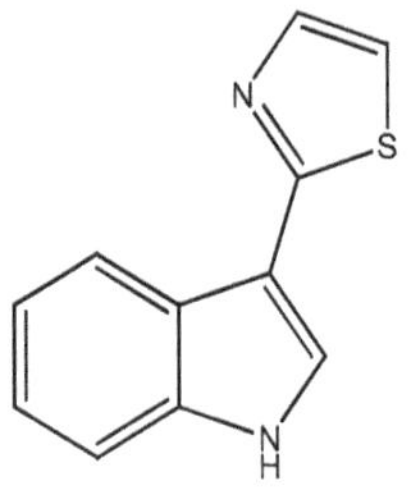

It has been shown to possess moderate antifungal and bacteriostatic, as well as antiproliferative and cancer chemopreventive properties [7,8]. Our laboratory has shown that camalexin induced apoptosis in prostate cancer cells (PCa) through the generation of Reactive Oxygen Species (ROS), and that the more aggressive prostate cancer cells with higher levels of endogenous ROS displayed greater sensitivity to camalexin treatments evidenced by decreased viability and increased apoptosis as compared to the less aggressive prostate cancer cells, while normal epithelial cells were unaffected [9]. The generation of intracellular ROS activates several signal transduction pathways leading to inflammation, cell cycle progression, apoptosis, migration, and invasion in cancer [10]. Thus, excessive production of ROS or inadequacy in a cell's antioxidant defense system (or both) induces oxidative stress with consequent initiation of cellular processes associated with initiation and development of many cancers including prostate cancer [11]. Hydrogen peroxide has been shown to increase progressively in the lesser aggressive LNCaP prostate cancer cells to its more aggressive sublines C4, C4-2 and C4-2B, with concomitant increase in tumorigenic and metastatic potential [12]. Hence, several studies were conducted on cancer treatments focusing on ROS inhibition with limited success [13]. Conversely, in many preclinical models of cancer, excessive ROS generation triggers pro-apoptotic pathways with subsequent apoptosis [14]. Thus, the therapeutic use of pro-oxidants to target prostate cancer cells is gaining traction in cancer research [15]. Therefore, compounds that are capable of inducing ROS especially in cancer cells warrant further investigation.

Apoptosis induced by oxidative stress has been observed in several studies and includes activation of cell cycle genes such as p53 and p21 [16–18]. Key players in the regulation of apoptotic cell death are the mitochondria, where they coordinate caspase activation through release of cytochrome C. Lysosomes, the major cell digestive organelles containing numerous hydrolases that degrade intracellular and extracellular materials delivered, has been implicated in the regulation of cell death [19]. Common to both mitochondria and lysosomes is their increased membrane permeability resulting in release of their contents in the early phase of apoptosis [20]. Varying stimuli induces lysosomal membrane permeabilization (LMP) with translocation of enzymes from lysosomal compartment to cytosol, such as oxidative stress [21,22], TNF-α [23] and p53 [24]. Among the hydrolytic enzymes released, the cathepsins were found to participate in apoptosis following their release into the cytosol. In oxidative stress-induced apoptosis, release of cathepsin D (CD) from lysosomes and induction of apoptosis could be prevented by the antioxidant α-tocopherol [21]. Deiss *et al.* have shown that CD anti-sense RNA protected Hela cells from IFN-γ- and Fas-induced cell death [25]. Additionally, it has been demonstrated that p53 accumulates rapidly after oxidative stress and has two binding sites located at the CD promoter gene, and that CD participates in p53-dependent apoptosis [26]. The role of CD in apoptosis has been linked to lysosomal release of mature CD into the cytosol, in turn leading to mitochondrial release of cytochrome c into the cytosol [27–30], activation of pro-caspases-9 and -3 [31,32], *in vitro* cleavage of Bid [33], or Bax activation independent of Bid cleavage [34]. Pepstatin A, an aspartate protease inhibitor of CD, could partially delay apoptosis induced by oxidative stress [28–31], or even when it was co-microinjected with CD [32]. Therefore, CD could play a key role in apoptosis mediated by its catalytic activity.

Previously, we have shown that prostate cancer cells expressing high levels of ROS (C4-2 and ARCaPE cells stably overexpressing Snail) displayed further increase in ROS upon camalexin treatment which led to decreased viability and increased apoptosis through activation of caspase-3 and -7 [9]. Interestingly, the less aggressive cells (LNCaP and ARCaPE with empty vector) were less responsive to camalexin and could be induced to be more responsive by addition of exogenous hydrogen peroxide [9] thus showing that camalexin mediates its response via ROS. In this current study, we have dissected the mechanism of camalexin-induced (apoptosis) decreased-cell viability further and shown for the first time that it is mediated through CD. Hence, in our experiments, we utilized two prostate cancer progression models, LNCaP/C4-2 and ARCaPE/ARCaPM and found that camalexin reduced cell viability in PCa cells that involved relocation of CD from lysosomes to cytosol, and increased protein expression of p53, mature CD, Bax, and cleaved PARP. Moreover, pepstatin A, the peptide inhibitor of CD activity was able to reverse the effects of camalexin. Targeting lysosomal proteases such as CD may therefore provide a great therapeutic potential in especially metastatic prostate cancer.

2. Results and Discussion

2.1. Camalexin Treatments Decreases Cell Proliferation in the More Aggressive Prostate Cancer Cells as Compared to the Lesser Aggressive Cells

Previously, we have shown that camalexin was more potent in reducing cell viability in C4-2 as compared to LNCaP cells suggesting that camalexin was more potent in the more aggressive cell line [9]. Confirming these results utilizing CellTiter 96® AQueous One Solution Cell Proliferation Assay (MTS assay) we observed that at day 0 for both LNCaP and C4-2 cells, viability was unaffected but on day 3, only 50 μM camalexin decreased cell viability in LNCaP by approximately 40% ± 2% ($p < 0.01$), while camalexin decreased C4-2 cell viability by approximately 40% ± 2% ($p < 0.001$) for 10 and 25 μM camalexin and 30% ± 5% ($p < 0.01$) for 50 μM camalexin, respectively (Figure 2A). We also tested the effect of camalexin on ARCaPE (epithelial) and ARCaPM (mesenchymal) cell lines that are derived from the same parental ARCaP but represent an EMT progression model [35,36]. CellTiter 96® AQueous One Solution Cell Proliferation Assay (MTS assay) was utilized and the results for day 0 and day 3 represented. For day 0, both ARCaPE and ARCaPM cells showed no change in viability as expected, however, on day 3, camalexin treatment of

ARCaPE decreased cell viability by approximately 23% ($p < 0.05$) at 25 μM treatment only, while for ARCaPM cells 10, 25 and 50 μM decreased viability by approximately 22% ± 5% ($p < 0.05$), 28% ± 1% ($p < 0.01$) and 47% ± 1% ($p < 0.001$), respectively (Figure 2B). Therefore, we show that the more aggressive C4-2 and ARCaPM cells displayed greater sensitivity to camalexin treatment than the lesser aggressive LNCaP and ARCaPE cells.

Figure 2. The more aggressive C4-2 and ARCaPM prostate cancer cells are more sensitive to camalexin as compared to LNCaP and ARCaPE cells. Viability was determined using MTS proliferation assay at day 0 or day 3 for LNCaP and C4-2 (**A**), and ARCaPE and ARCaPM (**B**) cells treated with 10, 25 and 50 μM camalexin. Statistical analysis was done using ANOVA and Tukey's Multiple Comparison as Post Hoc (* $p < 0.05$, ** $p < 0.01$, *** $p < 0.001$). Values were normalized to untreated controls and expressed as mean ± S.E.M (N = 3).

A

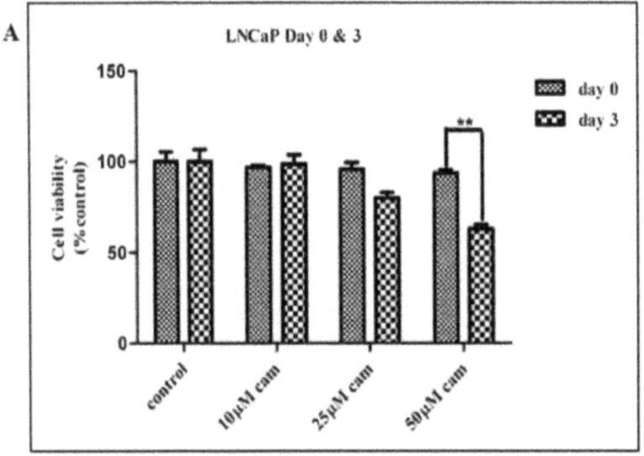

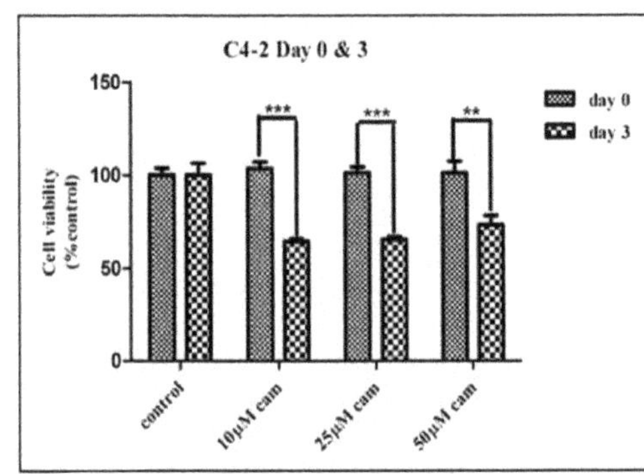

B

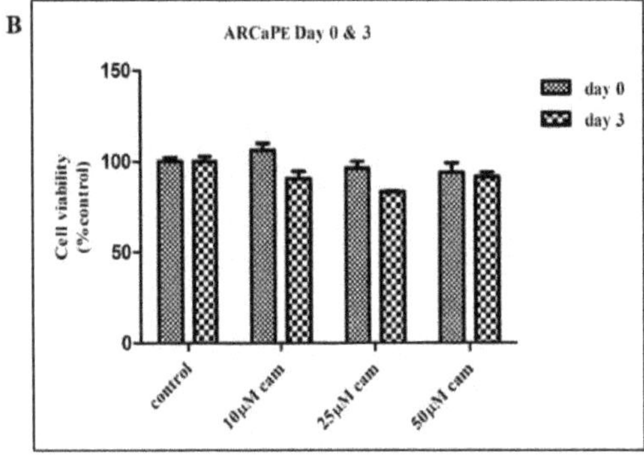

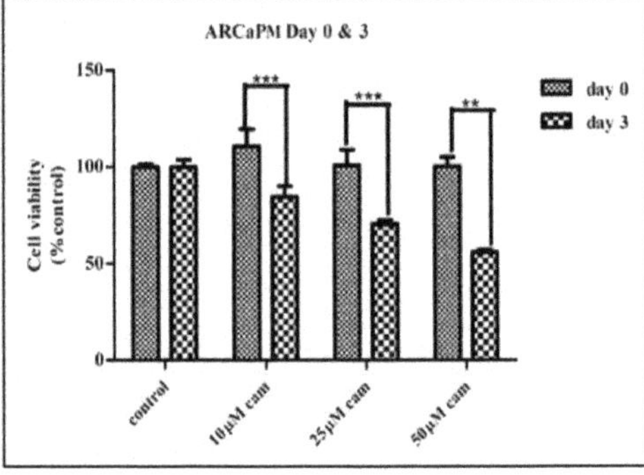

2.2. Camalexin Treatment of Prostate Cancer Cells Increases Protein Expression of the Lysosomal Protease CD, Bax and p53 Transcription Factor

Previously, our laboratory has shown that camalexin treatment of prostate cancer cells produces oxidative stress-induced apoptosis with consequent increased caspase 3 activity and PARP cleavage [9]. Survey of the literature revealed that several agents and molecules of endogenous origin can induce lysosomal membrane permeabilization, among which ROS are the most important [37,38]. As a result, CD is translocated from the lysosomes to the cytosol and triggers a rapid change in Bax conformation together with insertion of this protein to the outer mitochondrial membrane [34]. CD and BAX protein expression was analyzed by western blot analysis in camalexin-treated (10, 25 and 50 μM) ARCaPE, ARCaPM, LNCaP and C4-2 cells. Camalexin treatments of LNCaP cells showed a tendancy towards increased protein expression of CD at the 10, 25 and 50 μM camalexin treatments, whilst Bax showed no significance in protein expression levels of treated *versus* untreated control cells (Figure 3A). However, for C4-2 cells, camalexin treatments significantly increased protein expression of CD (25 and 50 μM treatments, *** $p < 0.001$, ** $p < 0.01$ respectively) and BAX protein (10, 25 and 50 μM treatments, ** $p < 0.01$, *** $p < 0.001$, *** $p < 0.001$ respectively) as compared to the untreated control cells (Figure 3B). In ARCaPE cells, no significant levels of CD protein expression were noted for untreated control, 10 and 25 μM camalexin treatments, while at 50 μM treatment there was a tendency toward increased expression although there was no statistical significance noted (Figure 3C). At 25 μM camalexin treatment only, significant increase in Bax protein expression was noted (* $p < 0.05$) in ARCaPE cells (Figure 3C), although at 50 μM camalexin treatment there is a strong tendency towards increased expression as compared to untreated control cells (Figure 3C). In ARCaPM cells, however, 10 and 25 μM camalexin treatments induced significant increased protein expression of CD (*** $p < 0.001$) and BAX (10 and 50 μM treatments, ** $p < 0.01$, * $p < 0.05$ respectively) *vs.* the untreated control cells (Figure 3D). At 25 μM camalexin treatment of ARCaPM cells, the data showed increased Bax protein expression level when compared to the untreated control cells (Figure 3D). Wu *et al.*, has demonstrated that p53 accumulates in cells subsequent to oxidative stress and can bind the CD promoter gene, and thus CD can participate in p53-dependent apoptosis [26]. Because we saw increased CD protein expression levels subsequent to camalexin treatments in our cells we decided to further investigate whether its trans-gene p53 may also be involved in the induced stress response. Western blot analysis showed that LNCaP and C4-2 cells at 25 and 50 μM camalexin treatments produced increased p53 protein expression when compared with the untreated control cells (Figure 3A,B). ARCaPE cell lysates did not display significant alterations in expression of p53 protein subsequent to camalexin treatments (Figure 3C). However, ARCaPM cell lysates showed significant increase in p53 protein expression at 10, 25 and 50 μM camalexin-treatments (*** $p < 0.001$, * $p < 0.05$, and ** $p < 0.01$) respectively (Figure 3D). Overall, our data shows that camalexin increases p53, CD and BAX expression, which are proteins involved in apoptosis and the effect is more pronounced in the ARCaPM and C4-2 cells as compared to ARCaPE and LNCaP cells, respectively.

Figure 3. Camalexin leads to increased expression of CD, Bax, and p53 protein in prostate cancer cell lines. Western Blot analysis was performed to examine CD, Bax and p53 protein expression following exposure of LNCaP (**A**) and C4-2 (**B**), ARCaPE (**C**), ARCaPM (**D**) cells to various doses of camalexin for 3 days. β-actin was used as loading control. Densitometry was performed for each Western Blot using Image J Software (National Institutes of Health). Data are representative of at least 3 independent experiments and statistical analysis was done using ANOVA and Tukey's Multiple Comparison as Post Hoc (* $p < 0.05$, ** $p < 0.01$, *** $p < 0.001$).

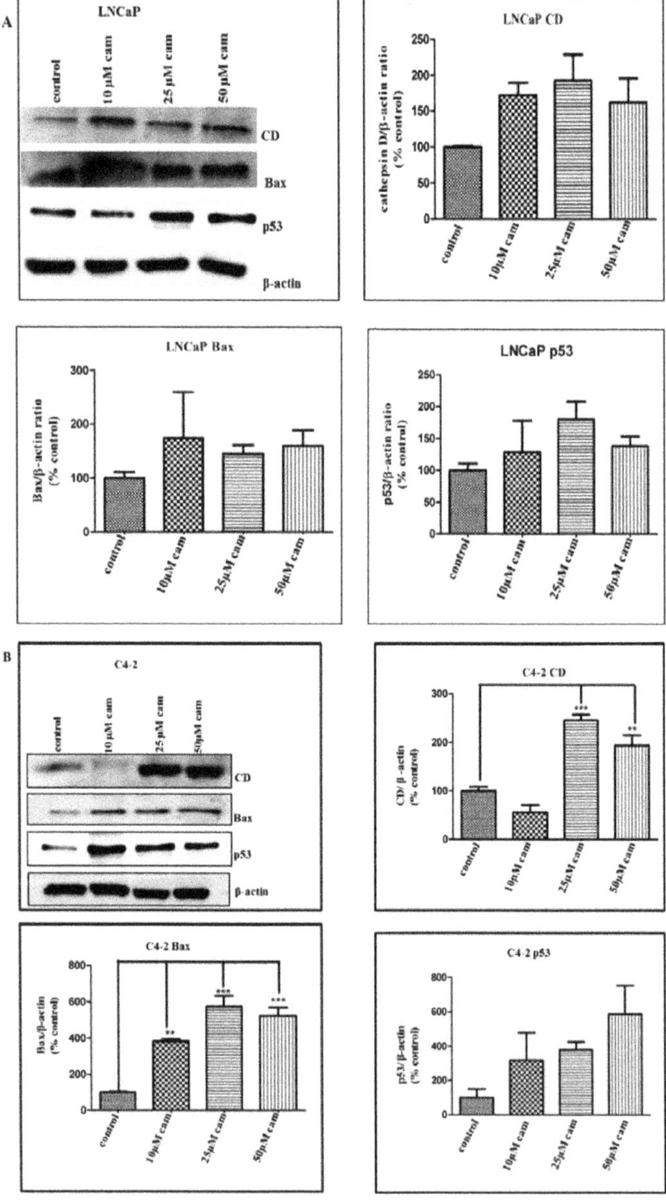

Figure 3. *Cont.*

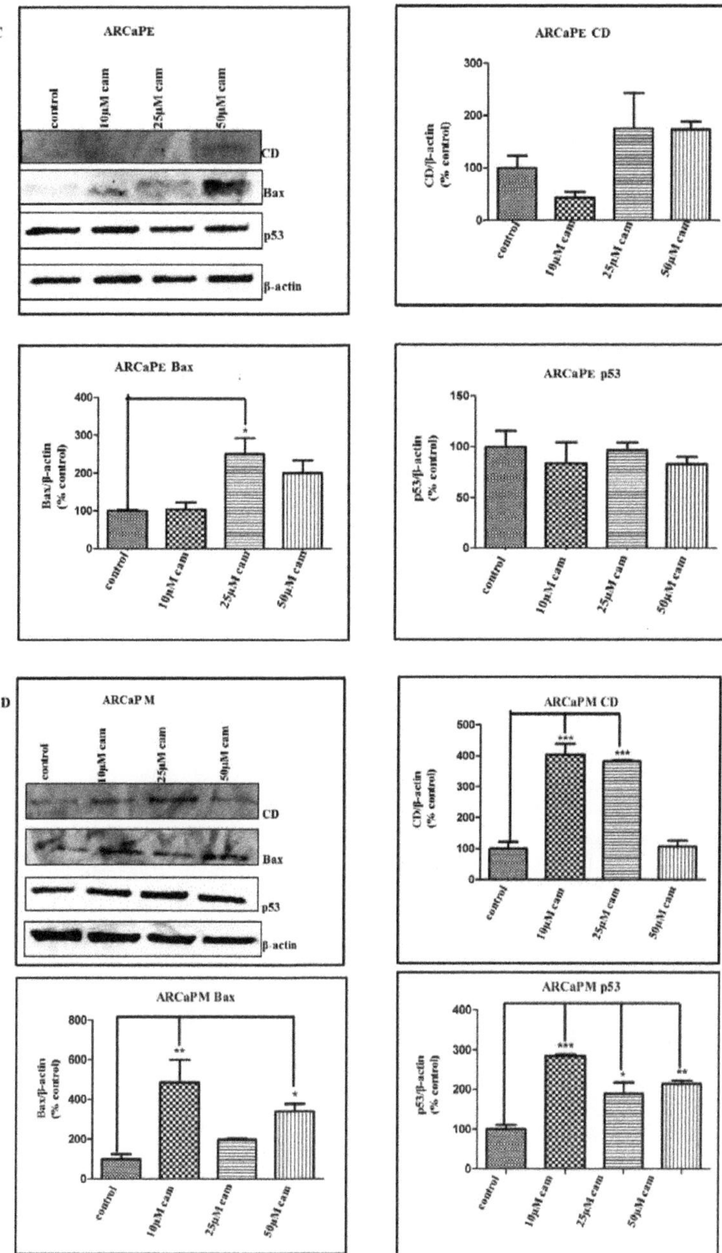

2.3. *Camalexin Shifts Subcellular Localization of CD from the Lysosome to the Cytosol*

A significant amount of hydrogen peroxide is produced in the mitochondria subsequent to oxidative stress and is able to diffuse into the lysosomes where ferruginous material delivered by autophagy is accumulated, resulting in Fenton-type reaction and production of highly reactive hydroxyl radicals [39]. These radicals cause lipid peroxidation of lysosomal membrane and subsequent leakage of proteases into the cytosol. In this way oxidative stress from the mitochondria is amplified with the help of redox-active iron-rich lysosomes [40,41]. The role of CD in apoptosis has been defined previously [25,26] and using immunocytochemistry, we analyzed CD localization within cells following camalexin treatments. We detected granular staining of CD in C4-2 untreated control cells depicting its location within intact lysosomes (Figure 4A). After 24 h camalexin treatment only, fluorescence staining for CD was more diffuse, indicating its translocation from the lysosomes to the cytosol of C4-2 cells (Figure 4B). Co-treatment of camalexin with the ROS scavenger, N-acetylcysteine (NAC), produced increased granular staining almost similar to the untreated cells suggesting that it could reverse in part, the effects of camalexin (Figure 4C). Therefore, camalexin not only increases CD expression but also leads to its leakage from the lysosome to the cytoplasm through increased ROS production.

Figure 4. Camalexin exposure in C4-2 cells induces release of lysosomal enzymes to the cytosol. CD localization was analyzed by Immunofluorescence analysis in C4-2 untreated (**A**), treated with 25 μM camalexin (**B**) or 25 μM camalexin and 10 mM NAC for 24 h (**C**). CD in untreated C4-2 cells appears punctuated in intact lysosomes while camalexin treatment led to more diffuse staining depicting CD release into cytosol. Co-treatment with NAC increased punctuated staining similar to untreated control cells.

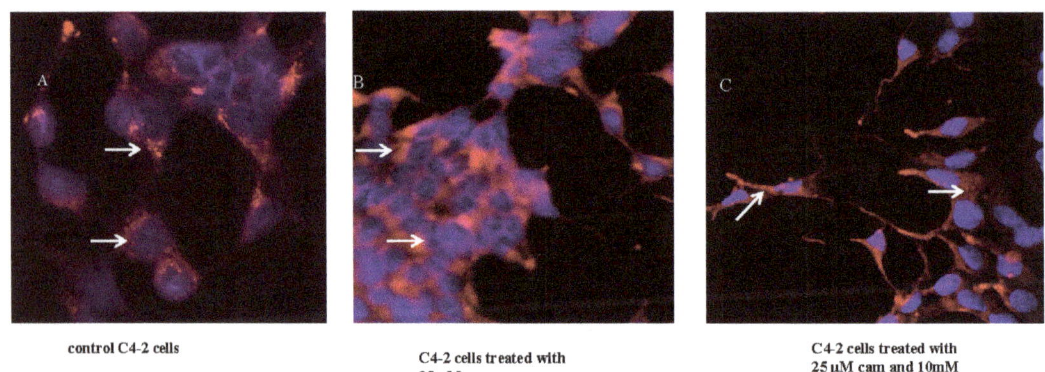

2.4. *NAC Abrogates Camalexin-Mediated Increased p53, Bax and CD Protein Expression in Prostate Cancer Cells*

Oxidative stress has been shown to be responsible for lysosomal membrane destabilization with consequent permeabilization and leakage of proteases into the cytosol [40]. The ROS scavenger NAC was added to camalexin-treated C4-2 cells and p53, Bax and CD protein expression assessed *via* western blot analysis. The results strongly indicated decreased expression of p53 and Bax, and significant abrogation of CD protein expression (** $p < 0.01$) by co-treatment of cells with NAC (Figure 5). Hence, camalexin utilizes oxidative stress to induce p53, Bax and CD protein expression.

Figure 5. NAC abrogates the effects of camalexin-mediated p53, Bax and CD protein expression. C4-2 cells were treated with 25 μM camalexin or 25 μM camalexin plus 10 mM NAC for 3 days. Western blot analysis was done to examine p53, Bax and mature CD protein expression. Treatment for 3 days indicated that NAC could inhibit p53, Bax and CD protein expression when combined with 25 μM camalexin. β-Actin was used as loading control and densitometry of protein expression assessed for treated *versus* untreated control cells. Data are representative of at least 3 independent experiments. Statistical analysis was done using ANOVA and Tukey's Multiple Comparison as *Post Hoc* (** $p < 0.01$).

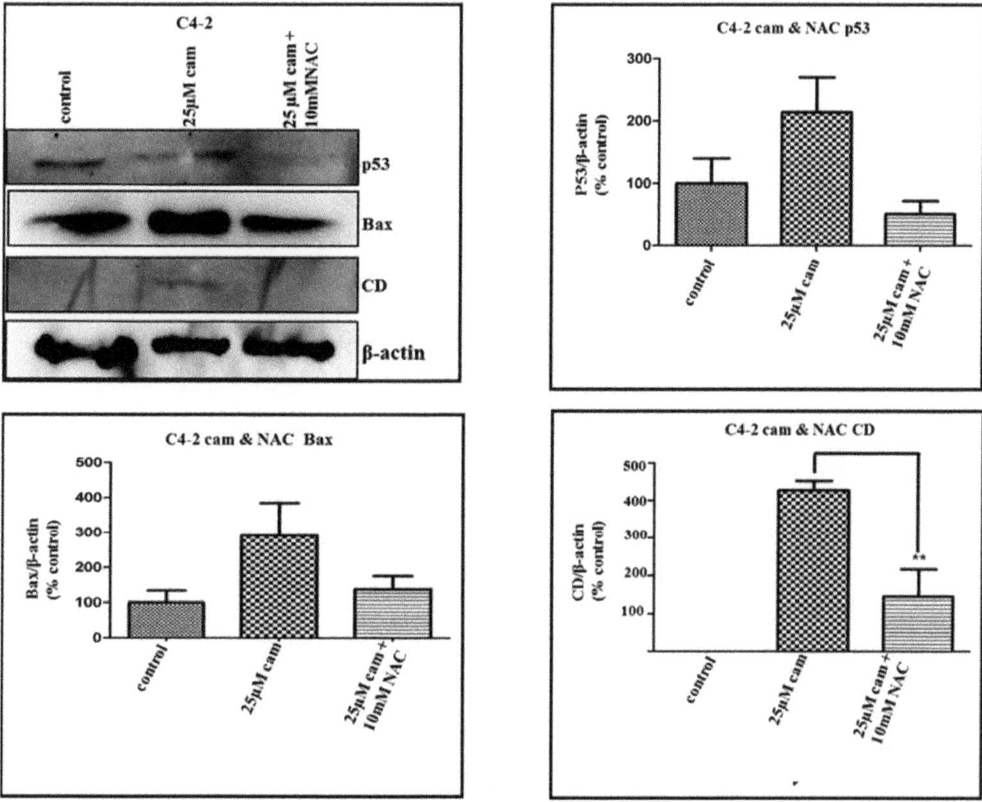

2.5. Pepstatin A, a CD Inhibitor, Antagonizes Camalexin-Mediated Decrease in Cell Viability, and Its Promotion of Pro-Apoptotic Protein Expression

We utilized the potent peptide inhibitor of CD activity, pepstatin A (Pep A), to confirm the role of this protease in camalexin-induced apoptosis in prostate cancer cells. We observed after 3 days that camalexin-mediated decrease in cell viability was blocked significantly by co-treatment with 100 μM Pep A plus 25 or 50 μM camalexin in C4-2 cells (Figure 6A). Additionally, treatment of C4-2 cells with Pep A significantly antagonized camalexin-mediated increase in Bax and PARP cleavage protein expression without significantly affecting p53 levels (Figure 6B). Hence, Pep A functions downstream of CD and our studies imply that p53 must be acting upstream of CD since it is not significantly affected by Pep A.

Figure 6. Pepstatin A abrogates the effects of camalexin-mediated decrease in cell viability and PARP cleavage. C4-2 cells were treated with 25 and 50 μM camalexin only, or 25 and 50 μM camalexin plus 100 μM Pep A for 3 days and cell viability assayed using the MTS proliferation assay **(A)**. Western blot analysis was done to examine p53, CD and cleaved PARP protein expression in untreated, camalexin-treated and camalexin plus Pep A- treated C4-2 cells, along with the densitometry of protein expression (**B**). Treatment for 3 days indicated that Pep A could significantly inhibit camalexin-mediated decrease in cell viability and increase in Bax and cleaved PARP protein expression, but does not significantly alter p53 and CD protein expression. Statistical analysis was done using ANOVA and Tukey's Multiple Comparison as a *Post Hoc* Test. Values were expressed as mean ± S.E.M normalized to untreated controls, and 25 μM camalexin *vs.* 25 μM camalexin + 100 μM Pep A (* $p < 0.05$, ** $p < 0.01$). (N = 3). β-Actin was used as loading control and data are representative of at least three independent experiments.

A

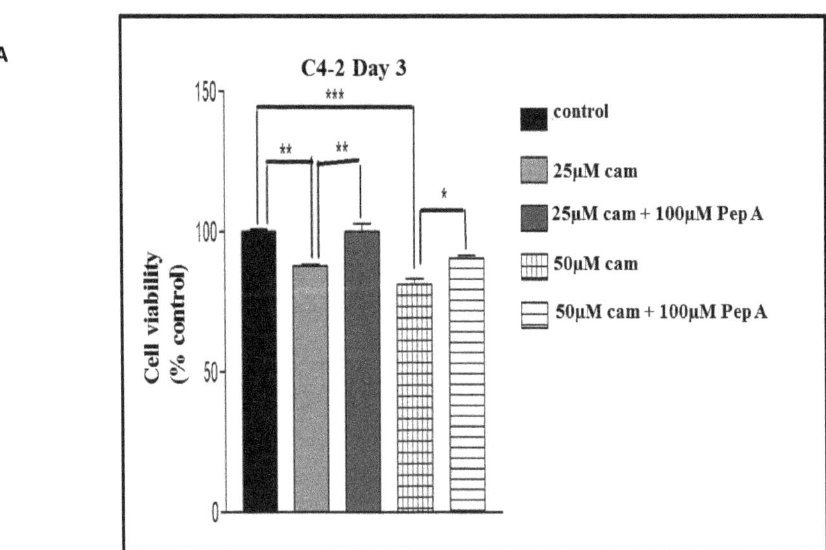

Figure 6. *Cont.*

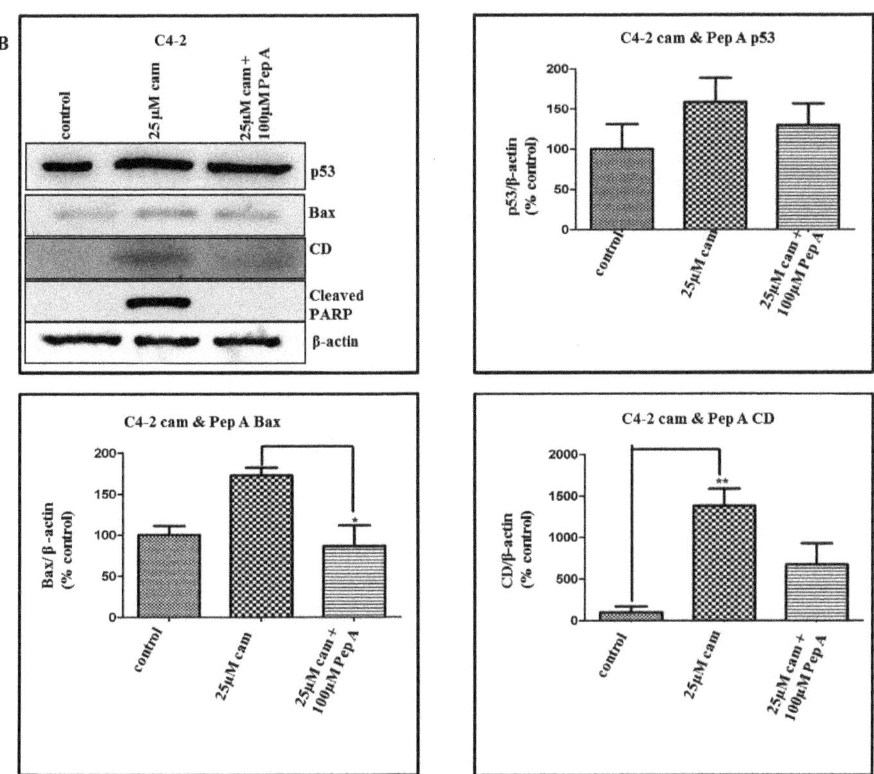

2.6. Discussion

The lysosomal pathway of apoptosis involves partial lysosomal membrane permeabilization (LMP) with subsequent release of cathepsins into the cytosol [19,42]. Resistance of cancer cells to chemotherapy is often considered to involve decreased sensitivity to apoptosis, and therefore alternate pathways to induce cell death in resistant tumors are highly demanded [43]. Hence, lysosomal cell death represents an alternate pathway where released cathepsins can activate apoptosis or execute apoptosis-like cell death independent of the caspases [44–46]. We have previously shown that camalexin decreases cell viability in prostate cancer cells by induction of oxidative stress leading to apoptosis [9]. Similarly, Mezencev *et al.* has shown that camalexin-induced apoptosis in Jurkat T leukemia cells occurred via increased ROS and involve mitochondrial superoxide generation evidenced by increased MitoSOX™ staining [47]. In this report we show that camalexin may also utilize the lysosomal pathway involving CD. We showed that camalexin decreased cell viability in ARCaPE and ARCaPM prostate cancer cell lines in a concentration-dependent manner and was more potent in ARCaPM cells. This would make camalexin a potential therapeutic agent for aggressive prostate cancer and we have previously shown that it does not affect normal prostate epithelial cells [9]. Camalexin-induced cell death in ARCaPE, ARCaPM, LNCaP and C4-2 prostate cancer cells is accompanied by Lysosomal Membrane Permeability as evidenced by increased CD protein expression in treated *vs.* untreated control cells. Additionally, immunocytochemical analysis of camalexin-treated C4-2 cells displayed diffuse staining for CD *vs.* punctuated staining in lysosomes for untreated control, indicating this protease release into the cytosol. The addition of NAC to the camalexin-treated cells, displayed increased punctated staining verifying increased CD retention within lysosomes and hence lysosomal membrane stabilization. CD has been discovered as a key mediator of apoptosis induced by several apoptotic agents including oxidative stress [27–31,48]. Its role in apoptosis has been linked to lysosomal release of its mature form (33 kDa) into the cytosol, activation of Bax [34], and in turn leading to mitochondrial release of cytochrome c into the cytosol [27–30].

Our data reveals increased Bax protein expression in camalexin-treated prostate cancer cell lines. Mounting evidence points to Bax and other pro-apoptotic family members as central regulators of the release of proteins from the mitochondrial intermembrane space. Bax overexpression in cells or the addition of purified recombinant Bax directly to isolated mitochondria triggers release of cytochrome c into the cytosol [49,50] with subsequent activation of procaspases 9 and 3 [28]. Immuno-electron microscopy has confirmed that Bax can directly insert into the lysosomal membrane during staurosporine treatment of fibroblasts [20]. Additionally, CD has two binding sites at its promoter region for the p53 transcription factor and can participate in p53-dependent apoptosis [27]. Moreover, p53 can localize to the mitochondrial membrane and can trigger mitochondrial membrane permeabilization (MMP) by direct interaction with Bcl-2 proteins [51–54], and also can promote Bax-mediated MMP through transcriptional upregulation of BH3-only-domain proteins Puma and Noxa [55]. Hence in our model of camalexin-induced apoptosis in prostate cancer cells via oxidative stress-induction involving lysosomal protease CD, increased p53 protein expression may function as an amplification loop in the death pathway.

In order to further demonstrate that camalexin is exerting its effects on CD through ROS, co- treatment of C4-2 cells with the antioxidant NAC decreased CD translocation to the cytosol and protein expression. We further attempted to highlight the role of CD in camalexin-induced apoptosis by pretreatment of cancer cells with its known peptide inhibitor pepstatin A (Pep A). Camalexin-mediated decrease in cell viability, increase in Bax protein levels and PARP cleavage was antagonized by cotreatment of cells with Pep A. Again, studies have shown that Pep A partially delayed the apoptosis induced by oxidative stress [30] or when Pep A was co-microinjected with CD in fibroblasts [32]. These authors have therefore evidenced CD's key role in apoptosis via its catalytic activity.

3. Experimental

3.1. Reagents and Antibodies

Growth media RPMI 1640 (1 × with L-glutamine, and without L-glutamine and phenol red), and penicillin-streptomycin were from Mediatech Inc. (Manassas, VA, USA). Fetal bovine serum (FBS) and charcoal/dextran treated FBS (DCC-FBS) were from Hyclone (South Logan, UT, USA). T-media was from Gibco by Life Technologies Corporation (Grand Island, NY, USA). Trypsin/EDTA was from Mediatech, Inc. MTS Cell Titer 96® Aqueous One Solution reagent was from Promega

Corporation (Madison, WI, USA). Camalexin (40 mM in absolute ethanol) was kindly provided by our collaborator Roman Mezencev of Georgia Institute of Technology (Atlanta, GA, USA) and was synthesized as described previously [8]. N-acetylcysteine (NAC) and mouse monoclonal anti-human actin antibody were from Sigma-Aldrich Inc. (St. Louis, MO, USA). Rabbit monoclonal anti-cleaved PARP and Bax primary antibody were from Cell Signaling Technology, Inc. (Danvers, MA, USA). Goat polyclonal anti-cathepsin D and p53 primary antibody, and, donkey anti-goat HRP and were from Santa Cruz Biotechnology (Santa Cruz, CA, USA). HRP conjugated sheep anti-mouse, donkey anti-rabbit secondary antibodies, and the Enhanced Chemiluminescence (ECL) western blotting detection reagent were from GE Healthcare UK Ltd. (Buckinghamshire, UK). Nitrocellulose membranes were from Bio-Rad Life Sciences Research (Hercules, CA, USA).

3.2. Cell Lines and Culture

Human prostate cancer cell line ARCaPE, an epithelial cell line and ARCaPM the more highly metastatic and mesenchymal cell line were all derived from parental ARCaP cells [35] and utilized in these experiments (a kind gift from Dr Leland Chung, Cedars-Sinai Medical Center, Los Angeles, CA, USA). They were cultured in T-media supplemented with 10% (*v/v*) fetal bovine serum (FBS), 2 mM L-glutamine, 50 μg/mL penicillin and 100 μg/mL streptomycin. Additionally, we utilized LNCaP cells from American Type Culture Collection (Manassas, VA, USA) with lesser metastatic potential and its more aggressive subline C4-2 cells (a kind gift from Dr Leland Chung). LNCaP and C4-2 cells were routinely cultured in RPMI 1640 medium supplemented with 10% (*v/v*) fetal bovine serum (FBS), 2 mM L-glutamine, 50 μg/mL penicillin and 100 μg/mL streptomycin. They were grown to 70% confluence in 95% air, 5% CO_2 humidified incubator at 37 °C, and routinely passaged using 0.05% Trypsin/EDTA solution. For all experimental conditions RPMI 1640 without L-glutamine and phenol red and supplemented with 5% dextran charcoal stripped FBS (DCC-FBS) was used.

3.3. Cell Viability Assay

ARCaPE, ARCaPM, LNCaP and C4-2 cells were plated at a density of 2,000 cells per well in 96-well plates and allowed to attach overnight. Camalexin (10, 25 and 50 μM) were added to cells and allowed to incubate at 37 °C in a humidified 5% CO_2 atmosphere for 0 through 5 days and then viability assessed daily using the CellTiter 96® Aqueous One Solution Cell Proliferation Assay according to supplier's protocol. For cathepsin D activity inhibition studies, C4-2 cells were treated with 25 or 50 μM camalexin alone or in combination with 100 μM pepstatin A for 0 and 3 days and then cell viability assessed.

3.4. Western Blot Analysis

Western blot was performed as described previously [36]. Briefly, cells were plated at a density of 2×10^6 cells in 75 cm^2 flasks, and allowed to attach overnight. Prior to treatments, ARCaPE and ARCaPM cells were serum-starved in serum-free and phenol red free RPMI medium overnight, while LNCaP and C4-2 cells were serum-starved for 4 h. Cells were then lysed with buffer containing

1 × RIPA, with protease inhibitors (aprotinin 0.1 mg/mL, PMSF 1 mM, leupeptin 0.1 mM, pepstatin A 0.1 mM) and phosphatase- inhibitor (10 mM sodium orthovanadate) and centrifuged at 13,000 rpm. Total protein content was determined using Pierce® BCA Protein Assay Kit (Thermo Scientific, Rockford, IL, USA) following protocol as per manufacturer's specifications. Cell lysates (40–50 μg) were subjected to SDS-Polyacrylamide gel electrophoresis (SDS-PAGE) and then subsequently transferred to pure nitrocellulose membrane. Western blot analysis for CD, Bax, cleaved-PARP and p53 protein expression was then performed. Anti-mouse, anti-goat and anti-rabbit IgG horseradish peroxidase in blocking solution (5% nonfat milk in TBS-Tween 20) was employed as secondary antibodies for chemiluminescent detection. Blots were visualized with chemiluminescence ECL detection system (Pierce, Rockford, IL, USA) and analyzed using FUJIFILM LAS 3000 imager and dark room x-ray film to observe protein expression. To evaluate protein loading, membranes were immediately stripped with Restore ™ Western Blot Stripping Buffer (Thermo Scientific) and reprobed for β-actin as a loading control. Densitometry of protein expression relative to untreated control cells was evaluated using ImageJ Software (National Institutes of Health, Bethesda, MD, USA).

3.5. Immunocytochemistry (ICC)

When C4-2 cells reached 60%–80% confluence, 2×10^3 cells were plated into 16 well chamber slides (Bio-Tek, Nunc, Winooski, VT, USA). Cells were serum-starved for 4 h, and then left untreated, treated with 25 or 50 μM camalexin, or 25 or 50 μM camalexin plus 10 mM NAC for 24 h. The media was removed and 200 μL of equal parts per volume of methanol/ethanol was added for 5 min at room temp for fixation. Methanol/ethanol was gently removed and the slides were washed 3 × 5 min with 1 × PBS and 50 μL of protein blocking solution (Dako, Camarillo, CA, USA) was added per well for 10 min at room temp. Blocking solution was drained off with a tissue and 50 μL of goat anti-CD primary antibody (1:50 or 1:100) (in Dako antibody diluent solution,) was added per well for 1 h at room temp. Slides were washed 3 × 5 min with commercial 1 × TBS-T (Dako), and then incubated with Texas red anti-goat secondary antibody in the dark for 1 h at room temp. Slides were briefly dipped into double deionized water, incubated with DAPI (1 μg/mL) for 5 min at room temp in the dark. Slides were washed 3 × 5 min with 1 × TBS-T mounted using Fluorogel mounting medium (Electron Microscopy Sciences, Hatfield, PA, USA). Slides were placed in a container covered with foil and left to dry overnight in the dark. Fluorescence microscopy was performed using AxioZeiss (Rel 4.8) fluorescence microscopy and images were captured at 20x magnification.

3.6. Statistical Analysis

Statistical analysis was performed using ANOVA and Tukey's Multiple Comparison Test as *Post-Hoc* test from Graph Pad Prizm3 software (GraphPad Software Inc., San Diego, CA, USA); * $p < 0.05$, ** $p < 0.01$, *** $p < 0.001$ were considered significant. Multiple experiments were performed in replicates of 3.

4. Conclusions

Considerable progress has been made in understanding the role of lysosomes and lysosomal proteases in the apoptotic pathways [56]; however, the exact mechanisms of lysosomal membrane permeabilization and its consequences are still not well understood. Although there are supporting data for camalexin ability to generate ROS in cells [9,47], the mechanism(s) involved are yet to be elucidated. We have attempted here to show the role of mature CD in camalexin-induced apoptosis via oxidative stress-induction in prostate cancer cells. Therefore, targeting lysosomes can provide a great therapeutic strategy in cancer cells because not only do they trigger apoptosis, but also can suppress autophagic flux which starves these cells and thus can sensitize them to chemotherapeutic agents [56]. Hence, camalexin treatment of especially metastatic prostate cancer cells in combination with current chemotherapeutic agents may provide a very vital therapeutic option.

Acknowledgments

This work was supported by grants to VOM 5P20MD002285 and Grant Number 8G12MD007590 from the NIH/National Institute on Minority Health and Health Disparities (NIMDH).

Authors Contributions

Conceived and designed the experiments: Basil Smith and Valerie Odero-Marah. Performed the experiments: Basil Smith, Diandra Randle, and Roman Mezencev. Analyzed the data: Basil Smith and Valerie Odero-Marah. Contributed reagents/materials/analysis tools: Valerie Odero-Marah, Basil Smith, Cimona Hinton, and Leeshawn Thomas. Wrote the manuscript: Basil Smith and Valerie Odero-Marah. All authors read and approved the final manuscript.

Conflicts of Interest

The authors declare that there was no conflict of interest.

References

1. Jemal, A.; Siegel, R.; Ward, E.; Murray, T.; Xu, J.; Thun, M.J. Cancer statistics. *CA-Cancer J. Clin.* **2007**, *57*, 43–66.
2. Crawford, E.D.; Eisenberger, M.A.; McLeod, D.G. A controlled trial of leuprolide with or without flutamide in prostatic carcinoma. *N. Engl. J. Med.* **1989**, *321*, 419–420.
3. Beer, T.M.; El Geneidi, M.; Eilers, K.M. Docetaxel (taxotere) in the treatment of prostate cancer. *Expert Rev. Anticancer Ther*. **2003**, *3*, 261–268.
4. Browne, L.M.; Conn, K.L.; Ayer, W.A.; Tewari, J.P. The Camalexins: New phytoalexins produced in leaves of Camelina sativa (Cruciferae). *Tetrahedron* **1991**, *47*, 3903–3914.
5. Jimenez, L.D.; Ayer, W.A.; Tewari, J.P. Phytoalexins produced in the leaves of Capsella bursa pastoris (Shepherd's purse). *Phytoprotection* **1997**, *78*, 99–103.

6. Tsuji, J.; Jackson, E.P.; Gage, D.A.; Hammerschmidt, R.; Somerville, S.C. Phytoalexin accumulation in Arabidopsis Thaliana during the hypersensitive reaction to Pseudomonas syringae pv syringae. *Plant. Physiol.* **1992**, *98*, 1304–1309.
7. Jeandet, P.; Clément, C.; Courot, E.; Cordelier, S. Modulation of phytoalexin biosynthesis in engineered plants for disease resistance. *Int. J. Mol. Sci.* **2013**, *14*, 14136–14170.
8. Mezencev, R.; Galizzi, M.; Kutschy, P.; Docampo, R. Trypanosoma cruzi: Antiproliferative effect of indole phytoalexins on intracellular amastigotes *in vitro*. *Exp. Parasitol.* **2009**, *122*, 66–69.
9. Moody, C.J.; Roffey, J.R.A.; Stephens, M.A.; Stratford, I.J. Synthesis and cytotoxic activity of indole thiazoles. *Anticancer Drugs* **1997**, *8*, 489–499.
10. Smith, B.A.; Neal, C.L.; Chetram, M.; Vo, B.; Mezencev, R.; Hinton, C.; Odero-Marah, V.A. The phytoalexin camalexin mediates cytotoxicity towards aggressive prostate cancer cells via reactive oxygen species. *J. Nat. Med.* **2013**, *67*, 607–618.
11. Wu, W.S. The signaling mechanism of ROS in tumor progression. *Cancer Metast. Rev.* **2006**, *25*, 695–705.
12. Khandrika, L.; Kumar, B.; Koul, S.; Maroni, P.; Koul, H.K. Oxidative stress in prostate cancer. *Cancer Lett.* **2009**, *28*, 125–135.
13. Lim, S.D.; Sun, C.; Lambeth, J.D.; Marshall, F.; Amin, M.; Chung, L.; Petros, J.A.; Arnold, R.S. Increased Nox1 and hydrogen peroxide in prostate cancer. *Prostate* **2005**, *62*, 200–207.
14. Gaziano, J.M.; Glynn, R.J.; Christen, W.G.; Kurth, T.; Belanger, C.; MacFadyen, J.; Bubes, V.; Manson, J.E.; Sesso, H.D.; Buring, J.E. Vitamins E and C in the prevention of prostate and total cancer in men: the Physicians' Health Study II randomized controlled trial. *JAMA* **2009**, *301*, 52–62.
15. Trachootham, D.; Alexandre, J.; Huang, P. Targeting cancer cells by ROS-mediated mechanisms: A radical therapeutic approach? *Nat. Rev. Drug. Discov.* **2009**, *8*, 579–591.
16. Kachadourian, R.; Day, B.J. Flavonoid-induced glutathione depletion: Potential implications for cancer treatment. *Free Radic. Biol. Med.* **2006**, *41*, 65–76.
17. Buttk, T.M.; Sandstrom, P.A. Oxidative stress as a mediator of apoptosis. *Immunol. Today* **1994**, *15*, 7–10.
18. Gardner, A.M.; Xu, F.; Fady, C.; Jacoby, F.J.; Duffey, D.C.; Tu, Y.; Lichtenstein, A. Apoptotic *vs.* non-apoptotic cytotoxicity induced by hydrogen peroxide. *Free Radic. Biol. Med.* **1997**, *22*, 73–83.
19. Forrest, V.J.; Kang, Y.H.; McClain, D.E.; Robinson, D.H.; Ramakrishnan, N. Oxidative stress- induced apoptosis prevented by trolox. *Free Radic. Biol. Med.* **1994**, *16*, 675–684.
20. Kagedal, K.; Johansson, A.C.; Johansson, U.; Heimlich, G.; Roberg, K.; Wang, N.S.; Jürgensmeier, J.M.; Ollinger, K. Lysosomal membrane permeabilization during apoptosis-involvement of Bax? *Int. J. Exp. Pathol.* **2005**, *86*, 309–321.
21. Mathiasen, I.S.; Jaattela, M. Triggering caspase independent cell death to combat cancer. *Trends Mol. Med.* **2002**, *8*, 212–220.
22. Roberg, K.; Ollinger, K. Oxidative stress causes relocation of the lysosomal enzyme cathepsin D with ensuing apoptosis in neonatal rat cardiomyocytes. *Am. J. Pathol.* **1998**, *152*, 1151–1156.

23. Kagedal, K.; Zhao, M.; Svensson, I.; Brunk, U. Sphingosine induced apoptosis is dependent on lysosomal proteases. *Biochem. J.* **2001**, *359*, 335–343.
24. Guicciardi, M.E.; Deussing, J.; Miyoshi, H.; Bronk, S.F.; Svingen, P.A.; Peters, C.; Kaufmann, S.H.; Gores, G.J. Cathepsin B contributes to TNF-alpha-mediated hepatocyte apoptosis by promoting mitochondrial release of cytochrome c. *J. Clin. Invest.* **2000**, *106*, 1127–1137.
25. Yuan, X.M.; Li, W.; Dalen, H.; Lotem, J.; Kama, R.; Sachs, L.; Brunk, U.T. Lysosomal destabilization in p53-induced apoptosis. *Proc. Natl. Acad. Sci. USA* **2002**, *99*, 6286–6291.
26. Deiss, L.P.; Galinka, H.; Berissi, H.; Cohen, O.; Kimchi, A. Cathepsin D protease mediates programmed cell death induced by interferon gamma, Fas/APO-1 and TNF-alpha. *EMBO J.* **1996**, *15*, 3861–3870.
27. Wu, G.S.; Saftig, P.; Peters, C.; El-Deiry, W.S. Potential role for cathepsin D in p53-dependent tumor suppression and chemosensitivity. *Oncogene* **1998**, *16*, 2177–2183.
28. Ollinger, K. Inhibition of cathepsin D prevents free-radical induced apoptosis in rat cardiomyocytes. *Arch. Biochem. Biophys.* **2000**, *373*, 346–351.
29. Roberg, K. Relocalization of cathepsin D and cytochrome c early in apoptosis revealed by immunoelectron microscopy. *Lab. Invest.* **2001**, *81*, 149–158.
30. Kagedal, K.; Johansson, U.; Ollinger, K. The lysosomal protease cathepsin D mediates apoptosis induced by oxidative stress. *FASEB J.* **2001**, *15*, 1592–1594.
31. Takuma, K.; Kiriu, M.; Mori, K.; Lee, E.; Enomoto, R.; Baba, A.; Matsuda, T. Roles of cathepsins in reperfusion-induced apoptosis in cultured astrocytes. *Neurochem. Int.* **2003**, *42*, 153–159.
32. Roberg, K.; Kagedal, K.; Ollinger, K. Microinjection of cathepsin d induces caspase-dependent apoptosis in fibroblasts. *Am. J. Pathol.* **2002**, *161*, 89–96.
33. Heinrich, M.; Neumeyer, J.; Jakob, M.; Hallas, C.; Tchikov, V.; Winoto-Morbach, S.; Wickel, M.; Schneider-Brachert, W.; Trauzold, A.; Hethke, A.; *et al.* Cathepsin D links TNF-induced acid sphingomyelinase to Bid-mediated caspase-9 and -3activation. *Cell Death Differ.* **2004**, *11*, 550–563.
34. Bidere, N.; Lorenzo, H.K.; Carmona, S.; Laforge, M.; Harpe, F.; Dumont, C.; Senik, A. Cathepsin D triggers Bax activation, resulting in selective apoptosis-inducing factor (AIF) relocation in T lymphocytes entering the early commitment phase to apoptosis. *J. Biol. Chem.* **2003**, *278*, 31401–31411.
35. Xu, J.; Wang, R.; Xie, Z.H.; Odero-Marah, V.; Pathak, S.; Multani, A.; Chung, L.W.; Zhau, H.E. Prostate cancer metastasis: Role of the host microenvironment in promoting epithelial to mesenchymal transition and increased bone and adrenal gland metastasis. *Prostate* **2006**, *66*, 1664–1673.
36. Odero-Marah, V.A.; Wang, R.; Chu, G.; Zayzafoon, M.; Xu, J.; Shi, C.; Marshall, F.F.; Zhau, H.E.; Chung, L.W. Receptor activator of NF-KappaB Ligand (RANKL) expression is associated with epithelial to mesenchymal transition in human prostate cancer cells. *Cell Res.* **2008**, *18*, 858–870.
37. Boya, P.; Kroemer, G. Lysosomal membrane Permeabilization in cell death. *Oncogene* **2008**, *27*, 6434–6451.

38. Stova, V.; Turk, V.; Turk, B. Lysosomal cysteine cathepsins: signaling pathways in apoptosis. *Biol. Chem.* **2007**, *388*, 555–560.
39. Kurz, T.; Terman, A.; Gustafsson, B.; Brunk, U.T. Lysosomes and oxidative stress in aging and apoptosis. *Biochem. Biophys. Acta* **2008**, *1780*, 1291–1303.
40. Terman, A.; Kurz, T.; Gustafsson, B.; Brunk, U.T. Lysosomal labilization. *IUBMB Life* **2006**, *58*, 531–539.
41. Antunes, F.; Cadenas, E.; Brunk, U.T. Apoptosis induced by exposure to low steady state concentration of H_2O_2 is a consequence of lysosomal rupture. *Biochem. J.* **2001**, *356*, 549–555.
42. Guicciardi, M.E.; Leist, M.; Gores, G.J. Lysosomes in cell death. *Oncogene* **2004**, *23*, 2881–2890.
43. Gimenez-Bonafe, P.; Tortosa, A.; Perez-Thomas, R. Overcoming drug resistance by enhancing apoptosis by tumor cells. *Curr. Cancer Drug Targets.* **2009**, *9*, 320–340.
44. Kroemer, G.; Jaattela, M. Lysosomes and autophagy in cell death control. *Nat. Rev. Cancer.* **2005**, *5*, 886–897.
45. Li, W.; Yuan, X.; Nordgren, G.; Dalen, H.; Dubowchik, G.M.; Firestone, R.A.; Brunk, U.T. Induction of cell death by the lysosomotropic detergent MSDH. *FEBS Lett.* **2000**, *470*, 35–39.
46. Groth-Pederson, L.; Jaattela, M. Combating apoptosis and multidrug resistant cancers by targeting lysosomes. *Cancer Lett.* **2013**, *332*, 265–274.
47. Mezencev, R.; Updegrove, T.; Kutschy, P.; Repovska, M.; McDonald, J.F. Camalexin induces apoptosis in Jurkat T-leukemia cells by increased concentration of reactive oxygen species and activation of caspases-8 and -9. *J. Nat. Med.* **2011**, *65*, 488–499.
48. Fehrenbacher, N.; Jaattela, M. Lysosomes as targets for cancer therapy. *Cancer Res.* **2005**, *65*, 2993–2995.
49. Rosse, T.; Olivier, R.; Monney, L.; Rager, M.; Conus, S.; Fellay, I.; Jansen, B.; Borner, C. Bcl-2 prolongs cell survival after Bax-induced release of cytochrome c. *Nature* **1998**, *391*, 496–499.
50. Eskes, R.; Antonsson, B.; Osen-Sand, A.; Montessuit, S.; Richter, C.; Sadoul, R.; Mazzei, G.; Nichols, A.; Martinou, J.C. Bax-induced cytochrome C release from mitochondria is independent of the permeability transition pore but highly dependent on Mg2+ ions. *J. Cell. Biol.* **1998**, *143*, 217–224.
51. Johansson, A.C.; Appelqvist, H.; Nilsson, C.; Kagedal, K.; Roberg, K.; Ollinger, K. Regulation of apoptosis-associated lysosomal membrane Permeabilization. *Apoptosis* **2010**, *15*, 527–540.
52. Werneburg, N.W.; Guciarddi, M.E.; Brunk, S.F.; Kaufmann, S.H.; Gores, G.J. Tumor necrosis factor-related apoptosis inducing ligand activates a lysosomal pathway of apoptosis that is regulated by Bcl-2 proteins. *J. Biol. Chem.* **2007**, *282*, 28960–28970.
53. Chipuk, J.E.; Kuwana, T.; Bouchier-Hayes, L.; Droin, N.M.; Newmeyer, D.D.; Schuler, M.; Green, D.R. Direct activation of Bax by p53 mediates mitochondrial membrane permeabilization and apoptosis. *Science* **2004**, *303*, 1010–1014.
54. Mihara, M.; Erster, S.; Zaika, A.; Petrenko, A.; Chittenden, T.; Pancoska, P.; Moll, U.M. p53 has a direct apoptogenic role at the mitochondria. *Mol. Cell* **2003**, *11*, 577–590.

55. Oda, E.; Ohk, R.; Murasaw, H.; Nemoto, J.; Shibue, T.; Yamashita, T.; Tokin, T.; Taniguchi, T.; Tanaka, N. Noxa, a BH3-only member of the Bcl-2 family and candidate mediator of p53-induced apoptosis. *Science* **2000**, *288*, 1053–1058.
56. Repnik, U.; Stoka, V.; Turk, V.; Turk, B. Lysosomes and lysosomal cathepsins in cell death. *Biochem. Biophys. Acta* **2012**, *1824*, 22–33.

Sample Availability: Samples of the compound camalexin, were obtained from Roman Mezencev, and can also be obtained commercially from AfferChem Inc., New Brunswick, NJ.

Resveratrol and Calcium Signaling: Molecular Mechanisms and Clinical Relevance

Audrey E. McCalley, Simon Kaja, Andrew J. Payne and Peter Koulen

Abstract: Resveratrol is a naturally occurring compound contributing to cellular defense mechanisms in plants. Its use as a nutritional component and/or supplement in a number of diseases, disorders, and syndromes such as chronic diseases of the central nervous system, cancer, inflammatory diseases, diabetes, and cardiovascular diseases has prompted great interest in the underlying molecular mechanisms of action. The present review focuses on resveratrol, specifically its isomer *trans*-resveratrol, and its effects on intracellular calcium signaling mechanisms. As resveratrol's mechanisms of action are likely pleiotropic, its effects and interactions with key signaling proteins controlling cellular calcium homeostasis are reviewed and discussed. The clinical relevance of resveratrol's actions on excitable cells, transformed or cancer cells, immune cells and retinal pigment epithelial cells are contrasted with a review of the molecular mechanisms affecting calcium signaling proteins on the plasma membrane, cytoplasm, endoplasmic reticulum, and mitochondria. The present review emphasizes the correlation between molecular mechanisms of action that have recently been identified for resveratrol and their clinical implications.

Reprinted from *Molecules*. Cite as: McCalley, A.E.; Kaja, S.; Payne, A.J.; Koulen, P. Resveratrol and Calcium Signaling: Molecular Mechanisms and Clinical Relevance. *Molecules* **2014**, *19*, 7327-7340.

1. Introduction

1.1. Effects and Mechanisms of Action of Resveratrol as the Basis for Its Therapeutic Potential in Various Diseases

Resveratrol is a stilbenoid commonly found in the roots of Japanese Knotweed and the skin of red grapes. Plants produce phytoalexin as a response to harmful stimuli such as UV radiation, infection, or other pathogenic threats [1,2]. Increased curiosity about resveratrol and its potential health benefits began with a phenomenon termed the "French Paradox" [3] that became a fascination of health professionals and researchers. The "French Paradox" describes the extremely low incidence rate of cardiovascular disease in France compared with other European countries despite a high fat diet with little to no exercise [4]. Epidemiological studies indicated that polyphenols present in red wine could be responsible for the cardioprotective effects experienced by the French [4–6]. Resveratrol then initially became a subject of study for potential cardiovascular benefits but additional benefits are becoming apparent in recent studies [7]. Some of the proposed mechanisms include antioxidant, anti-carcinogenic, anti-inflammatory, anti-aging [8], and anti-nociceptive functions indicating the potential of resveratrol as therapeutic agent for preventing and ameliorating a wide range of pathologies [9–14], including neurodegenerative diseases such as Alzheimer's disease (AD), Huntington's disease (HD), Parkinson's disease (PD), and amyotrophic lateral sclerosis (ALS) [8,15,16]. Cardiovascular diseases that may benefit from resveratrol treatment include hypertension, ventricular arrhythmia, myocardial infarction (MI) induced ventricular tachycardia, ventricular fibrillation, arteriosclerosis, arteriolosclerosis, and restenosis particularly after angioplasty [7,13,17–20]. In addition, cerebral ischemia, diabetes mellitus, cancer, autoimmune diseases, and other inflammatory-related issues have been discussed as potentially benefitting from resveratrol intervention [13,15,21,22]. Resveratrol also provides protection from UV radiation explaining its potential use in the prevention of age-related macular degeneration (AMD) [23,24]. Potential mechanisms underlying resveratrol's actions are its ability to control protein activity via interaction with transmembrane and intracellular enzymes [25].

1.2. Key Signaling Proteins Controlling Cellular Calcium Homeostasis

Resveratrol is a potent regulator of genomic and non-genomic processes including regulation of membrane potential, DNA transcription, enzyme activity, secretion, apoptosis, mitochondrial activity, and intracellular ion homeostasis, including the modulation of the intracellular calcium concentration [3,7,25].

Resveratrol can act as a ligand for trans-membrane proteins [23] including voltage-gated calcium channels (VGCC) and plasma membrane calcium ATPase (PMCA) [26,27]. Intracellularly, calcium release activated channels (CRAC), sarco-/endoplasmic reticulum calcium ATPase (SERCA), and intracellular calcium channels (ICC) contribute to calcium ion homeostasis [28–32]. Resveratrol potentially increases endoplasmic reticulum (ER) calcium concentrations through modulation of SERCA but may also contribute to the decrease or stabilization of the release of calcium from intracellular stores by modulating ICCs [31,33–36]. Most importantly, there is significant evidence to conclude that resveratrol contributes to overall calcium homeostasis during states of cellular dysfunction [37]. While resveratrol does not exhibit cellular toxicity, mild systemic yet reversible gastrointestinal disturbances were detected when resveratrol was used for extended periods of time in high concentrations [38,39].

2. Medical Relevance of Modulation of Cellular Calcium Signaling Mechanisms by Resveratrol

2.1. Modulation of Cellular Calcium Signaling Mechanisms by Resveratrol in Excitable Cells

Resveratrol potently modulates the intracellular calcium concentration in excitable cells through a variety of mechanisms that control calcium influx, store filling, release from intracellular stores, and downstream activation of calcium sensitive molecules. In myocytes, resveratrol increases the refractory period and decreases the threshold for membrane depolarization [17]. This results in restored calcium homeostasis evident by a reduction of delayed after depolarization (DAD, occurs with cytoplasmic calcium influx) and triggered activity (TA, ensues via irregular calcium release) in myocytes [37,40,41]. Resveratrol, therefore, can ameliorate cardiac arrhythmia and prevent premature atrial contraction (PAC) [17].

Similar membrane hyperpolarization in response to resveratrol also occurs in smooth muscle cells (SMC) leading to vasodilation, making resveratrol a candidate for the treatment of a wide variety of disorders including hypertension, ventricular arrhythmia, myocardial infarction (MI) induced ventricular tachycardia, and ventricular fibrillation [18,41,42]. Furthermore, it has been suggested that the ability of resveratrol to induce endothelium-dependent hyperpolarization of SMCs may compensate for pathologic absence or dysfunction of endothelium nitric oxide synthase (eNOS) and cyclooxygenase-1 (COX-1) [43]. Endothelial cells line the walls of the lymphatic and cardiovascular vessels and are critical modulators of contractility and vessel dimensions. Vasorelaxation is the most extensively studied effect of resveratrol on vascular endothelia. Most studies focus on two interlinked mechanisms of action, the direct effect on the intracellular calcium concentration and the downstream indirect effects on cellular potassium regulation [44]. A resveratrol dependent increase in nitric oxide biosynthesis was shown to result from calcium mobilization rather than direct agonism of eNOS by the drug [45]. The major plasma membrane targets of resveratrol include the L-type VGCC, which is directly inhibited by resveratrol, and the large conductance calcium-activated potassium (BK) channel, that resveratrol indirectly effects [19,37]. Under physiological conditions, calcium influx via L-type VGCCs stimulates BK channels that facilitate cytosolic potassium ion efflux, which results in membrane hyperpolarization. Resveratrol directly attenuates calcium influx, indirectly decreasing potassium efflux, and leading to a state of vasorelaxation in an endothelium-dependent manner [20,46,47]. Due to these effects, resveratrol may be beneficial in preventing or mitigating the physiological effects of hypertension, arteriosclerosis, arteriolosclerosis, and restenosis (particularly after angioplasty). Interestingly, diabetes mellitus is an important, clinically relevant risk factor for all four conditions [48].

In neuronal cells, resveratrol reduces the action potential threshold and both delays and prolongs calcium entry (Figure 1) [16]. This effect causes rapid sodium-induced action potentials that increase neuronal conduction. This mechanism of action has the potential to regulate and increase neural impulses which may provide a basis for the prevention of conductance disorders or to retard neurodegenerative processes. Resveratrol has also been shown to protect against cerebral ischemic damage via indirect BK channel activation, thereby preventing neuronal hyperexcitability and the ensuing cell death [15,16]. While resveratrol can prevent neuronal death and provide protection against oxygen-glucose deprivation [49,50], the potential of resveratrol for therapeutic intervention in axonopathy or other conduction block disorders remains to be elucidated [8,51]. The pleiotropic effects of resveratrol with its respective molecular mechanisms, and the pharmacological modulation of those mechanisms, affect separate components of membrane depolarization and repolarization, resulting in a distinct modulation of spatio-temporal properties of electrical signaling by resveratrol in excitable cells.

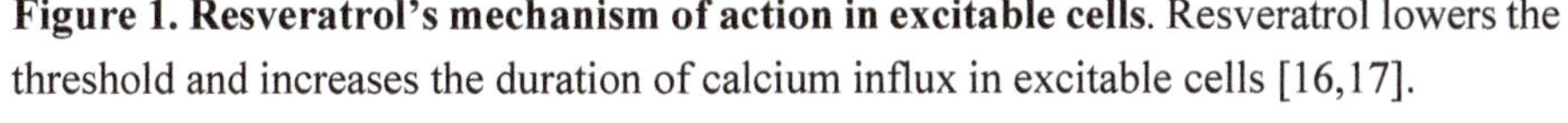

Figure 1. Resveratrol's mechanism of action in excitable cells. Resveratrol lowers the threshold and increases the duration of calcium influx in excitable cells [16,17].

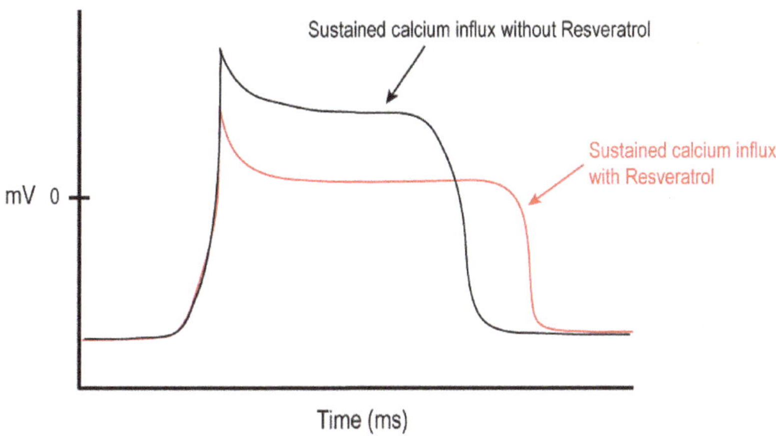

2.2. Modulation of Cellular Calcium Signaling Mechanisms by Resveratrol in Cancer and Immune Cells

Resveratrol shows great promise as a therapeutic agent for a variety of cancers [52]. Given its effects on calcium homeostasis, resveratrol may be useful as a latent remedy for the primary stages of some malignant cancers [53]. Calcium signaling pathways are necessary for the motility and invasive characteristics of malignant tumors, as it controls cell restructuring and dedifferentiation [54]. Another potential mechanism of action for resveratrol targets the excessive energy demand of tumor cells, lowers calcium thresholds, and prolongs intracellular calcium surges, all of which ultimately lead to cell death as a result of mitochondrial dysfunction [55]. Previously, resveratrol has been shown to suppress $CD4^{+}CD25^{+}$ regulatory T cells (T_{reg} cells) in a preclinical cancer model [56]. T_{reg} are critical mediators of self-tolerance and inhibit the proliferation of $CD4^{+}CD25^{-}$ conventional T-cells (T_{con}) through a recently identified mechanism involving the suppression of intracellular calcium release [57]. Further studies are needed to identify how resveratrol modulates intracellular calcium homeostasis during the early phases of tumor growth.

Resveratrol is a powerful modulator of the immune system response via both pro- and anti-inflammatory pathways [58,59]. Resveratrol has the ability to activate immune cells at low doses while exerting an inhibitory effect at high doses [21,52]. The mechanism of action of resveratrol in immune cells likely involves a combination of proteasome inhibition [60] and modification of the cytosolic calcium concentration through blocking calcium entry and increasing calcium storage [31]. It has been suggested that resveratrol can reversibly modify disturbed system responses such as mast cell granulation [56]. Resveratrol decreases mast cell hypersensitivity by reducing calcium influx thereby reducing granulation (Figure 2) [12]. Resveratrol has also been shown to decrease inflammatory responses in autoimmune disorders, such as multiple sclerosis, through potentiation of active T-cell apoptosis [56,61]. It appears that resveratrol exerts effect on intracellular calcium by modulating sequestration and release of intracellular stores but further studies are needed to elucidate the exact mechanism of action in lymphocytes and other cell types of the immune system.

Figure 2. Resveratrol attenuates immune responses related to degranulation. Resveratrol decreases calcium influx, inhibiting granule release in mast cells [14,58].

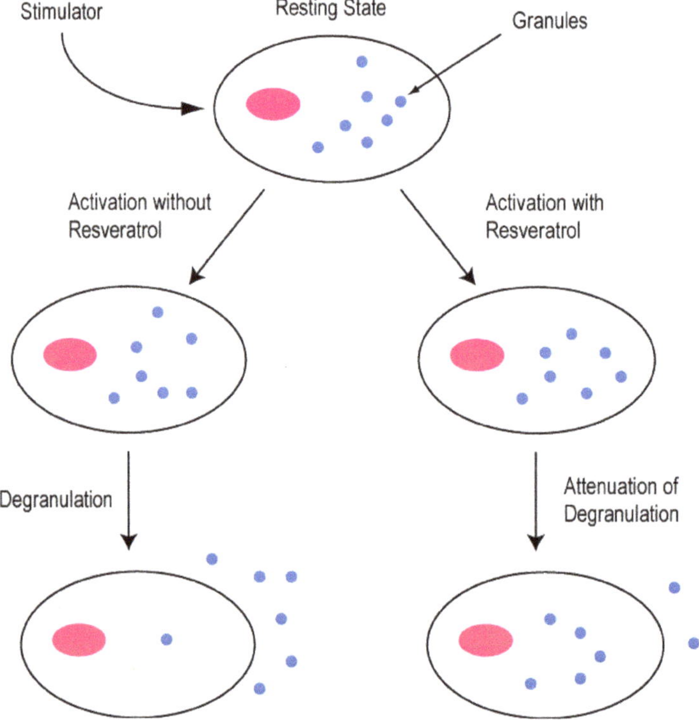

2.3. Modulation of Cellular Calcium Signaling Mechanisms by Resveratrol in Human Retinal Pigment Epithelial Cells

Trans-resveratrol is susceptible to photoisomerization to *cis*-resveratrol when exposed to ultraviolet irradiation [62]. When present in retinal pigment epithelial (RPE) cells, or any cell subject to UV ray exposure, resveratrol absorbs damaging UV light, thus reducing exposure of other structures to damaging wavelengths and ultimately enhancing the protective function of RPE cells (Figure 3) [23]. One key mechanism in RPE cells is phagocytosis, an important renewal mechanism that is susceptible to UV damage [23]. In experimental studies, resveratrol preserved RPE cell phagocytosis when applied prior to insult but had no protective effect when used as a post-insult treatment [23]. It has been proposed that this protective effect is mediated with BK channels [23]. Chronic resveratrol supplementation, rather than acute administration, would therefore be required for pharmaceutical intervention studies.

Another protection mechanism includes the anti-oxidative effect of resveratrol [63,64]. Pretreatment of RPE cells with resveratrol ameliorated the oxidative stress related damage caused by hydrogen peroxide by preventing oxidation related RPE phagocytic impairment, which may be linked to BK channel activity [23]. Furthermore, the clinical course of age-related macular degeneration (AMD), and potentially other vascular diseases of the retina, may be slowed by resveratrol when given as a nutritional additive [24,65].

Figure 3. Resveratrol protects RPE cells from UV damage. *Trans*-resveratrol absorbs UV light and is converted to *cis*-resveratrol thereby preventing RPE cells from the deleterious effects of UV damage [23,62–64].

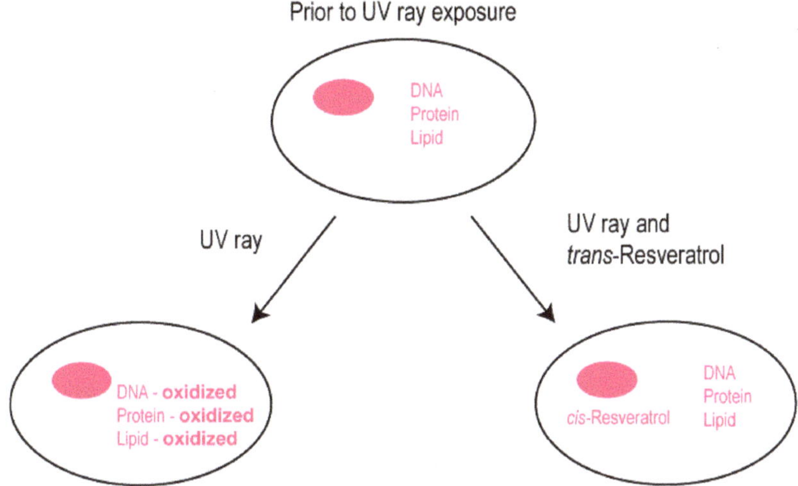

3. Therapeutic Potential of Resveratrol via the Modulation of Cellular Calcium Signaling

3.1. Intracellular Calcium Channels

Two main types of intracellular calcium channels (ICC), inositol 1,4,5-trisphosphate receptors (IP_3Rs) and ryanodine receptors (RyRs), are vital for the release of calcium from the endoplasmic and sarcoplasmic reticulum (ER & SR). Release of intracellular calcium through IP3Rs increases the intracellular calcium concentration which in turn activates RyRs to amplify the calcium signal—a process termed calcium induced calcium release (CICR). ICCs are involved in fundamental cellular processes such as cell proliferation, signaling, excitability, gene expression, and apoptosis [66]. Thus, ICCs are considered a potential target for neuroprotective and cytoprotective strategies for a number of pathologies [28,66,67] although, to date, no studies have been targeted at identifying the specific effects of resveratrol on ICCs (Table 1). Understanding the potential effects of resveratrol on ICCs could identify novel treatment strategies for neurodegenerative diseases such as Alzheimer's disease (AD), Huntington's disease (HD), amyotrophic lateral sclerosis (ALS), and other neurodegenerative diseases.

3.2. Store-Operated Calcium Channel

Calcium-release activated channels (CRACs) are store-operated calcium channels located at the plasma membrane that are activated by low intracellular calcium store concentrations, resulting in a sustained increase of the cytosolic calcium concentration. Immune cells rely on this mechanism and CRACs are potential drug targets for anti-inflammatory diseases [68]. Resveratrol has not been shown to exert a direct effect on CRAC channels [69] (Table 1) although it has been suggested that resveratrol acts indirectly on store-operated calcium entry (SOCE). However, the potency of resveratrol stimulation on CRAC is far lower than that of other known neuroprotective and neuromodulatory compounds including nonsteroidal estrogens [33].

3.3. Voltage-Gated Calcium Channels

Plasma membrane voltage-gated calcium channels (VGCCs) are classified into high or low voltage-activated channels. VGCCs are activated by a depolarization of the membrane potential and allow extracellular calcium flux into the cytosol, thereby translating an electric signal into a secondary chemical signal. The class of high VGCCs include Cav1 (L-type) and Cav2 (P/Q-, N-, and R-type) channels. The low VGCCs includes Cav3 (T-type) channels. Cav1, Cav2, and Cav3 have slow, intermediate, and fast inactivation responses, respectively. In excitatory cells, resveratrol decreases cellular excitability by attenuating calcium influx through VGCCs [70]. Diseases related to VGCC dysregulation include neurodegenerative disorders such as AD, PD, cerebral ischemia, cardiovascular diseases, chronic and acute inflammatory diseases, many cancers, and even mood disorders [28,45,71]. The primary mechanism of action of how VGCC dysregulation leads to these disorders is hyperexcitability resulting in a chronic increase in the cytosolic calcium concentration. Therefore, VGCCs, especially the slow inactivating Cav1 channels, have become a focus of neuroprotective strategies [28,72]. Resveratrol has been shown to dose-dependently inhibit both L-type and T-type calcium channels and to increase the time period for reactivation [29]. More specifically, resveratrol inhibits L-type currents during phase 2 of an action potential, leading to vasorelaxation [26]. Also, it has been suggested that resveratrol-induced vasorelaxation is partially due to a stimulation of NOS and the ensuing chronic increase of the nitric oxide concentration [30,41].

3.4. Calcium-Activated Potassium Channels

Calcium-activated potassium (K_{Ca}) channels include three major types of channels- large (BK), intermediate (IK), and small conductance (SK) channels. K_{Ca} channels help regulate the action potential activity of excitable cells via the modulation of membrane hyperpolarization. These channels play a critical role in restricting calcium influx [73]. Resveratrol does not appear to act directly on K_{Ca} channels but rather exerts its effects through other types of calcium channels. This higher level regulation of calcium homeostasis in turn indirectly regulates potassium flux [19,47,74]. While there is no general consensus, most studies suggest that the modulation of K_{Ca} channels by resveratrol's inhibitory effects illustrates an indirect partial reduction of potassium efflux due to moderation of calcium influx [9,75].

Table 1. Resveratrol's interactions with cellular proteins and its effects on components of the cellular calcium signaling machinery.

Protein	Modulatory Action	Therapeutic Application	References
Intracellular calcium channels	TBD	Potential for excitatory neuronal, cardiac, inflammatory and autoimmune diseases	---
Calcium-release activated channels	No direct effects	No direct disease amelioration	[60]
Store-operated calcium entry	Activation of store-operated calcium entry	Diseases of prolonged calcium influx such as immune and inflammatory diseases	[33]
Voltage-gated calcium channels	Dose-dependent inhibition of L- and T-type channels	Prevention of uncontrolled excitability	[26,27]
Calcium-activated potassium channels	Indirect inhibition, likely via modulation of voltage-gated calcium channels	Modulation of action potentials (particularly in cardiac and neurological disorders)	[19,47,74]
SERCA	Indirect up-regulation via SIRT1 activation	SIRT1 down regulation disorders, cancer	[26,27]
PMCA	Indirect PMCA degradation via calpain activation	Cancer	[27]

3.5. SERCA and PMCA

Calcium ATPases are calcium transport proteins that return the cellular calcium concentrations back to the transmembrane electrochemical gradient state prior to renewed depolarization. The two key proteins that mediate this mechanism are the plasma membrane calcium ATPase (PMCA), which pumps calcium out of the cell when present in high concentrations, and the SR calcium ATPase (SERCA), which replenishes the SR calcium store from the cytoplasm. To our knowledge there is no study investigating potential effects of resveratrol on PMCA. However, it appears that resveratrol upregulates SERCA via SIRT1 activation, further replenishing calcium stores and reducing intracellular calcium influx [62]. Resveratrol may prove useful to attenuate diseases related to SIRT1 downregulation [26]. Additionally, resveratrol facilitates apoptosis in cancerous cells by induction of a calpain-dependent PMCA degradation mechanism. This strategy may be further elucidated through experiments targeted at identifying resveratrol's mechanism of action in tumor cells [27].

3.6. Mitochondrial Calcium Signaling

Mitochondria are temporary and rapidly releasing calcium stores, capable of generating calcium spikes or waves to trigger second messenger systems. Mitochondria are critical initiators of apoptotic pathways via calcium signaling. Resveratrol has been shown to induce mitochondrial apoptosis by alteration of the mitochondrial membrane potential and eliciting a mitochondrial permeability transition (MPT) pore by lowering the calcium threshold necessary for MPT opening [27,55,76]. This mechanism is currently intensively investigated as a potential for anti-cancer drug therapy [77,78]. For instance, it has been proposed that resveratrol can induce this apoptosis in cancerous cells due to differences in mitochondrial function and that it can be injected directly into a tumor mass to initiate tumor regression prior to resection [79].

4. Conclusions

Resveratrol is a potent modulator of many cellular calcium signaling pathways (Table 1). The number of different targets for modulation by resveratrol (Table 1) confound the interpretation of *in vivo* analyses and result in cell type- and organism-specific responses. While L-type and T-type VGCCs have been identified as direct targets of resveratrol modulation, further studies are required to identify the effects on intracellular calcium channels and other neurotransmitter receptor systems. Yet there is significant evidence to suggest that the beneficial effects of resveratrol cannot be attributed only to its immunomodulatory function but also to a resveratrol dependent lowering of membrane hyperexcitability and cellular calcium toxicity. Thus resveratrol bears great promise for a number of multifactorial pathologies including neurodegeneration, autoimmune and cardiovascular disease, and cancer. Further, resveratrol as should be considered both an immunomodulator and a modulator of intracellular calcium signaling.

Acknowledgments

This publication was supported in part NIH grants EY014227, EY022774, AG010485, AG022550, and AG027956 and RR027093 (PK). The content is solely the responsibility of the authors and does not necessarily represent the official views of the National Institutes of Health. Additional support by the Felix and Carmen Sabates Missouri Endowed Chair in Vision Research, a Challenge Grant from Research to Prevent Blindness, and the Vision Research Foundation of Kansas City (PK) is gratefully acknowledged. The authors thank Margaret, Richard and Sara Koulen for generous support and encouragement.

Author Contributions

All authors have performed bibliographic searches, have written and reviewed the manuscript.

Conflicts of Interest

The authors declare no conflict of interest.

References

1. Jeandet, P.; Delaunois, B.; Conreux, A.; Donnez, D.; Nuzzo, V.; Cordelier, S.; Clément, C.; Courot, E. Biosynthesis, metabolism, molecular engineering and biological functions of stilbene phytoalexins in plants. *BioFactors* **2010**, *36*, 331–341.
2. Jeandet, P.; Clément, C.; Courot, E.; Cordelier, S. Modulation of phytoalexin biosynthesis in engineered plants for disease resistance. *Int. J. Mol. Sci.* **2013**, *14*, 14136–14170.

3. Neves, A.R.; Lucio, M.; Lima, J.L.; Reis, S. Resveratrol in medicinal chemistry: A critical review of its pharmacokinetics, drug-delivery, and membrane interactions. *Curr. Med. Chem.* **2012**, *19*, 1663–1681.
4. Richard, J.L. Coronary risk factors. The French paradox. *Arch. Mal. Coeur Vaiss.* **1987**, *80*, 17–21.
5. St. Leger, A.S.; Cochrane, A.L.; Moore, F. Factors associated with cardiac mortality in developed countries with particular reference to the consumption of wine. *Lancet* **1979**, *1*, 1017–1020.
6. Criqui, M.H.; Ringel, B.L. Does diet or alcohol explain the French paradox? *Lancet* **1994**, *344*, 1719–1723.
7. Kroon, P.A.; Iyer, A.; Chunduri, P.; Chan, V.; Brown, L. The cardiovascular nutrapharmacology of resveratrol: Pharmacokinetics, molecular mechanisms and therapeutic potential. *Curr. Med. Chem.* **2010**, *17*, 2442–2455.
8. Bagatini, P.B.; Saur, L.; Rodrigues, M.F.; Bernardino, G.C.; Paim, M.F.; Coelho, G.P.; Silva, D.V.; de Oliveira, R.M.; Schirmer, H.; Souto, A.A.; *et al.* The role of calcium channel blockers and resveratrol in the prevention of paraquat-induced parkinsonism in Drosophila melanogaster: A locomotor analysis. *Invertebr. Neurosci.* **2011**, *11*, 43–51.
9. Granados-Soto, V.; Arguelles, C.F.; Ortiz, M.I. The peripheral antinociceptive effect of resveratrol is associated with activation of potassium channels. *Neuropharmacology* **2002**, *43*, 917–923.
10. Park, S.J.; Ahmad, F.; Philp, A.; Baar, K.; Williams, T.; Luo, H.; Ke, H.; Rehmann, H.; Taussig, R.; Brown, A.L.; *et al.* Resveratrol ameliorates aging-related metabolic phenotypes by inhibiting cAMP phosphodiesterases. *Cell* **2012**, *148*, 421–433.
11. Athar, M.; Back, J.H.; Tang, X.; Kim, K.H.; Kopelovich, L.; Bickers, D.R.; Kim, A.L. Resveratrol: A review of preclinical studies for human cancer prevention. *Toxicol. Appl. Pharm.* **2007**, *224*, 274–283.
12. Yuan, M.; Li, J.; Lv, J.; Mo, X.; Yang, C.; Chen, X.; Liu, Z.; Liu, J. Polydatin (PD) inhibits IgE-mediated passive cutaneous anaphylaxis in mice by stabilizing mast cells through modulating Ca(2)(+) mobilization. *Toxicol. Appl. Pharm.* **2012**, *264*, 462–469.
13. Venkatesan, B.; Valente, A.J.; Reddy, V.S.; Siwik, D.A.; Chandrasekar, B. Resveratrol blocks interleukin-18-EMMPRIN cross-regulation and smooth muscle cell migration. *Am. J. Physiol. Heart Circ. Physiol.* **2009**, *297*, H874–H886.
14. Naylor, J.; Al-Shawaf, E.; McKeown, L.; Manna, P.T.; Porter, K.E.; O'Regan, D.; Muraki, K.; Beech, D.J. TRPC5 channel sensitivities to antioxidants and hydroxylated stilbenes. *J. Biol. Chem.* **2011**, *286*, 5078–5086.
15. Wang, Q.; Xu, J.; Rottinghaus, G.E.; Simonyi, A.; Lubahn, D.; Sun, G.Y.; Sun, A.Y. Resveratrol protects against global cerebral ischemic injury in gerbils. *Brain Res.* **2002**, *958*, 439–447.
16. Zhang, H.; Schools, G.P.; Lei, T.; Wang, W.; Kimelberg, H.K.; Zhou, M. Resveratrol attenuates early pyramidal neuron excitability impairment and death in acute rat hippocampal slices caused by oxygen-glucose deprivation. *Exp. Neurol.* **2008**, *212*, 44–52.
17. Chen, W.P.; Su, M.J.; Hung, L.M. *In vitro* electrophysiological mechanisms for antiarrhythmic efficacy of resveratrol, a red wine antioxidant. *Eur. J. Pharmacol.* **2007**, *554*, 196–204.

18. Chen, Y.R.; Yi, F.F.; Li, X.Y.; Wang, C.Y.; Chen, L.; Yang, X.C.; Su, P.X.; Cai, J. Resveratrol attenuates ventricular arrhythmias and improves the long-term survival in rats with myocardial infarction. *Cardiovasc. Drug. Ther.* **2008**, *22*, 479–485.
19. Calderone, V.; Martelli, A.; Testai, L.; Martinotti, E.; Breschi, M.C. Functional contribution of the endothelial component to the vasorelaxing effect of resveratrol and NS 1619, activators of the large-conductance calcium-activated potassium channels. *Naunyn-Schmiedeberg's Arch. Pharm.* **2007**, *375*, 73–80.
20. Gojkovic-Bukarica, L.; Novakovic, A.; Kanjuh, V.; Bumbasirevic, M.; Lesic, A.; Heinle, H. A role of ion channels in the endothelium-independent relaxation of rat mesenteric artery induced by resveratrol. *J. Pharmacol. Sci.* **2008**, *108*, 124–130.
21. Gao, X.; Xu, Y.X.; Janakiraman, N.; Chapman, R.A.; Gautam, S.C. Immunomodulatory activity of resveratrol: Suppression of lymphocyte proliferation, development of cell-mediated cytotoxicity, and cytokine production. *Biochem. Pharmacol.* **2001**, *62*, 1299–1308.
22. Sharma, S.; Anjaneyulu, M.; Kulkarni, S.K.; Chopra, K. Resveratrol, a polyphenolic phytoalexin, attenuates diabetic nephropathy in rats. *Pharmacology* **2006**, *76*, 69–75.
23. Sheu, S.J.; Wu, T.T. Resveratrol protects against ultraviolet A-mediated inhibition of the phagocytic function of human retinal pigment epithelial cells via large-conductance calcium-activated potassium channels. *Kaohsiung J. Med. Sci.* **2009**, *25*, 381–388.
24. Sheu, S.J.; Bee, Y.S.; Chen, C.H. Resveratrol and large-conductance calcium-activated potassium channels in the protection of human retinal pigment epithelial cells. *J. Ocul. Pharmacol.* **2008**, *24*, 551–555.
25. Pirola, L.; Frojdo, S. Resveratrol: One molecule, many targets. *IUBMB Life* **2008**, *60*, 323–332.
26. Sulaiman, M.; Matta, M.J.; Sunderesan, N.R.; Gupta, M.P.; Periasamy, M.; Gupta, M. Resveratrol, an activator of SIRT1, upregulates sarcoplasmic calcium ATPase and improves cardiac function in diabetic cardiomyopathy. *Am. J. Physiol. Heart Circ. Physiol.* **2010**, *298*, H833–H843.
27. Sareen, D.; Darjatmoko, S.R.; Albert, D.M.; Polans, A.S. Mitochondria, calcium, and calpain are key mediators of resveratrol-induced apoptosis in breast cancer. *Mol. Pharmacol.* **2007**, *72*, 1466–1475.
28. Duncan, R.S.; Goad, D.L.; Grillo, M.A.; Kaja, S.; Payne, A.J.; Koulen, P. Control of intracellular calcium signaling as a neuroprotective strategy. *Molecules* **2010**, *15*, 1168–1195.
29. Jakab, M.; Lach, S.; Bacova, Z.; Langeluddecke, C.; Strbak, V.; Schmidt, S.; Iglseder, E.; Paulmichl, M.; Geibel, J.; Ritter, M. Resveratrol inhibits electrical activity and insulin release from insulinoma cells by block of voltage-gated Ca^+ channels and swelling-dependent Cl- currents. *Cell Physiol. Biochem.* **2008**, *22*, 567–578.
30. Zhang, L.P.; Yin, J.X.; Liu, Z.; Zhang, Y.; Wang, Q.S.; Zhao, J. Effect of resveratrol on L-type calcium current in rat ventricular myocytes. *Acta Pharmacol. Sin.* **2006**, *27*, 179–183.
31. Dobrydneva, Y.; Williams, R.L.; Blackmore, P.F. trans-Resveratrol inhibits calcium influx in thrombin-stimulated human platelets. *Br. J. Pharmacol.* **1999**, *128*, 149–157.
32. Tian, X.M.; Zhang, Z.X. [Resveratrol promote permeability transition pore opening mediated by Ca2+]. *Yao Xue Xue Bao* **2003**, *38*, 81–84.

33. Dobrydneva, Y.; Williams, R.L.; Blackmore, P.F. Diethylstilbestrol and other nonsteroidal estrogens: Novel class of store-operated calcium channel modulators. *J. Cardiovasc. Pharmacol.* **2010**, *55*, 522–530.
34. Dobrydneva, Y.; Williams, R.L.; Katzenellenbogen, J.A.; Ratz, P.H.; Blackmore, P.F. Diethylstilbestrol and tetrahydrochrysenes are calcium channel blockers in human platelets: Relationship to the stilbene pharmacophore. *Thromb. Res.* **2003**, *110*, 23–31.
35. Wang, L.; Ma, Q.; Chen, X.; Sha, H.; Ma, Z. Effects of resveratrol on calcium regulation in rats with severe acute pancreatitis. *Eur. J. Pharmacol.* **2008**, *580*, 271–276.
36. Buluc, M.; Demirel-Yilmaz, E. Resveratrol decreases calcium sensitivity of vascular smooth muscle and enhances cytosolic calcium increase in endothelium. *Vascul. Pharmacol.* **2006**, *44*, 231–237.
37. Campos-Toimil, M.; Elies, J.; Alvarez, E.; Verde, I.; Orallo, F. Effects of trans- and cis-resveratrol on Ca^{2+} handling in A7r5 vascular myocytes. *Eur. J. Pharmacol.* **2007**, *577*, 91–99.
38. Sangeetha, M.K.; Eazhisai Vallabi, D.; Sali, V.K.; Thanka, J.; Vasanthi, H.R. Sub-acute toxicity profile of a modified resveratrol supplement. *Food Chem. Toxicol.* **2013**, *59*, 492–500.
39. Cottart, C.H.; Nivet-Antoine, V.; Beaudeux, J.L. Review of recent data on the metabolism, biological effects, and toxicity of resveratrol in humans. *Mol. Nutr. Food Res.* **2013**, *58*, 7–21.
40. Zhang, L.P.; Ma, H.J.; Zhao, J.; Wang, Q.S. Effects of resveratrol on delayed afterdepolarization and triggered activity induced by ouabain in guinea pig papillary muscles. *Sheng Li Xue Bao* **2005**, *57*, 361–366.
41. Liu, Z.; Zhang, L.P.; Ma, H.J.; Wang, C.; Li, M.; Wang, Q.S. Resveratrol reduces intracellular free calcium concentration in rat ventricular myocytes. *Sheng Li Xue Bao* **2005**, *57*, 599–604.
42. Campos-Toimil, M.; Elies, J.; Orallo, F. Trans- and cis-resveratrol increase cytoplasmic calcium levels in A7r5 vascular smooth muscle cells. *Mol. Nutr. Food Res.* **2005**, *49*, 396–404.
43. Li, H.; Xia, N.; Forstermann, U. Cardiovascular effects and molecular targets of resveratrol. *Nitric Oxide* **2012**, *26*, 102–110.
44. Baur, J.A.; Sinclair, D.A. Therapeutic potential of resveratrol: The *in vivo* evidence. *Nat. Rev. Drug Discovery* **2006**, *5*, 493–506.
45. Elíes, J.; Cuíñas, A.; García-Morales, V.; Orallo, F.; Campos-Toimil, M. Trans-resveratrol simultaneously increases cytoplasmic Ca^{2+} levels and nitric oxide release in human endothelial cells. *Mol. Nutr. Food Res.* **2011**, *55*, 1237–1248.
46. Protic, D.; Beleslin-Cokic, B.; Novakovic, R.; Kanjuh, V.; Heinle, H.; Scepanovic, R.; Gojkovic-Bukarica, L. Effect of Wine Polyphenol Resveratrol on the Contractions Elicited Electrically or by Norepinephrine in the Rat Portal Vein. *Phytother. Res.* **2013**, *27*, 1685–1693.
47. Novakovic, A.; Bukarica, L.G.; Kanjuh, V.; Heinle, H. Potassium channels-mediated vasorelaxation of rat aorta induced by resveratrol. *Basic Clin. Pharmacol. Toxicol.* **2006**, *99*, 360–364.
48. Su, H.C.; Hung, L.M.; Chen, J.K. Resveratrol, a red wine antioxidant, possesses an insulin-like effect in streptozotocin-induced diabetic rats. *Am. J. Physiol. Endocrinol. Metab.* **2006**, *290*, 1339–1346.

49. Raval, A.P.; Dave, K.R.; Perez-Pinzon, M.A. Resveratrol mimics ischemic preconditioning in the brain. *J. Cereb. Blood Flow Metab.* **2006**, *26*, 1141–1147.
50. Raval, A.P.; Lin, H.W.; Dave, K.R.; Defazio, R.A.; Della Morte, D.; Kim, E.J.; Perez-Pinzon, M.A. Resveratrol and ischemic preconditioning in the brain. *Curr. Med. Chem.* **2008**, *15*, 1545–1551.
51. Li, M.; Wang, Q.S.; Chen, Y.; Wang, Z.M.; Liu, Z.; Guo, S.M. Resveratrol inhibits neuronal discharges in rat hippocampal CA1 area. *Sheng Li Xue Bao* **2005**, *57*, 355–360.
52. Kundu, J.K.; Surh, Y.J. Cancer chemopreventive and therapeutic potential of resveratrol: Mechanistic perspectives. *Cancer Lett.* **2008**, *269*, 243–261.
53. Le Ferrec, E.; Lagadic-Gossmann, D.; Rauch, C.; Bardiau, C.; Maheo, K.; Massiere, F.; le Vee, M.; Guillouzo, A.; Morel, F. Transcriptional induction of CYP1A1 by oltipraz in human Caco-2 cells is aryl hydrocarbon receptor- and calcium-dependent. *J. Biol. Chem.* **2002**, *277*, 24780–24787.
54. Chen, Y.F.; Chen, Y.T.; Chiu, W.T.; Shen, M.R. Remodeling of calcium signaling in tumor progression. *J. Biomed. Sci.* **2013**, *20*, doi:10.1186/1423-0127-20-23.
55. Ma, X.; Tian, X.; Huang, X.; Yan, F.; Qiao, D. Resveratrol-induced mitochondrial dysfunction and apoptosis are associated with Ca^{2+} and mCICR-mediated MPT activation in HepG2 cells. *Mol. Cell. Biochem.* **2007**, *302*, 99–109.
56. Yang, Y.; Paik, J.H.; Cho, D.; Cho, J.A.; Kim, C.W. Resveratrol induces the suppression of tumor-derived CD4+CD25+ regulatory T cells. *Int. Immunopharmacol.* **2008**, *8*, 542–547.
57. Bishayee, A. Cancer prevention and treatment with resveratrol: From rodent studies to clinical trials. *Cancer Prev. Res.* **2009**, *2*, 409–418.
58. Feng, Y.H.; Zhou, W.L.; Wu, Q.L.; Li, X.Y.; Zhao, W.M.; Zou, J.P. Low dose of resveratrol enhanced immune response of mice. *Acta Pharmacol. Sin.* **2002**, *23*, 893–897.
59. Falchetti, R.; Fuggetta, M.P.; Lanzilli, G.; Tricarico, M.; Ravagnan, G. Effects of resveratrol on human immune cell function. *Life Sci.* **2001**, *70*, 81–96.
60. Shin, D.H.; Seo, E.Y.; Pang, B.; Nam, J.H.; Kim, H.S.; Kim, W.K.; Kim, S.J. Inhibition of Ca^{2+}-release-activated Ca^{2+} channel (CRAC) and K^+ channels by curcumin in Jurkat-T cells. *J. Pharmacol. Sci.* **2011**, *115*, 144–154.
61. Singh, N.P.; Hegde, V.L.; Hofseth, L.J.; Nagarkatti, M.; Nagarkatti, P. Resveratrol (trans-3,5,4'-trihydroxystilbene) ameliorates experimental allergic encephalomyelitis, primarily via induction of apoptosis in T cells involving activation of aryl hydrocarbon receptor and estrogen receptor. *Mol. Pharmacol.* **2007**, *72*, 1508–1521.
62. Nour, V.; Trandafir, I.; Muntean, C. Ultraviolet irradiation of trans-resveratrol and HPLC determination of trans-resveratrol and cis-resveratrol in Romanian red wines. *J. Chromatogr. Sci.* **2012**, *50*, 920–927.
63. Bertram, K.M.; Baglole, C.J.; Phipps, R.P.; Libby, R.T. Molecular regulation of cigarette smoke induced-oxidative stress in human retinal pigment epithelial cells: Implications for age-related macular degeneration. *Am. J. Physiol. Cell Physiol.* **2009**, *297*, C1200–C1210.
64. King, R.E.; Kent, K.D.; Bomser, J.A. Resveratrol reduces oxidation and proliferation of human retinal pigment epithelial cells via extracellular signal-regulated kinase inhibition. *Chem. Biol. Interact.* **2005**, *151*, 143–149.

65. Nagaoka, T.; Hein, T.W.; Yoshida, A.; Kuo, L. Resveratrol, a component of red wine, elicits dilation of isolated porcine retinal arterioles: Role of nitric oxide and potassium channels. *Invest. Ophthalmol. Vis. Sci.* **2007**, *48*, 4232–4239.
66. Berridge, M.J. Neuronal calcium signaling. *Neuron* **1998**, *21*, 13–26.
67. Wehrens, X.H.; Lehnart, S.E.; Marks, A.R. Intracellular calcium release and cardiac disease. *Annu. Rev. Physiol.* **2005**, *67*, 69–98.
68. Feske, S. Calcium signalling in lymphocyte activation and disease. *Nat. Rev. Immunol.* **2007**, *7*, 690–702.
69. Dorrie, J.; Gerauer, H.; Wachter, Y.; Zunino, S.J. Resveratrol induces extensive apoptosis by depolarizing mitochondrial membranes and activating caspase-9 in acute lymphoblastic leukemia cells. *Cancer Res.* **2001**, *61*, 4731–4739.
70. Wallace, C.H.; Baczko, I.; Jones, L.; Fercho, M.; Light, P.E. Inhibition of cardiac voltage-gated sodium channels by grape polyphenols. *Br. J. Pharmacol.* **2006**, *149*, 657–665.
71. Anekonda, T.S.; Quinn, J.F.; Harris, C.; Frahler, K.; Wadsworth, T.L.; Woltjer, R.L. L-type voltage-gated calcium channel blockade with isradipine as a therapeutic strategy for Alzheimer's disease. *Neurobiol. Dis.* **2011**, *41*, 62–70.
72. Cain, S.M.; Snutch, T.P. Voltage-gated calcium channels and disease. *Biofactors* **2011**, *37*, 197–205.
73. Vergara, C.; Latorre, R.; Marrion, N.V.; Adelman, J.P. Calcium-activated potassium channels. *Curr. Opin. Neurobiol.* **1998**, *8*, 321–329.
74. Chang, Y.; Wang, S.J. Inhibitory effect of glutamate release from rat cerebrocortical nerve terminals by resveratrol. *Neurochem. Int.* **2009**, *54*, 135–141.
75. Chen, W.-P.; Chi, T.-C.; Chuang, L.-M.; Su, M.-J. Resveratrol enhances insulin secretion by blocking KATP and KV channels of beta cells. *Eur. J. Pharmacol.* **2007**, *568*, 269–277.
76. Kalra, N.; Roy, P.; Prasad, S.; Shukla, Y. Resveratrol induces apoptosis involving mitochondrial pathways in mouse skin tumorigenesis. *Life Sci.* **2008**, *82*, 348–358.
77. Sun, W.; Wang, W.; Kim, J.; Keng, P.; Yang, S.; Zhang, H.; Liu, C.; Okunieff, P.; Zhang, L. Anti-cancer effect of resveratrol is associated with induction of apoptosis via a mitochondrial pathway alignment. *Adv. Exp. Med. Biol.* **2008**, *614*, 179–186.
78. Schmidt, A.; Oberle, N.; Weiß, E.M.; Vobis, D.; Frischbutter, S.; Baumgrass, R.; Falk, C.S.; Haag, M.; Brügger, B.; Lin, H.; *et al.* Human regulatory T cells rapidly suppress T cell receptor-induced Ca^{2+}, NF-kappaB, and NFAT signaling in conventional T cells. *Sci. Signal.* **2011**, *4*, doi:10.1126/scisignal.2002179.
79. Qureshi, A.A.; Guan, X.Q.; Reis, J.C.; Papasian, C.J.; Jabre, S.; Morrison, D.C.; Qureshi, N., Inhibition of nitric oxide and inflammatory cytokines in LPS-stimulated murine macrophages by resveratrol, a potent proteasome inhibitor. *Lipids Health Dis.* **2012**, *11*, 76.

Inhibition of Cancer Derived Cell Lines Proliferation by Synthesized Hydroxylated Stilbenes and New Ferrocenyl-Stilbene Analogs. Comparison with Resveratrol

Malik Chalal, Dominique Delmas, Philippe Meunier, Norbert Latruffe and Dominique Vervandier-Fasseur

Abstract: Further advances in understanding the mechanism of action of resveratrol and its application require new analogs to identify the structural determinants for the cell proliferation inhibition potency. Therefore, we synthesized new *trans*-resveratrol derivatives by using the Wittig and Heck methods, thus modifying the hydroxylation and methoxylation patterns of the parent molecule. Moreover, we also synthesized new ferrocenylstilbene analogs by using an original protective group in the Wittig procedure. By performing cell proliferation assays we observed that the resveratrol derivatives show inhibition on the human colorectal tumor SW480 cell line. On the other hand, cell viability/cytotoxicity assays showed a weaker effects on the human hepatoblastoma HepG2 cell line. Importantly, the lack of effect on non-tumor cells (IEC18 intestinal epithelium cells) demonstrates the selectivity of these molecules for cancer cells. Here, we show that the numbers and positions of hydroxy and methoxy groups are crucial for the inhibition efficacy. In addition, the presence of at least one phenolic group is essential for the antitumoral activity. Moreover, in the series of ferrocenylstilbene analogs, the presence of a hidden phenolic function allows for a better solubilization in the cellular environment and significantly increases the antitumoral activity.

Reprinted from *Molecules*. Cite as: Chalal, M.; Delmas, D.; Meunier, P.; Latruffe, N.; Vervandier-Fasseur, D. Inhibition of Cancer Derived Cell Lines Proliferation by Synthesized Hydroxylated Stilbenes and New Ferrocenyl-Stilbene Analogs. Comparison with Resveratrol. *Molecules* **2014**, *19*, 7850-7868.

1. Introduction

Polyphenolic compounds, including stilbenes, anthocyans, catechins and their oligomers, are widespread in a large number of plants. Polyphenolic stilbenoids have been discovered in numerous species, for instance, in the roots of the Asiatic plant *Polygonum cuspidatum* [1], in the South African plant *Erythrophleum lasianthu* [2], in red fruit, including grapes [3–5], in red wine [6,7], in Itadori green tea [8], in peanuts [9], and in rhubarb [10]. The common feature of these different plants is the presence of a phytoalexin, *trans*-resveratrol or *trans*-3,5,4'-trihydroxystilbene (**RSV**, Figure 1a) [1,3–5,8,9,11,12]. This well-known polyphenol proves to be a true (Swiss Army knife) molecule [13] in the therapeutic and biological fields [14–16]. Indeed, numerous publications and reviews report about *trans*-resveratrol's antitumoral [17,18], anti-inflammatory [19], antiviral [20], antimicrobial [21], and antifungal [22–24] activities. In addition, *trans*-resveratrol is a neuroprotective agent [25,26] and can also prevent heart disease [27–29]. The antioxidant features of *trans*-resveratrol may partly explain these numerous activities [30–32]. In cancer research, it has been shown that involvement of *trans*-resveratrol in antitumoral activity is also due to its ability to bind different cellular targets [33,34]. However, several derivatives of *trans*-resveratrol show a better activity than the parent molecule towards specific types of cancer [35]. The modifications of the chemical structure of *trans*-resveratrol involve the number and the position of the phenolic groups [35–37], the presence on the aromatic rings of methoxy groups [38–41], long alkyl chains [38,42], or functionalized chains [43]. These structural modifications improve mostly the lipophilicity of the stilbenes in the cellular environment and thus their biological effects inside the cell [44]. However, the methoxylated derivatives of *trans*-resveratrol seem to have a different way of delaying cancer growth. Indeed, our group has studied the biological activities of *E*- and *Z*-methoxylated stilbenes against the human colorectal tumor SW480 cell line and has reported that the methoxy group is a determinant substitution for the molecules bearing a *Z* configuration in inhibition of this cell line (compounds **A**, Figure 1) [45].

Figure 1. (**a**) Structure of *trans*-resveratrol (**RSV**). (**b**) Structure of *cis* and *trans*-resveratrol derivatives.

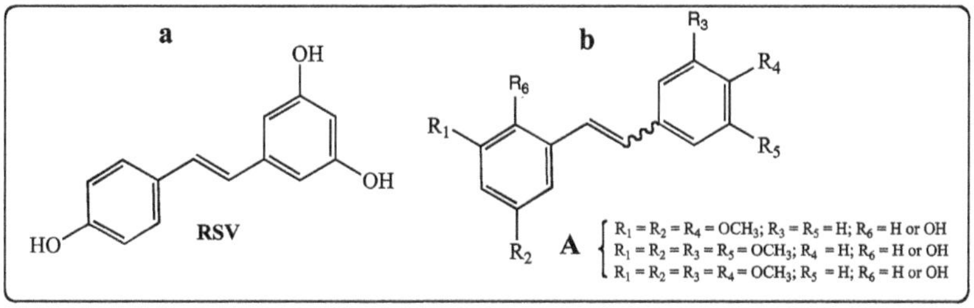

A: $R_1 = R_2 = R_4 = OCH_3$; $R_3 = R_5 = H$; $R_6 = H$ or OH
$R_1 = R_2 = R_3 = R_5 = OCH_3$; $R_4 = H$; $R_6 = H$ or OH
$R_1 = R_2 = R_3 = R_4 = OCH_3$; $R_5 = H$; $R_6 = H$ or OH

Zhang *et al.* have confirmed that *trans*-resveratrol was known to be active only in its *E* configuration while some methoxylated derivatives proved to be active in the *Z* configuration [41]. In order to deepen our understanding of the mechanism of action and to highlight compounds with enhanced effects on colorectal tumor SW480 and hepatoblastoma HepG2 cell lines, we synthesized a series of *E*-stilbenes, including three new original ferrocenylstilbene analogs, by improved Wittig and Heck methods [46]. Each compound was submitted to evaluation for biological properties (antiproliferative activity and cell cycle disturbance of SW480 colon cancer and hepatic HepG2 cancer cells). To obtain an inhibitory effect, the chemical parameters studied are the following: (a) the presence of a hydroxy group in position 4; (b) the increased effect due to the presence of a methoxy group (a decrease of the polar character leading to an increase in lipophilic property); (c) the lack (or masked form) of other hydroxy groups. In the series of ferrocenylstilbene analogs, the presence of a phenolic function as an ester greatly increases the antitumoral activity. Most of synthetic compounds are more efficient towards colorectal SW480 cells than liver-derived HepG2 cells. Furthermore, the lack of effects on non-tumor cells (IEC18 intestinal epithelium cells) demonstrates the selectivity of these molecules for cancer cells, which is an important aspect for possible therapeutic applications.

2. Results and Discussion

2.1. Chemical Results

2.1.1. Synthesis of *E*-4-Hydroxystilbenes

Given the importance of the free phenolic function in position 4 [30,31], we focused on the preparation of derivatives bearing a free phenolic group in position 4 and substituents on the ring B of the stilbenes (compounds **1**–**6**; Figure 2a) or on the A and B rings of the stilbenes (compounds **7**–**9**; Figure 2b). The methoxy group was often chosen as a substituent to improve the membrane permeability of the stilbenes. To highlight the importance of the presence and the position of the phenolic function in the activity of the stilbenes towards tumor cell lines, one derivative with OH group in position 3 was prepared (compound **10**; Figure 2c) and four resveratrol analogs without a free phenolic function were synthesized (compounds **11**–**14**; Figure 2d). Compound **10** was already studied by Zhang *et al.* for its effects on NQO1 induction in hepatoma cells, but its synthesis was not described [41].

On the contrary, compounds **1**–**4**, **6**, **7**, **12** and **13** were already synthesized by different method, including Horner-Emmons-Haworth [35,47,48], Perkin [49–51] and Mizoroki-Heck reactions [52]. Previously, our group has reported the synthesis of compounds **1**–**14** by two standard methods [46]. Stilbenes **4**, **7**–**13** were prepared by palladium-catalyzed Heck coupling using ferrocenylphosphane ligands. In our protocol, the hydroxylated stilbenes were obtained without the need of protection/deprotection steps on the phenolic functions. Stilbenes **1**–**3**, **5**, **6** and **14** were prepared by Wittig reactions; the protection on the hydroxy groups of aromatic aldehydes was achieved using the labile trimethylsilyl group, rarely used in this case. This protective group was easily cleaved during the aqueous work-up following the Wittig reaction.

Figure 2. Molecular structure of synthetic stilbene derivatives. (**a**) 4-OH stilbenes bearing substituents on cycle B. (**b**) 4-OH stilbenes bearing substituents on cycle A and/or cycle B. (**c**) 3-hydroxy-4'-methoxystilbene (**10**). (**d**) Stilbenes without free phenolic function.

a
1 R' = H
2 R' = 4'-OMe
3 R' = 3',5'-diOMe
4 R' = 4'-OH,3'-OMe
5 R' = 4'-vinyl
6 R' = 4'-Br

b
7 R = 3-OMe, R' = 3',4',5'-triOMe
8 R = 3-CH_2OH, R' = 4'-OMe
9 R = 3-OMe, R' = 4'-OMe

c
10

d
11 $R_1 = R_3 = R_4 = R_5 = R_7 = H$, $R_2 = R_6 = OMe$
12 $R_1 = R_3 = R_4 = R_6 = R_7 = H$, $R_2 = R_5 = OMe$
13 $R_1 = R_3 = R_5 = R_6 = R_7 = H$, $R_2 = R_4 = OMe$
14 $R_2 = R_4 = R_6 = H$, $R_1 = R_3 = R_5 = R_7 = OMe$

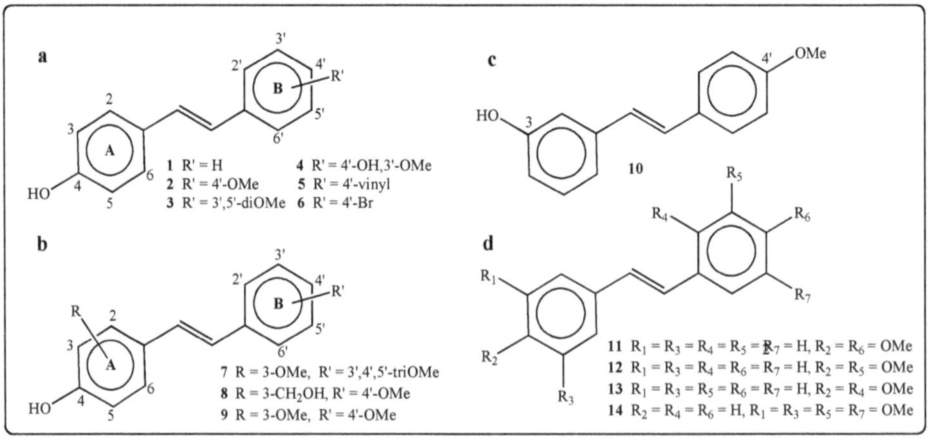

2.1.2. Synthesis of Stilbenes Bearing Ferrocenylstilbene Analogs

In addition to these stilbenes bearing classical substituents, we developed original ferrocenyl-analogs of stilbenes **15–17** (Figure 3).

Figure 3. Molecular structure of ferrocenyl-stilbene analogs **15–17**.

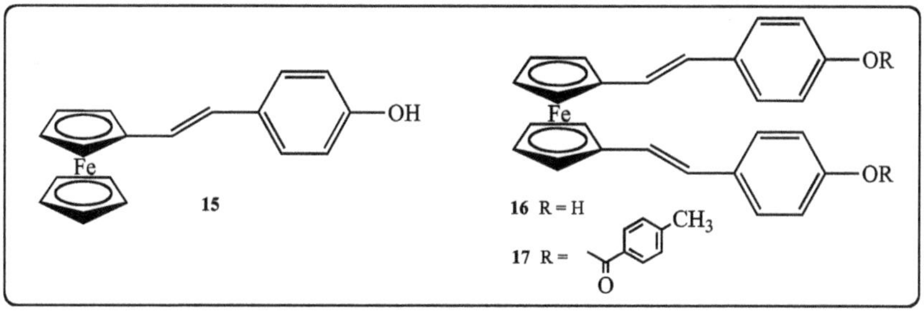

Indeed, since the discovery of the antitumoral properties of cisplatin [53], the therapeutic interests in metallic complexes and organometallic compounds has increased steadily [54], especially for ferrocenyl derivatives [55]. Several organometallic compounds bearing a ferrocenyl group display better biological properties than their organic counterparts, such as chloroquine and ferroquine used in the treatment of malaria [56]. A key example of an anticancer ferrocene derivative is the anti-breast cancer ferrocifen series. Jaouen's group has synthesized different derivatives of the ferrocen complexes of tamoxifen and has shown complementary activities of these compounds [57,58]. Therefore, in the aim to improve the antitumoral activities of the polyphenols, we have targeted the synthesis of an original stilbene molecular structure wherein a ferrocenyl ring replaced a benzenic ring; the position 4 of the remaining benzenic ring was substituted by a free phenolic function. The proposed strategy to access this series of ferrocenylstilbene analogs is to react under Wittig reaction conditions ferrocenecarbaldehyde (**18**) or ferrocene-1,1'-dicarbaldehyde (**19**) [59] with a benzylphosphonium bromide bearing a protected phenolic function **20** (Figure 4).

Figure 4. Starting reagents for the preparation of ferrocenyl-stilbene analogs **18–20**.

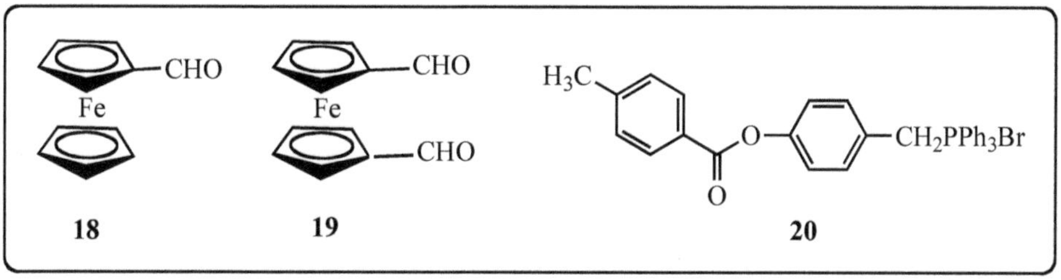

The precursor of **20** is 4-hydroxybenzylic alcohol (**21**), the corresponding bromide **22** is not commercially available and cannot be prepared by bromination of **21** because of its instability [60] (Scheme 1). Thus, the protection of the phenolic function has to be carried out before the bromination of the benzylic alcohol and in addition, the protective group should be stable to the bromination reagent.

These conditions preclude the use of the trimethylsilyl group [46]. Therefore, the phenolic function has been protected as an ester function by reacting **21** with *para*-toluoyl chloride in the presence of K_2CO_3 and acetone as a solvent [61]. The benzylphosphonium bromide **20** was obtained by reacting benzylic alcohol **23** successively with N-bromosuccinimide in CH_2Cl_2 [62] and triphenylphosphine in toluene (Scheme 1).

Scheme 1. Synthesis of benzylphosphonium bromide **20**.

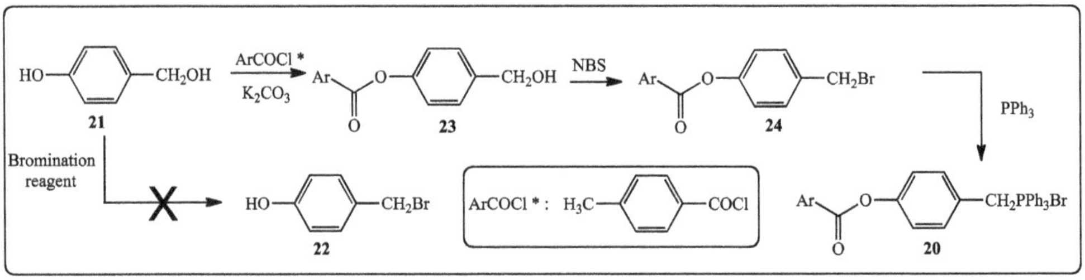

Finally, the benzylphosphonium bromide **20** was reacted with ferrocenecarbadehyde (**18**) in the presence of butyl lithium in THF. The cleavage of phenolic esters was carried out by KOH in methanol [63] and the ferrocenylstilbene analog **15** was recovered in 52% yield. In the same manner, the ferrocenyl derivative was obtained from **20** and ferrocene-1,1'-dicarbaldehyde (**19**) in 47% yield (Scheme 2).

2.2. Biological Effects

We compared the potency of the new resveratrol synthetic analogs towards the human colorectal tumor cell line SW480, the human hepatoblastoma HepG2 cell line and the rat normal intestine epithelium IEC18 cell, comparing their effect with the natural reference molecule, *i.e.*, *trans*-resveratrol.

Scheme 2. Synthesis of ferrocenyl-stilbene analogs **15–17**.

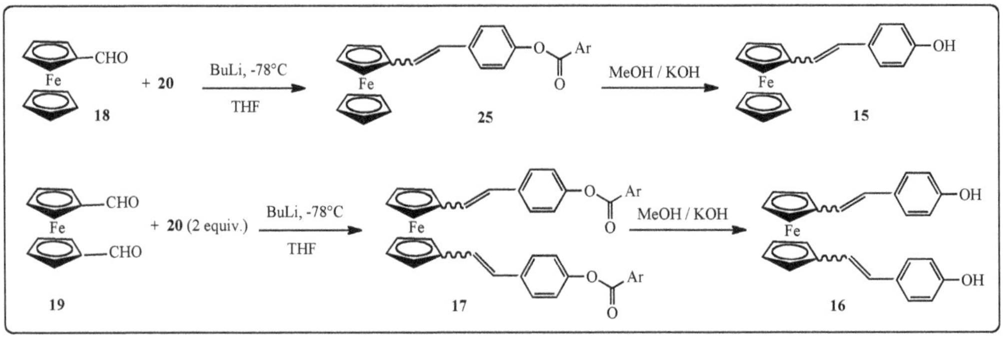

2.2.1. Effect of Stilbene Derivatives on Human Colorectal Tumor SW 480 Cell Line Proliferation

Firstly, we have determined the sensitivity of human tumoral colorectal cell line SW480 towards the newly synthesized stilbene derivatives and compared them to resveratrol, the parent molecule.

Figure 5 shows, as expected and in agreement with the literature [64], that resveratrol at 30 μM decreases drastically cell viability which is of 40% compared to the control (Figure 5).

Figure 5. Effect of stilbene derivatives on human cancerous colorectal SW480 cell viability. Cells were grown for 48 h in the presence of 30 μM resveratrol (or no RSV in a control experiment) or 30 μM stilbene derivatives (numbered on the x-axis). Cell viability was determined by counting cells using the trypan blue test (Co: cells control test). Data correspond to the mean of two independent experiments.

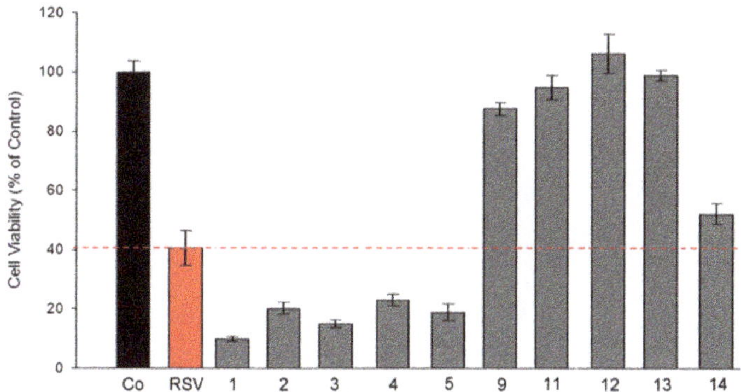

Interestingly, compounds **1–5** exhibit higher cytotoxicity than resveratrol. These derivatives bear, like resveratrol, at least one phenol group in the para position of the stilbene ring. The only structural differences between these molecules are the positions and numbers of methoxy groups. The efficiency of compound **1** indicates that its activity is due to the phenolic group, despite the absence of methoxy groups on its skeleton. Compound **14**, a tetramethoxylated derivative, shows similar activity as resveratrol, suggesting that these substituents are not essential for the activity. However, the fact that compounds **9**, **11–13** have only weak effects seems to indicate that a free phenolic group in the *para* position of the aromatic ring is needed for toxicity.

2.2.2. Effect of Stilbene Derivatives on the Cell Cycle Phase of the SW480 Cell Line

To further explore the mechanisms by which the most efficient compounds exert their antiproliferative potencies, we studied their effects on the cell cycle distribution of SW480 cells (Figure 6). The treatment of cells with compound **2**, which bears a hydroxy group in position 4 and a methoxy group in position 4', induces an accumulation of SW480 cells in S phase in the same manner as resveratrol (Figure 6). Interestingly, compound **4**, bearing hydroxy groups at positions 4 and 4' and a methoxy group at position 3, leads to an increase of S phase which is better than that of resveratrol and compound **2**. In contrast, pterostilbene (**3**) does not show any effect on the cell cycle, while it inhibits cell proliferation. This derivative has been reported to induce a blockade of HL60 intestine cancer cells in the G_1 phase, and to induce apoptosis [65]. The distribution of cells in the different cell cycle phases is reported in Figure 1 of the supplementary material.

One of the mechanisms by which resveratrol modulates carcinogenesis is the blockage of cells in S phase [66]. However, these effects at the cell cycle are complex and depend on the cell type, the resveratrol concentration and the duration of the treatment. Indeed, a low concentration of resveratrol induces accumulation of cells in S phase while at higher concentrations it leads to cell accumulation in G_1 or G_2/M phases [67]. Moreover, many cytotoxic agents also induce cell death by apoptosis. We have previously shown in SW480 and in HepG2 cell lines that resveratrol induces accumulation of cells in early S phase by action on the p21 protein and on the cyclin/cdk complexes formation and activity [68]. In the structural core of resveratrol, the phenol group in position 4 would be responsible for the antiproliferative effect by its action on DNA polymerases alpha and gamma [69,70]. Indeed, the increase of number of hydroxy groups on the stilbene moiety of resveratrol derivatives led to an increase of inhibition of tumor cell proliferation [71]. On the other hand, She *et al.* [72] have shown that *trans*-3,3',4',5-tetrahydroxystilbene and *trans*-3,3',4',5,5'-pentahydroxystilbene exhibit a higher apoptotic effect than resveratrol on the epidermal JB6 cell line.

Figure 6. Influence of stilbene derivatives on the cell cycle phases of the SW480 cells line. Cells were grown for 48 h in the presence of 30 μM resveratrol (or no RSV in a control experiment) or 30 μM stilbene derivatives (numbered on the x-axis). After treatment, nuclear DNA was labeled with propidium iodide. The cell cycle effect of the tested compounds was done analysing cell distribution in the different phases of the cell cycle (mean ± standard deviation of two independent experiments).

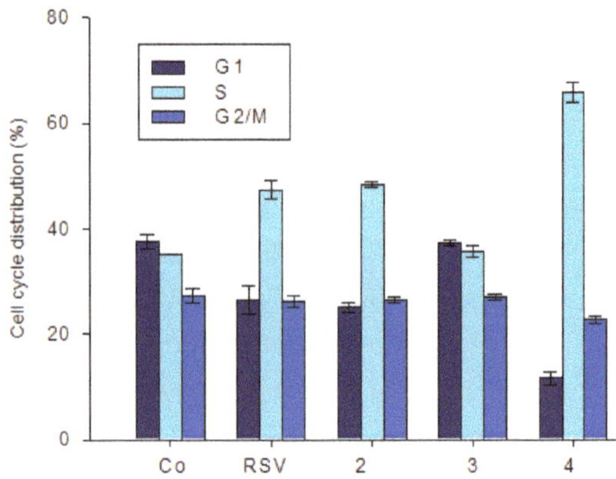

2.2.3. Evaluation of Toxicity Level of Stilbene Derivatives Towards Non-Cancerous Intestinal Epithelial Cells

With the aim of possible therapeutic applications using resveratrol derivatives in mind it was important to evaluate the specificity of cytoxicity towards normal cells. Hence, we evaluated the effect of potent derivatives on the proliferation of intestine epithelium IEC18 cells. The results shown in Figure 7 indicate no significant toxic effect of compounds **2**–**4** at 30 μM, except for compound **5** (presence of vinyl group in position 4). At higher concentration (100 μM) all compounds, including resveratrol, slightly inhibit cell proliferation, but much less than with the tumor SW480 cell line.

Figure 7. Effect of stilbenes derivatives on the proliferation of non-transformed IEC18 cells. Cells were grown for 48 h in the presence of 30 µM resveratrol (or no RSV in a control experiment) or 30 µM and 100 µM stilbene derivatives (numbered on the x-axis). Cell viability was determined by counting cells using the trypan blue exclusion. Data correspond to the mean ± standard deviation of two independent experiments.

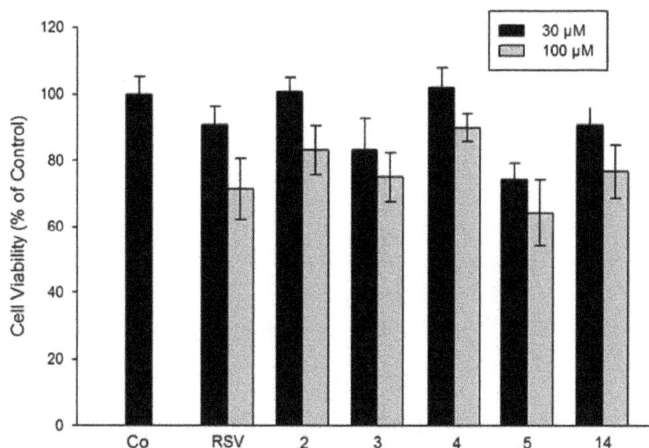

2.2.4. Comparison of Resveratrol Analogs on Cytotoxicity of Colorectal Tumor Cells and on Hepatoblastoma Cells

To have an overall view of the mechanisms involved in the inhibitory effect of the compounds, we performed a concentration-dependent analysis of the cytotoxicity evaluated by the crystal violet method. The crystal violet assay was chosen for the screening of the dose-effect of numerous molecules despite its lower sensitivity compared to some other cytotoxicity methods [73]. The results are presented as IC_{50} values. These IC_{50} values have been determined both on human tumor colorectal SW480 cell line and on human hepatoblastoma HepG2 cell line (Table 1). All tested molecules have lower IC_{50} than resveratrol towards SW480 cell line. Compounds **2** and **4** show a similar activity, indicating that the additional hydroxy group does not increase the activity of the stilbene. Comparison of the IC_{50} values between compounds **2** and **10** confirm the importance of the position 4 of the phenolic group [30,31]. In the series of ferrocenylstilbene analogs, compound **17** without a free phenolic function is the most active. This may be explained by a better lipophilicity due to the ester group while the antitumor activity can be attributed to the ferrocenyl moiety. Five of the most active derivatives (compounds **1**, **2**, **5**, **6** and **8**) have been subsequently tested on the HepG2 cell line (Table 1). Compounds **1**, **2**, **5** and **6** exhibit a lower potency on HepG2 than on SW480 cell line. Compounds **7** and **10** are the least active towards SW480 cells. Interestingly, compounds **5** (vinyl group in position 4') and **8** (carbinol group in position 3 and methoxy in position 4') exhibit a higher activity towards SW480 cell lines than HepG2 cell lines, while the bromine in position 4' (compound **6**) has an opposite effect. In the case of compound **8**, its metabolism by HepG2 cells may explain its weaker activity towards these cells. The difference between the resveratrol IC_{50}

cytotoxicity value (68.1 μM), (Table 1) and its inhibitory efficiency (30 μM) on cell proliferation (Figure 5) towards SW480 cell line would be attributed to the difference in the experimental approaches.

Table 1. Compared IC_{50} values of stilbene and ferrocenyl derivatives towards cell proliferation of SW480 and of HepG2 cell lines. For technical informations, see experimental procedure (Cell proliferation assays).

Compound Number	Compound Name	SW480 IC_{50} (μM)	HepG2 IC_{50} (μM)
	E-resveratrol	68.1 ± 5.5	57.3 ± 8.1
1	*E*-4-hydroxystilbene	18.6 ± 3.2	27.6 ± 5.0
2	*E*-4-hydroxy-4'-methoxystilbene	14.7 ± 2.1	26.3 ± 3.2
3	*E*-4-hydroxy-3',5'-dimethoxystilbene	16.1(± 2.9	Not Tested
4	*E*-4,4'-dihydroxy-4'-methoxystilbene	15.0 ± 0.9	Not Tested
5	*E*-4-hydroxy-4'-vinylstilbene	21.4 ± 0.3	33.2 ± 6.2
6	*E*-4-bromo-4'-hydroxystilbene	25.3 ± 2.4	18.6 ± 0.2
7	*E*-4-hydroxy-3,3',4',5'-tetramethoxystilbene	38.2 ± 0.7	Not Tested
8	*E*-3-carbinol-4-hydroxy-4'methoxystilbene	25.7 ± 2.1	77.7 ± 4.1
10	*E*-3-hydroxy-4'-methoxystilbene	81.7 ± 3.7	Not Tested
—	Ferrocene	>100	>100
15	*E*-(4-vinylphenol)-ferrocene	25.5 ± 1.6	40.2 ± 4.3
16	*(E,E)*-1,1'-bis(4-vinylphenol)-ferrocene	>100	>100
17	*(E,E)*-1,1'-bis[(1-*p*-toluoyloxy-4-vinyl)benzene]-ferrocene	5.9 ± 0.1	5.1 ± 0.2

2.2.5. Effect of Resveratrol Isosteres Bearing a Ferrocenyl Moiety. Determination of IC_{50} Values

Ferrocenyl derivatives were tested on cancerous SW480 and HepG2 cell lines and the IC_{50} values are reported in Table 1. Compound **17** shows the highest inhibitory activity in both cell lines with a very low IC_{50} value (5.9 μM), more than 10-fold higher compared to the resveratrol activity. Ferrocene used as a control does not induce any cytotoxic effect against SW480 cell line. Compound **16** (a deprotected version of compound **17**) shows a higher IC_{50} value (IC_{50} > 100 μM) than compound **17**. This data can be explained by the low solubility of **16** in DMSO in the cell medium. *E*-(4-vinylphenol)ferrocene (**15**), the closest isostere of resveratrol presented in this study shows a similar antiproliferative activity to resveratrol despite a lower solubility in the medium.

3. Experimental

3.1. General Experimental Procedures

Wittig reactions were performed under an inert atmosphere of argon using conventional vacuum-line and glasswork techniques. THF was degassed and distilled by refluxing over sodium and benzophenone under argon. The organic reagents were received from commercial sources and used without further purification. Separations by flash chromatography were performed on silica gel (230–400 mesh). ^{1}H-NMR, ^{13}C-NMR and ^{31}P-NMR spectra (δ, ppm) were recorded in $CDCl_3$

solutions on a Bruker 300 MHZ spectrometer, HRMS on MicroTOF Q-Bruker (ESI ionization). Spectroscopic analyses were performed at the Pôle de Chimie Moléculaire de l'Université de Bourgogne.

3.2. Precursors of Ferrocenyl-Stilbene Analogs

4-Toluoyloxybenzylic alcohol (**23**): To a mixture of 4-hydroxybenzylic alcohol (**21**, 100 g, 80.65 mmol) and potassium carbonate (13.4 g, 96.6 mmol) in acetone (300 mL) was added over 30 min at 0 °C a solution of *para*-toluoyl chloride (16 mL, 121 mmol) in acetone (100 mL). Then, the mixture was refluxed for 6 h. After cooling, the inorganic salts were filtrated and washed with acetone. The solvent was removed under vacuum and the crude product was purified by chromatography (EtOAc/heptane: 1/4) to give pure 4-toluoyloxybenzylic alcohol (**23**) in 47% yield. ^{1}H-NMR δ (ppm): 2.48 (s, 3H, CH_3), 4.75 (d, 2H, CH_2), 7.23 (d, 2H, Ar-H), 7.33 (d, 2H, Ar-H), 7.45 (d, 2H, Ar-H), 8.11 (d, 2H, Ar-H); ^{13}C-NMR δ (ppm): 21.75 (CH_3), 64.87 (CH_2), 117.46–144.52 (Ar-C).

4-Toluoyloxybenzylic bromide (**24**): To a mixture of **23** (9 g, 37.70 mmol) and triphenylphosphine (14.9 g, 56.53 mmol) in CH_2Cl_2 (150 mL) was added a solution of N-bromosuccinimide (10 g, 56.53 mmol) in CH_2Cl_2 (100 mL). After stirring for one hour, the mixture was poured into a separatory funnel and was washed with water. The organic phase was dried over $MgSO_4$. After removal of the solvent, the crude product was crystallized from ethanol (64%). ^{1}H-NMR δ (ppm): 2.39 (s, 3H, CH_3), 4.45 (d, 2H, CH_2), 7.12 (d, 2H, Ar-H), 7.24 (d, 2H, Ar-H), 7.38 (d, 2H, Ar-H), 8.01 (d, 2H, Ar-H); ^{13}C-NMR δ (ppm): 21.76 (CH_3), 32.74 (CH_2), 122.14, 126.61, 129.32, 129.78, 130.24, 135.29, 144.57, 150.96 (Ar-C), 165.04 (C=O).

4-Toluoyloxybenzyltriphenylphosphonium bromide (**20**): A mixture of **24** (18.7 g, 33 mmol) and triphenylphosphine (9.7 g, 36.3 mmol) in toluene (50 mL) was refluxed for five hours. The reaction mixture was cooled down to room temperature and a first crop of product was collected by filtration. The filtrate was then refluxed for five additional hours and a second crop of product precipitated. Two other crops were then collected and the combined fractions were crystallized from ethanol (86%). ^{1}H-NMR δ (ppm): 2.37 (s, 3H, CH_3), 5.47 (d, 2H, CH_2), 6.90 (d, 2H, Ar-H), 7.12 (d, 2H, Ar-H), 7.22 (d, 2H, Ar-H), 7.66 (m, 15H, Ar-H phosphonium), 7.96 (d, 2H, Ar-H); ^{13}C-NMR δ (ppm): 21.13 (CH_3), 60.48 (CH_2), 126.55 (Ar-C), 129.45, 130.28 (Ar-C phosphonium), 132.81, 134.54, 134.68, 135.05, 135.09, 144.76, 151.20 (Ar-C), 165.04 (C=O); ^{31}P-NMR δ (ppm): 23.50 (s, 1P).

3.3. Ferrocenyl-Stilbene Analogs **15–17** *and* **25**

E-[(1-paratoluoyloxy-4-vinyl)benzene]-ferrocene (**25**): Under argon atmosphere, butyllithium (1.6 M, 2.8 mL, 4.48 mmol) was slowly added to a solution of 4-toluoyloxybenzyltriphenylphosphonium bromide (**20**, 2.5 g, 4.41 mmol) in THF (40 mL) at −78 °C. The resulting solution was allowed to warm at room temperature. A solution of ferrocenecarbaldehyde [59] (**18**, 0.95 g, 4.41 mmol) in THF (15 mL) was added dropwise and the reaction mixture was then stirred overnight. Ice-cold water (500 mL) was added and the mixture stirred for an additional hour. The aqueous layer was extracted with ethyl acetate; the combined organic layers were washed with water and dried over $MgSO_4$. After evaporating the solvent, 52% of a crude mixture of isomers *Z* and *E* was isolated. The *E* isomer was isolated by chromatography (heptane/EtOAc: 9/1), yield 34%. ^{1}H-NMR δ (ppm): 3.33 (s, 3H, CH_3), 4.00 (s, 5H, Fc-H), 4.14 (t, 2H, Fc-H), 4.38 (d, 2H, Fc-H), 6.67 (d, 1H, ^3J = 16.65 Hz, =CH), 6.85 (d, 1H, ^3J = 16.65 Hz, =CH), 7.05 (d, 2H, Ar-H), 7.23 (d, 2H, Ar-H), 7.38 (d, 2H, Ar-H), 7.94 (d, 2H, Ar-H); ^{13}C-NMR δ (ppm): 21.4 (CH_3), 60.0, 65.9, 66.8 (Fc-C), 119.8, 124.2, 124.4, 125.1, 127.1, 127.9, 131.2, 135.09, 143.3, 148.3 (Ar-C), 165.3 (C=O); $C_{26}H_{22}FeO_2$ (MW 422.01). HRMS (ESI): *m/z* 422.09629 $[M]^+$, calculated mass 422.09637 (σ = 0.2 ppm).

(E,E)-1,1'-bis[(1-paratoluoyloxy-4-vinyl)benzene]-ferrocene (**17**): Under an argon atmosphere, butyl lithium (1.6 M, 5.6 mL, 8.96 mmol) was slowly added to a solution of 4-toluoyloxy-benzyltriphenylphosphonium bromide (**20**, 5 g, 8.82 mmol) in THF (80 mL) at −78 °C. The resulting solution was allowed to warm at room temperature. A solution of ferrocene-1,1'-dicarbaldehyde [59] (**19**, 0.95 g, 4.41 mmol) in THF (15 mL) was added dropwise and the reaction mixture was stirred overnight. Ice-cold water (500 mL) was added and the mixture was stirred for an additional hour. The aqueous layer was extracted with ethyl acetate; the combined organic layers were washed with water and dried over $MgSO_4$. After evaporating the solvent, 47% of a crude mixture of *EE*/*EZ*/*ZZ* isomers was obtained. The *EE* isomer was isolated by chromatography (heptane/EtOAc: 9/1), yield 25%. ^{1}H-NMR δ (ppm): 3.41 (s, 6H, CH_3), 4.28 (t, 4H, Fc-H), 4.48 (d, 4H, Fc-H), 6.63 (d, 2H, ^3J= 15.09 Hz, =CH), 6.81 (d, 2H, ^3J= 15.09 Hz, =CH), 7.11 (d, 4H, Ar-H), 7.28 (d, 4H, Ar-H), 7.41 (d, 4H, Ar-H), 8.09 (d, 4H, Ar-H); ^{13}C-NMR δ (ppm): 22.3 (CH_3), 67.9, 68.1, 70.4 (Fc-C), 121.7, 124.1, 124.5, 125.1, 127.2, 127.7, 131.3, 143.5, 148.0 (Ar-C), 164.3 (C=O); $C_{42}H_{34}FeO_4$ (MW 657.18). HRMS (ESI): m/z 658.17693 $[M]^+$, calculated mass 658.18018 (σ = 4.8 ppm).

E-(4-Vinylphenol)-ferrocene (**15**): To a solution of **25** (0.51 g, 1.1 mmol) in MeOH (15 mL) were added pellets of KOH (0.17 g, 3.2 mmol). The mixture was stirred for one hour at 30 °C. The reaction was quenched by addition of water (15 mL) and the solution was stirred for four hours. The solution was acidified to pH = 2 by concentrated HCl and then treated with aqueous $NaHCO_3$ solution (5%) to reach pH = 4. The ferrocene derivative **15** was extracted with ether. The combined organic layers were dried over $MgSO_4$ and after removal of the solvent, the compound **15** was isolated, yield 92%. ^{1}H-NMR δ (ppm): 4.25 (d, 4H, Fc-H), 4.27 (t, 2H, Fc-H), 4.43 (t, 2H, Fc-H), 4.60 (t, 1H, Fc-H), 4.74 (t, 1H, Fc-H), 4.79 (t, 1H, Fc-H), 6.36 (d, 1H, ^3J = 16.08 Hz, =CH), 6.71 (d, 1H, ^3J = 16.08 Hz, =CH), 6.79 (d, 2H, Ar-H), 7.28 (d, 2H, Ar-H); ^{13}C-NMR δ (ppm): 68.7, 69.1, 69.6, 73.3 (Fc-C), 115.8, 125.3, 127.6, 130.6, 157.8, (Ar-C); $C_{18}H_{16}FeO$ (MW 304.05). HRMS (ESI): *m/z* 304.05368 $[M]^+$, calculated mass 304.05452 (σ = 2.7 ppm).

(E,E)-1,1'-bis(4-Vinylphenol)ferrocene (**16**): Following the procedure described above, compound **16** was obtained from **17**; 88%. ^{1}H-NMR δ (ppm): 4.74 (d, 4H, Fc-H), 4.25 (t, 4H, Fc-H), 6.49 (s, 4H, =CH), 6.56 (d, 4H, Ar-H), 7.06 (d, 4H, Ar-H), 8.13 (s, 2H, OH); ^{13}C-NMR δ (ppm): 67.1, 69.3, 82.3, (Fc-C), 114.4, 126.7, 127.3, 131.5, 159.1, (Ar-C); $C_{26}H_{22}FeO_2$ (MW 422.09). HRMS (ESI): *m/z* 422.09588 $[M]^+$, calculated mass 422.09765 (σ = 4.2 ppm).

3.4. Biological Methods

3.4.1. Cell Culture

The human colon carcinoma cell line SW480 obtained from ATCC (American Type Culture Collection, Manassas,VA, USA) was cultured in RPMI-Medium with 10% fetal bovine serum (FBS) and 1% antibiotics. Human derived hepatoblastoma cell line HepG2 was obtained from the ECACC (European collection of cell culture, Salisbury, UK) and non-cancerous IEC18 cells from ileum epithelium of *Rattus norvegicus* (ATCC) were grown in monolayer culture system and maintained in phenol-red Dulbecco's Modified Eagle's Medium (DMEM) supplemented with 2 mM L-glutamine, 1% non-essential amino-acids, and 10% FBS (v/v) in a humidified atmosphere of 5% CO_2 at 37 °C.

3.4.2. Cell Viability Assays

Proliferation inhibition assays were performed in 24-well plates in triplicate, and each experiment was conducted two to three times. 30,000 cells were seeded per well, after 24 h cells were incubated in medium containing either 0.1% dimethylsulfoxide-solubilized *trans*-resveratrol, resveratrol derivatives, or 0.1% dimethylsulfoxide (DMSO) only as control. After 48 h, cells were harvested and the number of live cells was quantified using the trypan blue exclusion test which is based on the ability of a viable cell with an intact membrane to exclude trypan blue dye using a haemocytometer in microscopic counting. Results were expressed as percentage of control values.

3.4.3. Cell Proliferation Assays

After 48 h of incubation at 37 °C, medium was carefully removed from wells and the plates were washed gently with PBS 1X warmed at room temperature. Then the crystal violet solution was added and incubated for 10 min. Thereafter, plates were washed several times with tap water. The nucleus-incorporated crystal violet was dissolved using a sodium citrate solution and plates were agitated on orbital shaker until the color became uniform with no areas of dense coloration at the bottom of wells. The absorbance was read on each plate at 540 nm with a spectrophotometer (Dynex MRX-TC Revelation, Manassas, VA, USA). The absorbance is proportional to the relative density of cells adhering to multi-well dishes in regard to the absorbance of control well-plate (5% DMSO). After 48 h, IC_{50} values were determined by performing 0.75 to 100 μM treatments and the IC_{50} values were obtained after parametric regressions on the percentages of viable cells *versus* the control.

3.4.4. Cell Cycle Analysis

Cell cycle analysis was performed as described previously [67,74,75]. Briefly, cells were seeded 24 h before treatment into 25 cm^2 flasks. After treatment, the detached and adherent cells were pooled, fixed with ethanol, and stained with propidium iodide (PI) for subsequent analyses with a CyFlow Green flow cytometer and the fluorescence of PI was detected above 630 nm. For each sample 20,000 cells were acquired. Furthermore, data were analyzed with the MultiCycle software (Phoenix Flow Systems, San Diego, CA, USA); the x-axis corresponds to the DNA content and the y-axis to the number of cycling cells. The maximum value on the y-axis is inversely proportional to the altered cells level (non-cycling cells) which is excluded by gating.

4. Conclusions

While *trans*-resveratrol is considered a promising molecule for fighting cancer [76], a wide range of synthetic resveratrol analogs are potentially more active than *trans*-resveratrol. Some of these new synthetic molecules have interesting effects. Compounds **2** and **17** are the most active, while compounds **10** and **16** show the lowest activity. The comparison between compounds **16** and **17** indicates that the presence of a protecting group lead to a better efficacy which could be due to a better solubilisation in DMSO. It appears that the lack of substituents at position 3 and 5 (compound **1**) leads to a better inhibitory effect. Moreover, a limited number of methoxy groups (compounds **2**, **3** and **4**) provides better lipophilic properties. In most cases, the efficacy of the synthetic compounds is lower towards liver derived HepG2 cells than towards colorectal SW480 cells, except for compound **6** and mostly **17**, which is the most powerful derivative. These differences can be explained by the high xenobiotic metabolizing activities of HepG2 cells. Furthermore, the lack of effect on non-tumor cells (IEC18 intestinal epithelium cells) demonstrates the selectivity of these molecules for cancer cells, which is an important aspect for potential therapeutic applications. Concerning the possible targets of resveratrol analogs, an inhibition of the TNF alpha-induced activation NFkB by polyhydroxylated resveratrol derivatives *i.e.*, the hexahydroxystilbene in leukemia HL60 cells has been reported [70]. In terms of the structure-activity relationship, it appears that in order to obtain an inhibitory effect, the chemical parameters are the following: (a) the presence of a hydroxy group in position 4; (b) an increased inhibitory effect by the presence of a methoxy group (a decrease of the polar character leading to an increase in lipophilicity); (c) the lack (or masked form) of other hydroxy groups. In addition, *(E,E)*-1,1'-bis[(1-*para*-toluoyloxy-4-vinyl)benzene]ferrocene (**17**) a new compound, shows the highest efficacy.

Supplementary Materials

Supplementary materials can be accessed at: http://www.mdpi.com/1420-3049/19/6/7850/s1.

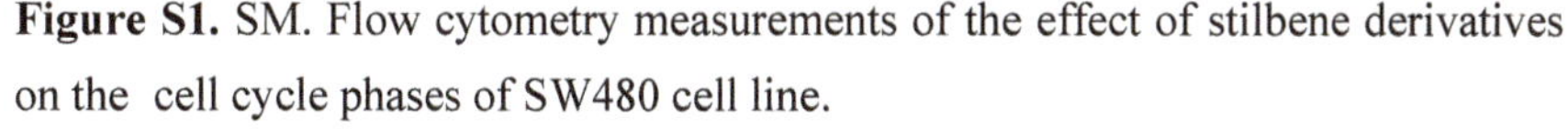

Figure S1. SM. Flow cytometry measurements of the effect of stilbene derivatives on the cell cycle phases of SW480 cell line.

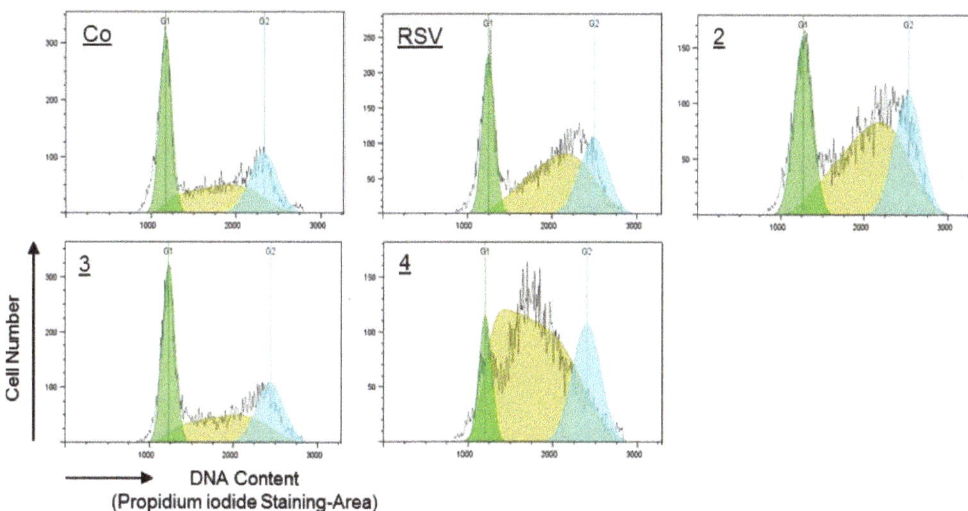

Cells were grown for 48 h in the presence or not (control) of resveratrol (RSV) at 30 μM or of stilbene derivatives (30 μM). After treatment, nuclear DNA was labeled with propidium iodide, then cell cycle phases were analysed by flow cytometry.

Acknowledgments

Université de Bourgogne, CNRS, INSERM UMR 866, are gratefully acknowledged for their financial support. Virginie Aires, Frédéric Mazué and M. Emeric Limagne are acknowledged for their technical help and advises as well as Richard Decreau and M. François Jacquin for English corrections. Malik Chalal was supported by a PhD grant from the Algerian government (Ministère de la Recherche et de l'Enseignement), which is sincerely acknowledged.

Author Contributions

The chemical work (syntheses and spectroscopic characterization of the compounds) as the biological study were performed by M. Chalal during his PhD work under the direction of P. Meunier and D. Vervandier-Fasseur for the chemical part and under the direction of N. Latruffe and D. Delmas for the biological part. The manuscript was written by N. Latruffe (N.L.) and D. Vervandier-Fasseur (D.V.-F.) and revised by the co-corresponding authors (N.L and D.V.-F.).

Conflicts of Interest

The authors declare no conflict of interest.

References and Notes

1. Arichi, H.; Kimura, Y.; Okuda, H.; Baba, M.; Kozowa, K.; Arichi, S. Effects of stilbene compounds of the roots of *Polygonum cuspidatum Sieb et Zucc* on lipid metabolism. *Chem. Pharm. Bull.* **1982**, *30*, 1766–1770.
2. Watt, J.M.; Breyer-Brandwijk, M.G. Medicinal and poisonous plants. *Nature* **1962**, *196*, 609–610.
3. Langcake, P.; Pryce, R. The production of Resveratrol by *Vitis vinifera* and other members of the *Vitaceae* as a response to infection or injury. *Physiol. Plant. Pathol.* **1976**, *9*, 77–86.
4. Langcake, P.; Pryce, R.A. New class of phytoalexins from grapevines. *Experientia* **1977**, *33*, 151–152.
5. Pawlus, A.D.; Sahli, R.; Bisson, J.; Rivière, C.; Delaunay, J.C.; Richard, T.; Gomes, E.; Bordenave, L.; Waffo-Teguo, P.; Mérillon, J.M. Stilbenoid profiless of canes from *Vitis* and *Muscadinia* species. *J. Agric. Food Chem.* **2013**, *61*, 501–511.
6. Siemann, E.H.; Creasy, L.L. Concentration of the phytoalexin resveratrol in wine. *Am. J. Enol. Vitic.* **1992**, *43*, 49–52.
7. Boutegrabet, L.; Fekete, A.; Hertkorn, N.; Papastamoulis, Y.; Waffo-Téguo, P.; Mérillon, J.M.; Jeandet, P.; Gougeon, R.D.; Schmitt-Koplin, P. Determination of stilbene derivatives in Burgundy red wines by ultra-high pressure liquid chromatography. *Anal. Bioanal. Chem.* **2011**, *401*, 1517–1525.
8. Burns, J.; Yokota, T.; Ashihara, H.; Lean, M.E.J.; Crozier, A. Plant foods and herbal sources of resveratrol. *J. Agric. Food Chem.* **2002**, *50*, 3337–3340.
9. Ingham, J.L. 3,5,4'-Trihydroxystilbene as a phytoalexin from groundnuts (*Arachis hypogaea*). *Phytochemistry* **1976**, *15*, 1791–1793.
10. Jeandet, P.; Delaunois, B.; Conreux, A.; Donnez, D.; Nuzzo, V.; Cordelier, S.; Clément, C.; Courot, E. Biosynthesis, metabolism, molecular engineering and biological functions of stilbene phytoalexins in plants. *Biofactors* **2010**, *36*, 331–341.
11. Jeandet, P.; Clément, C.; Courot, E.; Cordelier, S. Modulation of phytoalexin biosynthesis in engineered plants for disease resistance. *Int. J. Mol. Sci.* **2013**, *14*, 14136–14170.
12. Matsuda, H.; Morikawa, T.; Toguchida, I.; Park, J.Y.; Harima, S.; Yoshikawa, M. Antioxidant constituents from rhubarb: Structural requirements of stilbenes for the activity and structures of two new anthraquinone glucosides. *Bioorg. Med. Chem.* **2001**, *9*, 41–50.
13. Andrus, M.B.; Liu, J.; Meredith, E.L.; Nartey, E. Synthesis of resveratrol using a direct decarbonylative Heck approach from resorcylic acid. *Tetrahedron Lett.* **2003**, *44*, 4819–4822.
14. Frémont, L. Biological effects of resveratrol. *Life Sci.* **2000**, *66*, 663–673.
15. Saiko, P.; Szakmary, A.; Jaeger, W.; Szekeres, T. Resveratrol and its analogs: Defense against cancer, coronary disease and neurodegenerative maladies or just a fad? *Mutat. Res.* **2008**, *658*, 68–94.
16. Tomé-Carneiro, J.; Larrosa, M.; Gonzales-Sarrias, A.; Tomas-Barberan, F.A.; Garcia-Conesa, M.T.; Espin, J.C. Resveratrol and clinical trials: The crossroad from *in vivo* studies to human evidence. *Curr. Pharm. Des.* **2013**, *19*, 6064–6093.

17. Jang, M.; Cai, L.; Udeani, G.O.; Slowing, K.V.; Thomas, C.F.; Beecher, C.W.W.; Fong, H.S.; Farnsworth, N.R.; Kinghorn, A.D.; Mehta, R.G.; *et al.* Cancer, chemopreventive activity of resveratrol, a natural product derived from grapes. *Science* **1997**, *275*, 218–220.
18. Boyer, J.Z.; Jandova, J.; Janda, J.; Vleugels, F.R.; Elliott, D.A.; Sligh, J.E. Resveratrol-sensitized UVA induced apoptosis in human keratinocytes through mitochondrial oxidative stress and pore opening. *J. Phytochem. Photobiol. B* **2012**, *113*, 42–50.
19. Tili, E.; Michaille, J.J.; Adair, B.; Alder, H.; Limagne, E.; Taccioli, C.; Ferracin, M.; Delmas, D.; Latruffe, N.; Croce, C.M. Resveratrol decreases the levels of miR-155 by upregulating miR-663, a microRNA targeting JunB and JunD. *Carcinogenesis* **2010**, *31*, 1561–1566.
20. Docherty, J.J.; Fu, M.M.H.; Stiffler, B.S.; Limperos, R.J.; Pokabla, C.M.; de Lucia, A.L. Resveratrol inhibition of herpes simplex virus replication. *Antiviral. Res.* **1999**, *43*, 145–155.
21. Yim, N.; Ha, D.T.; Trung, T.N.; Kim, J.P.; Lee, S.; Na, M.; Jung, H.; Kim, H.S.; Kim, Y.H.; Bae, K. The antimicrobial activity of compounds from the leaf and stem of *Vitis amurensis* against two oral pathogens. *Bioorg. Med. Chem. Lett.* **2010**, *20*, 1165–1168.
22. Adrian, M.; Jeandet, P.; Veneau, J.; Weston, L.A.; Bessis, R. Biological activity of resveratrol, a stilbenic compound from grapevines, against *Botrytis cinerea*, the causal agent for gray mold. *J. Chem. Ecol.* **1997**, *23*, 1689–1702.
23. Adrian, M.; Jeandet, P. Effects of resveratrol on the ultrastructure of *Botrytis cinerea conidia* and biological significance in plant/pathogen interactions. *Fitoterapia* **2012**, *83*, 1345–1350.
24. Lambert, C.; Bisson, J.; Waffo-Téguo, P.; Papastamoulis, Y.; Richard, T.; Corio-Costet, M.-F.; Mérillon, J.-M.; Cluzet, S. Phenolics and their antifungal role in grapevine wood decay: Focus on the *Botryosphaeriaceae* family. *J. Agric. Food Chem.* **2012**, *60*, 11859–11868.
25. Rivière, C.; Richard, T.; Quentin, L.; Krisa, S.; Mérillon, J.M.; Monti, J.P. Inhibitory activity of stilbenes on Alzeimer's β-amyloid fibrils *in vitro*. *Bioorg. Med. Chem.* **2007**, *15*, 1160–1167.
26. Singh, N.; Agrawal, M.; Doré, S. Neuroprotective properties and mechanisms of resveratrol in *in vitro* and *in vivo* experimental cerebral stroke models. *ACS Chem. NeuroSci.* **2013**, *4*, 1151–1162.
27. Stef, G.; Csiszar, A.; Lerea, K.; Ungvari, Z.; Veress, G. Resveratrol inhibits aggregation of platelets from high-risk cardiac patients with aspirin resistance. *J. Card. Pharm.* **2006**, *48*, 1–5.
28. Li, H.; Xia, N.; Förstermann, U. Cardiovascular effects and molecular targets of resveratrol. *Nitric Oxide* **2012**, *26*, 102–110.
29. Mohar, D.S.; Malik, S. The sirtuin system: The holy grail of resveratrol? *J. Clin. Exp. Cardiolog.* **2012**, *3*, 1000216/1–1000216/4.
30. Wang, M.; Jin, Y.; Ho, C.T. Evaluation of resveratrol derivatives as potential antioxidants and identification of a reaction product of resveratrol and 2,2-diphenyl-1-picrylhydrazyl radical. *J. Agric. Food Chem.* **1999**, *47*, 3974–3977.
31. Fukuhara, K.; Nagakawa, M.; Nakanishi, I.; Ohkubo, K.; Imai, K.; Urano, S.; Fukuzumi, S.; Ozawa, T.; Ikota, N.; Mochizuki, M.; *et al.* Structural basis for DNA-cleaving activity of resveratrol in the presence of Cu(II). *Bioorg. Med. Chem.* **2006**, *14*, 1437–1443.
32. Athar, M.; Back, J.H.; Kopelovich, L.; Bickers, D.R.; Kim, A.L. Multiple targets of resveratrol: Anti-carcenogic mechanisms. *Arch. Biochem. Biophys.* **2009**, *486*, 95–102.

33. Pirola, L.; Frödjö, S. Resveratrol: One molecule, many targets. *IUBMB Life* **2008**, *60*, 323–332.
34. Latruffe, N.; Delmas, D.; Lizard, G.; Tringali, C.; Spatafora, C.; Vervandier-Fasseur, D.; Meunier, P. Resveratrol against major pathologies: From diet prevention to possible alternative chemotherapies with new structural analogues. In *Bioactive Compounds from Natural Sources*, 2nd ed.; Corrado, T., Ed.; Taylor & Francis Group: Boca Raton, FL, USA, 2011; pp. 340–378.
35. St John, S.E.; Jensen, K.C.; Kang, S.; Chen, Y.; Calamini, B.; Mesecar, A.D.; Lipton, M.A. Design, synthesis, biological and structural evaluation of functionalized resveratol analogues as inhibitors of quinone reductase 2. *Bioorg. Med. Chem.* **2013**, *21*, 6022–6037.
36. Stivala, L.A.; Savio, M.; Carafoli, F.; Perucca, P.; Bianchi, L.; Maga, G.; Forti, L.; Pagnoni, U.M.; Albini, A.; Prosperi, E.; *et al.* Specific structural determinants are responsible for the antioxidant activity and the cell cycle effects of resveratrol. *J. Biol. Chem.* **2001**, *276*, 22586–22594.
37. Cheng, J.C.; Fang, J.G.; Chen, W.-F.; Zhou, B.; Yang, L.; Liu, Z.L. Structure-activity relationship studies of resveratrol and its analogues by the reaction kinetics of low density lipoprotein peroxidation. *Bioorg. Chem.* **2006**, *34*, 142–157.
38. Cardile, V.; Lombardo, L.; Spatafora, C.; Tringali, C. Chemo-enzymatic synthesis and cell-growth inhibition activity of resveratrol analogues. *Bioorg. Chem.* **2005**, *33*, 22–33.
39. Horvath, Z.; Marihart-Fazekas, S.; Saiko, P.; Grusch, M.; Ozsüy, M.; Harik, M.; Handler, N.; Erker, T.; Jaeger, W.; Fritzer-Szekeres, M.; *et al.* Novel resveratrol derivatives induce apoptosis and cause cell cycle arrest in prostate cancer cell lines. *Anticancer Res.* **2007**, *27*, 3459–3464.
40. Pan, M.H.; Lin, C.L.; Tsai, J.H.; Ho, C.T.; Chen, W.J. 3,5,3',4',5'-Pentamethoxystilbene (MR-5), a synthetically methoxylated analogue of resveratrol, inhibits growth and induces G1 cell cycle arrest of human breast carcinoma MCF-7 cells. *J. Agric. Food Chem.* **2010**, *58*, 226–234.
41. Zhang, W.; Go, M.L. Methoxylation of resveratrol: Effects on induction of NAD(P)H quinone oxidoreductase 1 (NQO1) activity and growth inhibitory properties. *Bioorg. Med. Chem. Lett.* **2011**, *21*, 1032–1035.
42. Das, J.; Pany, S.; Mahji, A. Chemical modification of resveratrol for improved protein kinase C alpha activity. *Bioorg. Med. Chem.* **2011**, *19*, 5321–5333.
43. Biasutto, L.; Mattarei, A.; Marotta, E.; Bradaschia, A.; Sassi, N.; Garbisa, S.; Zoratti, M.; Paradisi, C. Development of mitochondria-targeted derivatives of resveratrol. *Bioorg. Med. Chem. Lett.* **2008**, *18*, 5594–5597.
44. Huang, X.F.; Ruan, B.F.; Wang, X.T.; Xu, C.; Ge, H.M.; Zhu, H.L.; Tan, R.X. Synthesis and cytotoxic evaluation of a series of resveratrol derivatives modified in C2 position. *Eur. J. Med. Chem.* **2007**, *42*, 263–267.
45. Mazué, F.; Colin, D.; Gobbo, J.; Wegner, M.; Rescifina, A.; Spatafora, C.; Fasseur, D.; Delmas, D.; Meunier, P.; Tringali, C.; *et al.* Structural determinants of resveratrol for cell proliferation inhibition potency: Experimental and docking studies of new analogs. *Eur. J. Med. Chem.* **2010**, *45*, 2972–2980.
46. Chalal, M.; Vervandier-Fasseur, D.; Meunier, P.; Cattey, H.; Hierso, J.C. Syntheses of polyfunctionalized resveratrol derivatives using Wittig and Heck protocols. *Tetrahedron* **2012**, *68*, 3899–3907.

47. Heynekamp, J.J.; Weber, W.M.; Hunsaker, L.A.; Gonzales, A.M.; Orlando, R.A.; Deck, L.M.; Vander Jagt, D.L. Substituted *trans*-stilbenes, including analogues of the natural product resveratrol, inhibit the human tumor necrosis factor alpha-induced activation of transcription factor nuclease factor kappaB. *J. Med. Chem.* **2006**, *49*, 7182–7189.
48. Shang, Y.J.; Qian, Y.P.; Liu, X.D.; Dai, F.; Shang, X.L.; Jia, W.Q.; Liu, Q.; Fang, J.G.; Zhou, B. Radical-scavenging activity and mechanism of resveratrol-oriented analogues: Influence of the solvent, radical, and substitution. *J. Org. Chem.* **2009**, *74*, 5025–5031.
49. Chandra, V.; Srivastava, V.B. Condensation of p-bromophenylacetic acid with aldehydes. *J. Indian Chem. Soc.* **1967**, *44*, 675–678.
50. Sinha, A.K.; Kumar, V.; Sharma, A.; Sharma, A.; Kumar, R. An unusual, mild and convenient one-pot two-step access to *E.*-stilbenes from hydroxy-substituted benzaldehydes and phenylacetic acids under microwave activation: A new facet of the classical Perkin reaction. *Tetrahedron* **2007**, *63*, 11070–11077.
51. Sinh, A.K.; Kumar, V.; Sharma, A. A Single Step Microwave Induced Process for the Preparation of Substituted Stilbenes and Analogs from Arylaldehydes and Phenylacetic Acids. WO2007/110883, 2007.
52. Albert, S.; Horbach, R.; Deising, H.B.; Siewert, B.; Csuk, R. Synthesis and antimicrobial activity of *E.*-stilbene derivatives. *Bioorg. Med. Chem.* **2011**, *19*, 5155–5166.
53. Rosenberg, B.; van Camp, L.; Trosko, J.E.; Mansour, V.H. Platinium compounds: A new class of potent antitumor agents. *Nature* **1969**, *222*, 385–386.
54. Hartinger, C.G.; Metzler-Note, N.; Dyson, P.J. Challenges and opportunities in the development of organometallic anticancer drugs. *Organometallics* **2012**, *31*, 5677–5685.
55. Fouda, M.F.R.; Abd-Elzaher, M.M.; Abdelsamaia, R.A.; Labib, A.A. On the medicinal chemistry of ferrocene. *Appl. Organomet. Chem.* **2007**, *21*, 613–625.
56. Navarro, M.; Castro, W.; Biot, C. Bioorganometallic compounds with antimalarial targets: inhibiting hemozoin formation. *Organometallics* **2012**, *31*, 5715–5727.
57. Top, S.; Vessières, A.; Cabestaing, C.; Laios, I.; Leclercq, G.; Provot, C.; Jaouen, G.J. Studies on organometallic selective estrogen receptor modulators. (SERMs) dual activity in the hydroxy-ferrocifen series. *Organomet. Chem.* **2001**, *637*, 500–506.
58. Messina, P.; Labbé, E.; Buriez, O.; Hillard, E.A.; Vessières, A.; Hamels, D.; Top, S.; Jaouen, G.; Frapart, Y.M.; Mansuy, D.; *et al.* Deciphering the activation sequence of ferrociphenol anticancer drug candidates. *Chem. Eur. J.* **2012**, *18*, 6581–6587.
59. Balavoine, G.G.A.; Doisneau, G.; Fillebeen-Khan, T. An improved synthesis of ferrocene-1,1'-dicarbaldehyde. *J. Organomet. Chem.* **1991**, *412*, 381–382.
60. We reacted 3,5-dihydroxybenzylic alcohol or 4-hydroxybenzylic alcohol with different brominating reagents: PBr_3, CBr_4 or HBr/AcOH but attempts to isolate by chromatography the corresponding bromides failed and the starting alcohols were recovered.
61. Vervandier-Fasseur, D.; Chalal, M.; Meunier, P. Method for the Production of *trans*-resveratrol and its Analogs via Wittig Reaction and Their Pharmaceutical and Cosmetic Compositions. *PCT Int. Appl.* WO 2013008175, 2013.

62. Firouzabadi, H.; Iranpoor, N.; Ebrahimzadeh, F. Facile conversion of alcohols into their bromides and iodides by N-bromo and N-iodosaccharins/triphenylphosphine under neutral conditions. *Tetrahedron Lett.* **2006**, *47*, 1771–1775.
63. Khurana, J.M.; Chauhan, S.; Bansal, G. Facile hydrolysis of esters with KOH-Methanol at ambient temperature. *Monatsch. Chem.* **2004**, *135*, 83–87.
64. Simoni, D.; Roberti, M.; Invidiata, F.P.; Aiello, E.; Aiello, S.; Marchetti, P.; Baruchello, R.; Eleopra, M.; di Cristina, A.; Grimaudo, S.; *et al.* Stilbene-based anticancer agents: Resveratrol analogues active toward HL60 leukemic cells with a non-specific phase mechanism. *Bioorg. Med. Chem. Lett.* **2006**, *16*, 3245–3248.
65. Pan, M.H.; Chang, Y.H.; Badmaev, V.; Nagabhushanam, K.; Ho, C.T. Pterostilbene induces apoptosis and cell cycle arrest in human gastric carcinoma cells. *J. Agric. Food Chem.* **2007**, *55*, 7777–7785.
66. Lee, S.K.; Zhang, W.; Sanderson, B.J.S. Selective growth inhibition of human leukemia and human lymphoblastoid cells by Resveratrol via cell cycle arrest and apoptosis induction. *J. Agric. Food Chem.* **2008**, *56*, 7572–7577.
67. Delmas, D.; Rébé, C.; Lacour, S.; Filomenko, R.; Athias, A.; Gambert, P.; Cherkaoui-Malki, M.; Jannin, B.; Dubrez-Daloz, L.; Latruffe, N.; *et al.* Resveratrol-induced apoptosis is associated with fas redistribution in the rafts and the formation of a death-inducing signaling complex in colon cancer cells. *J. Biol. Chem.* **2003**, *278*, 41482–41490.
68. Colin, D.; Gimazane, A.; Lizard, G.; Izard, J.C.; Solary, E.; Latruffe, N.; Delmas, D. Effects of resveratrol analogs on cell cycle progression, cell cycle associated proteins and 5-fluorouracil sensitivity in human derived colon cancer cells. *Int. J. Cancer* **2009**, *124*, 2780–2788.
69. Larrosa, M.; Tomas-Barberan, F.; Espin, J.C. Grape polyphenol resveratrol and the related molecule 4-hydroxystilbene induce growth inhibition, apoptosis, S-phase arrest, and upregulation of cyclins A, E and B1 in human SK-Mel-28 melanoma cells. *J. Agric. Food Chem.* **2003**, *51*, 4576–4584.
70. Fan, G.J.; Liu, X.D.; Qian, Y.P.; Shang, Y.J.; Li, X.Z.; Dai, F.; Fang, J.G.; Jin, X.L.; Zhou, B. 4,4'-Dihydroxy-*trans*-stilbene, a resveratrol analogue, exhibited enhanced antioxidant activity and cytotoxicity. *Bioorg. Med. Chem.* **2009**, *17*, 2360–2365.
71. Horvath, Z.; Murias, M.; Saiko, P.; Erker, T.; Handler, N.; Madlener, S.; Jaeger, W.; Grusch, M.; Fritzer-Szekeres, M.; Krupitza, G.; *et al.* Cytotoxic and biochemical effects of 3,3',4,4',5,5'-hexahydroxystilbenes, a novel resveratrol analog in HL-60 human promyelocytic leukemia cells. *Exp. Hematol.* **2006**, *34*, 1377–1384.
72. She, Q.B.; Ma, W.Y.; Wang, M.; Mingfu, K.; Kaji, A.; Ho, C.T.; Dong, Z. Inhibition of cell transformation by resveratrol and its derivatives: Differential effects and mechanism involved. *Oncogene.* **2003**, *22*, 2143–2150.
73. Nargi, F.E.; Yang, T.J. Optimization of the L-M cell bioassay for quantitating tumor necrosis factor alpha in serum and plasma. *J. Immunol. Methods* **1993**, *159*, 81–91.

74. Colin, D.; Lancon, A.; Delmas, D.; Lizard, G.; Abrossinow, J.; Kahn, E.; Jannin, B.; Latruffe, N. Antiproliferative activities of resveratrol and related compounds in human hepatocyte derived HepG2 cells are associated with biochemical cell disturbance revealed by fluorescence analyses. *Biochimie* **2008**, *90*, 1674–1684.
75. Marel, A.K.; Lizard, G.; Izard, J.C.; Latruffe, N.; Delmas, D. Inhibitory effects of trans-resveratrol analogs molecules on the proliferation and the cell cycle progression of human colon tumoral cells. *Mol. Nutr. Food Res.* **2008**, *52*, 538–548.
76. Delmas, D.; Lançon, A.; Colin, D.; Jannin, B.; Latruffe, N. Resveratrol as a chemoprotective agent: A promising molecule for fighting cancer. *Curr. Drug Targets* **2006**, *7*, 423–442.

Sample Availability: Samples of the compounds **1–17** are available from D. Vervandier-Fasseur (D.V.-F.).

MDPI AG
Klybeckstrasse 64
4057 Basel, Switzerland
Tel. +41 61 683 77 34
Fax +41 61 302 89 18
http://www.mdpi.com/

Molecules Editorial Office
E-mail: molecules@mdpi.com
http://www.mdpi.com/journal/molecules

www.ingramcontent.com/pod-product-compliance
Lightning Source LLC
Chambersburg PA
CBHW051926190326
41458CB00026B/6423

* 9 7 8 3 0 3 8 4 2 0 5 8 3 *